Introduction to
Organic
Mass Spectrometry

Second Edition

Introduction to
Organic
Mass Spectrometry

Second Edition

Anees Ahmad Siddiqui

MPharm PhD

Professor and Head
Department of Pharmaceutical Chemistry
School of Pharmaceutical Education and Research (SPER)
Jamia Hamdard, New Delhi

CBS

CBS Publishers & Distributors Pvt Ltd

New Delhi • Bengaluru • Chennai • Kochi • Kolkata • Mumbai
Hyderabad • Jharkhand • Nagpur • Patna • Pune • Uttarakhand

Introduction to
Organic
Mass Spectrometry
Second Edition

ISBN: 978-81-948986-7-2

Copyright © Author and Publisher

Second Edition: 2021

First Edition: 2011

Published by Satish Kumar Jain and produced by Varun Jain for

CBS Publishers & Distributors Pvt Ltd

4819/XI Prahlad Street, 24 Ansari Road, Daryaganj, New Delhi 110 002, India
Ph: 011-23289259, 23266861, 23266867   Website: www.cbspd.com
Fax: 011-23243014   e-mail: delhi@cbspd.com; cbspubs@airtelmail.in

Corporate Office: 204 FIE, Industrial Area, Patparganj, Delhi 110 092
Ph: 011-49344934   Fax: 011-49344935   e-mail: publishing@cbspd.com; publicity@cbspd.com

Branches

- **Bengaluru:** Seema House, 2975, 17th Cross, K.R. Road,
  Banasankari 2nd Stage, Bengaluru 560 070, Karnataka
  Ph: +91-80-26771678/79   Fax: +91-80-26771680   e-mail: bangalore@cbspd.com
- **Chennai:** 7, Subbaraya Street, Shenoy Nagar, Chennai 600 030, Tamil Nadu
  Ph: +91-44-26680620, 26681266   Fax: +91-44-42032115   e-mail: chennai@cbspd.com
- **Kochi:** 42/1325, 1326, Power House Road, Opposite KSEB, Kochi 682018, Kerala
  Ph: +91-484-4059061-65   Fax: +91-484-4059065   e-mail: kochi@cbspd.com
- **Kolkata:** 6/B, Ground Floor, Rameswar Shaw Road, Kolkata-700 014, West Bengal
  Ph: +91-33-22891126, 22891127, 22891128   e-mail: kolkata@cbspd.com
- **Mumbai:** PWD Shed, Gala No. 25/26, Ramchandra Bhatt Marg,
  Next to JJ Hospital, Gate No. 2, Opp. Union Bank of India, Noorbaug,
  Mumbai-400009, Maharashtra
  Ph: +91-22-66661880/89   e-mail: mumbai@cbspd.com

Representatives

| • Hyderabad | 0-9885175004 | • Jharkhand | 0-9811541605 | • Nagpur | 0-9421945513 |
| • Patna | 0-9334159340 | • Pune | 0-9623451994 | • Uttarakhand | 0-9716462459 |

Printed at: SRK Graphics, Shahdara, Delhi, India

Preface to the Second Edition

The first edition of the book titled *Introduction to Organic Mass Spectrometry* was compiled in 2011. It was found much useful at undergraduate as well as postgraduate levels in science, pharmacy, biotechnology, agriculture, food technology, forensic science, etc. courses. The book was very much appreciated due to its usefulness especially in pharmacy subjects. Hence, it becomes the responsibility of the author to revise and update the book. It also becomes more necessary with revision of syllabi recently.

There is no change in number of chapters but every chapter is revised and updated. Chapter 1 is completely changed in which application part is completely revised as per the recent syllabus in most of the courses of various disciplines. In Chapter 4, more examples to determine the molecular formulae are added. In Chapter 5, rules determining the fragmentation pattern are systematically arranged. In Chapter 6, fragmentation pattern of some more relevant compounds is included. Similarly, some more examples related to analytical mass spectrometry are incorporated. The chapters of tandem mass spectrometry, GC-MS, LC-MS and biological mass spectrometry are also revised with the inclusion of more informative figures. Appendices included in this book are also made more informative.

The author hopes that this book will be more useful in teaching the topic of mass spectrometry to the students in various disciplines.

I express my thanks to authors of those articles from which information has been collected for compiling this book. I am thankful to my colleagues and students for their fruitful suggestions. I am also thankful to Mr SK Jain (CMD) and Mr Varun Jain (MD), CBS Publishers & Distributors Pvt Ltd, for bringing out the second edition of this book.

All constructive criticism and comments regarding this edition from students and teachers are most welcome for further improvement as per requirement in further edition.

Prof Anees Ahmad Siddiqui

Preface to the First Edition

The first edition of the book titled **"Introduction to Organic Mass Spectrometry"** is compiled to guide the students of various scientific disciplines with a complete overview of the principles, theories and key applications of modern mass spectrometry. All instrumental aspects of mass spectrometry are clearly and concisely described: Sources, analyzers and detectors. Tandem mass spectrometry, GC-MS and LC-MS are explained in more detail in a separate chapter. Emphasis is placed throughout the text on optional utilization conditions. Various fragmentation patterns are described together with analytical information that derives from the mass spectra. Large number of mathematical equations and derivations has been avoided and compilation of book has been undertaken to present the field of mass spectrometry in a logical manner. I hope, this book will be an invaluable resource for all undergraduate and postgraduate students using this technique in departments of chemistry, biochemistry, medicine, pharmacology, agriculture, pharmacy and food science. It is also of interest for researchers looking for an overview of the latest techniques and development. Biological mass spectrometry is also included to illustrate the importance of mass spectrometry in clinical medical research. Appendices included in this book provide vide range of information at a glance.

The author hopes that this book will assist in teaching the subject of mass spectrometry to the students in different levels.

I express my thanks to those authors of articles from which information are collected for compiling this book. I am thankful to their staff members for their suggestions and Mr. S.K. Jain and Mr. Vinod K. Jain of M/s CBS Publishers & Distributors Pvt. Ltd. for bringing out the first edition of this book.

All constructive criticism and comments of this book from students and teachers are most welcome and shall form the base for future editions.

New Delhi   **Dr Anees Ahmad Siddiqui**

Contents

1

Mass Spectrometry:
Introduction and its Application

Mass spectrometry is a sophisticated analytical technique used to:

• quantify known materials,
• identify unknown compounds within a sample, and
• elucidate the structure and chemical properties of different molecules.

Mass spectrometer works by converting the molecule into gaseous ions, with or without fragmentation, which are separated and then characterized by their mass to charge ratios (m/e) and relative abundances. With the help of masses of fragment, the mass/original structure of molecule is predicted. Two key components in this process are the ion source, which generates the ions, and the mass analyzer, which sorts the ions. Each ion then goes to the detector and recorded its presence in the form streak line depending on its relative abundance.

To illustrate the function of mass spectrometry (Fig 1.1), imagine a stone being projected from a catapult towards a delicate vase (I). On impact, the vase is shattered (II). If the pieces are carefully collected (III), the vase can be reconstructed from the fragments (IV). In this example, the vase represents the molecule, the catapult and stone represent the device for making the fragments.

Mass spectrometry is unlike most other forms of spectroscopy or spectrometry that are concerned with non-destructive interactions between molecules and electromagnetic radiation,

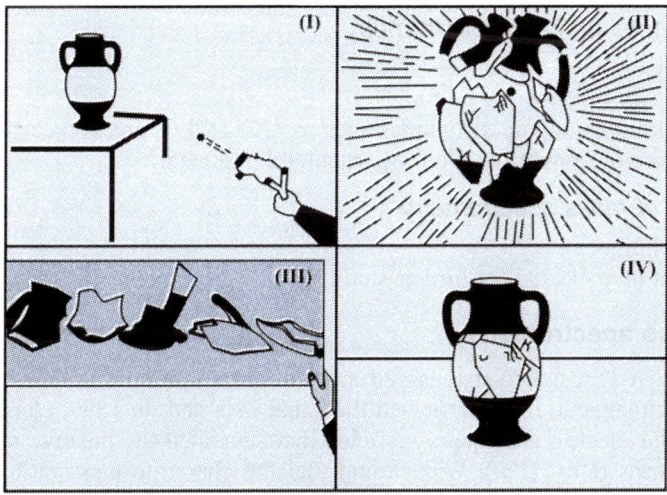

Fig. 1.1: Illustration of the principle of mass spectrometry by electron bombardment

is a destructive technique. This is because mass spectrometry is the study of the effect of ionizing energy on molecules. It depends upon chemical reactions in the gas phase in which sample molecules are consumed during the formation of ionic and neutral species. Although sample is consumed destructively by the mass spectrometer, the technique is very sensitive and only trace amounts of material are used in the analysis. A mass spectrometer converts sample molecules into ions in the gas phase, separates them according to their mass to charge ratio (*m/e*) and sequentially records the individual ion fragment in the form of the mass spectrum. The mass spectrum (Fig. 1.2) usually presented as a vertical bar graph in which bar represent an ion (fragment ion) having specific mass to charge ratio and length of bar indicates the relative abundance or relative intensity. The highest peak intensity is counted 100%, and rest of peaks is measured with respect to 100% intensity peak (base peak). For example, in mass spectrum of methyl-t-butyl ketone, the highest peak (base peak) at *m/e* 57 is taken as 100% intensity peak. The rest of fragment ions peaks, e.g. *m/e* 100, 85, 43, 41, 29 are measured with respect to the highest peak, *m/e* 57.

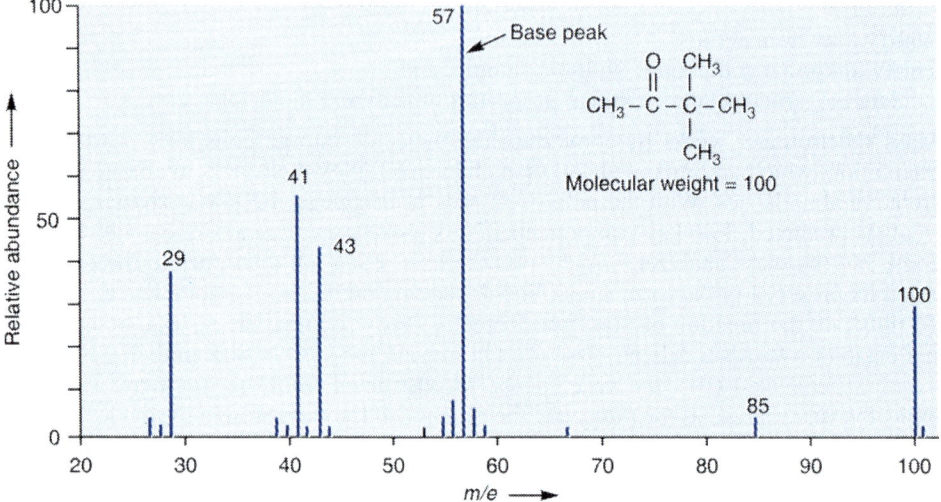

Fig. 1.2: Mass spectrum of methyl-*t*-butyl ketone

Special advantages of mass spectrometry
- High sensitivity
- High accuracy
- Coupling of chromatographic techniques such as GC, HPLC, etc. It means analysis can be done in a mixture or impure sample for particular component.

Disadvantages of mass spectrometry
- Destructive technique
- Samples cannot be collected for further study

History of mass spectrometry

Goldstein discovered the positively charged 'rays' in 1886. Later on, Wilhelm Wien found that strong electric or magnetic fields deflected the canal rays and, in 1899, constructed a device with perpendicular electric and magnetic fields that separated the positive rays according to their charge-to-mass ratio (*Q/m*). Wien found that the charge-to-mass ratio depends on the nature of the gas in the discharge tube. He deflected a beam of positive ions in electric and magnetic fields. In 1912, **J.J. Thomson** (Father of Mass Spectrometry) built his 'parabola mass spectrograph' to measure the charge-to-mass ratio (z/m) for several ionic species. He was able to demonstrate the existence of two isotopes of neon, masses 20 and 22, using a magnetic deflection instrument. In the expression z/m, z is the charge number, i.e. the total charge on an ion divided by the elementary charge (e), and m is the nucleon number, i.e. the sum of the total number of protons and neutrons in an atom, molecule or ion. In modern mass spectrometry, the parameter measured is m/z, rather than z/m.

Thomson's student, **Aston** continued the work at Cambridge and built instruments that helped him to establish the presence of isotopes. He was subsequently able to measure the atomic mass of most elements with sufficient accuracy to be able to calculate the 'packing fraction' of their atomic nuclei. The packing fraction is the difference between the accurate atomic mass of the isotope and the nearest whole number divided by the mass number, also known as the mass defect. His work on isotopes also led to his formulation of the **Whole Number Rule** which states that "the mass of the oxygen isotope being defined [as 16], all the other isotopes have masses that are very nearly whole numbers," a rule that was used extensively in the development of nuclear energy.

Aston also obtained accurate measurements of the ratios of the stable isotopes of many of the known elements. At the end of this exciting period of development, Aston was convinced that much of the potential of mass spectrometry had been exploited. It was not until the 1940s that the technique was put to work in elucidating organic structures in the petroleum industry. Ionization was effected by electron 'impact' [now called electron ionization (EI)] for those molecules that could withstand vaporization into the heated and evacuated ion source without decomposition. This limited the practical mass range to less than 1000 Daltons (Da)* but yielded useful fragmentations for structure elucidation (*see* Chapter 2).

By choosing to work with 70 electron volt (eve), many ions were formed with internal energies far in excess of the ionization energy (IE). These ions decompose rapidly to produce lower mass (fragment) ions and neutral radicals or molecules. During the 1950s, commercial instruments were being built and new applications discovered. One of the earliest of these was the identification of low molecular weight volatile food flavor compounds. Ten years later, the powerful combination of electron ionization mass spectrometry (EIMS) with gas chromatography (GC/MS) led to an explosion of applications where mass spectrometry was used in qualitative and quantitative, chemical and biochemical studies. GC/MS instruments produced enormous amounts of data, which were best handled by computers data acquisition methods. In 1966, **Munson and Field** described chemical ionization (CI) technique. This technique increased the yield of ions representative of the molecular weight of volatile molecules through interactions with reagent gas ions (e.g. CH_5^+ ions from methane) with little excess energy. Other 'soft' ionization techniques such as field desorption (FD) and particle desorption methods based upon ion generation by Cf-252 fast fission products [plasma desorption, (PDMS)] were introduced during the 1970s for nonvolatile compounds. At the same time (and in response to these developments) the instrumental mass range was increased to cope with the larger sample molecule ions now entering the gas phase. This process accelerated in the 1980s with the introduction of Fast Atom Bombardment (FAB) ionization. FAB was the first ionization technique to enable biologists and biochemists routinely to obtain molecular weight information on complex, labile biomolecules, including polypeptides and small proteins.

Ionization from the liquid state, followed by evaporation/desolvation of charged droplets, includes techniques such as ion spray, thermospray (TSP) and electrospray ionization (ESI). These methods differed mainly in the manner in which ionization was initiated. Multiple charged molecular ions could be formed under ESI, facilitating the measurement of high molecular masses, even on conventional instruments (i.e. those with a mass range up to 2000 or 4000 The).

---

* Dalton (Da) is the unit of mass (also known as the mass unit, u) and is 1/12 of the mass of C (defined as 12.000000).

More efficient pumping systems were required to cope with the increased gas volumes generated by vaporizing liquids.

Separation techniques such as liquid chromatography and capillary electrophoresis coupled to mass spectrometry (LC/MS and CE/MS respectively) have extended the advantages first associated with the analysis of volatile compounds by GC/MS to compounds of low volatility and high molecular weight. Tandem mass spectrometry (MS/MS) collision-induced dissociation (CID), focal-plane array detectors, ion traps and hybrid instruments are providing a high sensitivity structure elucidation facility for nonvolatile compounds similar to that provided by EIMS of volatiles. Recently, laser desorption (LD), and especially matrix assisted laser desorption ionization (MALDI), combined with time-of-flight (ToF) mass analysis has extended the practical application for biomolecules. These techniques require minimum ionization energy (minimum energy of excitation for an atom or molecule) to remove an electron in order to produce a positively charged ion.

Fourier transform mass spectrometry (FTMS), also known as Fourier transform ion cyclotron resonance (FT-CR) mass spectrometry is only slowly entering into the commercial area.

Historical development in mass spectrometry

19th century

1886   Eugen Goldstein observed canal rays.
1898   Wilhelm Wien demonstrated that canal rays can be deflected using strong electric and magnetic fields. He showed that the mass-to-charge ratio of the particles has opposite polarity and is much larger compared to the electron. He also realized that the particle mass is similar to the one of hydrogen particles.
1898   J.J. Thomson measured the mass-to-charge ratio of electrons.

20th century

1901   Walter Kaufmann used a mass spectrometer to measure the relativistic mass increase of electrons.
1905   JJ Thomson began his study of positive rays.
1906   Thomson was awarded the Nobel Prize in Physics "in recognition of the great merits of his theoretical and experimental investigations on the conduction of electricity by gases".
1913   Thomson was able to separate particles of different mass-to-charge ratios. He separates the ^{20}Ne and the ^{22}Ne isotopes, and he correctly identifies the m/z = 11 signal as a doubly charged ^{22}Ne particle.
1919   Francis Aston constructed the first velocity focusing mass spectrograph with mass resolving power of 130.
1922   Aston was awarded the Nobel Prize in chemistry "for his discovery, by means of his mass spectrograph, of isotopes, in a large number of non-radioactive elements, and for his enunciation of the whole-number rule".
1931   Ernest O. Lawrence invented the cyclotron.
1934   Josef Mattauch and Richard Herzog developed the double-focusing mass spectrograph.
1936   Arthur J. Dempster developed the spark ionization source.
1937   Aston constructed a mass spectrograph with resolving power of 2000.
1939   Lawrence received the Nobel Prize in Physics for the cyclotron.
1942   Lawrence developed the Calutron for uranium isotope separation.
1946   William Stephens presented the concept of a time-of-flight mass spectrometer.
1956   Fred McLafferty proposed a hydrogen transfer reaction that is known as the McLafferty rearrangement.
1959   Researchers at Dow Chemicals developed an interface to couple gas chromatograph to a mass spectrometer.
1966   FH Field and MSB Munson developed chemical ionization.
1968   Malcolm Dole developed electrospray ionization.

1969 HD Beckey developed field desorption.

1974 Comisarow and Marshall developed Fourier Transform Ion Cyclotron Resonance Mass Spectrometry.

1976 Ronald MacFarlane and co-workers developed plasma desorption mass spectrometry.

1984 John Bennett Fenn and co-workers used electrospray technique to ionize biomolecules.

1985 Franz Hillenkamp, Michael Karas and co-workers described and coined the term matrix-assisted laser desorption ionization (MALDI).

1987 Koichi Tanaka used the "ultra fine metal plus liquid matrix method" to ionize intact proteins.

1989 Wolfgang Paul received the Nobel Prize in Physics "for the development of the ion trap technique".

1999 Alexander Makarov presented the Orbitrap mass spectrometer.

21st century

2002 John Bennett Fenn and Koichi Tanaka were awarded one-quarter of the Nobel Prize in chemistry each "for the development of soft desorption ionisation methods ... for mass spectrometric analyses of biological macromolecules".

APPLICATIONS OF MASS SPECTROMETRY

1. Identification of unknown compounds

From mass spectra of any compound, we can get the information about:

- Molecular weight
- Structural characteristics
- Elemental composition of molecular ion and fragment ions

2. Monitoring of chemical reactions

During the last several decades, mass spectrometry (MS) has rapidly developed as a practical technique that can be used to monitor chemical reactions and investigate reaction mechanisms (Fig. 1.3). The real-time discovery of intermediates and products provides critical information regarding the reaction mechanism, which can facilitate the optimization of reaction conditions. With appropriate ionization methods, most chemical compounds can be ionized and detected with MS hybrid systems. Reactions can be monitored either by observing the disappearance of the reactants or appearance of the products in the mass spectrum.

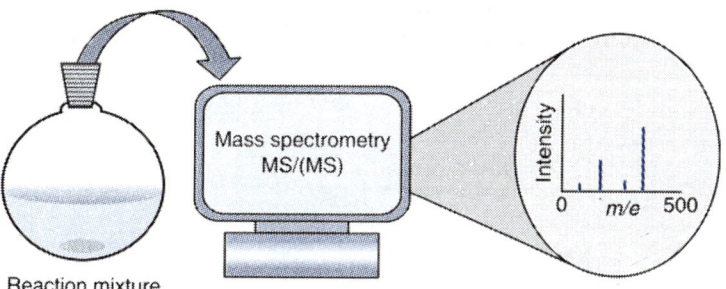

Fig. 1.3: Monitoring of chemical reactions

3. Determination of sequence of amino acids in peptide chain

There are three different types of bonds that can fragment along the amino acid backbone: the **NH–CH**, **CH–CO**, and **CO–NH** bonds (Fig. 1.4). Each bond breakage gives rise to two species, one neutral and the other one charged, and only the charged species is monitored by the mass spectrometer. The charge can stay on either of the two fragments depending on the chemistry and relative proton affinity of the two species. Hence, there are six possible fragment ions for

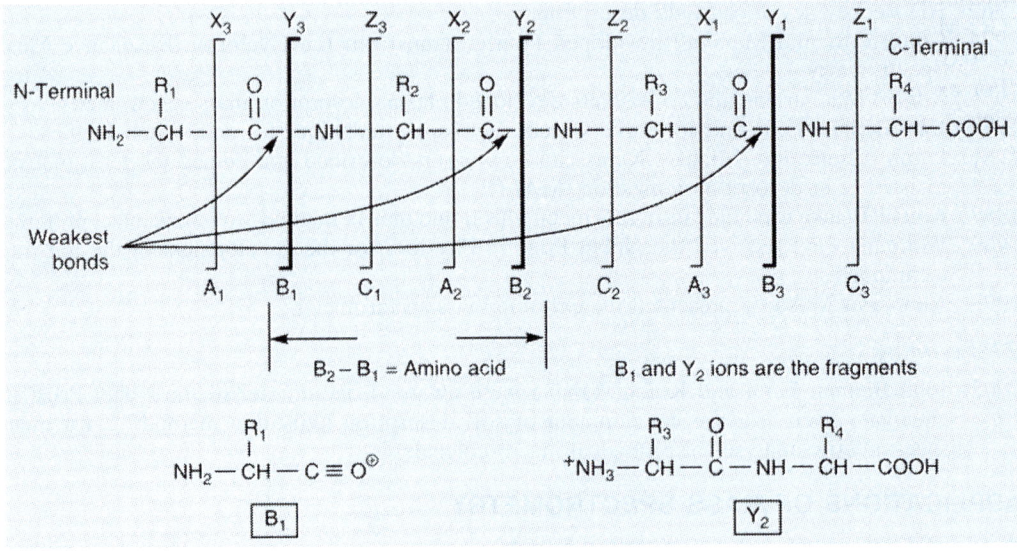

Fig. 1.4: Sequence of amino acids in peptide chain

each amino acid residue and these are labelled as in the diagram, with the **A**, **B**, and **C ions** having the charge retained on the **N-terminal fragment**, and the **X**, **Y**, and **Z ions** having the charge retained on the **C-terminal fragment**. The most common cleavage sites are at the CO–NH bonds which give rise to the B and/or the Y ions. The mass difference between two adjacent B ions, or Y ions, is indicative of a particular amino acid residue. Software are also available to determine the peptide structure (example: SEQUEST).

4. Determination of oligosaccharide structure

Structural elucidation of complex carbohydrates requires determination of monosaccharide composition, sequence, branching pattern, glycosidic linkages, and anomeric configuration (Fig. 1.5). One of the efficient methods for the derivatization of oligosaccharides, wherein the oligosaccharide is efficiently ligated to a basic aminooxyacetyl peptide by oxime formation. The resulting glycopeptide yields much higher sensitivity in matrix-assisted laser desorption/ionization mass spectrometry than does the underivatized oligosaccharide.

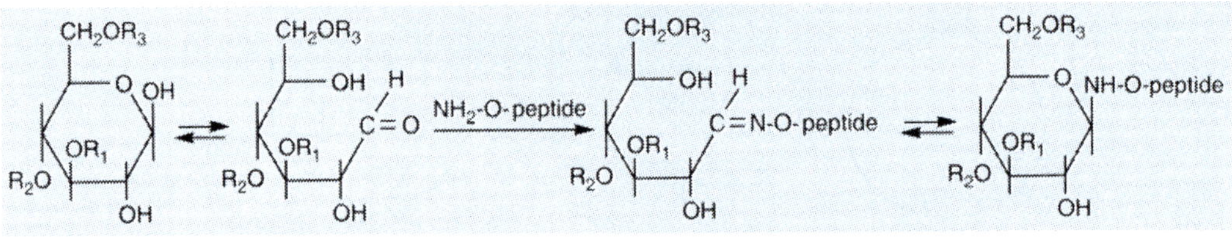

Fig. 1.5: Sequence of oligosaccharide in carbohydrate

5. Mass spectrometry in pesticides analysis

- Pesticides are indispensable chemicals and poisonous to mankind—adversely affects nerve functioning, direct exposure can cause eye problems like blurring of vision, reddening, retardation in fetal growth, etc.
- Residual analysis in food, water and environment samples is of paramount importance from viewpoint of preventive medicine.

- Frequently used pesticides are organophosphorous compounds (OP) and carbamate derivatives.
- GC-MS can determine meconium, cypermethrin, malathion, cyfluthin, etc. in the concentration of 0.01–4.15 μg/g.
- For example, in the mass spectrometry, peak at *m/e* 157 and 125 (Fig. 1.6) is focussed to detect the presence of malathion.

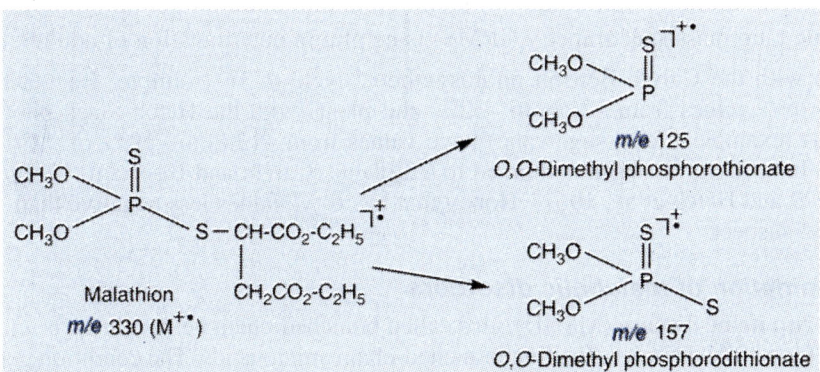

Fig. 1.6: Pesticide analysis by mass spectrometry

6. Mass spectrometry in determination of drug of abuse

Cocaine is a carboxylic acid ester that is rapidly hydrolyzed *in vivo* to the free acid benzoyl ecgonine (BEG) and detection of BEG is used to monitor cocaine abuse.

BEG itself has poor vapor phase properties because of the polar, ionizable carboxylate moiety, and it is therefore converted to a carboxylic acid ester, in this example, the pentafluoropropyl (PFP) ester, before GC/MS analysis. The molecular ion of PFP-BEG is *m/e* 421, and two characteristic fragment ions of *m/e* 316 and *m/e* 300 that arise from cleavages about an oxygen atom of the ester (Fig. 1.7). A ring-labeled [^{2}H$_3$]-BEG internal standard is added before extraction, processed along with the target analyte, and also derivatized (d$_3$-PFP-BEG). The ions in the mass spectrum of d$_3$-PFP-BEG analogous to those in the spectrum of PFP-BEG are *m/e* 424, 319, and 303, and in each case retain all three [^{2}H] atoms. GC/MS is then performed with selected ion monitoring of the ion pairs *m/e* 316 and 319, *m/e* 300 and 303, and *m/e* 421 and 424.

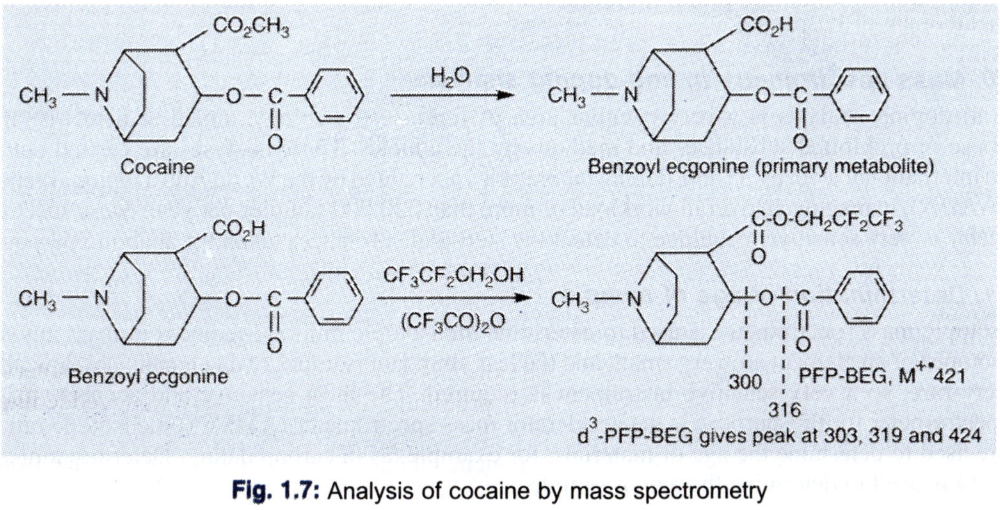

Fig. 1.7: Analysis of cocaine by mass spectrometry

7. Determination of adulteration of honey

Bee honey is a unique sweetening agent that can be used by humans without processing, and it provides significant nutritious and medical benefits. Because of its nutritional and medicinal value, honey continues to be a popular food. However, honey can easily be adulterated with various cheaper sweeteners, such as refined cane sugar, beet sugar, high fructose corn syrup and maltose syrup, resulting in higher commercial profits.

Isotopic ratio mass spectrometry (IRMS) is helpful in determination of adulteration.

Plants with the Calvin-Benson photosynthetic cycle (C3) (example: Beet, wheat) have $^{13}C/^{12}C = 8‰$ values from $-21‰$ to $-32‰$ and plants with the Hatch-Slack photosynthetic cycle (C4) (example: Corn, sugarcane) have values from $-12‰$ to $-19‰$ of $^{13}C/^{12}C = 8‰$; C4 plants have high ^{13}C when compared to C3 plants (Calvin and Bassham, 1962, Hatch and Slack, 1979 and Hatch et al., 1967). Honey that has $\delta^{13}C$ values less negative than $-23.5‰$ is considered suspect.

8. Determination of metabolic disorders

Maple syrup urine disease (MSUD), also called branched-chain ketoaciduria, is an autosomal recessive metabolic disorder affecting branched-chain amino acids. The condition gets its name from the distinctive sweet odor of affected infants' urine, particularly prior to diagnosis, and during times of acute illness.

MSUD is a metabolic disorder caused by a deficiency of the branched-chain alpha-keto acid dehydrogenase complex (BCKDC), leading to a buildup of the branched-chain amino acids (leucine, isoleucine, and valine) and their toxic by-products (keto acids) in the blood and urine.

Newborn screening for maple syrup urine disease involves analyzing the blood of 1–2 day-old newborns through tandem mass spectrometry. The blood concentration of leucine and isoleucine is measured relative to other amino acids to determine if the newborn has a high level of branched-chain amino acids.

9. Mass spectrometry in drug metabolic studies

Mass spectrometry plays a pivotal role in drug metabolism studies, which are an integral part of drug discovery and development nowadays. Metabolite identification is helpful in understanding the metabolic fate of drug candidates and to aid lead optimization with improved metabolic stability, toxicology and efficacy profiles. The mass spectrum of buspirone (antipsychotic drug) is shown in Fig. 1.8. The peak at m/e 122 can be monitored to know the metabolism of buspirone.

10. Mass spectrometry in anti-doping analysis

Anti-doping analysis is a very peculiar area of forensic toxicology, aimed at detecting the abuse of prohibited substances and methods by the athletes. These analyses are carried out at an international level by 33 anti-doping laboratories accredited by the World Anti-Doping Agency (WADA), managing an overall workload of more than 220,000 samples per year. Mass spectrometry is very sensitive technique to detect the steroidal substances taken for anabolic purpose.

11. Determination of age of sample

Isotopic mass spectrometry is used to determine the isotopic ratio. Differences in mass among isotopes of an element are very small, and the less abundant isotopes of an element are typically very rare, so a very sensitive instrument is required. The most sensitive and accurate mass spectrometer for this purpose is the accelerator mass spectrometer (AMS). Some isotope ratios are used to determine the age of materials, for example, as in carbon dating. Determination of C-14 is used to determine the age of sample.

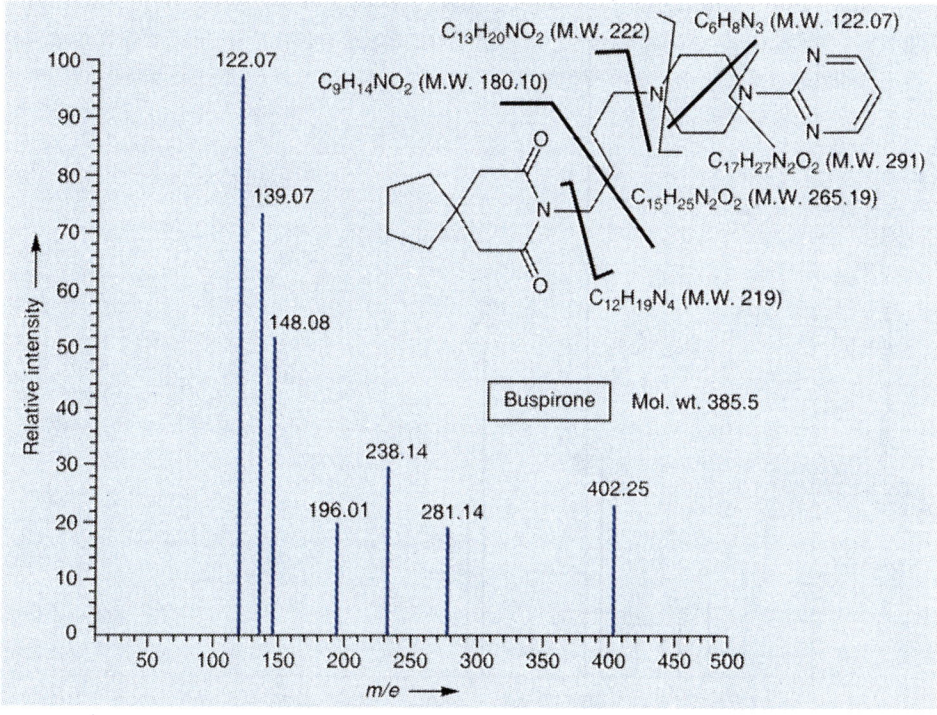

Fig. 1.8: Mass spectrum of buspirone with its metabolite

12. Monitoring of anesthesia

Anesthesia adequacy was assessed with mass-spectrometric method by monitoring the ratio of mass concentrations of end-tidal CO_2 and inhaled O_2 in every respiratory cycle during surgery. For real-time monitoring, we used a mass spectrometer with electron ionization connected to the respiratory contour of inhalation anesthesia machine. The study has demonstrated advantages of the novel method in real-time assessment of adequacy of the total intravenous anesthesia.

13. Mass spectrometry in fermentation industry

Ethanol produced by yeast fermentation, called bioethanol, accounts for approximately 95% of the ethanol production. Ethanol is produced from various kinds of substrates. The substrate used for ethanol production is chosen based on the regional availability and economical efficiency.

They monitor the composition of gas streams into and out of fermentors and bioreactors continuously, accurately and reliably. Ethanol in the vent gas is linearly related to the concentration in the fermentor broth they give a continuous monitor of the ethanol production which is particularly important for detecting the start of ethanol production and also for monitoring changes in ethanol production.

Although the molecular weight of ethanol is 46 it can be seen that the molecular ion $(CH_3CH_2OH^+)$ peak at mass 46 is not the largest peak, in fact it is not even the second largest peak. The ethanol molecule tends to fragment during ionization and the largest peak is actually at mass 31 due to CH_2OH^+. Also there is considerable interference from the CO_2 in the vent gas at masses 45 and 46, due to the 13C, 17O and 18O isotopes. Therefore, we have to use mass 31 to analyze ethanol (Fig. 1.9). However, we need to consider the presence of a very large peak at mass 32 from the percentage levels of O_2 in the vent gas. We need to correct for the tail from the 32 peak if we are to make an accurate measurement of ethanol at low concentrations (ppm) at the start of ethanol production.

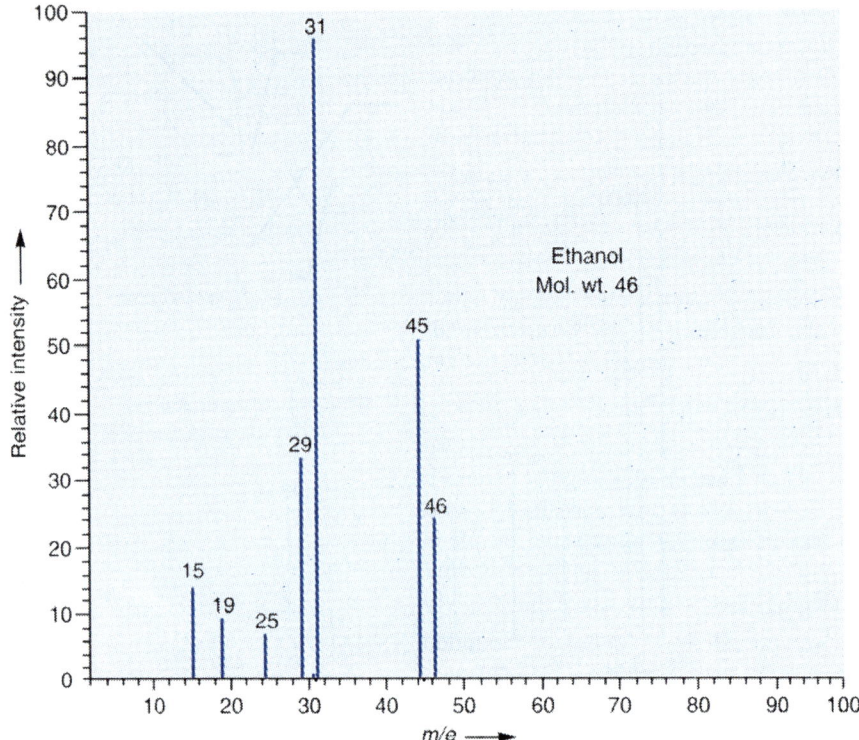

Fig. 1.9: Mass spectrum of ethanol

14. Determination of composition of crude oil (geological importance)

Crude oil is a natural multicomponent mixture. Its major part is composed of hydrocarbons (alkanes, naphthenes, and aromatics). Their content in oils ranges between 30% and 100%. The emergence of analytical techniques such as the gas chromato-mass spectrometry enabled scientists to obtain new information on the composition and structure of petroleum hydrocarbons, study in detail their homological series, and determine the distribution patterns of normal and branched alkanes, methylalkanes, and isoprenoid alkanes in oils.

15. Detection of dioxins by mass spectrometry

Dioxins are not created intentionally, but they can be produced through industrial processes, including combustion, chlorine bleaching of pulp and paper and through certain types of chemical manufacturing and processing. These are toxic, effective in low concentration and cause cancer in human beings.

Polychlorinated dibenzodioxins (PCDDs), polychlorinated dibenzofurans (PCDFs) and polychlorinated biphenyls (PCBs) constitute a group of polyhalogenated aromatic compounds that have become known as "dioxin" and "dioxin-like" compounds.

Gas chromatography (GC) with high resolution mass spectrometry (HRMS) is a sensitive method of detection for these compounds.

16. Mass spectrometry in environmental monitoring

Monitoring environmental pollutants is a major application of GC-MS. It is widely used in the detection of dibenzofurans, dioxins, herbicides, sulfur, pesticides, phenols, and chlorophenols in air, soil, and water.

17. Food and fragrance analysis

Aromatic compounds such as fatty acids, esters, aldehydes, alcohols, and terpenes present in food and beverages can be easily analyzed using GC-MS. The technique can also be used to detect the spoilage or contamination of food. The analysis of a wide range of oils such as lavender oil, olive oil, spearmint oil, and essential oils, perfumes, fragrances, allergens, menthol, and syrups is also possible using GC-MS.

18. Cancer diagnosis by mass spectrometry

Why to use mass spectrometry for cancer diagnosis?

The big reason is that mass spectrometry can in theory conclusively prove the presence, identify, and report the concentration of all of the small peptides and proteins in an unknown sample, such as blood. No other diagnostic protocol can do this.

Why is this important?

Cancer is often diagnosed by an abnormal amount of a particular peptide or protein in blood. These may act as **Cancer biomarkers**. These biomarkers may be (i) a mediator of the disease pathology, (ii) present at low and stable expression levels in healthy individuals and higher expression levels in patients, and (iii) simple and quick to evaluate. Such a biomarker can be assayed and linked to cancer using a defined mechanism. For example, the prostate cancer can be diagnosed by determining Zn-alpha $(\alpha)_2$ glycoprotein in serum by LC-MS/MS technique.

The mass spectrometric technique has additional advantage. Traditional cancer diagnoses are specific to one type of cancer. In theory, mass spectrometry can instead be used as a universal cancer screening.

19. Miscellaneous

Mass spectrometric techniques can be used for:

(a) Determination of gene damage due to environmental causes. DNA-damaging agents generate a plethora of products in the DNA of living organisms. The DNA damage can lead to numerous diseases including carcinogenesis. Both gas chromatography-mass spectrometry (GC-MS) or liquid chromatography-mass spectrometry (LC-MS), in single or tandem versions, have been used for the measurement of numerous DNA damage resulting products.

(b) Location of petroleum deposit by testing rock samples.

(c) Testing the purity of semiconductor material used in making microchips for computers.

2

Introduction to
Mass Spectrometer

INTRODUCTION

Mass spectrometer (MS) is a kind of instrument which uses an analytical technique to measure the mass-to-charge ratio of ions in the analytical technique, known as Mass spectrometry. For analysis, the mass spectrometer involves the following sequences:

Sequence 1: Ionization

The molecule is ionized by knocking one or more electrons off to give a positive ion. This is true even for things which you would normally expect to form negative ions (chlorine, for example). Mass spectrometers always work with positive ions. Note: *It is easy to handle the charged particle in comparison to neutral molecule.*

Sequence 2: Acceleration

The ions are accelerated so that they all have the same kinetic energy.

Sequence 3: Deflection

The ions are then deflected by a magnetic field according to their masses. The lighter they are, the more they are deflected. The amount of deflection also depends on the number of positive

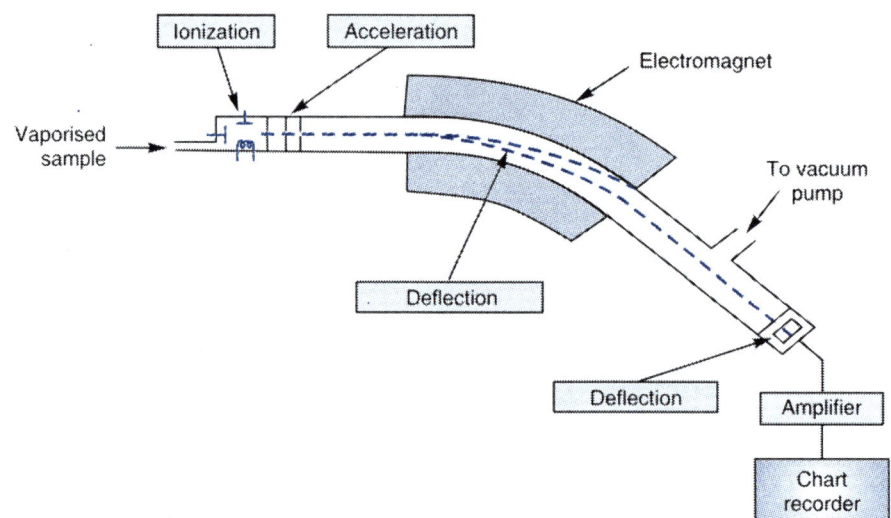

Fig. 2.1: Sequences of mass spectrometry technique

charges on the ion—in other words, on how many electrons were knocked off in the first stage. The more the ion is charged, the more it gets deflected.

Sequence 4: Detection

The beam of ions passing through the machine is detected electrically.

The need for a vacuum

It is important that the ions produced in the ionization chamber have a free run through the machine without hitting air molecules. For this reason, vacuum is needed.

GENERAL STRUCTURE OF MASS SPECTROMETER

Generally, a typical mass spectrometer consists of three parts: An ion source, a mass analyzer and a detector.

The function of the ion source is to produce ions from the sample. The function of the mass analyzer is to separate ions with different mass-to-charge ratios. Then the numbers of different ions are detected by the detector. Finally, the mass spectrum is generated after all the data have been collected. Fig. 2.2 is a schematic graph of the mass spectrometer.

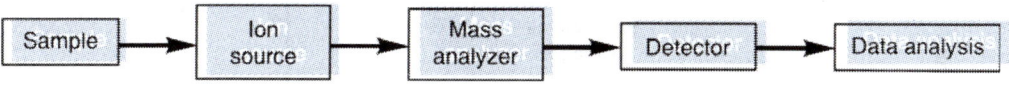

Fig. 2.2: Schematic diagram of mass spectrometer

Apart from the above components, there is a sample inlet system and vacuum system.

Ion source

A mass spectrometer works by using magnetic and electric fields to exert forces on charged particles (ions) for separation in a vacuum. Therefore, a compound must be charged or ionized to be analyzed by a mass spectrometer. Furthermore, the ions must be introduced in the gas phase into the vacuum system of the mass spectrometer. This is easily done for gaseous or heat-volatile samples. However, many (thermally labile) analytes decompose upon heating. These kinds of samples require either desorption or desolvation methods if they are to be analyzed by mass spectrometry. Although ionization and desorption/desolvation are usually separate processes, the term "ionization method" is commonly used to refer to both ionization and desorption (or desolvation) methods.

The choice of ionization method depends on the nature of the sample and the type of information required from the analysis. So-called 'soft ionization' methods such as field desorption and electrospray ionization tend to produce mass spectra with little or no fragment-ion content. Comparatively, *hard ionization source* causes the extensive fragmentation. Hence, nature of mass spectrum depends on methods of ionization. For example, decanol when subjected to hard source ionization shows more fragments (more peaks) in mass spectrum (Fig. 2.3a) in comparison to mass spectrum (Fig. 2.3b) obtained by soft source ionization.

In the present text, the ionization methods are discussed into the gaseous phase and desorption ionization.

Gas-phase ionization

These methods rely upon ionizing gas-phase samples. The samples are usually introduced through a heated batch inlet, or heated direct insertion probe, or a gas chromatograph.

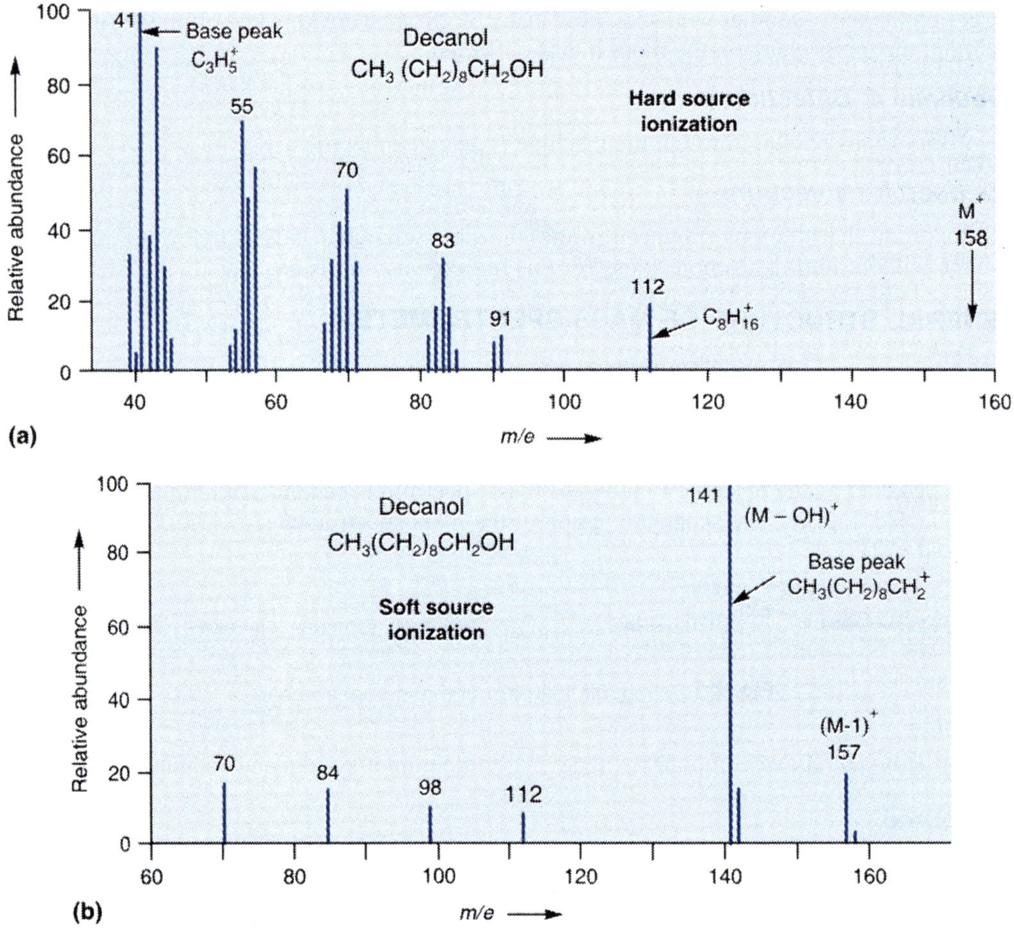

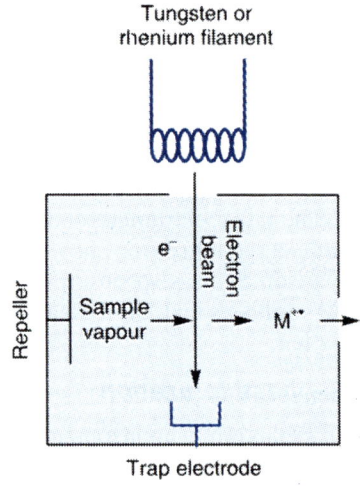

Fig. 2.3: Comparison of mass spectrum: (a) Hard source ionization; (b) Soft source ionization

Electron ionization (EI)

It, also referred to as electron impact ionization, is the oldest and best-characterized of all the ionization methods. The process is a relatively harsh form of ionization producing a wide range of molecular fragments. EI is best suited to relatively non-polar, volatile samples. The electron impact source (Fig. 2.4) consists of a heated tungsten or rhenium filament that produces electrons which are accelerated to another electrode called the ion trap. Sample vapor diffuses at the right angle into the electron beam and become ionized and fragmented. The size of fragment depends on the electron energy which is controlled by the accelerating potential on the ion trap electrode. Low energy electrons produce molecular ions and larger fragments, whereas high energy electrons produce many smaller fragments and possibly no molecular ions. The following gas phase reaction describes the electron ionization process.

Fig. 2.4: Electron impact ion source

$$M + e^- \rightarrow M^{+\bullet} + 2e^-$$

Here, M represents the analyzed molecule and $M^{+\bullet}$ is its molecular ion which may be further fragmented depending on the energy.

Note: Ionization potential is the energy required to remove the one electron from the molecule. When the transferred energy is equal to the ionization potential of the molecule, it loses an electron and ionizes. For most elements, the ionization potential is in the range of 5 to 15 eV. It varies from 8 to 12 eV for organic compounds. When more energy is transferred to the molecule by the bombarding electron, a molecule M can undergo the various types of electron impact induced reactions; fragmentation of the molecule takes place and so more positive ion fragments are formed. Generally, the spectrum is run at 70 eV (average energy of the electron beam).

After the ions are produced, they are driven by a potential applied to an ion-repeller electrode, away from the ion source into the accelerating region of the mass spectrometer, where mass analysis takes place.

Most mass spectrometers use electrons with an energy of 70 electron volts (eV) for EI. Decreasing the electron energy can reduce fragmentation, but it also reduces the number of ions formed.

Sample introduction

- Heated batch inlet.
- Heated direct insertion probe.
- Gas chromatograph.
- Liquid chromatograph (particle-beam interface).

Benefits

- Well-understood.
- Can be applied to virtually all volatile compounds.
- Reproducible mass spectra.
- Fragmentation provides structural information due to more fragmentation (hard source ionization).
- Libraries of mass spectra can be searched for EI mass spectral "fingerprint".

Limitations

Sample must be thermally volatile and stable. To increase the volatility, chemical derivatives can be prepared. For example, organic acids, fatty acids, etc. can be analyzed by EI method with the formation of their methyl esters. Many chemical classes can be rendered volatile through the preparation of a variety of silylated products that enable mass and structural information to be obtained. Special chemical derivatives can be made that yield characteristic fragmentation properties, thereby aiding analysis. An example is the preparation of *t*-butyldimethylsilyl derivatives of steroids. These yield an abundant $[M - 57]^+$ ion (generated by loss of C_4H_9) that is especially useful in quantitative measurements.

The molecular ion may be weak or absent for many compounds.

Mass range

Low. Typically less than 1,000 Da.

Chemical ionization

Chemical ionization (CI) is a soft ionization technique which produces ions with a little excess energy (Fig. 2.5). Chemical ionization uses ion-molecule reactions to produce ions from the analyte (M). Chemical ionization (CI) is applied to samples similar to those analyzed by EI and is primarily used to enhance the abundance of the molecular ion.

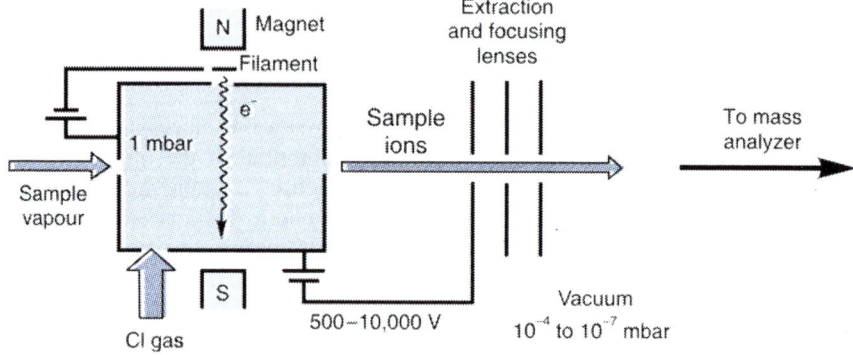

Fig. 2.5: Schematic diagram of chemical ion source

The chemical ionization process begins when a reagent gas such as methane, isobutane, or ammonia is ionized by electron impact. A high reagent gas pressure (or long reaction time) results in ion-molecule reactions between the reagent gas ions and reagent gas neutrals. Some of the products of these ion-molecule reactions can react with the analyte molecules to produce analyte ions.

For example, chemical ionization involving methane is shown in Fig. 2.6.

(A) EI of reagent gas to form ions:

$$CH_4 + e^- \longrightarrow CH_4^{+\bullet} + 2e^-$$

(B) Reaction of reagent gas ions to form adducts:

$$CH_4 + CH_4^{+\bullet} \longrightarrow CH_3^{+\bullet} + CH_5^+$$

or $$CH_4^{+\bullet} \longrightarrow CH_3^+ + H^\bullet$$

$$CH_3^+ + CH_4 \longrightarrow C_2H_5^+ + H_2$$

(C) Reaction of reagent gas ions with analyte molecule (M):

$$CH_5^+ + M \longrightarrow CH_4 + MH^+ (M+1)$$

$$C_2H_5^+ + M \longrightarrow C_2H_4 + MH^+ (M+1)$$

$$CH_3^+ + M \longrightarrow CH_4 + (M+H)$$

Fig. 2.6: Chemical ionization reactions with methane gas

Similarly, chemical ionization with isobutane and ammonia occurs. The isobutane loses an electron upon EI and yields the corresponding radical cation, which will fragment mainly through the loss of hydrogen radical to yield a *t*-butyl cation, and to a lesser extent through the loss of a methyl radical:

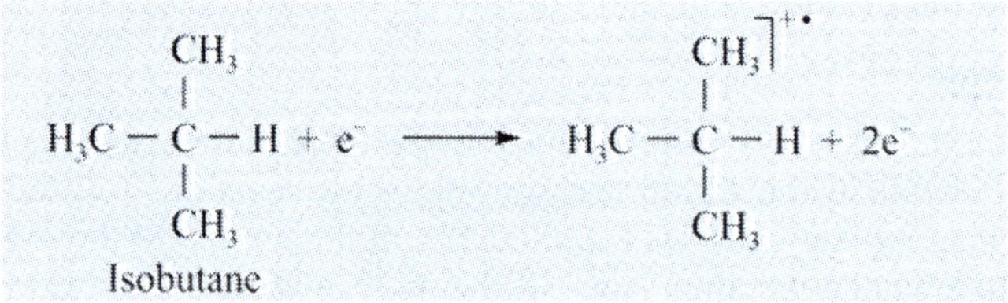

$$
\begin{array}{ccc}
& CH_3 & & CH_3 \rceil^{+\bullet} \\
& | & & | \\
H_3C-C-H + e^- & \longrightarrow & H_3C-C-H + 2e^- \\
& | & & | \\
& CH_3 & & CH_3 \\
\end{array}
$$

Isobutane

$$CH_3 \cdot C - H \xrightarrow{\hspace{2cm}} \begin{array}{l} H_3C - \overset{+}{\underset{CH_3}{\overset{CH_3}{\mid}}}C + H^\bullet \\[2em] H_3C - \overset{+}{\underset{CH_3}{C}} + H + CH_3^\bullet \end{array}$$

$$M + C_4H_9^+ \rightarrow (M + H)^+ + C_4H_8$$

With ammonia,

$$NH_3 + e^- \rightarrow NH_3^{+\bullet} + 2e^-$$

$$NH_3 + NH_3^{+\bullet} \rightarrow NH_4^+ + {}^\bullet NH_2$$

$$NH_4^+ + NH_3 \rightarrow N_2H_7^+$$

$$M + NH_4^+ \rightarrow (M + H)^+ + NH_3$$

Two factors determine the choice of the reagent gas to be used:
• Proton affinity (PA)
• Energy transfer

Sample introduction
• Heated batch inlet.
• Heated direct insertion probe.
• Gas chromatograph.
• Liquid chromatograph (particle-beam interface).

Benefits
• Often gives molecular weight information through molecular-like ions such as $[M + H]^+$, even when EI would not produce a molecular ion.
• Simple mass spectra, fragmentation reduced compared to EI.

Figures 2.7 (a and b) and Figs 2.8 (a and b) compare the mass spectra of *m*-nitrobenzyl alcohol and cyclophosphamide respectively.

The electron impact mass spectrum involves extensive fragmentation and gives $M^{+\bullet}$. The chemical ionization spectrum is much simpler and gives $[M + H^+]$.

Limitations
• Sample must be thermally volatile and stable.
• Less fragmentation than EI, fragment pattern not informative or reproducible enough for library search. Results depend on reagent gas type, reagent gas pressure or reaction time, and nature of sample.

Mass range
Low. Typically less than 1,000 Da.

Note: In order to rationalize the fragmentation of MH^+ ions, one must consider at which sites in the sample molecule the proton is attached. The spectrum may then be rationalised in terms of the fragmentation of the different types of MH^+ ions. In general, protonation occurs on hetero-atoms having lone pairs of electrons, such as O, N and Cl. This frequently followed by charge-induced elimination of a molecule containing the hetero-atom. Other possible protonation sites are aromatic rings and regions of unsaturation.

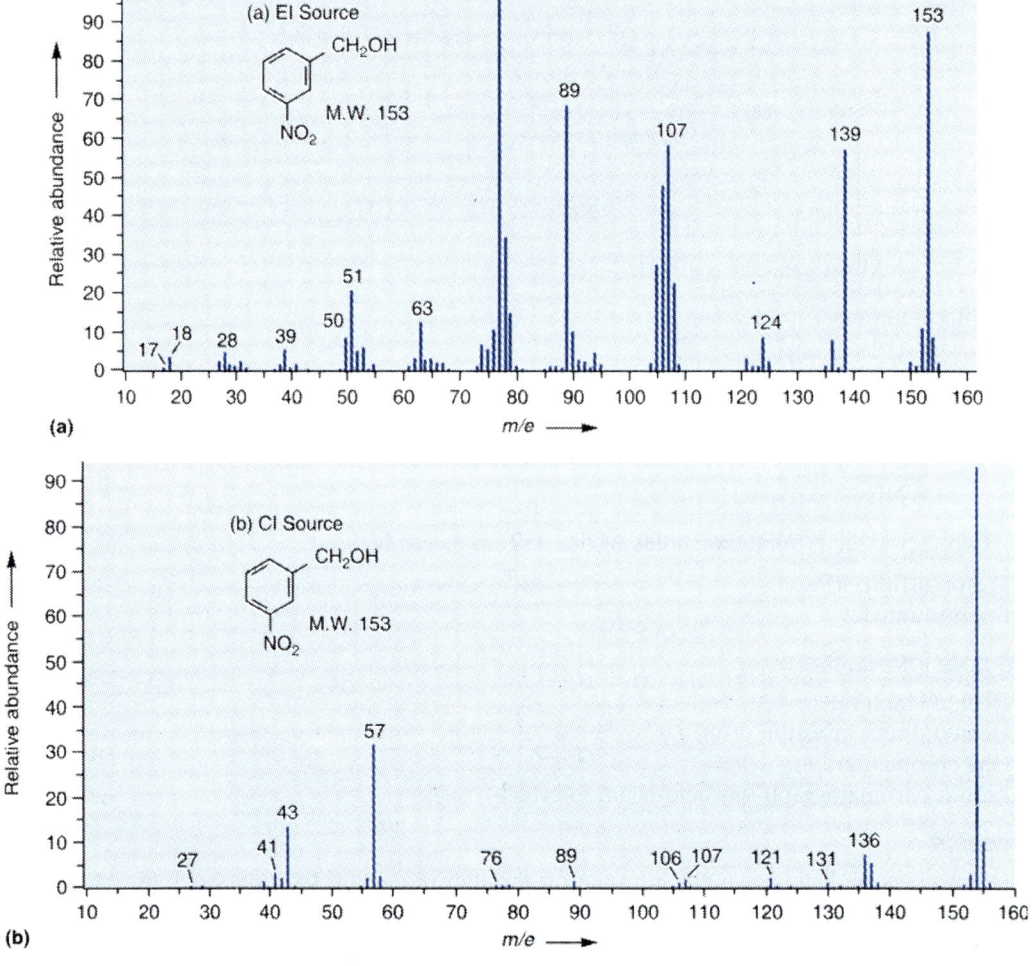

Fig. 2.7: Mass spectra of *m*-nitrobenzyl alcohol: (a) EI source; (b) CI source

Ephedrine ionised by methane CI may protonate, for example, on the O atom of the OH group resulting in loss of water molecule.

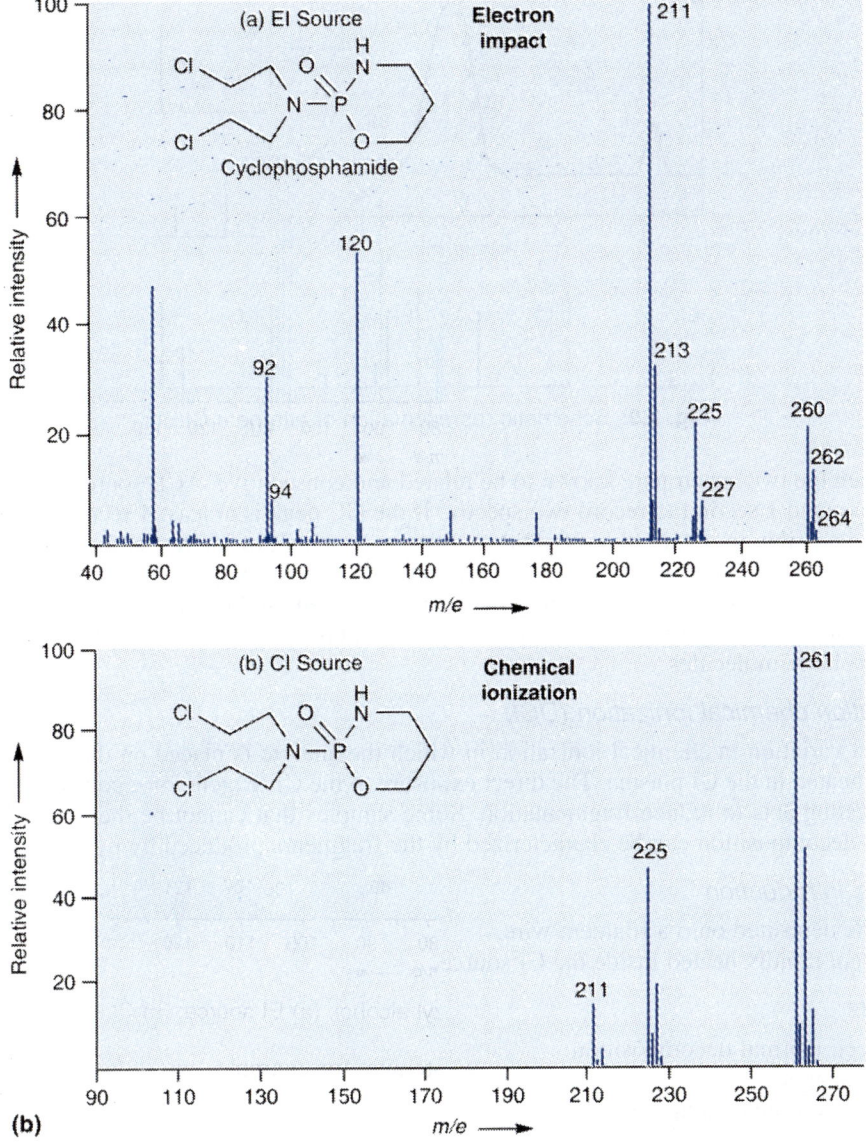

Fig. 2.8: Mass spectra of cyclophosphamide: (a) EI source; (b) CI source

Protonation on the N atom leads to the loss of CH_3NH_2 by a similar mechanism, yielding an ion of m/e 135. Both m/e 148 and 135 are observed in the CI spectrum, indicating the presence of OH and $HNCH_3$ groups in the molecule.

Alternate CI-EI (ACE)

Electron and chemical ionization produce complementary information on volatilisable samples. This becomes particularly useful when analyzing compounds in complex mixtures by high-resolution capillary column gas chromatography. Although, it is possible to run the same mixture twice, once by CI and once by EI, it can be difficult to replicate the chromatograms exactly. Furthermore, the time for each analysis is at least doubled. However, if the conditions in the source can be changed rapidly from EI to CI and back again during the elution time of the GC

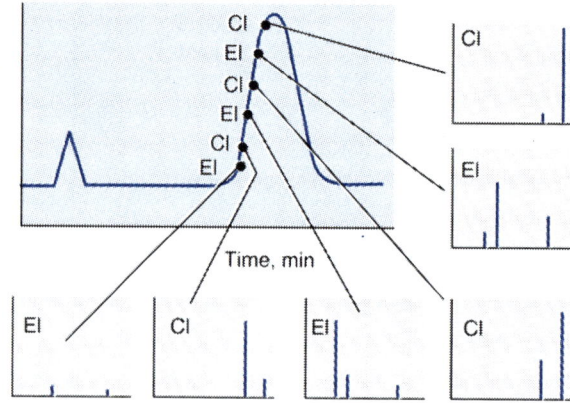

Fig. 2.9: Schematic representation of alternate CI/EI

peak, then the two spectra are known to be related and consecutive. ACE sources have a cycle time of around 1 second to record two spectra. If the GC peak is at least 4 seconds wide at the base it is possible to obtain two sets of data as the component of interest elutes (Fig. 2.9).

Chemical ionization can be performed with carefully selected reagent gases to reveal, through specific chemical reactions, something of the nature of the unknown sample. However, the sample must withstand transfer to the gaseous phase and is therefore only suitable for thermally stable, volatile molecules.

Desorption chemical ionization (DCI)

This is a variation in chemical ionization in which the analyte is placed on the lament that is rapidly heated in the CI plasma. The direct exposure to the CI reagent ions, combined with the rapid heating acts to reduce fragmentation. Some samples that cannot be thermally desorbed without decomposition can be characterized by the fragments produced by pyrolysis DCI.

Sample introduction

- Sample deposited onto a filament wire.
- Filament rapidly heated inside the CI source.

Benefits

- Reduced thermal decomposition.
- Rapid analysis.
- Relatively simple equipment.

Limitations

- Not particularly reproducible.
- Rapid heating requires fast scan speeds. Fails for large or labile compounds.

Mass range

Low. Typically less than 1,500 Da.

Negative-ion chemical ionization (NCI)

Various types of reaction can take place according to the nature of the reagent gas used, e.g. a mixture of hydrocarbon and water (95 : 5) can produce negatively-charged sample ions by the following reactions:

$$H_2O + e^- \rightarrow OH^- + H^+$$

$$AH + OH^- \rightarrow A^- H_2O$$

$$M + HO^- \rightarrow MOH^-$$

Other possible reagent gases include hydrocarbon/organic halide/oxygen (which produces Cl^- and O^- attachment ions) and fluorocarbons yielding negative ions by hydride abstraction and fluorine attachment.

Negative ions can also be formed from suitable sample molecules by electron capture processes. In this case, the reagent gas acts as a moderator, generating thermal electrons that can then attach to molecules with high electron affinities forming negative radical ions.

$$M + e^- \text{ (thermal)} \rightarrow M^{-\bullet}$$

Negative ion chemical ionization (CI), especially electron capture ionization, can be two to three orders of magnitude more sensitive than positive ion chemical ionization (CI) for electronegative molecules. It is, therefore, especially useful in the quantitative determination of trace substances that have, or that through the production of suitable derivatives are induced to have, electron-capturing properties. For example, testosterone (Fig. 2.10) is derivatized to enhance vaporization and ionization for NCI-GC/MS.

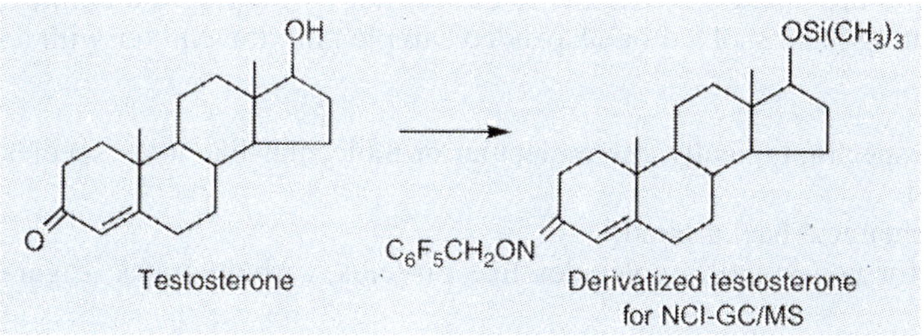

Testosterone

$C_6F_5CH_2ON$

Derivatized testosterone for NCI-GC/MS

Fig. 2.10: The pentafluorobenzyl trimethyl silyl ether derivatives of steroids make them more amenable to high sensitivity measurements using negative chemical ionization

Sample introduction
Same as for CI.

Benefits
- Efficient ionization, high sensitivity.
- Less fragmentation than positive-ion EI or CI.
- Greater selectivity for certain environmentally or biologically important compounds.

Limitations
- Not all volatile compounds produce negative ions.
- Poor reproducibility.

Mass range
Low. Typically less than 1,000 Da.

Field desorption ionization
These methods are based on electron tunneling from an emitter that is based at a high electrical potential. The emitter is a lament on which the crystalline 'whiskers' are grown. When a high potential is applied to the emitter, a very high electric field exists near the tips of the whiskers.

There are two kinds of emitters used on JEOL mass spectrometers: Carbon emitters and silicon emitters. Silicon emitters are robust, relatively inexpensive, and they can handle a higher current for field desorption. Carbon emitters are more expensive, but they can provide about an order of magnitude better sensitivity than silicon emitters.

Field desorption and ionization are soft ionization methods that tend to produce mass spectra with a little or no fragment-ion content.

Field desorption (FD)

This technique was developed by Beckly in 1969. The sample is deposited onto the emitter and the emitter is biased to a high potential (several kilovolts) and a current is passed through the emitter to heat up the filament. Mass spectra are acquired as the emitter current is gradually increased and at certain fixed temperature, the sample is evaporated from the emitter into the gas phase. The analyte molecules are ionized by electron tunneling at the tip of the emitter 'whiskers'. Characteristic positive ions produced are radical molecular ions and cation attached species such as $[M + H]^+$, $[M + Na]^+$ and $[M - Na]^+$. The latter are probably produced during desorption by the attachment of trace alkali metal ions present in the analyte.

Sample introduction

Direct insertion probe.

The sample is deposited onto the tip of the emitter by dipping the emitter into an analyte solution depositing the dissolved or suspended sample onto the emitter with a micro syringe

Benefits

* Simple mass spectra, typically one molecular or molecular-like ionic species per compound [Fig. 2.11(b)].
* Little or no chemical background.
* Works well for non-volatile molecules like proteins, carbohydrates, sugars and industrial polymers.

Limitations

* Sensitive to alkali metal contamination and sample overloading.
* Emitter is relatively fragile.
* Relatively slow analysis as the emitter current is increased.
* The sample must be thermally volatile to some extent to be desorbed.

Mass range

Low–moderate, depends on the sample. Typically less than about 2,000 to 3,000 Da.

Some examples have been recorded from ions with masses beyond 10,000 Da.

Field ionization (FI)

The removal of electrons from any species by interaction with a high electrical field. The sample is evaporated from a direct insertion probe, gas chromatograph, or gas inlet. As the gas molecules pass near the emitter, they are ionized by electron tunneling.

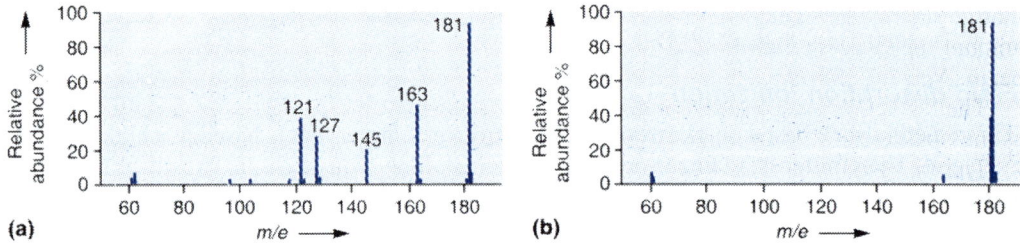

Fig. 2.11: (a) FI mass spectrum—D-glucose; (b) FD mass spectrum—D-glucose

Sample introduction

- Heated direct insertion probe.
- Gas inlet.
- Gas chromatograph.

Benefits

- Simple mass spectra, typically one molecular or molecular-like ionic species per compound [Fig. 2.11(a)].
- Little or no chemical background.
- Works well for small organic molecules and some petrochemical fractions.

Limitations

The sample must be thermally volatile. Samples are introduced in the same way as for electron ionization (EI). This method is not suitable for thermally unstable sample.

Mass range

Low. Typically less than 1,000 Da.

Comparison of field ionization and field desorption methods

- FI: Sample is heated in a vacuum so as to volatilize it onto an ionization surface. FI is suited for use with volatile, thermally stable compounds. FI sources are arranged to function also as FD sources.
- FD: The sample is placed directly onto the surface (dipping emitter in an analyte solution) before ionization but FD is needed for non-volatile and/or thermally labile substances.

Particle bombardment

In these methods, the sample is deposited on a target that is bombarded with atoms, neutrals, or ions. The most common approach for organic mass spectrometry is to dissolve the analyte in a liquid matrix with low volatility and to use a relatively high current of bombarding particles (FAB or dynamic SIMS). Other methods use a relatively low current of bombarding particles and no liquid matrix (static SIMS). The latter methods are more commonly used for surface analysis than for organic mass spectrometry.

The primary particle beam is the bombarding particle beam, while the secondary ions are the ions, produced from bombardment of the target.

$$Xe^+ + Xe \longrightarrow Xe + Xe^+$$
Fast atom

Fast atom bombardment (FAB)

FAB is a "soft" ionization method which overcomes many of the limits of EI and CI. FAB works well for polar, thermally labile compounds. A FAB mass spectrum often contains a protonated molecule, $[M + H]^+$, and a few fragment ions. The analyte is dissolved in a liquid matrix such as glycerol, thioglycerol, *m*-nitrobenzyl alcohol, or diethanolamine and a small amount (about 1 microliter) is placed on a target. The target is bombarded with a fast atom beam. *Note:* With MALDI, the matrix is typically a solid crystalline.

The fast atoms are generated by accelerating xenon ions to 6–9 keV, then neutralizing them as they pass xenon atoms at low pressure. Neutralisation takes place by electron transfer:

The fast atom (for example, xenon atoms), accelerated through a potential difference of 5000–35000 V, strike the sample molecule (Fig. 2.12). When the sample is struck by the fast atoms, it is desorbed from the surface of the probe by the transfer of momentum, usually as an ion. In common with chemical ionization the sample molecule in FAB is usually detected as

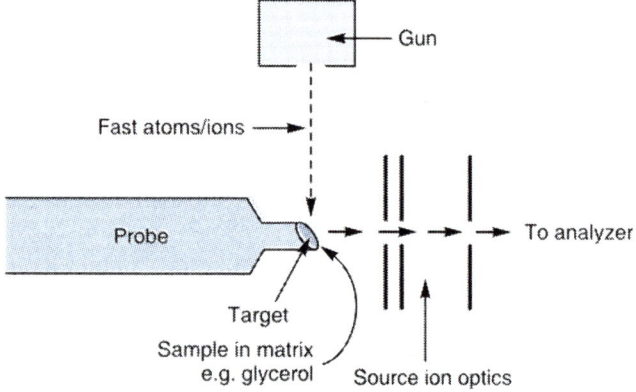

Fig. 2.12: Schematic representation of a FAB ion source

$(M + H)^+$ or as $(M - H)^-$. The ions produced are analyzed in the mass spectrometer in the same way as ions produced by other methods. It is important to notice that FAB is about 1000 times less sensitive than MALDI. Observed peaks in FAB mass spectrum are those of matrix cluster ions, analyte ions (M^+ and M^-), impurities, and ions of matrix modifiers (Fig. 2.13).

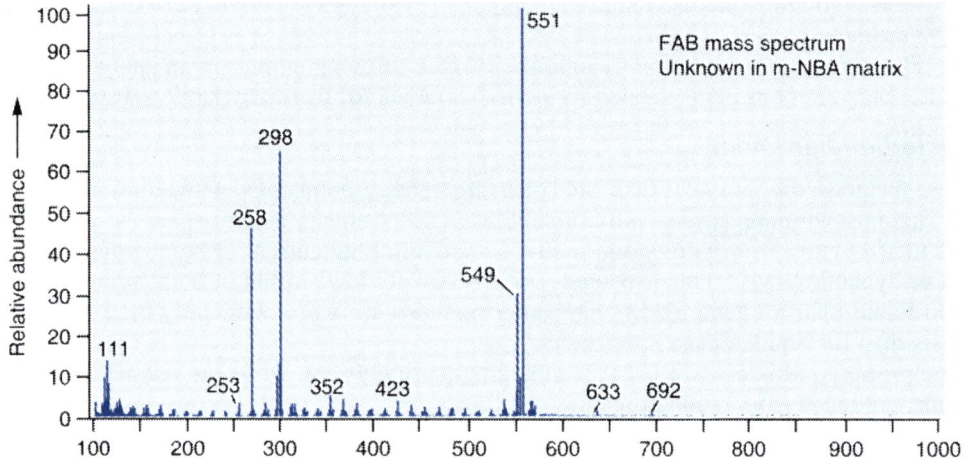

Fig. 2.13: FAB mass spectrum of unknown compound in *m*-nitrobenzyl alcohol

Function of matrix

- Facilitating the desorption and ionization process.
- Constantly replenish the surface with new sample as it is bombarded by the incident ion beam.
- By absorbing most of the incident energy, the matrix also minimizes sample degradation from the high energy particle beam.

Choice of matrix in FAB

- Sample MUST be soluble in matrix.
- Under vacuum conditions matrix must have low volatility (so that matrix/sample will maintain liquid nature).
- Matrix ions should not interfere with analyte ions.
- Matrix should not undergo unexpected chemical reactions with the sample ions.

Sample introduction

- Direct insertion probe.
- LC/MS (frit FAB or continuous-flow FAB).

Benefits

- Rapid, simple, sensitive and consumes less sample.
- Relatively tolerant of variations in sampling.
- Good for a large variety of compounds. Generally used to analyze polar, ionic, thermally unstable and high MW compounds.
- Strong ion currents—good for high resolution measurements.

Limitations

- High chemical background defines detection limits.
- May be difficult to distinguish low molecular weight compounds from chemical background.
- Analyte must be soluble in the liquid matrix.
- Not good for multiple charged compounds with more than 2 charges.
- Need skilled operator.

Mass range

Moderate. Typically ~300 Da to about 6,000 Da.

Secondary ion mass spectrometry (SIMS)

SIMS is nearly identical to FAB except that the primary particle beam is an ion beam (e.g. cesium ions) rather than a neutral beam. In SIMS, the surface of the sample is subjected to bombardment by high energy ions—this leads to the ejection (or *sputtering*) of both neutral and charged (+/–) species from the surface. The ejected species may include atoms, clusters of atoms and molecular fragments.

In traditional SIMS it is only the positive ions that are mass analyzed. This is primarily for practical ease but it does lead to problems with quantifying the compositional data since the positive ions are but a small, non-representative fraction of the total sputtered species. It should be further noted that the displaced ions have to be energy filtered before they are mass analyzed (i.e. only ions with kinetic energies within a limited range are mass analyzed).

The most commonly employed incident ions (denoted by I^+ in Fig. 2.14) used for bombarding the sample are argon ions (Ar^+) but other ions (e.g. alkali metal ions, Ga^+) are preferred for some applications. The mass analyzer is typically a quadrupole MS analyzer with unit mass resolution, but high specification time-of-flight (TOF) analyzers are also used and provide substantially higher sensitivity and a much greater mass range (at a higher cost).

The use of SIMS for moderate-size (3,000–13,000 Da) proteins and peptides has largely been supplanted by electrospray ionization.

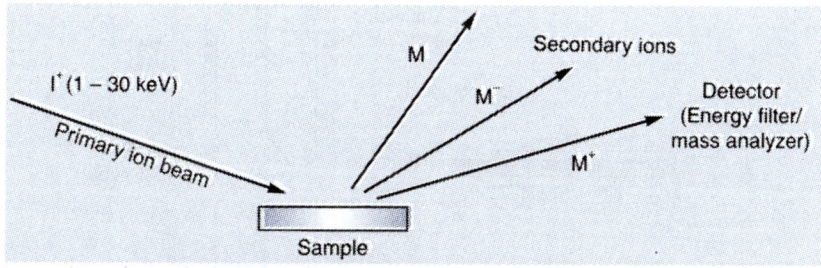

Fig. 2.14: Secondary ions mass spectrometry

Sample introduction

Same as for FAB.

Benefits

Same as for FAB, except sensitivity is improved for higher masses (3,000 to 13,000 Da).

Limitations

Same as for FAB except:

- Target can get hotter than in FAB due to more energetic primary beam.
- High-voltage arcs more common than FAB.
- Ion source usually requires more maintenance than FAB.

Mass range

Moderate. Typically 300 to 13,000 Da.

Atmospheric pressure ionization (spray methods)

As the name implies, APCI is a technique which creates ions at atmospheric pressure. A sample solution flows through a heated tube where it is volatilized and sprayed into a corona discharge with the aid of nitrogen nebulization. Ions are produced in the discharge and extracted into the mass spectrometer. APCI is best suited to relatively polar, semi-volatile samples. An APCI mass spectrum usually contains the quasi-molecular ion, $[M + H]^+$. These methods are well-suited for flow-injection and LC/MS techniques.

Electrospray ionization (ESI)

ESI allows for large, non-volatile molecules to be analyzed directly from the liquid phase. Large charged droplets are produced by 'pneumatic nebulization', i.e. the forcing of the analyte solution through a fine capillary tube at the end of which is applied a potential (Fig. 2.15). The potential used is sufficiently high to disperse the emerging solution into a very fine spray of charged droplets all at the same polarity. The solvent evaporates away, shrinking the droplet size and increasing the charge concentration at the droplet's surface. Eventually coulombic repulsion overcomes the droplet's surface tension and the droplet explodes. This 'Coulombic explosion' forms a series of smaller, lower charged droplets (Fig. 2.16). The process of shrinking followed by explosion is repeated until individually charged 'naked' analyte ions are formed. Increasing the rate of solvent evaporation, by introducing a drying gas flow countercurrent to the sprayed ions increases the extent of multiple-charging. Decreasing the capillary diameter and lowering the analyte solution flow rate, i.e. in nanospray ionization, will create ions with higher m/e ratios (i.e. it is a softer ionization technique). Methanol and aqueous formic acid make the best make-up solvent for ESI technique.

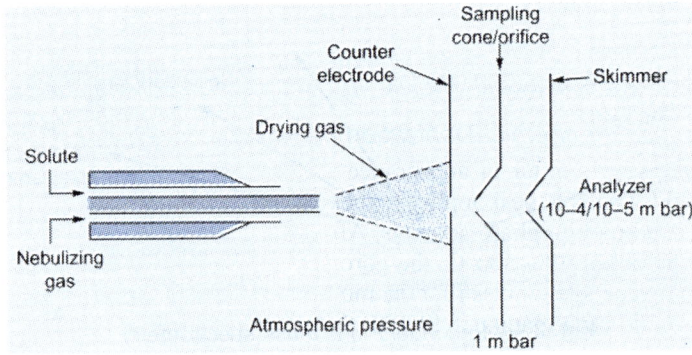

Fig. 2.15: Electrospray ionization

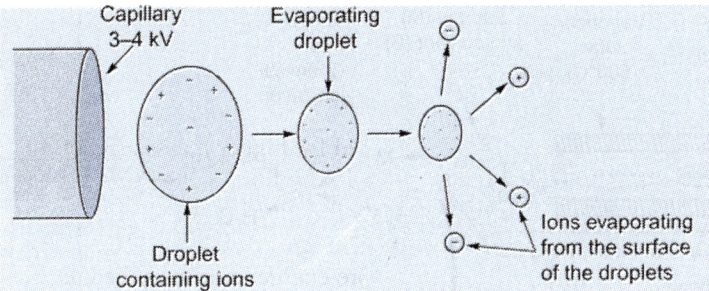

Fig. 2.16: The electrospray ionization process

Sample introduction
- Flow injection.
- LC/MS.
- Typical low rates are less than 1 microliter per minute up to about a milliliter per minute.

Benefits
- Soft ionization method, it provides **molecular weight information**.
- Used for ionizing non-volatile solutions:
  - Mass determination of biomolecules.
  - Analysis and sequencing of proteins and oligonucleotides.
  - Analyzing drugs, pesticides, and carbohydrates.
  - Long chain fatty acid.
- Permits the detection of high-mass compounds at mass-to-charge ratios that are easily determined by most mass spectrometers (m/e typically less than 2,000 to 3,000).
- Best method for analyzing multiple charged compounds.
- Very low chemical background leads to excellent detection limits.
- Can control presence or absence of fragmentation by controlling the interface lens potentials.
- Compatible with MS/MS methods.

Limitations
- Multiple charged species require interpretation and mathematical transformation (can sometimes be difficult).
- Complementary to APCI. Not good for uncharged, non-basic, low-polarity compounds (e.g. steroids).
- Very sensitive to contaminants such as alkali metals or basic compounds.
- Relatively low ion currents.
- Relatively complex hardware compared to other ion sources.

Mass range
Low–high. Typically less than 200,000 Da.

Atmospheric pressure chemical ionization (APCI)
APCI sources are very similar in appearance to their ESI counterparts, but the principle is different. Unlike ESI, no voltage is applied to the capillary, but there is a heater for the evaporation of both analytes and mobile phase solvents. After nebulization and solvent evaporation in the heated chamber at about 300–500°C, the corona discharge created by the high voltage (e.g. 3–4 kV) applied to a needle ionizes first the mobile phase components since they are present at much higher concentration than analytes (M). The formation of reactant gases is analogous to CI, but occurs at atmospheric pressure.

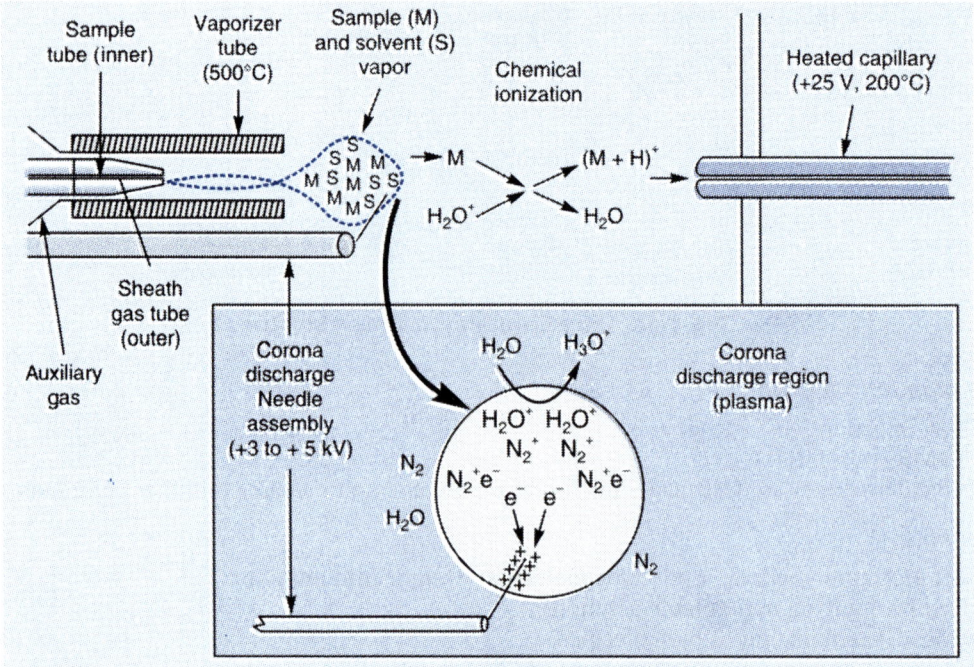

Fig. 2.17: Atmospheric pressure chemical ionization

Then, the ionized mobile phase species react with the analyte by ion-molecule reactions to generate analyte ions. As with ESI and other soft ionization techniques, mostly even-electron molecular adducts are formed, such as $[M + H]^+$, alkali metal adducts $[M + Na]^+$ and $[M + K]^+$ or $[M + NH_4]^+$ in case of the positive ion mode and mobile phase containing ammonium additives/buffers. In negative ion mode, $[M - H]^-$ is the predominate adduct ion. For example,

Ion primary formation

$$N_2 + e^- \rightarrow N_2^{+\bullet} + 2e^-$$

Ion secondary formation

$$N_2^{+\bullet} + H_2O \rightarrow H_3O^+ + HO^\bullet$$

Proton transfer

$$M + H_3O^+ \rightarrow (M + H)^+ + H_2O$$

Sample introduction

Same as for electrospray ionization.

Benefits

• Good for less-polar compounds.
• Excellent LC/MS interface.
• Compatible with MS/MS methods.

Limitations

Complementary to ESI.

Mass range

Low–moderate. Typically less than 2,000 Da.

Laser desorption

Laser desorption methods use a pulsed laser to desorb species from a target surface. Therefore, one must use a mass analyzer such as time-of-flight (TOF) or Fourier transform ion cyclotron resonance (FTICR) that is compatible with pulsed ionization methods. Magnetic sector mass spectrometers equipped with an array detector can also be used for the detection of ions produced by MALDI.

Direct laser desorption relies on the very rapid heating of the sample or sample substrate to vaporize molecules so quickly that they do not have time to decompose. This is good for low to medium-molecular weight compounds and surface analysis. The more recent development of matrix-assisted laser desorption ionization (MALDI) relies on the absorption of laser energy by a matrix compound. MALDI has become extremely popular as a method for the rapid determination of high-molecular-weight compounds.

Matrix-assisted laser desorption ionization (MALDI)

The analyte is dissolved in a solution containing an excess of a matrix such as sinapinic acid or dihydroxybenzoic acid (Table 2.1) that has a chromophore that absorbs at the ultraviolet laser wavelength (337 nm). A small amount of this solution is placed on the laser target (probe) (Fig. 2.18). The matrix absorbs the energy from the laser pulse and produces a plasma that results in vaporization and ionization of the analyte (Fig. 2.19).

Table 2.1: Matrices most frequently used for MALDI together with the usable wavelengths	
Matrix	Wavelength (nm)
Nicotinic acid	266, 220–290
Benzoic acid derivatives	
2,5-Dihydroxybenzoic acid	266, 337, 355
2-Aminobenzoic acid	266, 337, 355
Pyrazine carboxylic acid	266
3-Aminopyrazine-2-carboxylic acid	337
Cinnamic acid derivatives	
Ferulic acid	266, 337, 355
Sinapinic acid	266, 337, 355
Caffeic acid	266, 337, 355
3-Nitrobenzyl alcohol	266

In MALDI-TOF MS the specific process is as follows:

Put the sample proteins into the solvent, i.e. matrix mixed with water. The sample proteins will dissolve in the solvent. The matrix material also reacts with the proteins (analyte) to make the polymers become charged ions. No one really knows how this happens. Then, wait until the water in the solvent totally evaporates. When the water evaporates, the sample proteins are surrounded by the matrix, which forms a crystal lattice. Keep the sample proteins in some boxes which are formed by the matrix molecules. Finally, put the target, i.e. proteins and 'boxes', into the source, using a beam of ultraviolet laser to fire it. This results in each protein, picks up a proton and turns into gas phase ion.

Sample introduction

- Direct insertion probe.
- Continuous-flow introduction.

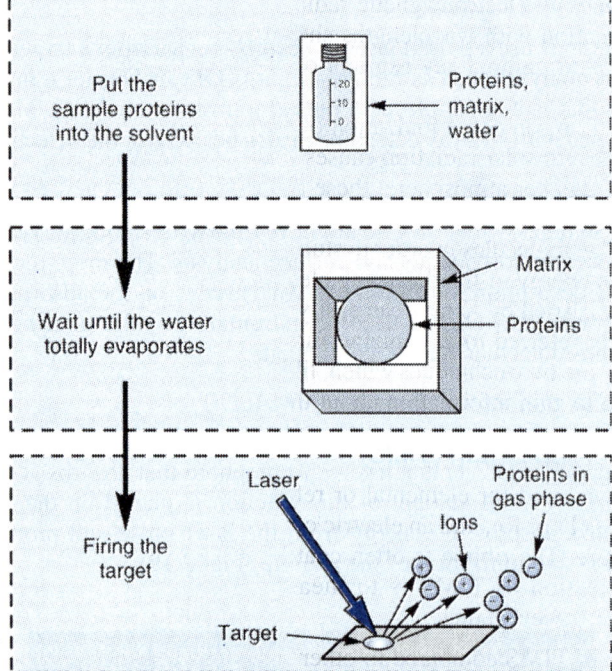

Fig. 2.18: Process in the source

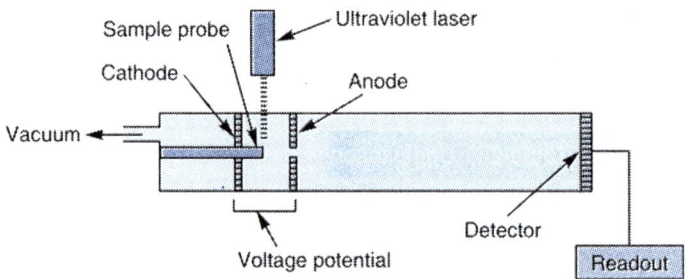

Fig. 2.19: Schematic diagram of MALDI instrument

Benefits

Rapid and convenient molecular weight determination.

Limitations

- MS/MS difficult.
- Requires a mass analyzer that is compatible with pulsed ionization techniques.
- Not easily compatible with LC/MS.

Mass range

Very high. Typically less than 500,000 Da.

Photoionization

If the ionization energy is supplied by electromagnetic radiation, the ionization is called photoionization, referring to the fact that a photon of radiation produces the ionization. This technique works well with non-polar or low polarity compounds not efficiently ionized by other ionization sources. However, not all electromagnetic radiation has sufficient energy to cause ionization. Generally, only radiation with wavelengths shorter than visible light, that is, radiation in the ultraviolet, X-ray, and gamma ray regions of the electromagnetic spectrum can produce ionization.

Ultraviolet radiation can cause ionization of many small molecules, including oxygen, O_2. In fact, short wavelength solar radiation causes ionization of molecular oxygen and molecular nitrogen found in the upper atmosphere; these processes are important to the chemistry of the earth's atmosphere. In the laboratory, ultraviolet light from special lamps, e.g. krypton lamp or lasers is used to ionize molecules in order to study them. Ultraviolet photoelectron spectroscopy (UPS) measures the energy of the departing electron.

The high energy carried by X-rays can easily cause ionization of isolated atoms. X-rays are, therefore, frequently referred to as ionizing radiation. X-ray photoelectron spectroscopy and Auger spectroscopy are two techniques which, like ultraviolet photoelectron spectroscopy, study the ejected electron to gain information about the atom from which it came.

Thermal (surface) ionization (TIMS)

Thermal ionization is used for elemental or refractory materials. A sample is deposited on a metal ribbon, such as Pt or Re, and an electric current is passed to heat the metal under vacuum to a high temperature. The ribbon is often coated with graphite to provide a reducing effect. The primary application of TIMS is to measure the isotope ratios of elements used in **geochronology** and **tracer** studies.

The advantages of TIMS compared to other isotope ratio techniques include:

• The chemical and physical stability of the measurement environment, which leads to highly precise measurements.
• The ability to ionize and evaporate samples at different temperatures by using multiple filament assemblies.
• Near 100% transmission of ions from source to collector.

Table 2.2: Comparison of ionization methods					
Ionization method	Type of ion formed	Analytes	Sample introduction	Mass limits (dalton)	Method type
EI	M^+, M^-	Small volatiles	GC, liquid or solid probe	10^3	Hard method
CI	$[M + H]^+$, $[M + X]^+$	Small volatiles	GC, liquid or solid probe	10^3	Soft method
APCI	$[M + H]^+$, $[M + X]^+$, $[M - H]^-$	Small volatiles (less polar species)	LC or syringe	2×10^3	Soft method
FI/FD	$[M + H]^+$, $[M + X]^+$	FI: Volatiles, FD: Non-volatiles	GC, liquid or solid probe	2×10^3	Soft method
ES	$[M + nH]^{n+}$, $[M - nX]^{n-}$	Peptides, proteins, non-volatiles	LC or syringe	2×10^5	Soft method multiple charged ion
FAB	$[M + H]^+$, $[M - H]^-$	Carbohydrates, organometallics, peptides, non-volatiles	In viscous matrix	6×10^3	Soft but harder than ESI or MALDI
MALDI	$[M + H]^+$, $[M + X]^+$	Peptides, proteins, nucleotides	In solid matrix	5×10^5	Soft method

Limitations

The disadvantages include:

- Not all elements are easily ionized, which restricts applications to elements with low ionization potentials.
- Ionization is not equally efficient for all elements, and is generally less than 1%.

Mass analyzers

Ions produced in the ionization chamber are accelerated by the application of acceleration potential (2 to 8 kV). These ions enter in the mass analyzer which differentiate them on the basis of mass to charge ratio by magnetic and electric field. There are two main functions of a mass analyzer (i) resolve the ion beam into its component-ions and (ii) to maximise the resolved ions intensities (dispersive focussing). The efficiency of mass spectrometer depends on type of mass analyzer. Mass spectrometer can be classified into two types depending on efficiency— low resolution and high resolution types. Two compounds, C_3H_6O and C_3H_8O, have nominal masses of 58 and 60, and can be distinguished by low-resolution MS. But the compounds, $C_2H_4O_2$ and C_3H_8O, having the same nominal mass of 60 cannot be differentiated by low resolution type mass spectrometer. It requires high-resolution MS.

Nominal mass: 60

Molecular formula	Precise mass
$C_2H_4O_2$	60.02112
C_3H_8O	60.05754

Similarly, for a mass of 58, eleven (11) molecular formulas are possible, as given here:

Nominal mass: 58

Molecular formula	Precise mass
N_3O	58.0042
$C_2H_2O_2$	58.0054
N_4H_2	58.0280
C_2H_4NO	58.0293
CNO_2	57.9929
$C_2H_6N_2$	58.0532
CH_2N_2O	58.0167
C_3H_6O	58.0419
CH_4N_3	58.0406
C_3H_6N	58.0657
C_4H_{10}	58.0783

In the low resolution mass spectrometer, the atomic weight of the most abundant isotopes are determined as a whole number, means $^1H = 1$; $^{16}O = 16$; $^{12}C = 12$; $^{14}N = 14$, etc.

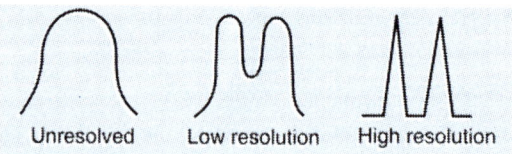

Unresolved   Low resolution   High resolution

Fig. 2.20: Diagram showing three states of resolution

Low resolution mass spectrometers cannot differentiates the CO, N_2, C_2H_4 as all will be recorded as whole number of which mass is 28.

From a physical point of view, only the atomic weight of most abundant isotope of carbon is equal to 12 exactly. The other atoms are not equal to whole numbers and their actual atomic weight can be determined to the sixth decimal place.

For example,

$^1H = 1.007825$	$^{16}O = 15.994915$	$^{28}Si = 27.976901$	$^{79}Br = 78.918301$
$^{12}C = 12.000000$	$^{19}F = 18.998405$	$^{31}P = 30.973800$	$^{126}I = 126.904500$
$^{14}N = 14.003074$	$^{32}S = 31.972074$	$^{35}Cl = 34.9689$	

From these values, it can be observed that CO, C_2H_4 and N_2 show a significant difference in their actual masses.

CO = 27.994914, N_2 = 28.006158, C_2H_4 = 28.031299.

The mass spectrometer with good resolving power can differentiate them. Resolving power can be calculated as:

$$\text{Resolving power (RP)} = \frac{m_1}{m_2 - m_1} = \frac{m_1}{\Delta m}$$

where m_1 and m_2 are the masses of two sample molecule.

For nitrogen and ethylene:

$$RP = \frac{28.006158}{28.031299 - 28.006158} = 1110$$

Hence, mass spectrometer possessing RP = 1,110 can differentiate the masses of nitrogen and ethylene.

The different types of mass analyzers can be classified on the basis of their working principle as follows:

Mass analyzers involving magnetic field

The ion optics in the ion-source chamber of a mass spectrometer extract and accelerate ions to a kinetic energy (KE) given by:

$$KE = 0.5\, mv^2 = eV \text{ (PE), means, Kinetic energy = Potential energy} \qquad \text{...(2.1)}$$

where m is the mass of the ion, v is its velocity, e is the charge of the ion and V is the applied voltage of the ion optics.

The ions enter the flight tube between the poles of a magnet and are deflected by the magnetic field, H. Only ions of mass-to-charge ratio that have equal centrifugal and centripetal forces pass through the flight tube:

$$\frac{mv^2}{r} = HeV \qquad \text{...(2.2)}$$

Centrifugal force = Centripetal force

where r is the radius of curvature of the ion path:

$$r = \frac{mv}{eH} \qquad \text{...(2.3)}$$

Substitute the value of v in Eq. (2.3):

$$r = \frac{1}{H}\sqrt{2V\frac{m}{e}} \qquad \text{...(2.4)}$$

On rearranging this equation:

$$\frac{m}{e} = \frac{H^2 r^2}{2V} \qquad \qquad ...(2.5)$$

This equation shows that the m/e of the ions that reach the detector can be varied by changing either H or V.

A. Single focussing mass analyzer

It is a directional focussing with the application of magnetic field. Hence, the instrument is defined as a 'direction' or 'single' focussing mass spectrometer (Fig. 2.21). Different ions are deflected by the magnetic field by different amounts. The amount of deflection depends on:

- The mass of the ion. Lighter ions are deflected more than heavier ones.
- The charge on the ion. Ions with 2 (or more) positive charges are deflected more than ones with only 1 positive charge.

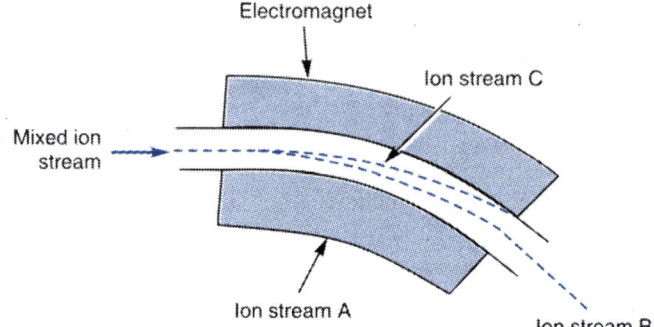

Fig. 2.21: Magnetic sector mass analyzer

These two factors are combined into the **mass/charge ratio**. Mass/charge ratio is given the symbol m/e (or m/z). For example, if an ion had a mass of 28 and a charge of 1+, its mass/charge ratio would be 28. An ion with a mass of 56 and a charge of 2+ would also have a mass/charge ratio of 28.

In Fig. 2.21, ion stream A is most deflected—it will contain ions with the smallest mass/charge ratio. Ion stream C is the least deflected—it contains ions with the greatest mass/charge ratio. The mass spectrum is obtained when ions of different mass arrive at collector, one after the other. The time of arrival will be proportional to the square root of the m/e ratio. They have lower mass ranges.

B. Double focussing mass analyzer

Double focussing mass spectrometers use a combination of magnetic and electrical fields to focus and sort ions. A common configuration for a sector instrument is the geometry shown in Fig. 2.22, in which a magnetic "sector" follows an electric "sector". The slit acts as a filter to select for a specific m/e value. The electric sector focuses the ions with respect to differences in kinetic energy that they may have as they exit the source region. "Double focussing," this combination of "angular" or "directional" focussing and energy focussing, provide mass resolution high enough to separate ions of the same nominal mass but different chemical formulae, such as C_2H_4, N_2 and CO at m/e 28. The so called "exact masses", more properly "high precision masses", of C_2H_4, N_2 and CO are 28.0313, 28.0061, and 27.9949 daltons, respectively.

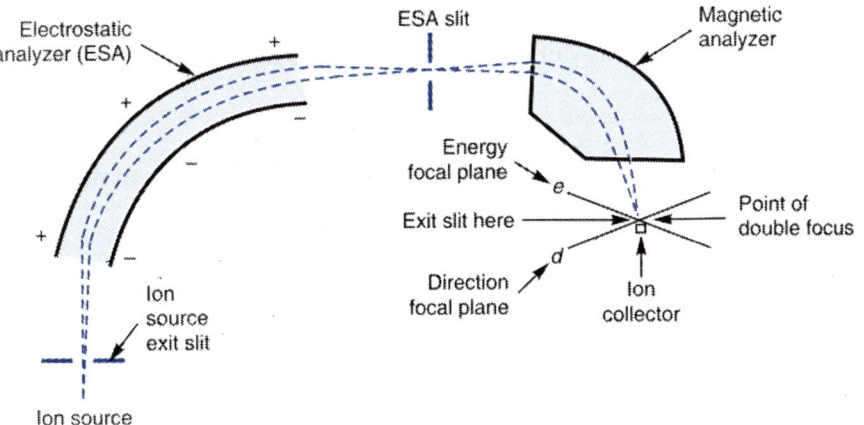

Fig. 2.22: Double focussing mass analyzer

Time-of-flight mass analyzers

The time-of-flight mass analyzers (Fig. 2.23a) separate ions by virtue of their different flight times over a known predetermined distance.

TOF analyzers separate ions by time without the use of an electric or magnetic field. In a crude sense, TOF is similar to chromatography, except there is no stationary/mobile phase, instead the separation is based on the kinetic energy and velocity of the ions through a fixed distance (in a tube of fixed length).

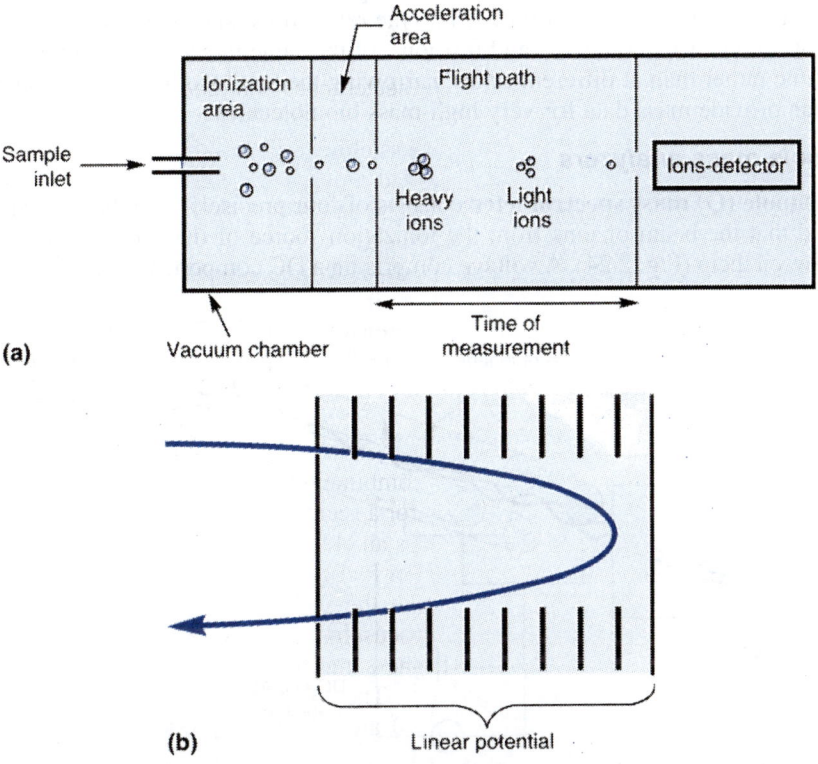

Fig. 2.23: (a) Time-of-flight analyzer; (b) reflectron

Ions of the same charges have equal kinetic energies; kinetic energy of the ion in the flight tube is equal to the kinetic energy of the ion as it leaves the ion source:

$$\text{Kinetic energy} = \frac{mv^2}{2} = eV \qquad \qquad \dots(2.6)$$

The time-of-flight (T_f), or time it takes for the ion to travel the length of the flight tube is:

$$T_f = \frac{L}{v} \qquad \qquad \dots(2.7)$$

where L is the length of tube, and v is the velocity of the ion.

Substituting Eq. (2.6) for kinetic energy in Eq. (2.7) for time-of-flight:

$$T_f = L\sqrt{\frac{m}{e}}\sqrt{\frac{1}{2V}} \propto \sqrt{\frac{m}{e}} \qquad \qquad \dots(2.8)$$

During the analysis, L, length of tube, the voltage from the ion source V are held constant, which can be used to say that time-of-flight is directly proportional to the root of the mass to charge ratio.

Unfortunately, at higher masses, resolution is difficult because flight time is longer. Also, at high masses, not all of the ions of the same m/e values reach their ideal TOF velocities. To fix this problem, often a **reflectron (optic device)** is added to the analyzer. The reflectron consists of a series of ring electrodes of very high voltage placed at the end of the flight tube (Fig. 2.23b). When an ion travels into the reflectron, it is reflected in the opposite direction due to the high voltage.

The reflectron increases resolution by narrowing the broadband range of flight times for a single m/e value. Faster ions travel further into the reflectrons, and slower ions travel less into the reflector. In this way, both slow and fast ions, of the same m/e value, reach the detector at the same time rather than at different times, narrowing the bandwidth for the output signal.

They can provide mass data for very high-mass biomolecules.

Quadrupole mass analyzers

The **quadrupole (Q) mass spectrometer** consists of four precisely straight and parallel rods, so arranged that the beam of ions from the ionization source of the spectrometer is directed axially between them (Fig. 2.24). A voltage comprising a DC component and a radio frequency

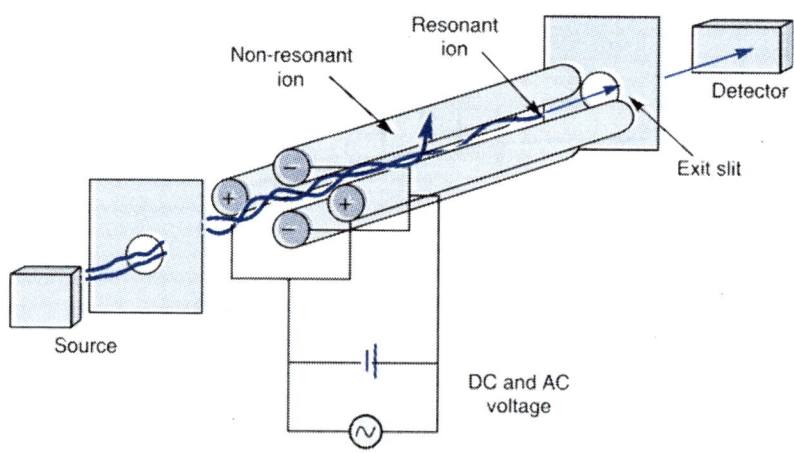

Fig. 2.24: Schematic diagram of quadrupole analyzer (Resonant ions will travel down through the poles toward the detector. Non-resonant ions will strike the poles and neutralized.)

component is applied between adjacent rods, opposite rods being electrically connected. Once inside the **quadrupole** the ions will oscillate in the (x) (y) direction as a result of the high frequency electric field. The oscillations are only stable for a certain function of frequency and the DC voltage, otherwise the ions strike the rods and become dissipated. The mass range of the oscillating ions is scanned by changing the DC voltage and the frequency, keeping the ratio of the DC voltage to the frequency constant. Each mass is detected by an appropriate ion sensor. The resolution of the spectrometer can be increased by either employing **eight poles** or by connecting two or three **quadrupoles** in series.

Quadrupole ion trap (QIT) mass analyzer

The quadrupole ion trap (QIT) mass analyzer was developed in parallel with the quadrupole mass analyzer by the third Nobel prize winning mass spectrometry pioneer, Wolfgang Paul. A schematic of the basic set up of a QIT mass analyzer is shown in Fig. 2.25. The ions, produced in the ionization source of the instrument, enter into the trap through the inlet and are trapped through action of the three hyperbolic electrodes: The ring electrode and the entrance and exit endcap electrodes. A combination of RF and DC voltages is applied to the electrodes to create a quadrupole electric field similar to the electric field for the quadrupole mass analyzer. This electric field traps ions in a potential energy well at the center of the analyzer. The mass spectrum is acquired by scanning the RF and DC fields to destabilize low mass to charge ions. These destabilized ions are ejected through a hole in one endcap electrode and strike a detector. The mass spectrum is generated by scanning the fields, so that ions of increasing *m/e* value are ejected from the cell and detected. The trap is then refilled with a new batch of ions to acquire the next mass spectrum. The mass resolution of the ion trap is increased by adding a small amount 0.1 Pa (10 torr) of helium as a bath gas. Collisions between the analyte ions and the inert bath gas dampen the motion of the ions and increases the trapping efficiency of the analyzer.

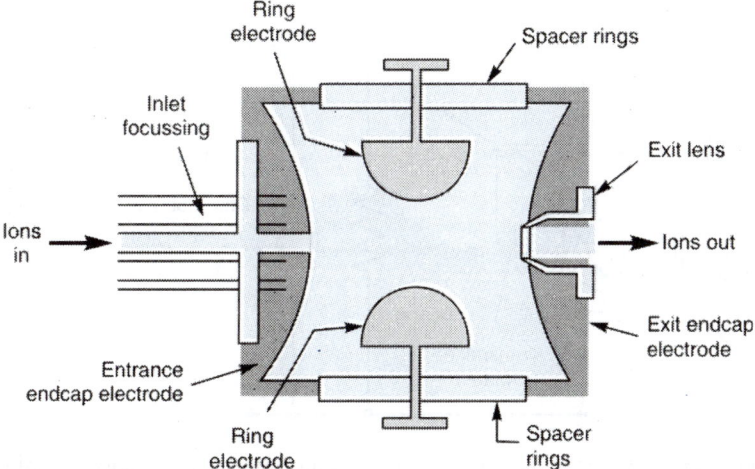

Fig. 2.25: A schematic (cutaway view) of a QIT mass analyzer

The very nature of trapping and ejection makes a quadrupolar ion trap especially suited to performing experiments in structural elucidation studies.

Ion cyclotron resonance (ICR)

Principle of operation

Ions move in a circular path in a magnetic field. The cyclotron frequency of the ion's circular motion is mass-dependent. By measuring the cyclotron frequency, one can determine an ion's mass. The working equation for ICR can be quickly derived by equating the centripetal force (mv^2/r) and the Lorentz force evH experienced by an ion in a magnetic field:

$$\frac{mv^2}{r} = evH$$

Solving for the angular frequency (omega), which is equal to v/r:

$$\text{Angular frequency}\,(\omega) = \frac{v}{r} = \frac{eH}{m}$$

A group of ions of the same mass-to-charge ratio will have the same cyclotron frequency, but they will be moving independently and out-of-phase at roughly thermal energies. If an excitation pulse is applied at the cyclotron frequency, the "resonant" ions will absorb energy and be brought into phase with the excitation pulse.

As ions absorb energy, the size of their orbit also increases. The packet of ions passes close to the receiver plates in the ICR cell and induces image currents that can be amplified and digitized.

The signal induced in the receiver plates depends on the number of ions and their distance from the receiver plates.

If several different masses are present, then one must apply an excitation pulse that contains components at all of the cyclotron frequencies. This is done by using a rapid frequency sweep ("chirp"), an "impulse" excitation, or a tailored waveform. The image currents induced in the receiver plates will contain frequency components from all of the mass-to-charge ratios. The various frequencies and their relative abundances can be extracted mathematically by using a Fourier transform which converts a time-domain signal (the image currents) to a frequency-domain spectrum (the mass spectrum). A schematic representation of a cubic FTICR cell is shown in Fig. 2.26.

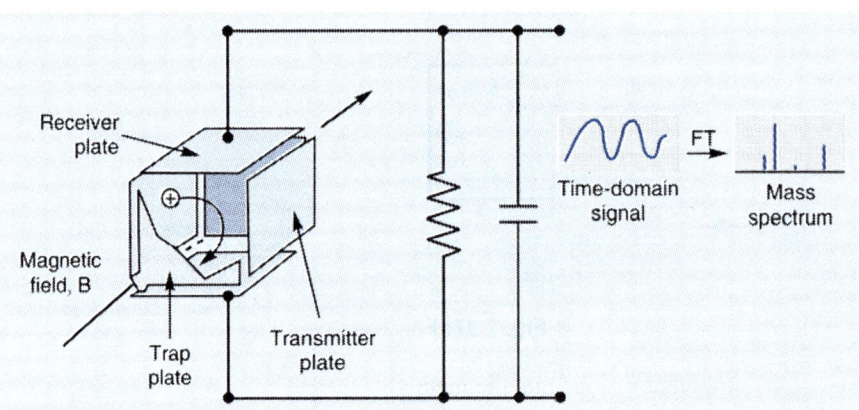

Fig. 2.26: Schematic diagram of ICR showing three pairs of parallel plates

Table 2.3: Comparison of different mass analyzers	
Analyzer	**System highlights**
Quadrupole	Unit mass resolution, fast scan, low cost
Sector (magnetic and/or electrostatic)	High resolution, exact mass
Time-of-flight (TOF)	Theoretically, no limitation for m/e maximum, high throughput
Ion cyclotron resonance (ICR)	Very high resolution, exact mass, perform ion chemistry

The ICR is also used as a Fourier Transform Mass Spectrometer (FT-MS). Instead of using a single excitation frequency, a fast RF pulse is applied to the transmitter electrodes. This simultaneously excites all the ions and produces a signal at the cyclotron frequency of each m/e ion present. This signal is similar to the Free Induction Decay (FID) produced in an FT-NMR experiment. A complete mass spectrum is obtained by using the Fourier transform to convert this signal from the time domain to the frequency domain.

The ICR has highest recorded mass resolution of all mass spectrometers, and has powerful capabilities for ion chemistry and MS/MS experiments but it has limited dynamic range.

Detectors

Several types of detectors are available for mass spectrometers. The choice of detector depends on the design of the instrument and the analytical applications that will be performed. Ion detectors can be divided into two classes. Some detectors are made to count ions of a single mass at a time and, therefore, they detect the arrival of all ions sequentially at one point (point ion collectors). Other detectors, such as photographic plates, image current detectors or array detectors, have the ability to count multiple masses and detect the arrival of all ions simultaneously along a plane (array collectors). The detector used for most routine experiments is the electron multiplier. Another type of detector is photographic plates coated with a silver bromide emulsion, it is sensitive to energetic ions. A photographic plate can give a higher resolution than an electrical detector.

Faraday cup

The Faraday cup is named after Michael Faraday who first theorized ions around 1830. A Faraday cup is made of a metal cup or cylinder with a small orifice. It is connected to the ground through a resistor, as illustrated in Fig. 2.27.

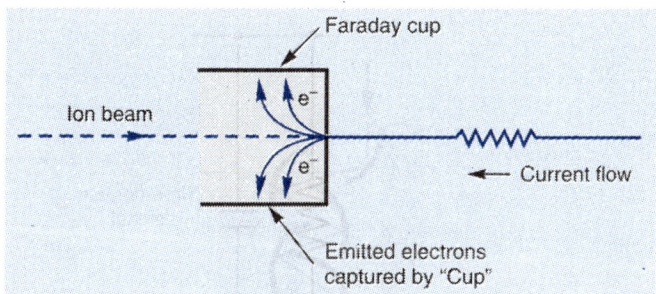

Fig. 2.27: Faraday cup

When a beam or packet of ions hits the metal it gains a small net charge while the ions are neutralized by either accepting or donating electron. This leads to a current flow through the resistor. The discharge current is then amplified and detected. It provides a measure of ion abundance. Essentially, the Faraday cup is part of a circuit where ions are the charge carriers in vacuum and the Faraday cup is the interface to the solid metal, where electrons act as the charge carriers (as in most circuits). Because the charge associated with an electron leaving the wall of the detector is identical to the arrival of a positive ion at this detector, secondary electrons that are emitted when an ion strikes the wall of the detector are an important source of errors, if they are not suppressed. In consequence, the accuracy of this detector can be improved by preventing the escape of reflected ions and ejected secondary electrons. Various devices have been used to capture ions efficiently and to minimize secondary electron losses. For instance, the cup is coated with carbon because it produces a few secondary ions. The shape of the cup and the use of a weak magnetic field prevent also any secondary electrons produced inside to exit. The disadvantages of this simple and robust detector are its low sensitivity and its slow response time.

The Faraday cup was widely used in the beginning of mass spectrometry due to its accuracy because of the direct relation between the measured current and number of ions.

Electron multiplier

An electron multiplier is used to detect the presence of ions emerging from the mass analyzer of a mass spectrometer. It is essentially the "eyes" of the instrument. The task of the electron multiplier is to detect every ion of the selected mass passed by the mass filter. How efficiently the electron multiplier carries out this task represents a potentially limiting factor on the overall system sensitivity. Consequently, the performance of the electron multiplier can have a major influence on the overall performance of the mass spectrometer.

The basic physical process that allows an electron multiplier to operate is called secondary electron emission. When a charged particle (ion or electron) strikes a photoemissive surface, it causes secondary electrons to be released from atoms in the surface layer. The number of secondary electrons released depends on the type of incident primary particle, its energy and characteristic of the incident surface. Ultimately, the released electrons are collected at anode with the generation of current. The strength of current is proportional to number of electrons (or number of ions striking to photoemissive surface).

There are two basic forms of electron multipliers that are commonly used in mass spectrometry:

1. Discrete-dynode electron multiplier; and
2. The continuous-dynode electron multiplier (often referred to as a channel electron multiplier or CEM).

Generally, electron multipliers are of the discrete-dynode type (Fig. 2.28).

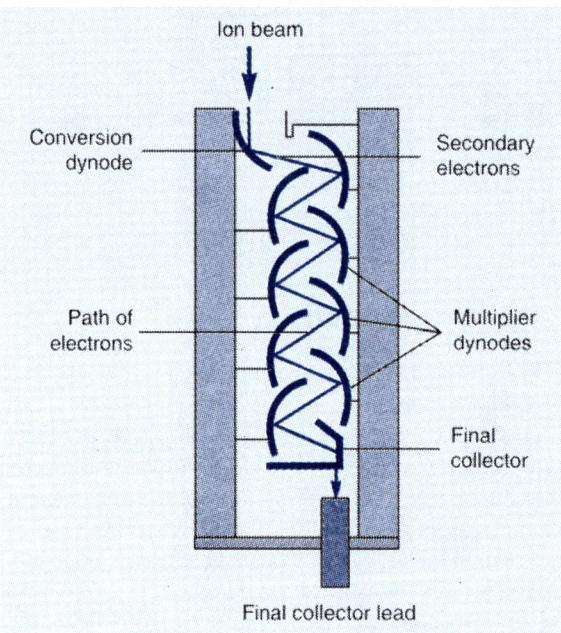

Fig. 2.28: Ion-optics of discrete-dynode electron multiplier showing the electron gain at each successive dynode. The electron cascading process results in gains up to 10^8 being achieved with ~21 dynodes

A typical discrete-dynode electron multiplier has between 12 and 24 dynodes and is used with an operating gain of between 10^4 and 10^8, depending on the application. It has Cu/Be photoemmisive surface. These detectors have the advantage of being able to produce high output currents (in excess of 100 mA), but suffer from the disadvantage of being relatively unstable when repeatedly exposed to atmosphere.

In recent years, new structures have been fabricated based on aluminum dynodes, which are reportedly less susceptible to degradation but are bulky and relatively expensive.

The second type of multiplier is the continuous dynode multiplier. The vast majority of these is fabricated of glass, although some are constructed from coated ceramic materials or are a combination of glass and ceramic. These detectors are, in general, much more suitable for applications requiring frequent exposure to atmosphere. Most of these detectors are of the single channel type; however, microchannel plates (two dimensional arrays of single channels) have been used in applications involving simultaneous imaging of an entire mass spectrum and time-of-flight mass apectrometry.

Channeltron® Single Channel Electron Multipliers (CEMs) are durable and efficient detectors for positive and negative ions as well as electrons and photons. Fig. 2.29 illustrates the basic structure and operation of the Channeltron® CEM. A glass tube having an inner diameter of approximately 1 mm and an outer diameter of 2, 3, or 6 mm is constructed from a specially formulated lead silicate glass. When appropriately processed, this glass exhibits the properties of electrical conductivity and secondary emission. An ion striking the input face of the device typically produces 2–3 secondary electrons. These electrons are accelerated down the channel by a positive bias. The electrons strike the channel walls, producing additional electrons (and so on) until, at the output end a pulse of 10^7 to 10^8 electrons emerges. Today, CEMs are the most widely used detector in quadrupole and ion trap mass spectrometers and find many applications in magnetic sector instruments as well.

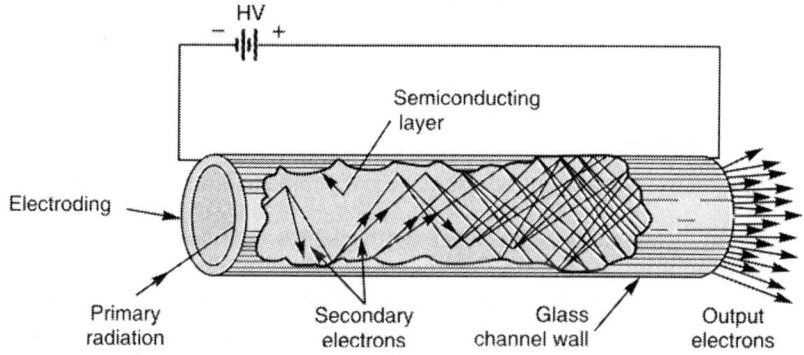

Fig. 2.29: Cutaway view of a CEM

Scintillator ('Daly' detector)

The combination of a scintillator and a light detector is called a scintillation (or Daly) detector. A fast ion causes electrons to be emitted and these are accelerated towards a second 'dynode'. In this case, the dynode consists of a substance (a scintillator) which emits photons (light). The emitted light is detected by a photomultiplier (PMT) and is converted into an electric current. Since photon multipliers are very sensitive, high gain amplification of the arrival of a single ion is achieved. These detectors are also important in studies on metastable ions.

Array detectors

Array detectors can attain very high sensitivities by collecting all the ions or ions over a large range of m/e ratios and determined simultaneously. This contrasts with conventional scanning

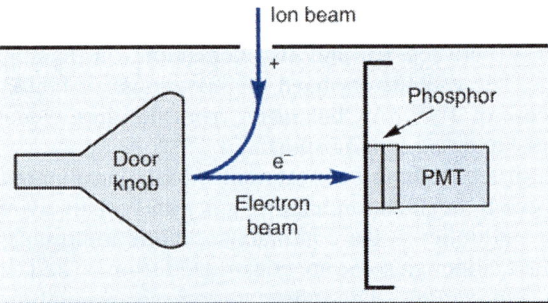

Fig. 2.30: Scintillation detector

methods using point detectors (e.g. electron multipliers) that collect only a small fraction of the ions in each mass channel as the spectrum is scanned. Photodiode arrays (PDA) are the most common form in which a microchannel plate detector is coupled to fibre optical channel via a phosphorescent screen (Fig. 2.31).

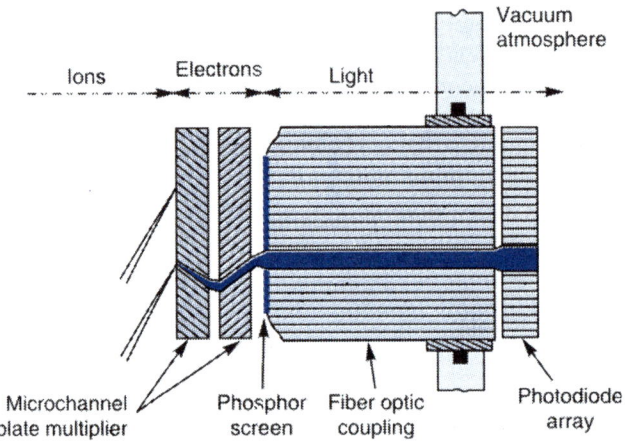

Fig. 2.31: Schematic representation of photodiode array detector

Ions enter the microchannel plates and the electrons emitted strike the phosphorescent screen. The screen consists of a layer of aluminium coated with crystalline phosphor often composed of caesium iodide (CsI) and thallium (Tl). This surface emits light in the form of photons that are transmitted down the fiber optic channels onto a charge or plasma-coupled, unlike the phosphor screen of the image intensifier, react independently of each other. Each pixel of the photodiode array is typically 25 μm by 2.5 mm in size and converts the photons to charge. The photodiode of the charge coupled device (CCD) array records the charge necessary to neutralize that accumulated in each pixel and these data are integrated and stored.

Sample inlet system

The sample has to pass from the atmosphere into the mass spectrometer under vacuum. To prevent interactions between the sample and residual molecules in the mass spectrometer, analysis has to be performed at a very low pressure. The inlet system has to control the flow of sample into the vacuum. This must happen at a relatively low flow rate to allow the vacuum pumps to maintain low pressure.

Samples are introduced into mass spectrometer via some sample inlet system. The choice of inlet system depends upon nature of sample and method of ionization. Following are the different types of inlet system.

Batch inlet systems

These systems are the simplest and simply involve the volatilization of the sample externally and then the gradual leakage of the volatilized sample into the evacuated ionization chamber. For gases, the sample is introduced into the metering volume container and then expanded into the reservoir flask where it is then leaked into the ionization chamber. For liquids, a small quantity of sample is introduced into the reservoir and the pressure of the system is reduced to about 10^{-5} torr. The inlet system is lined with glass to avoid losses of polar analytes by adsorption.

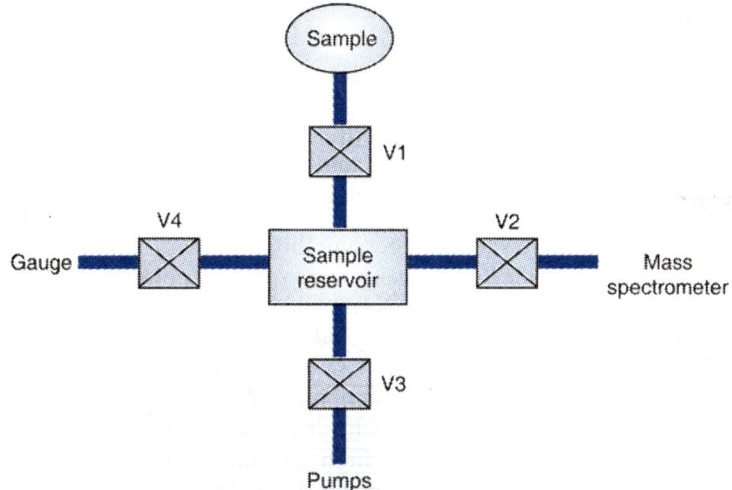

Fig. 2.32: Schematic representation of batch inlet system

Direct probe inlet

These systems are used for solids and non-volatile liquids and in these systems the sample is introduced into the ionization region by means of a sample holder, or probe, which is inserted through a vacuum lock. Probes are also used when the amount of the sample to be analyzed is small. With a probe, the sample is generally held on the surface of a glass or aluminum capillary tube and positioned within a few meters of the ionization source.

Chromatographic inlet system

Gas chromatography

It is probably the most common technique for introducing samples into a mass spectrometer. Complex mixtures are routinely separated by gas chromatography and mass spectrometry is used to identify and quantitate the individual component. Several different interface designs are used to connect these two instruments. The most significant characteristics of the inlets are the amount of GC carrier gas that enters the mass spectrometer and the amount of analyte that enters the mass spectrometer. If a large flow of GC carrier gas enters the mass spectrometer it will increase the pressure in the source region. Maintaining the required source pressure will require larger and more expensive vacuum pumps. The amount of analyte that enters the mass spectrometer is important for improving the detection limits of the instrument. Ideally all the analyte and none of the GC carrier gas would enter the source region.

The most common GC/MS interface now uses a capillary GC column. Since the carrier gas flow rate is very small for these columns, the end of the capillary is inserted directly into the source region of the mass spectrometer. The entire flow from the GC enters the mass spectrometer. Since capillary columns are now very common, this inlet is widely used. However, wide bore capillaries and packed GC columns have higher flow rates. This significantly increases the pressure in the mass spectrometer. Several inlet designs are available to reduce the gas flow into the source. The simplest design splits the GC effluent so that only a small portion of the total flow enters the mass spectrometer. Although this inlet reduces the gas load on the vacuum system, it also reduces the amount of analyte. Effusive separators and membrane inlets are more selective and transport a higher fraction of the analyte into the source region. Each of these methods has efficiency and resolution drawbacks but they are necessary for some experiments.

Liquid chromatography

Liquid chromatography inlets are used to introduce thermally labile compounds not easily separated by gas chromatography. These inlets have undergone considerable development and are now fairly routine. Because these inlets are used for temperature sensitive compounds, the sample is ionized directly from the condensed phase.

Vacuum system

All mass spectrometers operate at very low pressure (high vacuum). This reduces the chance of ions colliding with other molecules in the mass analyzer. Any collision can cause the ions to react, neutralize, scatter, or fragment. All these processes will interfere with the mass spectrum. To minimize collisions, experiments are conducted under high vacuum conditions, typically 10^{-2} to 10^{-5} Pa (10^{-4} to 10^{-7} torr) (1 Pa = 133.32 torr) depending upon the geometry of the instrument. This high vacuum requires two pumping stages. The first stage is a mechanical pump that provides rough vacuum down to 0.1 Pa (10^{-3} torr). The second stage uses diffusion pumps or turbomolecular pumps to provide high vacuum. Ion cyclotron resonance (ICR) instruments have even higher vacuum requirements and often include a cryogenic pump for a third pumping stage. The pumping system is an important part of any mass spectrometer but a detailed discussion is beyond the scope of this chapter

Data system

The final component of a mass spectrometer is the data system. It is the heart of the mass spectrometry instrument. Without it, it is difficult to deal vast number of information from mass spectrometric analysis. This part of the instrument has undergone revolutionary changes in the past twenty years. It has evolved from photographic plates and strip chart recorders to data systems that control the instrument, acquire hundreds of spectra in a minute and search tens of thousands of reference spectra to identify an unknown. Because these systems are evolving so rapidly, a through discussion is not included in this paper. Interested readers should study the manuals for their instrument.

3

Types of Ions Observed in Mass Spectrometry

The ions which form in mass spectrometry are molecular ions, isotopic ions, fragment ions, rearrangement ions, multiple charged ions, metastable ions, negative ions and ions formed by ion molecule interaction.

MOLECULAR ION

When a sample of compound is bombarded with electron of 9 to 15 eV energies, molecular ion is produced by loss of single electron.

$$M + e \rightarrow M^{+\bullet} + e^-$$

The molecular ion is often given the symbol M^+ or $M^{+\bullet}$—the dot in this second version represents the fact that somewhere in the ion there will be a single unpaired electron.

If some of the molecular ions remain intact long enough (about 10^{-6} seconds) to reach to the detector, a molecular ion peak is observed. This peak gives the molecular weight of the compound. The mass of this ion corresponds to the sum of the masses of the most abundant naturally occurring isotopes of the various atoms that make up the molecule (with a correction for the masses of the electron(s) lost or gained). For example, the mass of the molecular ion of ethyl bromide C_2H_5 ^{79}Br (Fig. 3.1) will be 2×12 plus 5×1.0078246 plus 78.91839 minus the mass of the electron (m_e). This is equal to 107.95751 u $- m_e$, u being the unified atomic mass unit based on the standard that the mass of the isotope $^{12}C = 12$ u exactly.

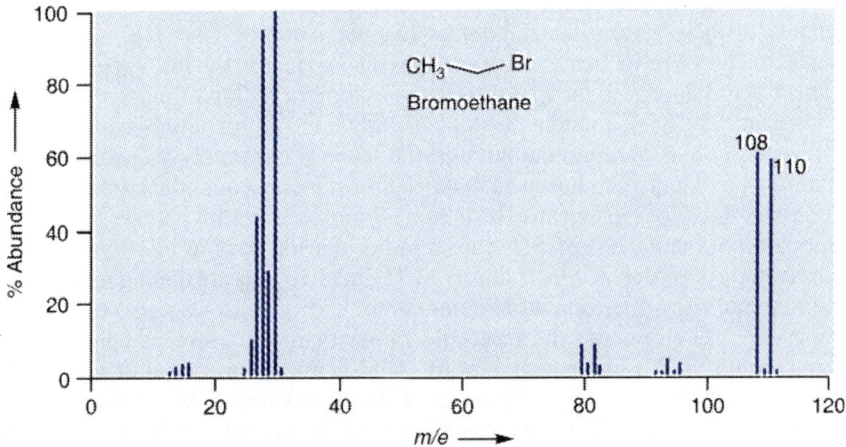

Fig. 3.1: Mass spectrum of bromoethane

It is the precursor of other ions appeared in the mass spectrum. Hence, it is called parent ion. This $M^{+\bullet}$ is not fragmented before collection at the detector and is recorded at m/e corresponding to the molecular weight of the compound, hence called molecular ion or *parent ion* peak. All other ions are generated from the fragmentation of molecular ion.

If the molecular ion appears, it will be the highest mass in an EI spectrum (except for isotope peaks discussed below).

Its appearance depends on the stability of the molecular ion. The molecular ion is stabilized by pi electron system which are capable of accommodating a loss of one electron more easily. Cyclic structure also gives the more intense molecular ion peak.

Generally, the stability of molecular ion decreases in the following order:

Aromatics > Conjugated Olefins > Cyclics > Unbranched Hydrocarbons > Mercaptans > Ketones > Amines > Esters > Ethers > Carboxylic Acids > Branched Hydrocarbons > Alcohols

The most stable molecular ions of purely aromatic system. The detection of molecular ion is difficult in aliphatic alcohols, nitrites, nitrates, nitro compounds, nitriles and highly branched compounds.

Important features of molecular ions are:

1. The molecular ion peak is intense in aromatic compounds due to the presence of pi (π) electron system.
2. Conjugated olefins show more intense molecular ion peak in compared to non-conjugated olefins having the same number of unsaturation. Conjugated olefins are more stable than unconjugated olefins.
3. Unsaturated compound shows more intense molecular ion peak than saturated or cyclic compound.
4. The molecular ion peak of saturated hydrocarbon is more intense than branched chain compound having the same number of carbon atom. For example, molecular ion peak is more intense than neopentane.
5. The substituent groups like –OH, NH_2 which lowers the ionization potential, increase the relative abundance in case of aromatic compounds.
   The substituent groups like $-NO_2$, –CN which increase the ionization potential, decrease the relative abundance in case of aromatic compounds.
6. Absence of molecular ion peak in the mass spectrum means that the compound is highly branched or tertiary alcohol.

ISOTOPIC IONS

Many elements in their natural states contain heavier isotopes, because mass spectrometer measures mass to charge ratio, hence these isotopes also appear in the mass spectrum as isotopic ion peak. These peaks appear in the mass spectrum in the last region of mass spectrum just after molecular ion peak. The most abundant isotope of carbon is ^{12}C but natural carbon also contains ^{13}C and ^{14}C. Although as natural beta emitter, the latter is extremely valuable for radiotracer work, its natural abundance is so low as to make it almost inconsequential to mass spectrometry. This case does not exist for ^{13}C because its natural abundance is 1.08 percent in carbon. Hence, the mass spectrum of methane (Fig. 3.2) shows molecular ion peak at m/e 16 due to $^{12}CH_4$ but also contains isotopic ion peak at m/e 17 due to $^{13}CH_4$, the two ions are having relative abundance of 99 : 1. As the number of carbon atoms in a compound increases so also do the chances of incorporating one ^{13}C atom into the molecule rather than ^{12}C atom. A compound with ten carbon atoms would yield a molecular ion, M^+, and isotopic ion one unit greater $(M + 1)^+$, which would be about $10 \times 1.08 = 10.8$ percent of the abundance of M^+. The chance of finding two ^{13}C atoms in the same molecule increases with the increasing number of carbon atoms, so that $(M + 2)^+$ starts to become more prominent. The chance of finding more than two ^{13}C atoms

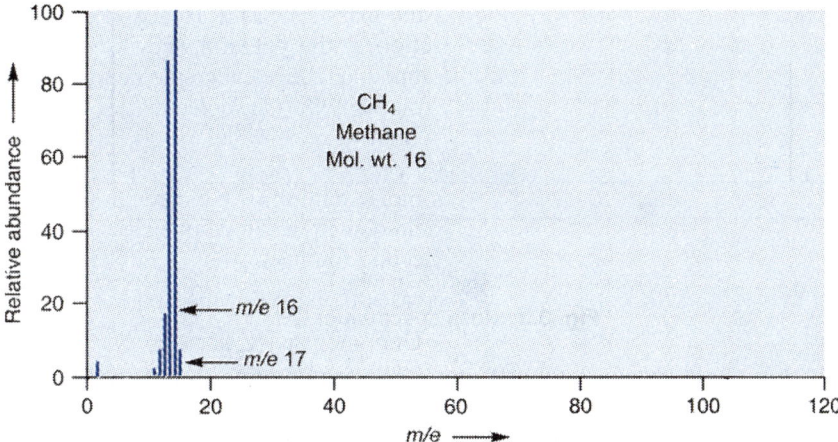

Fig. 3.2: Mass spectrum of methane

are very small except with large number of carbon atoms and $(M + 3)^+$ can safely ignored. For a compound of ten carbon atoms, the approximate relative height of M^+, $(M + 1)^+$, $(M + 2)^+$ and $(M + 3)^+$ are 100 : 10 : 0.45 : 0.01 from which it can be seen how unimportant are peaks greater than $(M + 2)^+$. Tables are available giving the relative heights of M^+, $(M + 1)^+$ and $(M + 2)^+$ peaks for various elemental compositions (Beynon and Williams 1963). All the carbon compounds contain molecular and fragment ion accompanied by isotopic ion one and two unit greater, making the appearance of mass spectrum complex. Figure 3.2 shows mass spectrum with ^{13}C isotope included. Table 3.1 gives, for some elements commonly met in mass spectrometry, the approximate natural abundances of most significant isotopes. From the table it should be noted that bromine, chlorine and sulphur have particularly abundant isotopes separated by two mass units. For these reasons, the bromine and chlorine containing compounds especially recognized by mass spectrometry and by examining the isotopic ion pattern in the molecular ion region, the number of bromine and chlorine atoms can be determined in an original molecule.

Table 3.1: Natural abundance of some elements						
Element	**Isotope**	**Relative abundance**	**Isotope**	**Relative abundance**	**Isotope**	**Relative abundance**
Carbon	^{12}C	100	^{13}C	1.11		
Hydrogen	^{1}H	100	^{2}H	0.016		
Nitrogen	^{14}N	100	^{15}N	0.38		
Oxygen	^{16}O	100	^{17}O	0.04	^{18}O	0.20
Sulphur	^{32}S	100	^{33}S	0.78	^{34}S	4.40
Chlorine	^{35}Cl	100			^{37}Cl	32.5
Bromine	^{79}Br	100			^{81}Br	98.0
Silicon	^{28}Si	100	^{29}Si	5.1	^{30}Si	3.4

Chlorine is an excellent example of how isotope distributions are useful for determining the presence of chlorine atom in a molecule. The molecular weight of chlorine is 35.45 u. This is calculated from the natural abundance of ^{35}Cl (75%) and ^{37}Cl (25%). The natural abundance of these two isotopes is observed in the mass spectrum as two peaks separated by m/e 2 with a relative intensity of 3 : 1. The mass spectrum of CH_3Cl (Fig. 3.3) clearly shows two peaks with the isotope distribution pattern for an ion with a single chlorine atom. CH_3 ^{35}Cl (m/e 50) and CH_3 ^{37}Cl (m/e 52) are separated by m/e 2 value and possess the 3 : 1 abundance ratio characteristic

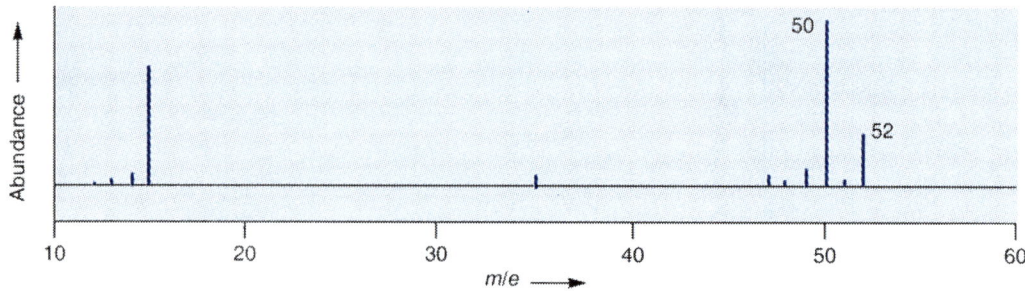

Fig. 3.3: Mass spectrum of CH_3Cl

of an ion with a single chlorine atom. If more than one chlorine atom is present, the isotope abundance is more complex. An ion with two chlorine atoms has three possible isotope combinations. This pattern is apparent in the mass spectrum of CH_2Cl_2 (Fig. 3.4). Ions are observed for $CH_2{}^{35}Cl_2{}^+$ (m/e 84), $CH_2{}^{35}Cl{}^{37}Cl^+$ (m/e 86), and $CH_2{}^{37}Cl_2$ (m/e 88). Based upon the probability of each combination of isotopes, the relative intensity of these peaks is 10 : 6 : 1. The 3 : 1 isotope ratio for an ion with a single chlorine atom is observed at m/e 49 and m/e 51. This corresponds to $CH_2{}^{35}Cl$ and $CH_2{}^{37}Cl$ fragments formed by loss of Cl from the molecular ion. Careful examination of the spectrum also shows ions produced by loss of H· and H_2.

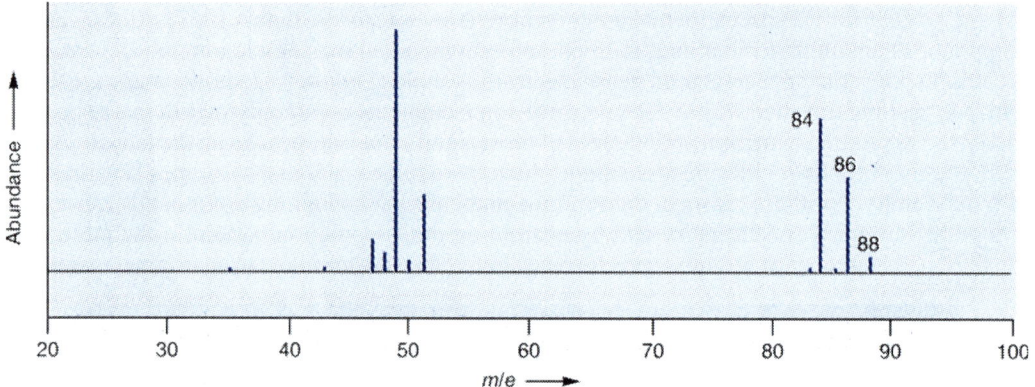

Fig. 3.4: Mass spectrum of CH_2Cl_2

Similarly, the mass spectrum of propyl bromide (Fig. 3.5) has two prominent peaks of equal intensity at 122 (M^+) and 124 $(M + 2)^+$ and then two more at 79 and 81 belonging to the bromine fragment. The relative height of the peaks at 122 and 124 is almost equal due to their isotopes ^{79}Br (100%) and ^{81}Br (98%).

Note: In general, the contribution of various isotopes of different elements can be calculated from the following expression.

$$(a + b)^n$$

where a = % natural abundance of the lighter isotope; b = % natural abundance of the heavy isotope; and n = number of atoms of the element present in a molecule.

In case of a molecule containing two atoms of chlorine, such as CH_2Cl_2,

$$(a + b)^n = (0.754 + 0.246)^2 = 0.568 + 0.371 + 0.060 = 100 : 65.65 : 10.60$$

The three ions' mass M^+ ($CH_2{}^{35}Cl_2$), $(M + 2)^+$ ($CH_2{}^{35}Cl^{37}Cl$), and $(M + 4)^+$ ($CH_2{}^{37}Cl_2$) have intensity ratios of 100 : 65.25 : 10.60, respectively.

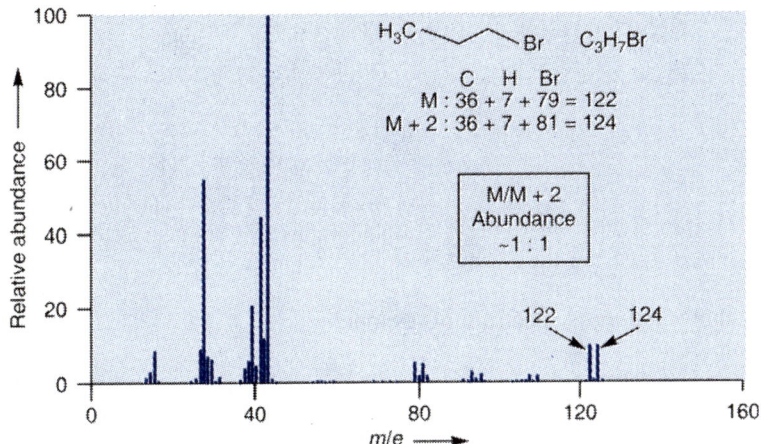

Fig. 3.5: Mass spectrum of propyl bromide

When a molecule is formed with two different polyisotopic elements (for example, Cl and Br), the following equation must be used.

$$(a + b)^n (c + d)^r$$

where a, b, c and d refer to the percentage abundance of ^{35}Cl, ^{37}Cl, ^{79}Br, and ^{81}Br, respectively, and n and r are the number of atoms of chlorine and bromine respectively.

FRAGMENT IONS

When a molecule is ionized by removal of an electron, a molecular ion ($M^{+\bullet}$, a cation radical) is produced. This molecular ion may contain sufficient excess internal energy to fragment by ejection of neutral particle (N) with the formation of fragment ion ($A^{+\bullet}$ or A^+). A neutral molecule gives a radical cation as the molecular ion, and the fragment ion may be either a radical cation or cation. The ejected neutral particle (N) may be a radical or neutral molecule.

$$M \xrightarrow{e^-} M^{+\bullet} \longrightarrow A^+ + N^\bullet$$

$$M \xrightarrow{e^-} M^{+\bullet} \longrightarrow A^{+\bullet} + N$$

If the fragment ion ($A^{+\bullet}$) has sufficient excess of internal energy, the further decomposition may occur with the formation of new fragment ions (B^+, C^+, etc.) until there is insufficient excess of internal energy in any one of ion for further fragmentation. Such a series of decomposition when elucidated from mass spectrum is referred to as fragmentation pathway. The molecular ion ($M^{+\bullet}$) and any of the fragment ions (A^+, B^+, C^+, etc.) may decompose by more than one pathway. The various fragmentation pathways together comprise the fragmentation pattern, characteristic of a compound under investigation.

Here, fragment ions of methane and acetone are shown below.

$$CH_4 + e^- \longrightarrow CH_4^{+\bullet} + 2e^-$$
(Molecular ion)

$$\downarrow$$

$$CH_3^+ + H^\bullet$$
(Fragment ion)

$$CH_3COCH_3 + e^- \longrightarrow CH_3COCH_3^{+\bullet} + 2e^-$$
$$\text{(Molecular ion)}$$

$$\downarrow$$

$$CH_3C \equiv O^+ + CH_3^\bullet$$

Acylium ion   Methyl radical
(fragment ion)   (not detected)

Figure 3.6 shows the mass spectrum of acetone.

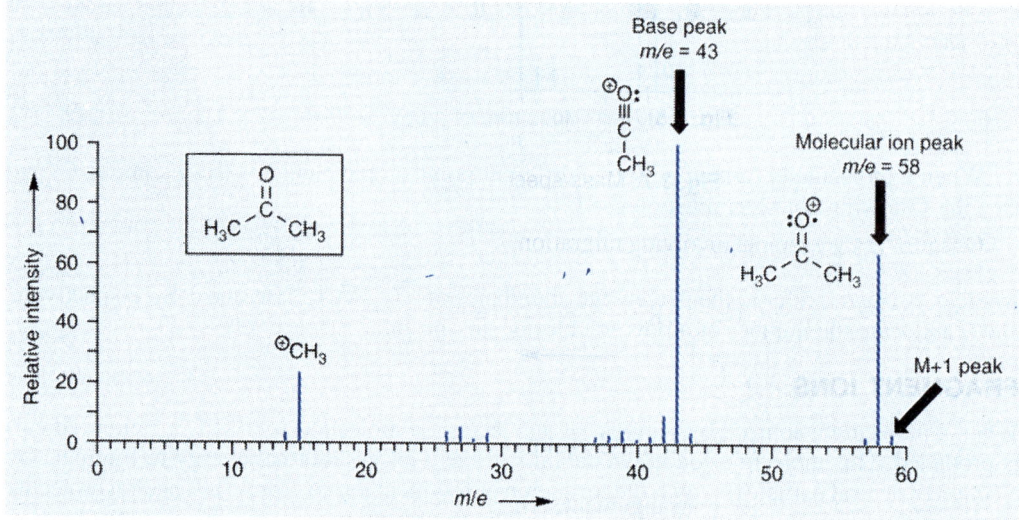

Fig. 3.6: Mass spectrum of acetone

REARRANGEMENT IONS

Rearrangement ions are fragments whose origin cannot be described by simple cleavage of bonds in the molecular ion, but are a result of intramolecular atomic rearrangement during fragmentation. Rearrangements involving migration of hydrogen atoms in molecules that contain a heteroatom are especially common. For example, diethyl ether shows a prominent peak at m/e 31 due to rearrangement (Fig. 3.7).

$$CH_3-CH_2-O-CH_2-CH_3 \rbrack^{+\bullet} \longrightarrow CH_3-CH_2-\overset{\bullet\bullet}{\underset{\bullet\bullet}{O}}-CH_2^+$$

Diethyl ether
m/e 74

$$CH_2 = CH_2 + H\overset{+}{\underset{\bullet\bullet}{O}} = CH_2 \longleftarrow \underset{\overset{\displaystyle\frown}{}}{CH_2} - CH_2 - \overset{+}{\underset{\bullet\bullet}{O}} = CH_2$$

m/e 31

m/e 59

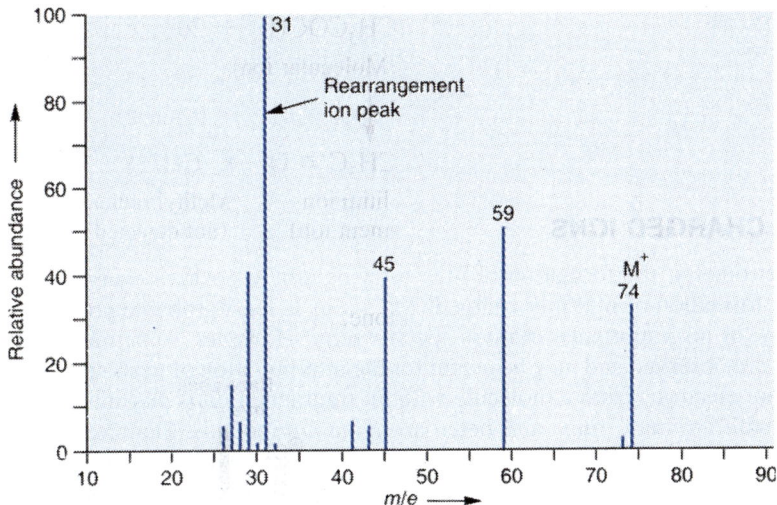

Fig. 3.7: Mass spectrum of diethyl ether

One important example involving migration of hydrogen atom is *McLafferty rearrangement*:

R may be H, alkyl group, *O*-alkyl, NH_2, Cl, OH, etc.

To undergo McLafferty rearrangement, a molecule must possess an appropriately located heteroatom (e.g. O), a π-system (usually a double bond) and an abstractable hydrogen ã to the C=O system. Such rearrangements often account for prominent characteristic peaks and are consequently very useful for our purpose. They can frequently be rationalized on the basis of low-energy transitions and increased stability of the products. Rearrangements resulting in elimination of stable neutral molecule are common (e.g. the olefin product in the McLafferty rearrangement). Rearrangement peaks can be recognized by considering the mass (*m/e*) number for fragment ions and for their corresponding molecular ions. A simple (no rearrangement) cleavage of an even-numbered molecular ion gives an odd numbered fragment ion and simple cleavage of an odd-numbered molecular ion gives an even-numbered fragment. Observation of a fragment ion mass different by 1 unit from that expected for a fragment resulting from simple cleavage (e.g. an even-numbered fragment mass from an even-numbered molecular ion mass) indicates rearrangement of hydrogen has accompanied fragmentation. Rearrangement peaks may be recognized by considering the corollary to the "nitrogen rule". Thus, even-numbered peak derived from an even-numbered molecular ion is a result of two cleavages, which may involve a rearrangement.

In some compounds, structural features also promote double McLafferty rearrangements.

$$CH_3 \quad \overset{O^{+\cdot}}{\quad} \quad \overset{H\ H}{\quad} CH_3 \xrightarrow{1.\ MR} CH_3 \quad \overset{\cdot^+OH}{\quad} \overset{}{CH_2} + \quad H_3C \quad \overset{}{CH_2}$$

5-Nonanone
m/e 142

m/e 100

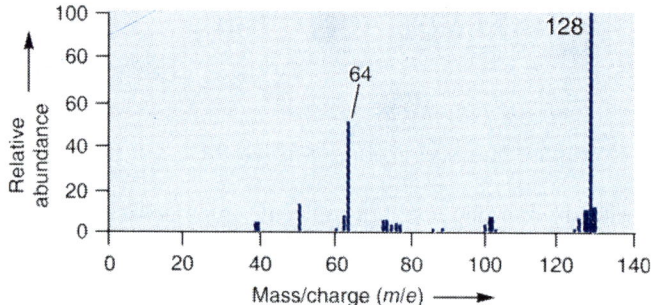

MULTIPLE CHARGED IONS

In mass spectrometry, the formation of ions carrying single positive charge is most probable process. The formation of multiply charged (M^{n+}) ions is least probable process. Hence, these are, generally, of no importance in mass spectrometry. However, sometimes doubly or triply charged ions are observed and may be useful for the interpretation of mass spectra. The removal of two or more electrons from a molecule without fragmentation is possible in case of organic compounds with aromatic rings and heteroaromatic compounds. Double and triple charged ions are formed by removal of two or more electrons respectively and they appear at 1/2 and 1/3 of the actual mass respectively. For example, naphthalene shows molecular ion peak at m/e 128 due to singly charged ion but peak at m/e 64 appears due to doubly charged (M^{2+}) (Fig. 3.8) (half value of molecular mass of naphthalene). The doubly charged ion undergoes fragmentation to give doubly charged ions which are recorded at m/e values numerically equal to one-half of their actual mass numbers.

Fig. 3.8: Mass spectrum of naphthalene

METASTABLE IONS

These ions are formed with sufficient excitation to dissociate spontaneously during its flight from the ion source to the detector. The presence of metastable ion is indicated by broad peaks, called "metastable peak", that do not usually appear at whole mass numbers.

If the average lifetime of ion (m_1) is $>5 \times 10^{-6}$ secs, the ion is accelerated, deviated and recorded as m_1. If the average lifetime of ion $m_1 < 5 \times 10^{-6}$ secs, it fragments before acceleration to give a new ion of mass m_2.

If the ion decomposed in the zone between the source and analyser, there will be an ion with the acceleration of m_1 that becomes deviated as m_2. The ion m_2 appears in the spectrum at m^*. Such decomposition is called "metastable transition" and the peak m^* (Fig. 3.9) is an example of metastable transition.

The numerical value of m^* is given by

$$m^* = (m_2)^2/m_1$$

where m_1, m_2 and m^* are the masses (strictly m/e) of the ions m_1^+, m_2^+ and m^* respectively and $m_1 > m_2$. For example, toluene shows intense peak at m/e 91 ($C_7H_7^+$) and m/e 65 ($C_5H_5^+$) and appearance of a broad peak at 46.4 ($= 65^2/91$) indicates that at least some of the ions at m/e 65 arise through ejection of C_2H_2 (26 mass unit) from the ions at m/e 91.

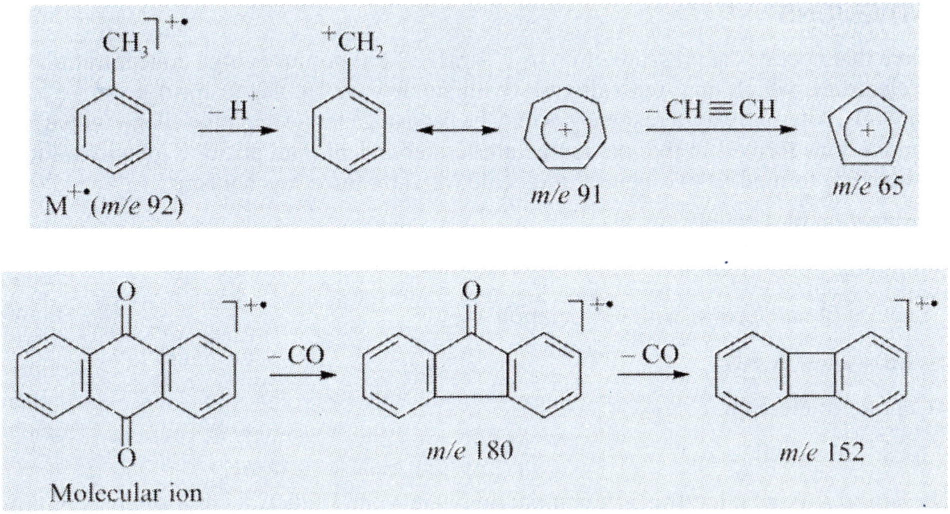

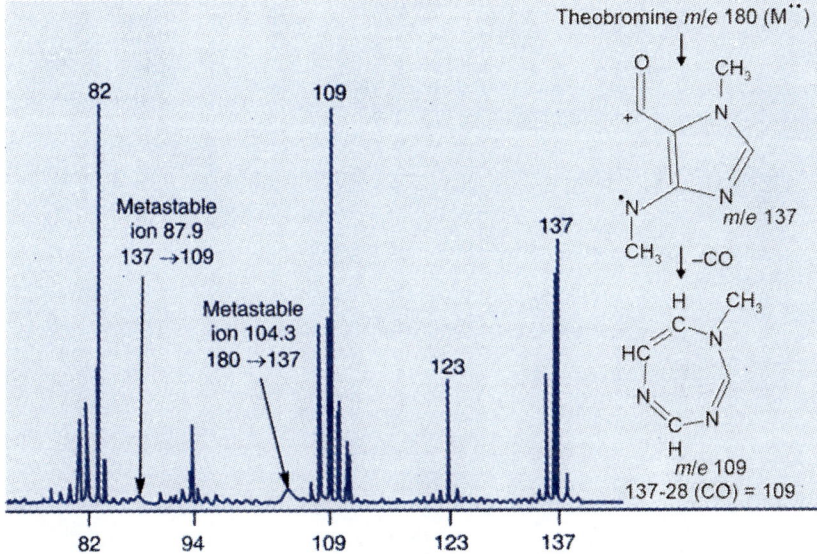

Fig. 3.9: Metastable ions

The mass spectrum of acetophenone also shows metastable peak at 92.1 and give evidence for the decomposition of the molecular ion of acetophenone (m/e 120) to $C_6H_5CO^+$ (m/e 105) in one step because the predicted mass

$$m^* = (105)^2/120 = 91.88$$

$$C_6H_5COCH_3^{+\bullet} \rightarrow C_6H_5CO^+ + {}^\bullet CH_3$$

Figure 3.10 shows an example of the spectrum of theobromine, the metastable fragmentation of which is shown in the scheme. The signal detected at an apparent m/e 87.9 comes from the metastable fragmentation of the ion at 137, which loses 28 and yields the fragment, m/e 109. Similarly, the signal at apparent m/e 104.3 comes from the metastable fragmentation 180→137.

Fig. 3.10: Mass spectrum of theobromine showing metastable peaks

NEGATIVE IONS

Ion source that operates at pressure of up to 1 Torr (133 Pa) produces high concentration of low energy electrons, which may react directly with suitable molecules to form negative ion and fragment ions. Alternatively, a reagent gas may be used to generate negative chemical ionization.

Negative ions formed in the ion source are less abundant than positive ions by a factor of 10^{-4}. These are formed from a neutral molecule (AB) by three mechanisms:

(i) Production of a couple of ions

$$AB + e^- \rightarrow A^+ + B^- + e$$

(ii) Capture of an electron with dissociation

$$AB + e^- \rightarrow A + B^-$$

(iii) Capture of electron

$$AB + e^- \rightarrow AB^-$$

The suitability of electron attachment may be very high for compounds containing electronegative elements and it may be enhanced for other compound by making the use of fluorinated derivatives.

IONS FORMED BY ION MOLECULE INTERACTION

The percentage of molecules of a compound ionized to the vapor in ionization chamber is very low. Therefore, there is possibility of collision between molecular ion and neutral molecule. In such a collision, the molecular ion can subtract an atom from the neutral molecule, forming a heavier ion. The simplest example is the protonation of the molecular ion, resulting in the formation of M + 1 peak. This peak can be differentiated from the isotopic peak with the help of high resolution data. This reaction occurs when the sample pressure is high.

Chapter

4

Determination of Molecular Formula

It is difficult to get the exact molecular formula solely on the basis of mass spectral data alone. A mass spectrum can only give the idea about the probable molecular formula. It needs further confirmation from the other spectral (IR and NMR) data.

The prime step to determine the molecular formula is determination of molecular ion peak. If the molecular ion appears, it will be the highest mass in an EI spectrum (except for isotope peaks discussed earlier). This peak will represent the molecular weight of the compound. Its appearance depends on the stability of the compound. Double bonds, cyclic structures and aromatic rings stabilize the molecular ion and increase the probability of its appearance and, thus, increase its abundance. There are two situations in which the identification of the molecular ion peak may be difficult.

(a) The molecular ion does not appear or is very weak. The obvious remedy in most cases is to run the spectrum at maximum sensitivity (and accept the resulting loss in resolution) and to use a larger sample. (Sometimes a large sample exaggerates the M + 1 peak) Still the molecular ion may not be evident and then other sources of information may be useful. The type of compound may be known, and the molecular mass may be deduced from the breakdown pattern. For example, alcohol usually gives a very weak parent molecular ion peak but often shows a pronounced peak resulting from loss of water (M-18). Similarly, loss of CH_3 radical (M-15), $-OCH_3$ (M-31), $-COOH$ (M-45), CH_3, H_2O (M-33) can be focussed to locate molecular ion peak if it is weak or not present in the mass spectrum.

(b) The molecular ion is present but is one of several peaks which may also be as prominent. In this situation, the question is that of purity. If the compound can be assumed to be pure, the usual problem is to distinguish the molecular ion peak from a more prominent M-1 peak. One good test is to reduce the energy of the bombarding electron beam to near the appearance potential. This will reduce the intensities of all peaks, but will increase the intensity of the molecular ion relative to other peaks, including fragmentation peaks (but not molecular ion peaks) of impurities. Another test frequently used is to increase the size of the sample, or increase the time the sample spends in the ionization chamber by decreasing the ion repeller voltage. In either case the net effect is to increase the opportunity for bimolecular collisions to occur in the ion chamber. The most common result of the bimolecular collision of a molecular ion containing a heteroatom (O, N or S) is a contribution to the M + 1 peak (i.e., the net effect is the transfer of a hydrogen atom from a neutral molecule to the molecular ion). Thus, an increase in peak size relative to other peaks, as sample size is increased or the repeller voltage is decreased, designates that peak as the M + 1 peak and affords an indirect identification of the molecular ion. Of course, the dependence of the M + 1 on a sample size must be kept in mind when the peak is used to establish a molecular formula of a compound containing a heteroatom. Many peaks can be ruled out as possible molecular ions simply on grounds of following reasonable structure requirements:

(i) **Nitrogen rule** is often helpful identifying the molecular ion peak. It states that a molecule of even-numbered molecular weight must contain no nitrogen or an even number of nitrogen atoms; an odd-numbered molecular weight requires an odd number of nitrogen atoms. This rule holds for all compounds containing carbon, hydrogen, oxygen, nitrogen, sulphur, and the halogens, as well as many of the less usual atoms such as phosphorous, boron, silicon, arsenic, and the alkaline earths.

Molecule:	NH_3	H_2NNH_2		
Formula:	NH_3	N_2H_4	$C_7H_5N_3O_6$	$C_8H_{10}N_4O_2$
m/e for M (amu):	17	32	227	194

This distinction is also illustrated nicely by the following two examples (Figs 4.1 and 4.2). The unsaturated ketone, 4-methyl-3-pentene-2-one, below has no nitrogen, so the mass of the molecular ion (*m/e* 98) is an even number. Most of the fragment ions have odd-numbered masses. Diethylmethylamine, on the other hand, has one nitrogen and its molecular mass (*m/e* 87) is an odd number. A majority of the fragment ions have even-numbered masses (ions at *m/e* 30, 42, 56 and 58 are not labelled). The weak even-electron ions at *m/e* 15 and 29 are due to methyl and ethyl cations (no nitrogen atoms).

(ii) A useful corollary states that fragmentation at a single bond gives an odd-numbered ion fragment from an even-numbered molecular ion, and an even-numbered ion fragment from an odd numbered molecular ion. For this corollary to hold, the fragment must contain all nitrogen (if any) of the molecular ion.

(iii) Consideration of the breakdown pattern coupled with other information also assists in identifying molecular ions. It should be kept in mind that the intensity of the molecular ion peak depends on the stability of the molecular ion. The most stable molecular ions are those of purely aromatic systems. If substituents that have a favour-

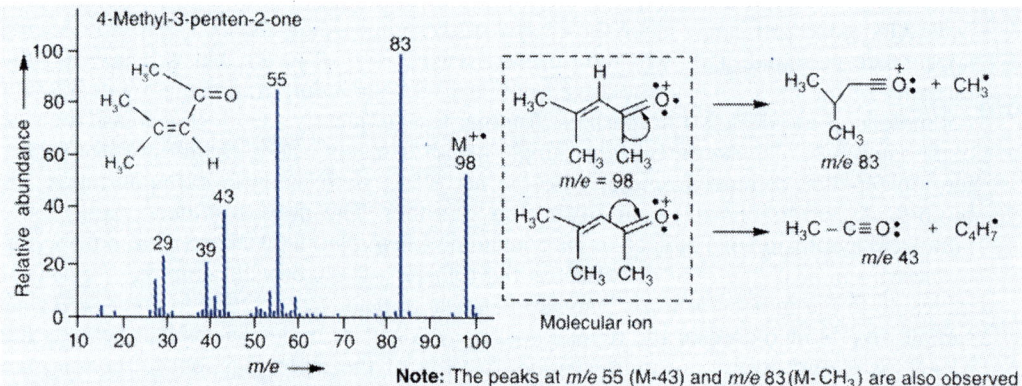

Note: The peaks at *m/e* 55 (M-43) and *m/e* 83 (M-CH₃) are also observed

Fig. 4.1: Mass spectrum of 4-methyl-3-penten-2-one

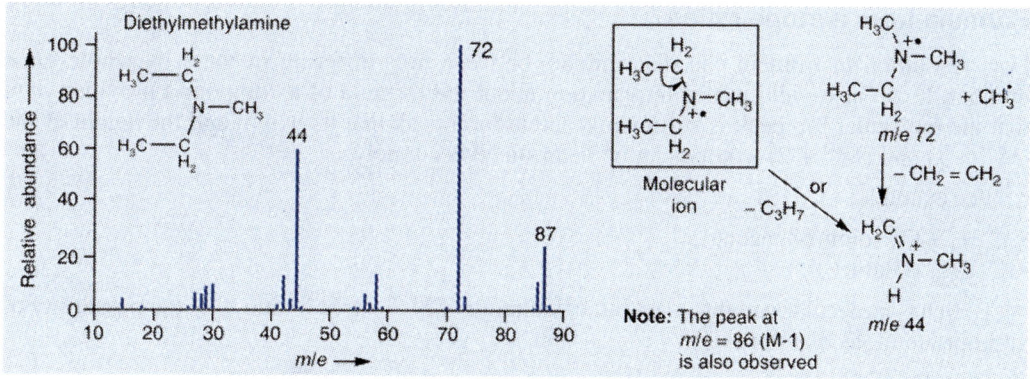

Fig. 4.2: Mass spectrum of diethylmethylamine

able mode of cleavage are present, the molecular ion peak will be less intense, and the fragment peaks relatively more intense. In general, the following group of compounds will, in order of decreasing ability, **give prominent molecular ion peaks:** aromatic compounds > conjugated alkenes > cyclic compounds > organic sulphides > short, normal alkanes.

Recognizable molecular ions are usually produced for these compounds in order of decreasing ability: Ketones > amines > esters > carboxylic acids-aldehydes-amides-halides. The **molecular ion is frequently not detectable** in aliphatic alcohol, nitrites, nitrates, nitro compounds, nitriles and in high.

The presence of M-15 peak (loss of CH_3) or an M-18 peak (loss of H_2O) or an M-31 peak (loss of OCH_3 from methyl esters), etc., is taken as confirmation of the molecular ion peak. An M-1 peak is common and occasionally an M-2 peak (loss of H_2 by either fragmentation or thermolysis or even a rare M-3 peak (from alcohol) is reasonable. Peaks in the range of M-3 to M-14, however, indicate that contaminants may be present or that the presumed molecular ion peak is actually a fragment ion peak. Loss of fragments of masses 19 to 25 are also unlikely (except for the loss of F = 19 or HF = 20 from flourinated compounds). Loss of 16 (O), 17 (OH), or 18 (H_2O) are likely only if an oxygen atom is in the molecule.

Another way of confirming a molecular ion is that all fragments in the spectrum (other than impurities) should come from the assigned molecular ion with logical loss of neutral fragments. If a fragment or two do not comply, this will mean either the assigned molecular ion is not really the molecular ion but a fragment ion or the fragment does not belong to the compound but to a background contaminant.

DETERMINATION OF MOLECULAR FORMULA

Formula from high resolution instrument

The probable molecular formula can often be derived from the exact mass of the molecular ion peak provided we have some structures in our mind. High resolution instrument is capable of detecting mass differences of a few thousands of a mass unit.

For example:

- Purine $C_5H_4N_4$ (M = 120.044)
- Benzimidine $C_7H_8N_2$ (M = 120.069)
- Ethyl toluene C_9H_{12} (M = 120.096)
- Acetophenone C_7H_8O (M = 120.058)

If the measured mass of molecular ion peak is 120.070 (± 0.005), then all except $C_7H_8N_2$ are excluded as possible formula.

Formula from isotope ratios

Low resolution instrument can discriminate between ions differing in mass by whole mass number. It can also yield useful information about the formula of a compound provided only that the molecular ion peak is sufficiently intense ion peak that its height and the height of the $(M^{+\bullet} + 1)$ and $(M^{+\bullet} + 2)$ isotope can be determined accurately.

For example:

- $C_6H_4N_2O_4$ (dinitrobenzene)
- $C_{12}H_{24}$ (Olefin)

Both has molecular weight ~ 168 but calculation of $M^{+\bullet} + 1/M^{+\bullet}$ ratio give the possibility of appropriate molecular formula.

$C_6H_4N_2O_4$	$C_{12}H_{24}$
^{13}C 6 × 1.08 = 6.48%	^{13}C 12 × 1.08 = 12.96%
^{2}H 4 × 0.015 = 0.060%	^{2}H 24 × 0.015 = **0.36%**
^{15}N 2 × 0.37 = 0.74%	$\dfrac{M^{+\bullet}+1}{M^{+\bullet}} = 13.32\%$
^{17}O 4 × 0.04 = **0.16%**	
$\dfrac{M^{+\bullet}+1}{M^{+\bullet}} = 7.44\%$	

In an approximate but rapid manner, ratios $(M^{+\bullet} + 1)/M^{+\bullet}$ and $(M^{+\bullet} + 2)/M^{+\bullet}$ can be determined in the following ways:

$$(M^+ + 1)/M^+ = 1.1\% \times \text{No. of carbon atoms} + 0.36\% \times \text{No. of nitrogen atoms}$$

$$\frac{M^{+\bullet}+2}{M^{+\bullet}} = \frac{(1.1\% \times \text{No. of carbon atoms})^2}{200} + 0.2\% \times \text{No. of nitrogen atoms}$$

Molecular formula from Beynon table

Generally, position of the molecular ion and isotopic ion peak along with their intensities are mentioned in mass spectral data. For example, in a compound following mass spectral data are recorded:

$\quad$ 150 $(M^{+\bullet})$ $\qquad$ 100%

$\quad$ 151 $(M^{+\bullet} + 1)$ $\quad$ 9.9%

$\quad$ 152 $(M^{+\bullet} + 2)$ $\quad$ 0.9%

To know the molecular formula of this compound, we select all the compounds of molecular weight 150 possessing nearly the same $M^+ + 1$ and $M^+ + 2$ peak intensities from Beynon table:

	$M^{+\bullet} + 1$	$M^{+\bullet} + 2$
(i) $C_7H_{10}N_4$	9.25	0.38
(ii) $C_8H_8NO_2$	9.23	0.78
(iii) $C_8H_{10}N_2O$	9.61	0.61
(iv) $C_8H_{12}N_3$	9.98	0.45
(v) $C_9H_{10}O_2$	9.96	0.84
(vi) $C_9H_{12}NO$	10.34	0.68
(vii) $C_9H_{14}N_2$	10.71	0.52

There are seven molecular formula at the molecular weight – 150 possessing same or nearly same $M^{+\bullet} + 1$ and $M^{+\bullet} + 2$ (intensities peak). According to nitrogen rule, molecular formulas, (ii), (iv) and (vi) are rule out (compound possessing even numbered molecular weight should possess even number nitrogen or no nitrogen atoms). Then, it is observed that molecular formula $C_9H_{10}O_2$ (V) possess the more close $M^{+\bullet} + 1$ (9.96) and $M^{+*} + 2$ (0.84) values. Hence, it is the possible molecular formula of the compound which should be further verified from the IR, NMR and mass fragmentation pattern.

How the M + 2 (and possibly M + 4) peak arises, and its use in showing the presence of chlorine or bromine in a compound.

Molecular formula from molecular ion and isotopic ion peaks and Beynon table

For example, molecular mass is 74. With the help of molecular mass, 74, and intensities of M + 1 and M + 2 peaks, following formulas come in consideration as per Beynon table.

Molecular ion peak ($M^{+\bullet}$) formulas	Exact mass	M + 1	M + 2
$CH_2N_2O_2$	74.0117	1.95	0.41
CH_4N_3O	74.0355	2.33	0.22
CH_6N_4	74.0594	2.70	0.03
$C_2H_2O_3$	74.0004	2.31	0.62
$C_2H_4NO_2$	74.0242	2.69	0.42
C_2H_6NO	74.0480	3.06	0.23
$C_3H_6O_2$	74.0368	3.42	0.44
$C_3H_{10}N_2$	74.0845	4.17	0.07
$C_4H_{10}O$	74.0003	4.52	0.28

From the above formulas, we assume that formula for compound is $C_4H_{10}O$. Calculate the % intensities of M + 1 and M + 2.

M + 1 peak → arises from different possibilities of additional amu = 75 amu.

Sum of possibilities for M + 1 peak ($^{13}C + {}^2H + {}^{16}O$) = $(0.044 + 0.0015 + 0.0004) \times 100 =$ 4.59%.

M + 2 peak → arises from different possibilities of two additional amu = 76 amu.

Sum of possibilities for M + 2 peak (two ^{13}C + two ^2H + one ^{16}O) = $(0.0007 + 0.0020 + 0.0001) \times 100 = 0.28$.

Now, it is observed that calculated M + 1 and M + 2 matches with the observed value in the molecular formula → $C_4H_{10}O$. Hence, it is required molecular formula.

Note: The molecular weight is even. The number of nitrogen atoms must be even (0, 2, 4, ...) or nitrogen should be absent.

Molecular formula from molecular and isotopic ion peaks

Mass → Formula

Example 1

Molecular ion	m/e	Relative abundance	Conclusions
M	102	100%	Mass = 102, even number of nitrogens
M + 1	103	6.9	$6.9 \div 1.1 = 6.3$, six carbons
M + 2	104	0.38	S, Cl or Br atom absent; oxygen atom?

$M - C_6 = 102 - (6 \times 12) = 30$ amu for oxygen, nitrogen, hydrogen.

Oxygens	Nitrogens	30 – O – N = H	Formula notes
0	0	30 – 0 – 0 = 30	C_6H_{30}, violates hydrogen rule
1	0	30 – 16 – 0 = 14	$C_6H_{14}O$, reasonable
2	0	30 – 32 – 0 = –2	Not possible
0	2	30 – 0 – 28 = 2	$C_6H_2N_2$, reasonable

Two molecular formula are possible and one formula can be selected/ruled out on the basis of other spectral data.

Example 2

Molecular ion	m/e	Relative abundance	Conclusions
M	157	100%	Mass = 157, odd number of nitrogens
M + 1	158	9.39	9.39 ÷ 1.1 = 8.5, eight or nine carbons
M + 2	159	34	One chlorine; no bromine; # of sulphur?

$M - C_8 - N - Cl = 157 - (8 \times 12) - 14 - 35 = 12$ amu for oxygen, nitrogen, hydrogen. Not enough for amu for oxygen or another nitrogen. Probable molecular formula = $C_8H_{12}NCl$.

Example 3

Molecular ion	m/e	Relative abundance	Conclusions
M	148	100%	Mass = 148, zero or even number of nitrogens
M + 1	149	11.25	11.25 ÷ 1.1 = 10.23, ten carbons
M + 2	159	34	One chlorine; no bromine; # of sulphur?

$148 - 120 (C_{10}) = 28$ amu (atomic mass units) for oxygen, nitrogen and hydrogen.

Oxygens	Nitrogens	28 – O – N = H formula	Comments
None	None	28 – 0 – 0 = 28	$C_{10}H_{28}$, violates hydrogen rule
One	None	28 – 16 – 0 = 12	$C_{10}H_{12}O$, reasonable
None	Two	28 – 0 – 28 = 0	$C_{10}N_2$, reasonable

Index of Hydrogen Deficiency (IHD)

It indicates the sum of the number of rings and pi bonds in a molecule. Once the formula is known, it is possible to calculate the hydrogen deficiency of the molecule. For instance, for a saturated hydrocarbon without rings, the formula is $C_nH_{2n + 2}$. The difference between the number of hydrogens in the formula and 2n + 2 divided by two is the hydrogen deficiency or a sum of the number of bonds and rings.

Each multiple bond and each ring contributes 1 degree of hydrogen deficiency to a compound:

Structural unit	Contribution to index of hydrogen deficiency
Double bond	1
Triple bond	2
Ring	1

The hydrogen deficiency in molecules containing C, H, N, O, Cl, Br or I for the molecule $C_cH_hN_nO_oX_x$ (where X represents any halogen and c, h, n and x represent the total number of carbon, hydrogen, nitrogen and halogen atoms respectively) can be calculated from the equation:

$$\text{Index of hydrogen deficiency}(U) = \frac{2c + 2 - h + n - x}{2}$$

Where does this equation come from?

- The **maximum number of hydrogen atoms for** "c" carbon atoms is $2c + 2$ (for example, saturated hydrocarbon, methane has one carbon atom, $2 \times 1 + 2 = 4$).
- From this number, subtract the "h" hydrogens that you have.
- Since, like hydrogen, a halogen only forms one bond, then they can be treated as if they are hydrogens, so subtract them as well.
- Oxygen forms two bonds, therefore, it has **no impact** (compare H count for methane, CH_4, and methanol, CH_3OH). Both have same index of hydrogen deficiency.
- Nitrogen forms three bonds. This means for "n" nitrogens, "n" extra hydrogen atoms are needed (compare the H count for methane, CH_4, and methyl amine, CH_3NH_2), therefore, add "n". Both have zero index of hydrogen deficiency.
- The factor of 0.5 accounts for us counting H **atoms**, but adding hydrogen, H_2, **molecules**.

Example 1

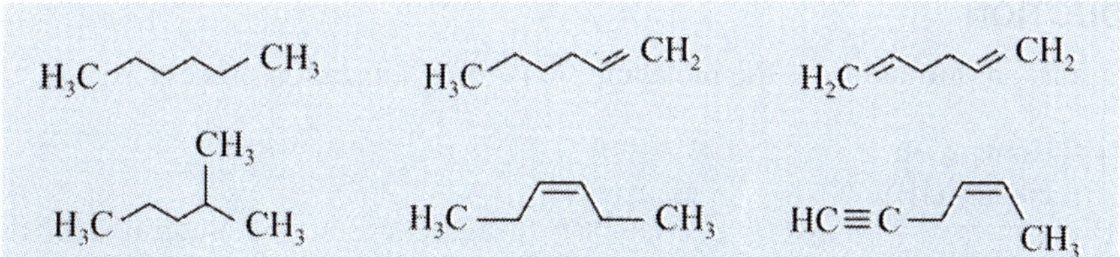

C_6H_{14}
IHD = 0

C_6H_{12}
IHD = 1

C_6H_{10}
IHD = 2

Example 2

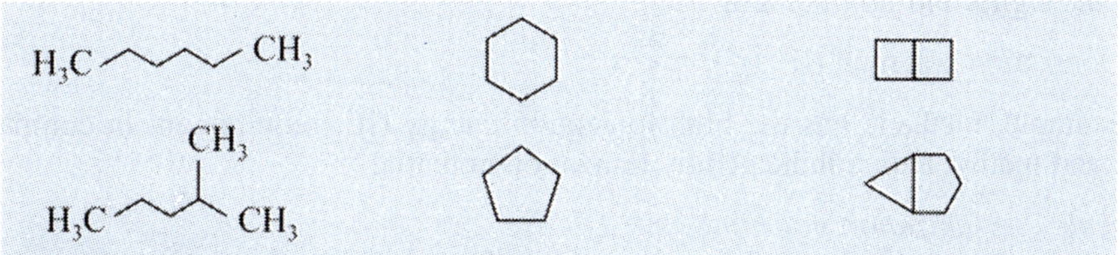

C_6H_{14}
IHD = 0

C_6H_{12}
IHD = 1
(Confirm the presence
of one ring)

C_6H_{10}
IHD = 2
(Confirm the presence
of two rings)

Example 3

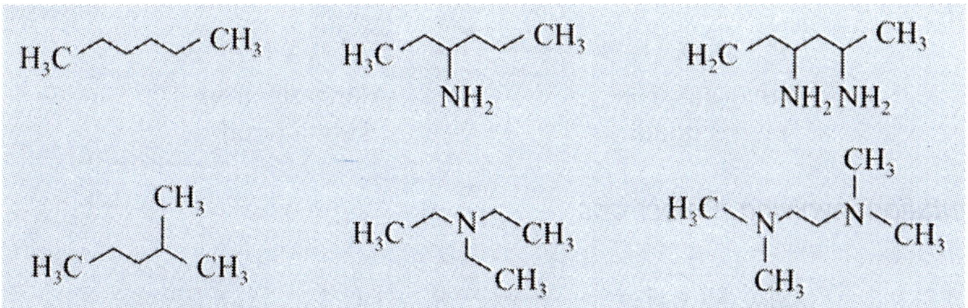

C_6H_{14}
IHD = 0

$C_6H_{15}N$
IHD = 0

$C_6H_{16}N_2$
IHD = 0

5

Fragmentation of Organic Compounds

INTRODUCTION

Fragment ions are formed from the molecule (M) during ionization process as:

$$M \xrightarrow[\text{70 eV}]{\text{Ionization energy (IE)}} M^{+\cdot} \xrightarrow{\text{Excess energy}} \text{Fragment ion}^{+\cdot} + \text{Neutral radicals}$$

Molecular ion

Electron will be removed from orbital having the lowest ionization potential (IP). In general, non-bonding (n) electrons possess lowest and the sigma (σ) have highest while pi (π) lies in between the sigma and non-bonding.

Ionization potential (IP): $n < \pi < \sigma$

For example, methane has the high ionization energy (IE) requirement in comparison to ethylene and methyl amine indicated by ionization potential.

Molecule	Ionization potential
CH_4	12.6 eV
C_2H_4	10.52 eV
CH_3NH_2	10.3 eV

Ionization mechanism

Fragmentation involving σ electrons

$$H_3C \bullet\bullet CH_3 \xrightarrow{IE} H_3C \bullet + CH_3 + 2e^-$$

A two-electron sigma bond → A one-electron sigma bond

Fragmentation involving π electrons

$$H_2C ::: CH_3 \xrightarrow{IE} H_2C :: CH_2 + 2e^-$$

A two-electron pi bond → A one-electron pi bond

Fragmentation involving non-bonding electrons

$$H_3C - \overset{..}{\underset{..}{O}}H \xrightarrow[\text{IE}]{e^-} H_3C - \overset{..+}{\underset{..}{O}}H + 2e^-$$

A lone pair
electron removed

The products of the ionization process are called molecular (M^+) ions. They are radical cations. It is the radical cations that fragment to give a cation, the observed species in the mass spectrum, and a neutral radical. Fragmentation occurs via homolytic cleavage, heterolytic cleavage, alpha cleavage, beta cleavage and rearrangement process depending on structural features of organic compound. This will be elucidated in the coming text.

STEVENSON'S RULE

Fragmentation often produces both a radical and a cation. This can be represented by the following equation:

$$R - R_1 + e^- \longrightarrow [R - R_1]^{+\bullet} \begin{cases} R^\bullet + R_1^+ \\ \text{Radical} \quad \text{Cation} \\[1em] R^+ + R_1^\bullet \\ \text{Cation} \end{cases}$$

Molecular ion
(Radical cation)
$+ 2e^-$

The factor that determines which of the fragments is a radical or a cation can be emphasized as a competition between two cations to capture the electron:

$$R^+ \dots\dots\ e^- \dots\dots\ R_1$$

According to **Stevenson's rule**, if two charged fragments are in competition to produce a neutral radical by electron attachment, the radical having the highest ionization energy will be produced. The other ion, whose corresponding neutral radical has lower ionization energy, will hold its charge and will thus be the observed fragment. Indeed, this reaction can be considered as a competition between two cations to carry away the electron:

$$R^+ \dots\dots\ e^- \dots\dots\ R_1$$

Overall, as per Stevenson's rule, the most probable fragmentation is the one that leaves positive charge on the fragment with lowest IE. In other words, fragmentation processes that lead to the formation of more stable ions are favoured over pathway that leads to less stable ions.

IONS WITH ODD OR EVEN NUMBER OF ELECTRONS

Organic molecules have an even number of electrons. Because ionization occurs by the removal of an electron, the molecular ion can be identified with radical-ion that possesses an odd number of electrons, or rather, an unpaired. A molecular ion, or any ion, with an odd number of electrons (i.o.e.) can decompose by homolytic or heterolytic processes. The *homolytic cleavage* occurs by the use transfer of a single electron, and this is indicated by the use of an arrow with one hook $\frown$.

$$CH_3 - CH_2 - CH^+ - {}^\bullet CH_2 \longrightarrow CH_3^+ + CH_2 = CH - CH_2^+$$

$$\updownarrow$$

$${}^+CH_2 - CH = CH_2$$

The *heterolytic cleavage* occurs by the transfer of a pair of electrons and indicated by a conventional arrow $\curvearrowright$.

$$CH_3CH_2 - \overset{\bullet}{Br} \longrightarrow CH_3CH_2^+ + :\overset{\bullet}{\underset{\bullet\bullet}{Br}}:$$

In both cases, an ion with an even number of electrons (i.e.e.) is obtained by the elimination of an uncharged fragment with an odd number of electrons (radical, r).

When there are two consecutive or simultaneous fragmentation of an ion with an odd number of electrons, an ion with an odd number of electrons (i.o.e.) and a neutral molecule with an even number of electrons (n.e.e.) are formed by either type of cleavage (homolytic or heterolytic).

$$R - CH - CH_2 \longrightarrow R - \overset{\bullet}{CH} - CH_2^+ + H_2O$$
$$\underset{H}{|} \quad \underset{OH}{|}$$

The majority of ions in a spectrum have an even number of electrons. Either they decompose by loss of a non-charged fragment or a neutral (n) molecule with an even number of electrons to give an ion with an even number of electrons, or they decompose by loss of a radical (r) to give an ion with an odd number of electrons (o.e.). The latter fragmentation is very rare.

$$CH_3 \overset{\frown}{-} CH_2CH_2^+ \longrightarrow {}^{\bullet}CH_3 + CH_2 = CH_2$$

$$CH_3 \overset{\frown}{-} CH_2CH_2^+ \longrightarrow {}^+CH_3 + \overset{\bullet}{C}H_2 - {}^+CH_2$$

The ions with an even number of electrons (e.e.) give rise to a significant series of peaks for many types of compounds, particularly in the lower part of mass spectrum. The aliphatic hydrocarbons form two series of alkyl ions, the aliphatic amines a series of $C_nH_{2n+2}N^+$, the alcohol a series of $C_nH_{2n+1}O^+$, the ketones a series of $C_nH_{2n}O^+$. In all cases, the series are useful for recognizing the class to which a substance belongs.

The concepts expressed above are summarized in Table 5.1.

Table 5.1: Types of cleavage	
i.o.e. → i.e.e. + r	Simple homo- and heterolytic cleavage
i.o.e. → i.o.e. + n.e.e.	Double cleavage
i.e.e. → i.e.e. + n.e.e.	Simple heterolytic cleavage
i.e.e. → i.o.e. + r	Simple homolytic cleavage

FACTORS INFLUENCING THE FRAGMENTATION

The mass spectrum of a molecule shows the numerous peaks, some of them are intense, whereas others are weak or scarcely visible. The formation of fragment ions depends on the (i) preferential breakage of some bonds in comparison to others, and (ii) on the stability of some fragments because of their structure. In conclusion, the formation of ions depends on three factors:

1. The relative strength of the bonds.
2. The stability of the fragmentation products. These may be neutral molecule, radicals and positive ions.
3. The relative spatial rearrangements of the atoms or groups.

Relative strength of the bonds

The energies of the most common bonds of an organic molecule are shown in Table 5.2. The weakest bond (C–I, C–Br, C–Cl) are preferentially broken in comparison to other. Similarly, breakage of single bond requires less energy in preference to double or triple bond.

Table 5.2: Energy profile of some bonds present in organic compounds (kcal/mole)					
Bond	**Energy**	**Bond**	**Energy**	**Bond**	**Energy**
C–H	97.8	C≡N	147	C–F	116
C–C	82.6	C–O	85.5	C–Cl	81
C=C	145.1	C=O	179	C–Br	68
C≡C	199.6	C–S	65	C–I	51
C–N	72.8	C=S	128	O–H	110.6

Stability of the fragments

Fragment ions are stabilized by following under-mentioned process. The stabilization favors the pathway which in turn increases the intensity of the peak.

Inductive effect

The formation of tertiary cations is favored with respect to secondary or primary ions. For example, in the mass spectra of n-butyl alcohol and t-butyl alcohol, the fragment ions are formed in both cases but more abundant in the case of t-butanol because tertiary butyl ion is stabilized by the inductive effect of the alkyl groups attached to carbon atom, carrying the positive charge. To some extent, it is also stabilized by hyperconjugation effect, that is, delocalisation of the electron of a C–H bond to form a p bond with an adjacent carbon atom having an empty orbital.

$$CH_3CH_2CH_2CH_2 - \overset{\cdot +}{O}H \longrightarrow CH_3CH_2CH_2CH_2^+ + \,{}^{\cdot}OH$$
$$\text{n-Butyl alcohol} \qquad\qquad \text{n-Butyl cation}$$

t-Butyl alcohol → t-Butyl cation

In the straight chain paraffins, the rupture of the alkyl chain takes place in one C–C bonds. In branched paraffins, the fragmentation occurs at the most substituted carbon atom to give the most stability to the ion thus formed. For example, in isopentane, the fragmentation occurs at the bonds of the carbon atom at the C–2 position and the various probable ions have the following order of stability:

The inductive effect also has an influence on the breaking of a bond. In a compound R–X in which X can be Cl, Br, O, S or N, the electronegativity of heteroatom lowers the electron density of the R–X bond. Formation of an alkyl ion occurs according to below mentioned reaction:

$$R - Cl \longrightarrow \overset{+}{R} + \overset{\bullet}{Cl}$$

$$R - O - R \longrightarrow \overset{+}{R} + \overset{\bullet}{O}R$$

The order of electron attraction is Cl > Br, O, S > I > N, C, H.

The unsaturated functional groups, such as carbonyl group, have a similar effect.

$$R - \overset{\overset{\displaystyle +\bullet}{\overset{\displaystyle O}{\|}}}{C} - R \longrightarrow R - CO^{\bullet} + \overset{+}{R}$$

The above three fragmentations are heterocyclic fragmentation directed by the ionic center present on the heteroatom.

Neighboring electron participation

The carbonyl compounds, such as ketones, aldehydes, esters, amides and the corresponding sulphur derivatives, give rise to ions of $C^+ = X^-$ stabilized in the canonical form $-C\equiv X^+$. For example, acetone shows the presence of fragment, CH_3CO^+ in its mass spectrum.

$$CH_3 - \overset{\overset{\displaystyle +\bullet}{\overset{\displaystyle O}{\|}}}{C} - CH_3 \longrightarrow CH_3 - C \equiv \overset{+}{O} + \overset{\bullet}{C}H_3$$

The canonical form is stabilized by heteroatom which possesses at least a couple of electrons not used in bonding. The fragmentation reaction can be represented by the homolytic cleavage directed by the radical center present on oxygen.

The non-bonding electrons of an atom can also stabilize a positive charge on carbon atom, as in the case of $C_4H_8Br^+$ and $C_4H_8Cl^+$ which are formed from the halide.

X = Br or Cl

Similarly, alcohol, ethers, thiols, thioethers and amines show the presence of fragment ions which are stabilized by heteroatom due to presence of non-binding electrons.

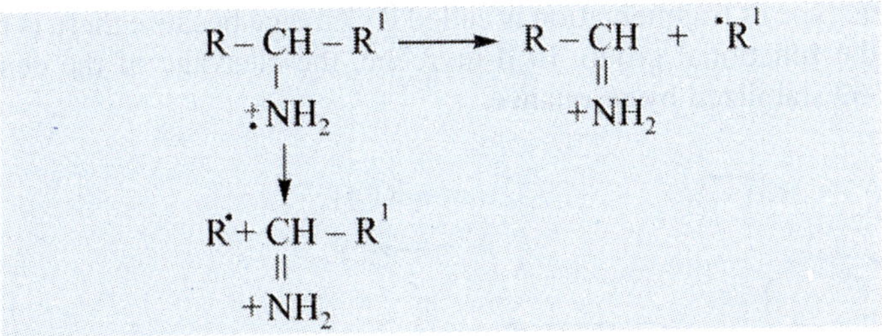

$$R - \underset{\underset{\overset{|}{:}NH_2}{|}}{CH} - R^1 \longrightarrow R - \underset{\overset{||}{+NH_2}}{CH} + {}^{\bullet}R^1$$

$$\downarrow$$

$$R^{\bullet} + \underset{\overset{||}{+NH_2}}{CH} - R^1$$

The fragmentation reaction is represented as a **homolytic reaction** directed by the radical center present on the heteroatom. It is indicated by an **alpha cleavage** because the bond is adjacent to that alpha carbon and the heteroatom.

The capacity of a heteroatom to stabilize an adjacent positive charge is very high for nitrogen atom and decreases gradually, passing through S, O and the halogen for which it is very low. In the following given example, the stabilizing capacity of various atoms, in the same molecule is compared. The numerical values indicated below the formulae represent the percentage abundance of the CH_2X ions.

$CH_2 - CH_2 - CH_2$		$CH_2 - CH_2 - CH_2$		$CH_2 - CH_2 - CH_2$	
NH_2	OH	SH	OH	OH	Cl
100%	9%	100%	60%	100%	12%

Resonance effect

Vinyl chloride and allyl chloride behave differently toward solvolysis. The first is extremely resistant to nucleophilic reagent, whereas the second reacts by the S_{N_1} mechanism to form ionic intermediate stabilized by resonance.

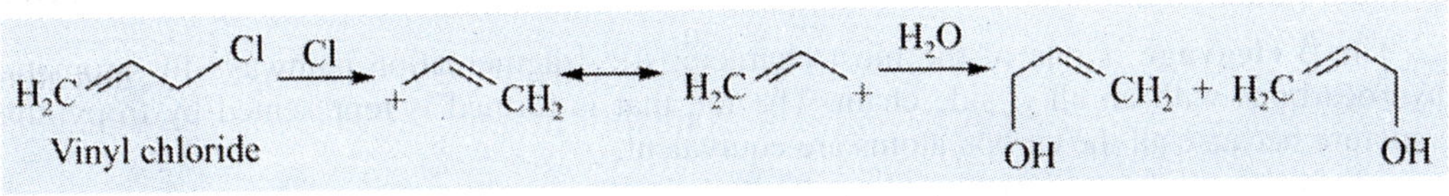

In the mass spectrometer, the vinylic bonds in the compound show allylic cleavage (favored fragmentation mode in unsaturated compound). The resonance stabilization of an allyl cation leads to increased probability of fragmentation of carbon-carbon bond to double bonds.

$$R - CH_2 - CH = CH_2{}^{+\bullet} \longrightarrow R^{\bullet} + CH_2 \overset{+}{=} CH = CH_2$$

In the case of alkenes, fragmentation favors the allylic cleavage because allylic carbocations are stabilized by two resonating structures as shown below:

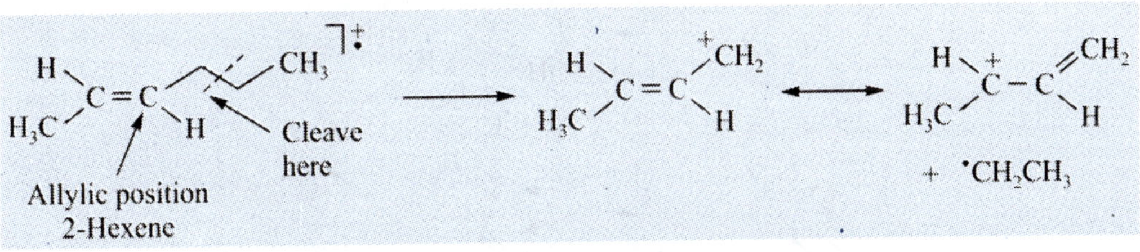

Allylic position
2-Hexene

This homolytic type of fragmentation is called β cleavage because there is rupture of β bond with respect to the functional group. In β-myrcene, the cleavage of the central allylic bond gives ion at m/e 69 stabilized by resonance.

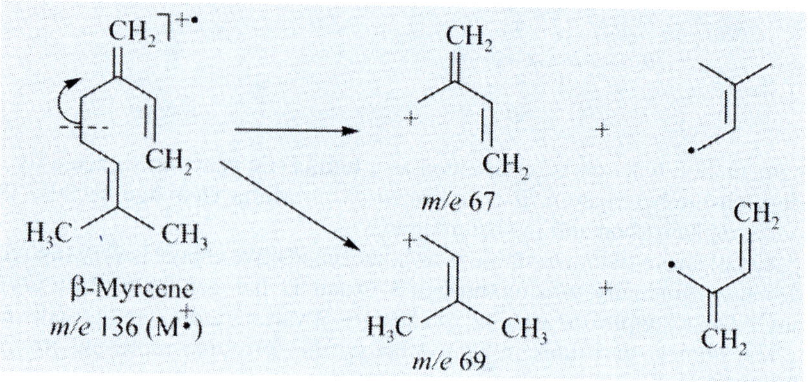

However, in allocimene, the ion at *m/e* 69 is not formed because there is no allylic bond in the molecule.

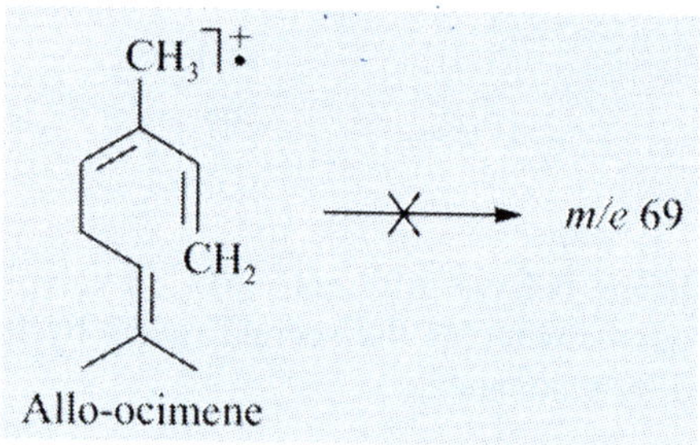

The **β cleavage** is one of the most characteristic fragmentation pathways for aromatic hydrocarbons with an alkyl side chain. The ion that is formed is represented by tropylium structure because all the carbon atoms are equivalent.

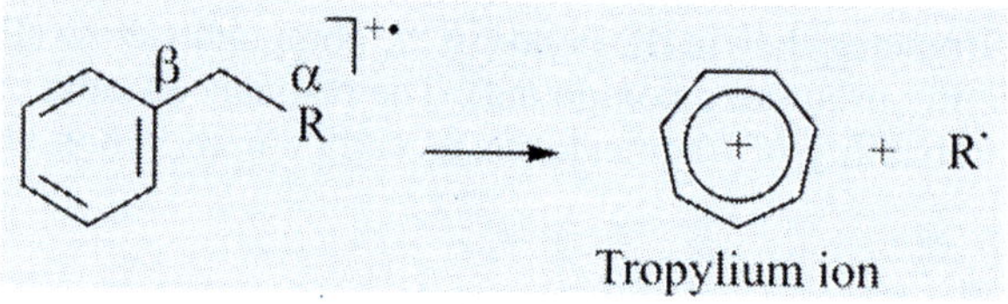

Tropylium ion

The **β cleavage** is also characteristic for five-member aromatic heterocyclic compounds with an alkyl side chain in the α or β position.

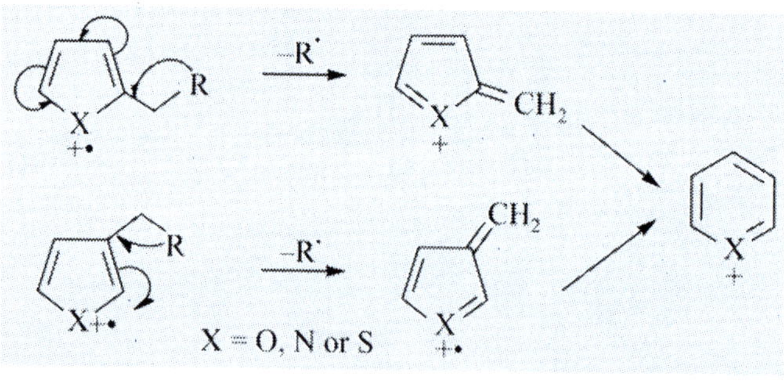

X = O, N or S

In the case of acetophenone, benzophenone, benzoic acid and substituted benzoates, the abundance of acyl ion, formed as in reaction decreases if the substituent X is an electron donor, whereas it increases if X is an electron acceptor.

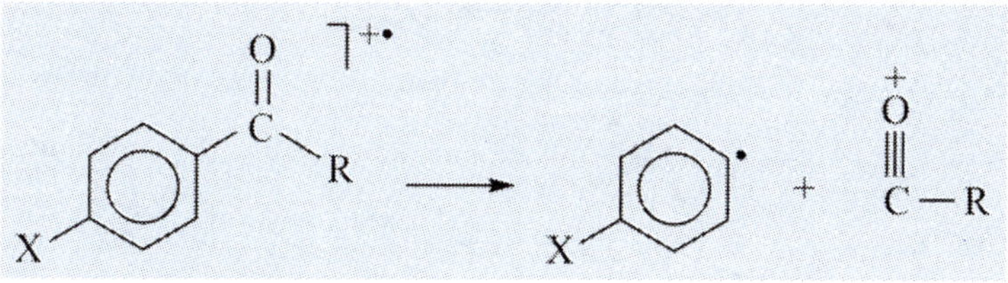

Multi-center fragmentations and stearic factors

The fragmentation so far observed involve the cleavage of only one bond. In a complex molecule, the interaction of various functional groups can give a complicated fragmentation reaction that involves the rupture of more than one bond. This is called a multi-center fragmentation.

A multi-center fragmentation is governed by a number of factors: The center involved in the fragmentation must have a suitable arrangement in space; the formation of a stable neutral molecule is the driving force of the process; and the fragmentation occurs by an energetically favorable cyclic transition state. In this, there may be hydrogen atom migration with the expulsion of neutral fragment (elimination and McLafferty rearrangement) and the reaction of internal rearrangement of bonds (Retro Diels-Alder).

Elimination reaction

This elimination process can be represented by the following scheme:

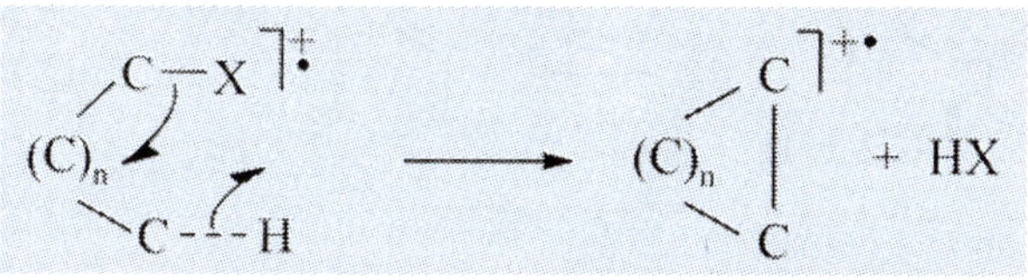

where X = Cl, Br, I, OH, OR, OCOR, NH_2, NR_2, SH.

The number of carbons atoms between the two groups undergoing elimination (H and X) can vary but for every atom of X there is a preferred value. In other words, the hydrogen atom that is eliminated must have a precise steric relationship with the X.

For example, alcohol shows the $M–H_2O$ peak due to elimination-reaction. The present reaction shows 1,4-elimination reaction to exhibit the loss of water.

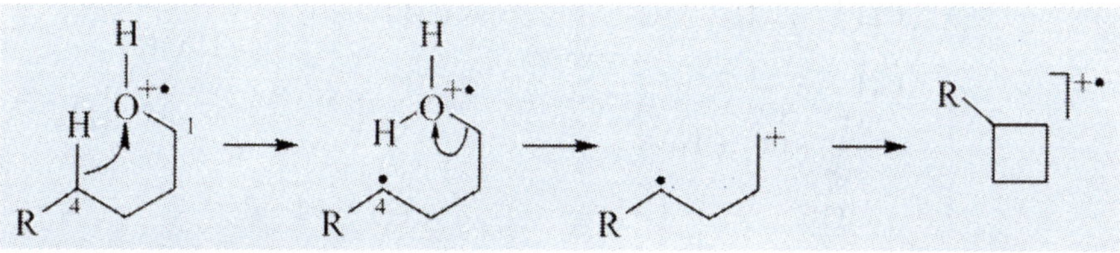

In the cyclic alcohol, the elimination of H_2O has much more complex route. In cyclohexanol (Scheme 5.1), for example, the introduction of a deuterium atom in positions 2, 3 or 4 demonstrate that the eliminated hydrogen atom comes from position 3 or 4 because of the conformation of the ring.

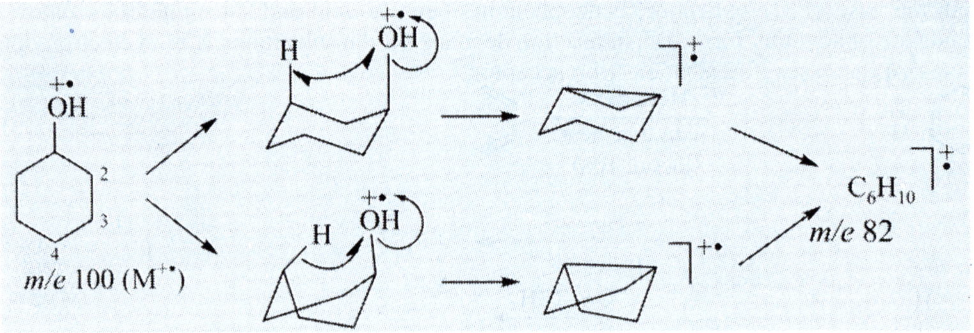

Scheme 5.1: Fragmentation pathway in cyclohexanol

Similarly, there is elimination of hydrogen chloride/bromide in alkyl chloride/bromide.

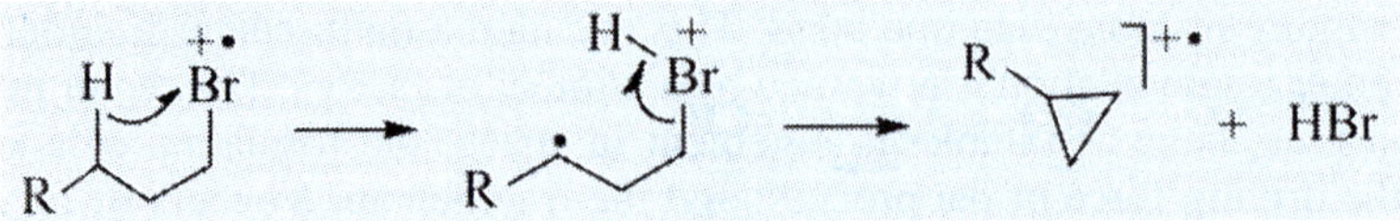

Ortho effect

Ortho disubstituted aromatic system or *cis* olefins can give rise to the specific migration of a hydrogen atom onto an atom or group that is eliminated in a form of neutral molecule.

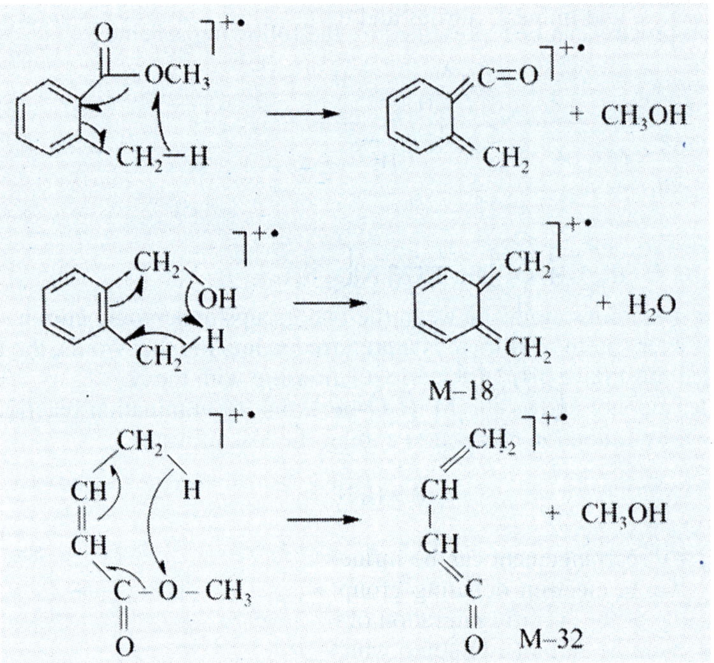

It is useful to distinguish between *cis* and *trans* compounds and o, m and p-isomers.

McLafferty rearrangement

This fragmentation mode is characteristic of carbonyl compound containing at least one gamma hydrogen atom like ketones, aldehydes, acids, esters, amides.

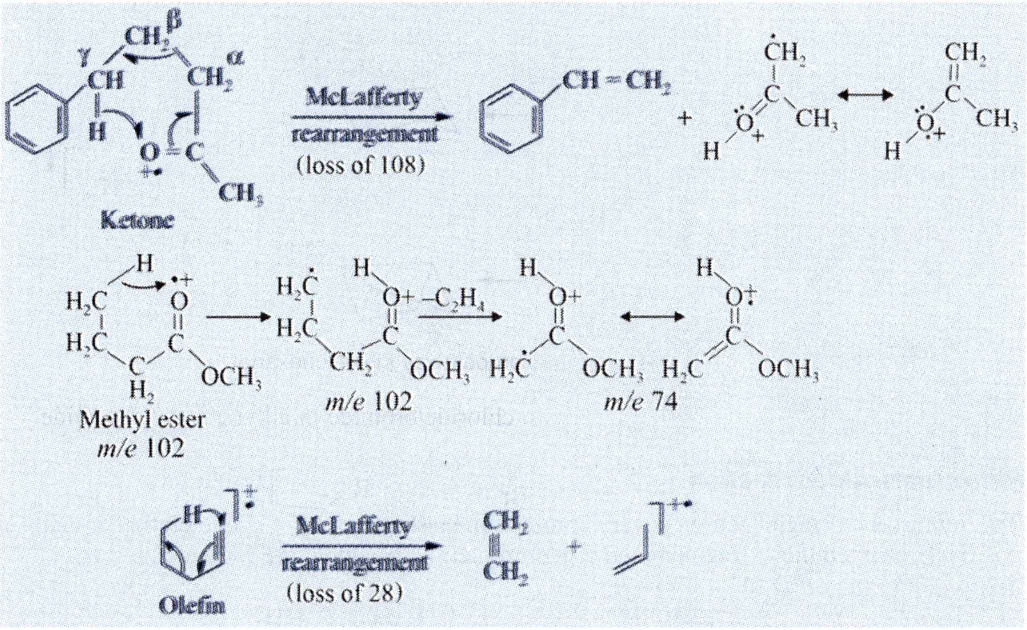

Other molecules that satisfy the structural requirements of McLafferty rearrangement are the olefins, the alkyl benzenes, and the alkyl aryl ether, nitrogen derivatives of carbonyl group (oximes, hydrazones, and imines), nitriles and epoxides.

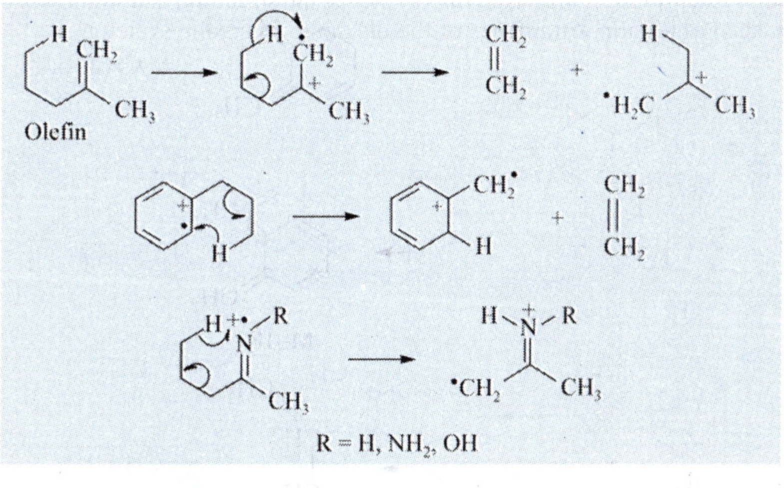

R = H, NH$_2$, OH

The McLafferty rearrangement can be influenced by electronic and steric factors. In alkyl benzenes that have an electron donating group at meta position, with respect to alkyl side chain, the tendency of the specific migration of hydrogen atom is opposed by the localization of positive charge on the carbon atom onto which the hydrogen atom must migrate. Similarly, in a compound, in which R=CH$_3$, the hydrogen migration is sterically hindered by the methyl group in the ortho position.

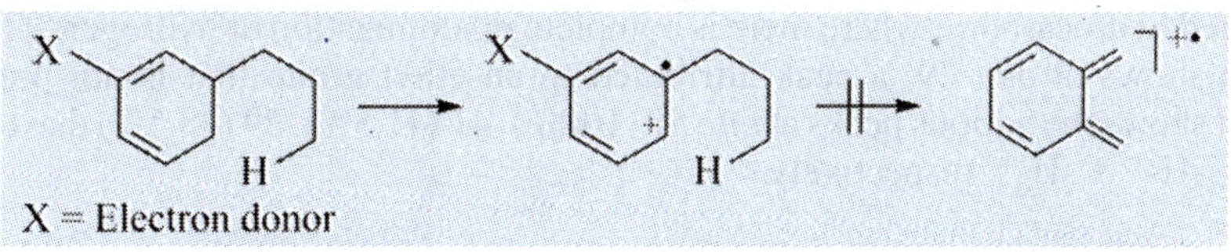

X = Electron donor

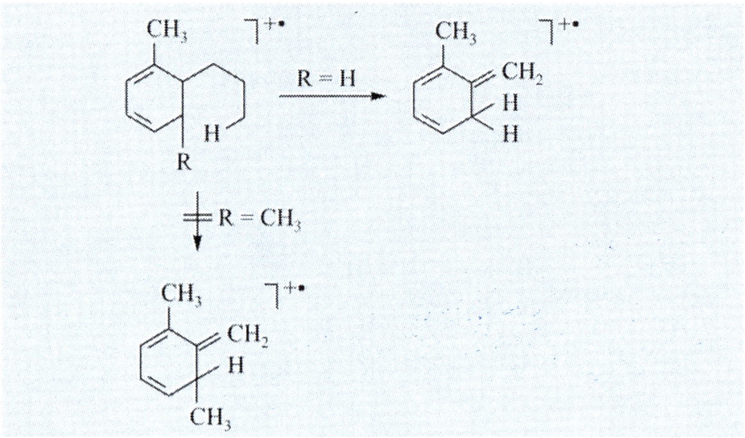

Retro Diels-Alder reaction

This multi-center fragmentation is very common in mass spectrum of cyclic olefins. Usually, a positively charged diene fragment and a neutral olefinic fragment are formed.

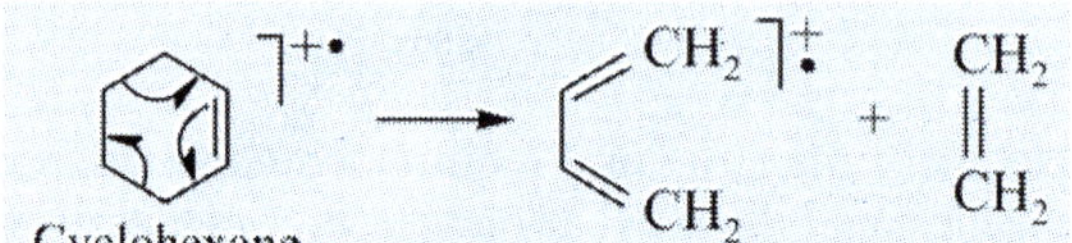

Cyclohexane

This type of reaction is considerable importance in the structural determination of numerous compounds, such as tetralin, triterpenes of the oleanolic or ursolic skeleton.

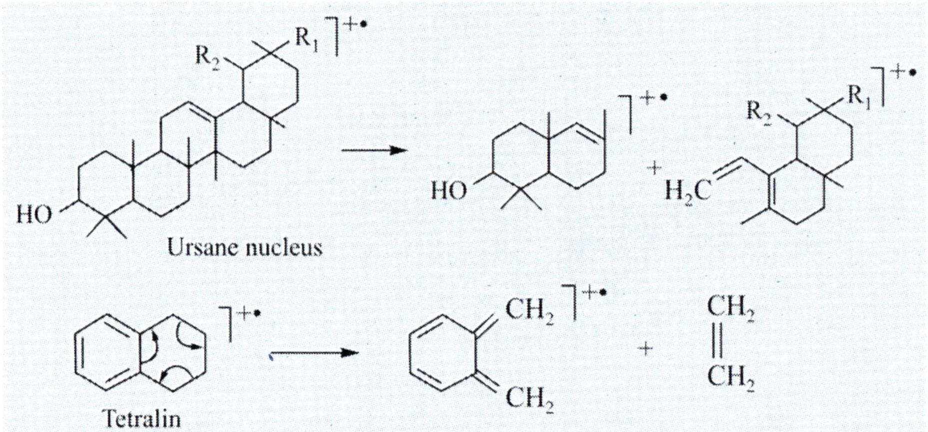

Ursane nucleus

Tetralin

Hydrogen migration

The McLafferty rearrangement and the elimination of HX are the examples of the specific migration of hydrogen atoms or groups of atoms with the expulsion of a neutral molecule.

In the mass spectrum of a complex molecule, the migration of a different type from those mentioned above is also a very common fragmentation process. Some processes are not specific, as in saturated hydrocarbons, where there is complete randomization of hydrogen. The spectrum neopentane shows at m/e 29, a peak attributed to an ethyl group that is not present in the molecule. It shows the various peaks at m/e 57 (100%), 41 (41.5%), 29 (38.5%) due to formation of $C_4H_9^+$, $C_3H_7^+$, $C_2H_5^+$, respectively.

Expulsion of a stable neutral molecule

Expulsion of neutral molecules like CO, CO_2, N_2, SO_2, $CH_2 = C=O$, $CH_2 = CH_2$ and $CH \equiv CH$ has a considerable importance in the course of a fragmentation. The expulsion of ketene and ethylene occurs in general through the hydrogen migration reaction, as already shown in McLafferty rearrangement for ethylene, and as indicated in the following reaction of an acetate to alcohol and to form ketene.

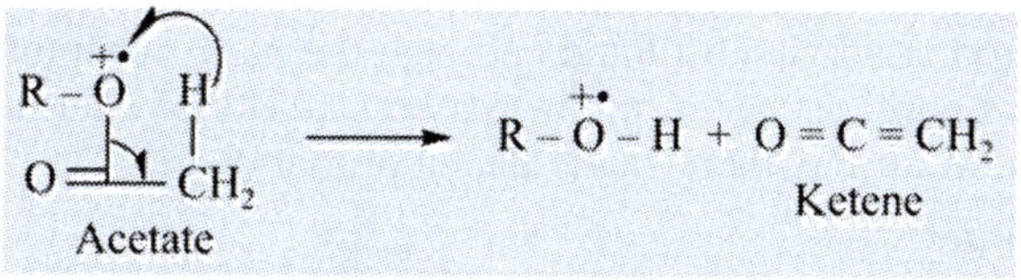

The elimination of the other groups requires the cleavage of two bonds without hydrogen transfer and the formation of new bond or a lone pair of electrons in the molecule that is eliminated.

The mass spectrum of phthalic anhydride shows the elimination first of CO_2 and then of CO to give an ion at m/e 76, designated as benzene.

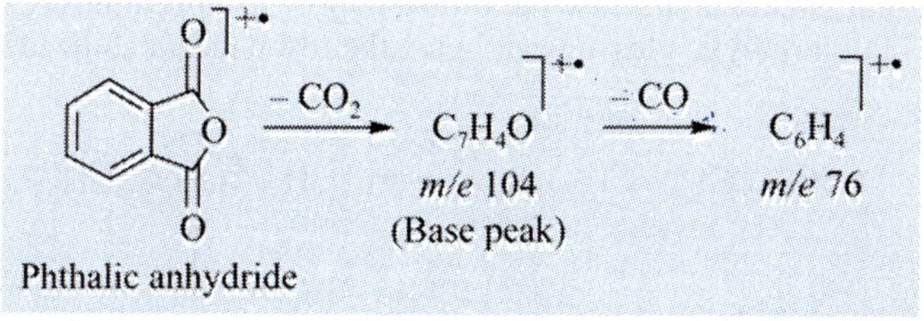

Anthraquinone loses two molecules of carbon monoxide to give a compound $C_{12}H_{10}$ to which is attributed the diphenyl structure.

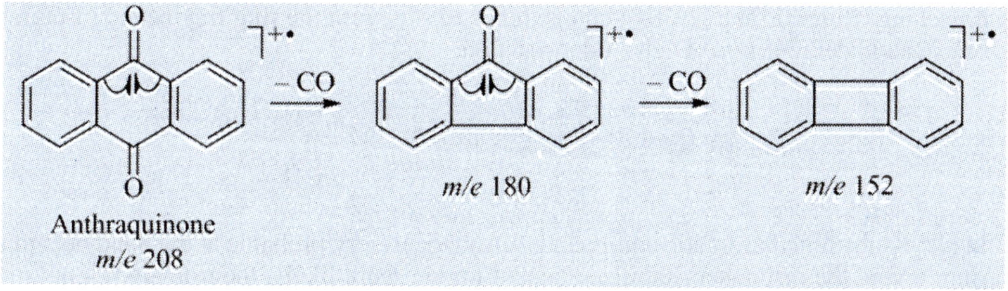

Acetylene is eliminated from aromatic compounds. Benzene, for example, loses a molecule of acetylene, but it is not possible to assign a reliable mechanism from this reaction. It seems that the molecular ion of benzene should be represented by the open chain form and not the cyclic form.

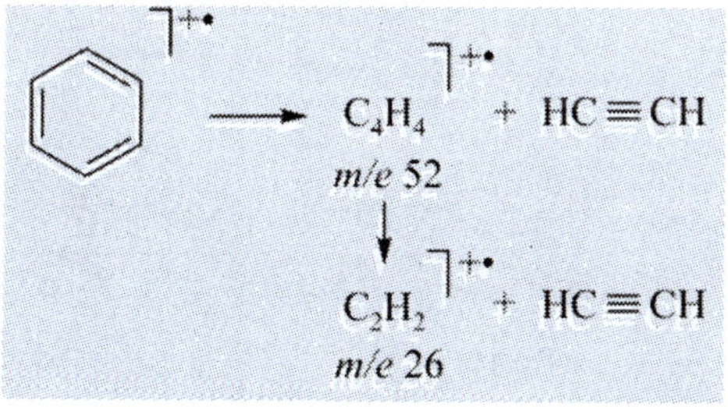

HANDY RULES OF FRAGMENTATION

A number of general rules for predicting prominent peaks in electron-impact spectra can be written and rationalized by using standard concepts of physical organic chemistry:

1. The relative height of the molecular ion peak is greatest for the straight-chain compound and decreases as the degree of branching increases (*see* Rule 3).
2. The relative height of the molecular ion peak usually decreases with increasing molecular weight in a homologous series. Fatty esters appear to be an exception.
3. Cleavage is favored at alkyl substituted carbons; the more substituted, the more likely is cleavage. This is a consequence of the increased stability of a tertiary carbocation over a secondary, which, in turn, is more stable than a primary:

Cation stability order:

$$CH_3^+ < RCH_2^+ < R_2CH^+ < R_3C^+$$

Generally, the largest substituent at a branch is eliminated most readily as a radical, presumably because a long-chain radical can achieve some stability by delocalization of the lone electron.

4. Double bonds, cyclic structures, and especially aromatic (or heteroaromatic) rings stabilize the molecular ion, and thus increase the probability of its appearance.
5. Double bonds favour allylic cleavage and give the resonance-stabilized allylic carbonium ion.

6. Saturated rings tend to lose side chains at the α-bond. This is merely a special case of branching (Rule 3). The positive-charge tends to stay with the ring fragment. Unsaturated rings can undergo a *retro*-Diels-Alder reaction:

7. In alkyl-substituted aromatic compounds, cleavage is very probable at the bond beta to the ring, giving the resonance-stabilized benzyl ion or, more likely, the tropylium ion:

8. C–C bonds next to a heteroatom (Y) are frequently cleaved, leaving the charge on the fragment containing the heteroatom whose non-bonding electrons provide resonance stabilization.

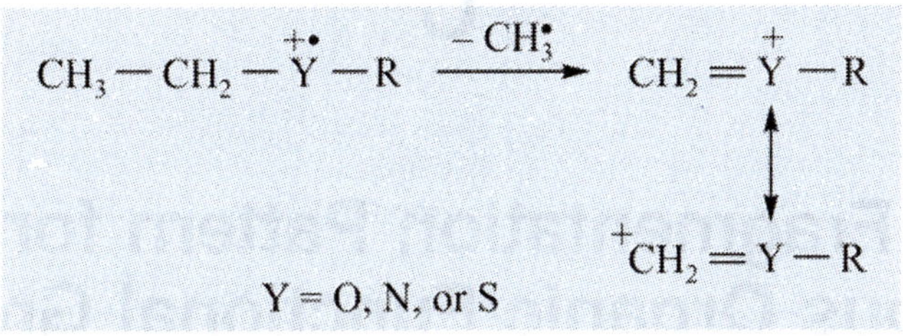

$$CH_3-CH_2-\overset{+\bullet}{Y}-R \xrightarrow{-CH_3^{\bullet}} CH_2=\overset{+}{Y}-R$$

$$\updownarrow$$

$$\overset{+}{C}H_2=Y-R$$

$$Y = O, N, \text{ or } S$$

For example:

In ether:

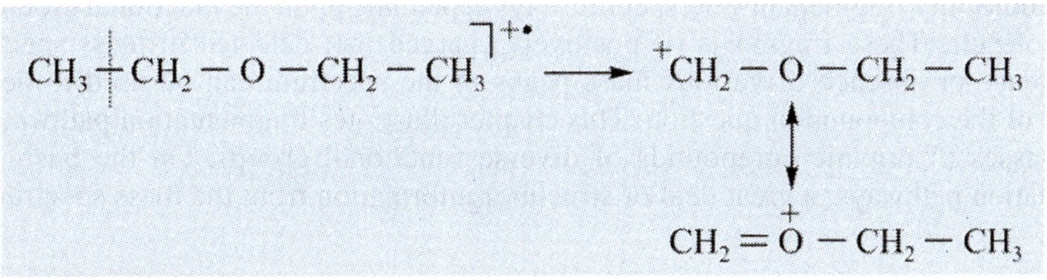

$$\left[CH_3\!\!\mid\!\!CH_2-O-CH_2-CH_3\right]^{+\bullet} \longrightarrow {}^{+}CH_2-O-CH_2-CH_3$$

$$\updownarrow$$

$$CH_2=\overset{+}{O}-CH_2-CH_3$$

In ester, acids and amides

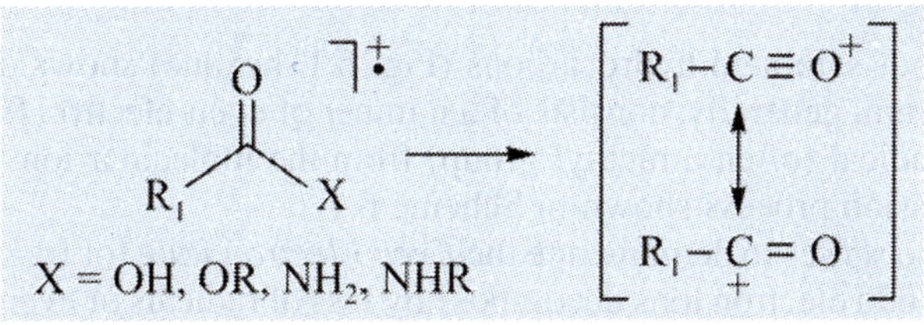

$$\left[\underset{R_1}{\overset{O}{\overset{\|}{\diagup}}\diagdown}X\right]^{+}_{\bullet} \longrightarrow \left[\begin{array}{c} R_1-C\equiv O^{+} \\ \updownarrow \\ R_1-\underset{+}{C}=O \end{array}\right]$$

$$X = OH, OR, NH_2, NHR$$

In ketones and aldehydes

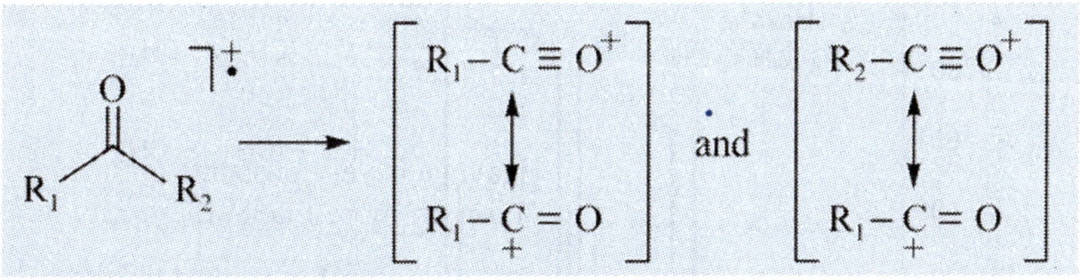

$$\left[\underset{R_1}{\overset{O}{\overset{\|}{\diagup}}\diagdown}R_2\right]^{+}_{\bullet} \longrightarrow \left[\begin{array}{c} R_1-C\equiv O^{+} \\ \updownarrow \\ R_1-\underset{+}{C}=O \end{array}\right] \quad \text{and} \quad \left[\begin{array}{c} R_2-C\equiv O^{+} \\ \updownarrow \\ R_1-\underset{+}{C}=O \end{array}\right]$$

9. Cleavage is often associated with elimination of small stable neutral molecules, such as carbon monoxide, olefins, water, ammonia, hydrogen sulphide, hydrogen cyanide, mercaptans, ketene, or alcohols. For example, anisole shows prominent peak at m/e 65 due to expulsion of carbon monoxide molecule.

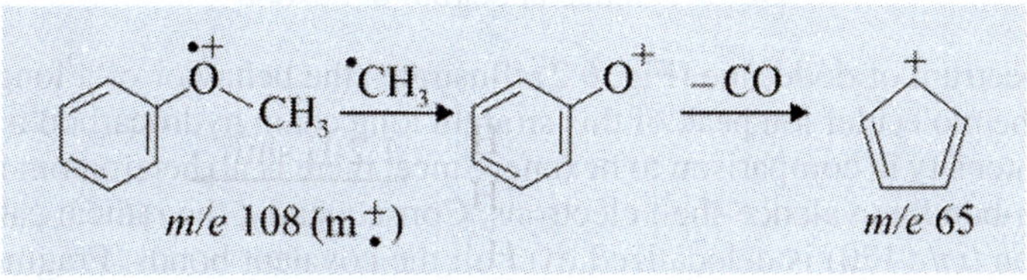

$$m/e\ 108\ (m^{+}_{\bullet}) \qquad\qquad\qquad m/e\ 65$$

6

Fragmentation Pattern for Various Organic Functional Groups

Organic molecules fragment in very specific ways depending upon the functional groups present in the molecule. These fragments (if positively charged) are detected in mass spectrometry. The presence or absence of various mass peaks in the spectrum can be used to deduce the structure of the compound in question. This chapter illustrates fragmentation pathways for the major classes of organic compounds of diverse functional groups. On the basis of these fragmentation pathways, a great deal of structural information from the mass spectrum can be deduced.

HYDROCARBON

Saturated hydrocarbon

The mass spectrum of saturated hydrocarbons (Fig. 6.1, hexane) shows detectable molecular ion peak. The spectrum, generally, consists of a number of even-electron fragment ions formed by expulsion of a radical (often a methyl group) from the molecular ion, followed by loss of ethene due to the fission process shown in Scheme 6.1

This example also serves to demonstrate the *Even Electron Rule* for fragmentation of cations and cation-radicals; odd electron ions decompose by loss of radicals or even electron molecules, while even-electron ions decompose by loss of even-electron molecules.

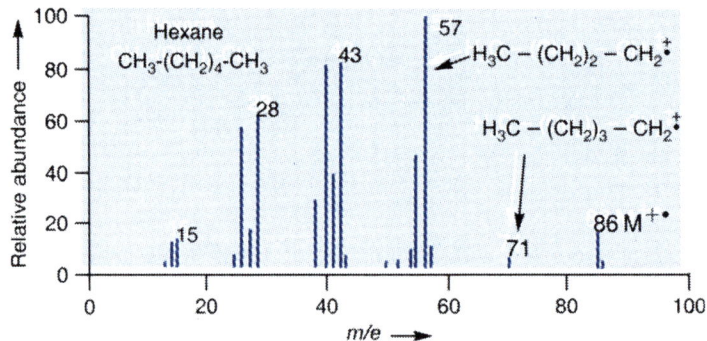

Fig. 6.1 : Mass spectrum of hexane

The mass spectrum of dodecane (Fig. 6.2) illustrates the behavior of a long chain aliphatic hydrocarbon. The molecular ion peak of this straight long chain hydrocarbon is always present, though of low intensity in comparison to hexane. Since, there is no heteroatom in this molecule, there are no non-bonding valence shell electrons. Consequently, the radical cation character of the molecular ion (m/e 170) is delocalized over all the covalent bonds. Fragmentation of C–C

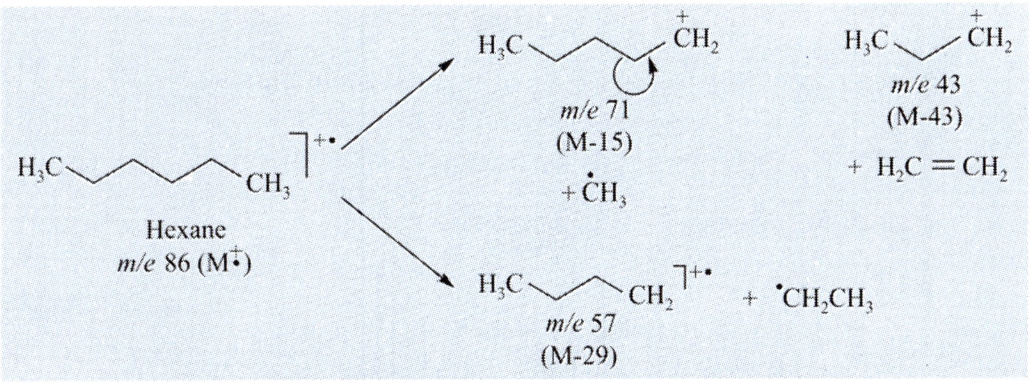

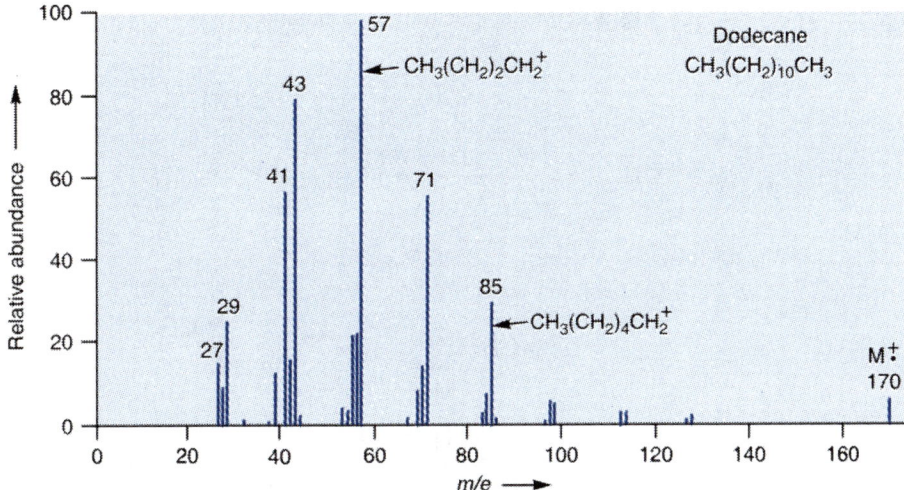

Scheme 6.1: Fragmentation pattern for hexane

Fig. 6.2: Mass spectrum of dodecane

bonds will take place to produce a mixture of alkyl radicals and alkyl cations. The positive charge is usually retained by the smaller fragment, so, we see a homologous series of hexyl (*m/e* 85), pentyl (*m/e* 71), butyl (*m/e* 57), propyl (*m/e* 43), ethyl (*m/e* 29) and methyl (*m/e* 15) cations. These are accompanied by a set of corresponding alkenyl carbo-cations (e.g. *m/e* 55, 41 and 27) formed by loss of 2 H. In most of the alkane spectra, the propyl (C3) and butyl (C4) ions are the most abundant.

Branched hydrocarbon

Spectra of branched saturated hydrocarbons are generally similar to those of straight-chain compounds, but the smooth curve of decreasing intensities is broken by the preferred fragmentation at each branch. This is due to the formation of stable secondary and tertiary carbocations. For this reason, the molecular ion is much less abundant than for straight-chain alkanes. The most important mode of fragmentation in branched alkanes usually occurs at the branch point. In the mass spectrum of isobutane (Fig. 6.3), there is reduced intensity of the molecular ion peak (*m/e* 58). Scheme 6.2 shows the mechanism of fragmentation of isobutane.

The effect of branching can be also shown with the example of isomeric hexanes (Fig. 6.4).

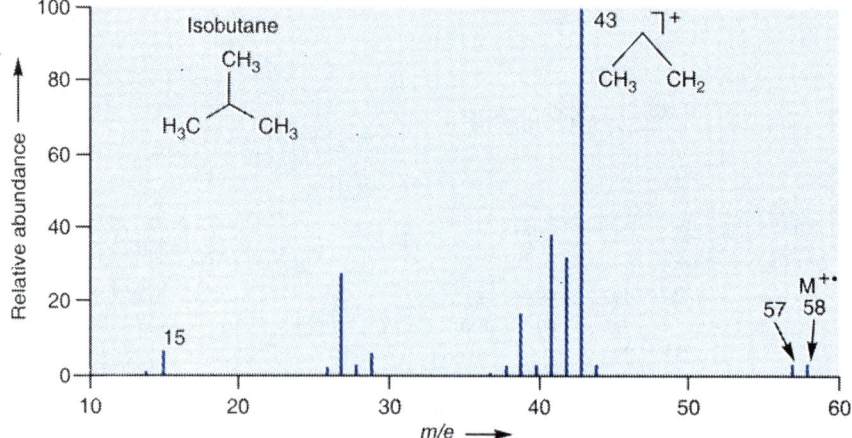

Fig. 6.3: Mass spectrum of isobutane

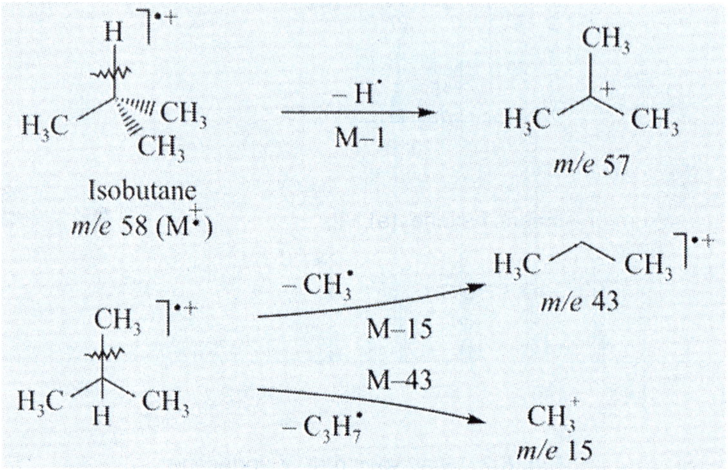

Scheme 6.2: Fragmentation pattern for isobutane

Hexane shows the same fragmentation pattern as other unbranched alkanes. Thus, alkyl carbocations at m/e 15, 29, 43 and 57 provide the dominant peaks in the spectrum. The m/e 57 butyl cation (M-29) is the base peak, and the m/e 43 and 29 ions are also abundant.

Chain branching clearly influences the fragmentation of this isomeric hexane. The molecular ion at m/e 86 is weaker than that for hexane itself in 2,3-dimethylbutane and the M-15 ion at m/e 71 is stronger. The m/e 57 ion is almost absent (try to find a simple cleavage that gives a butyl group). An isopropyl cation (m/e 43) is very strong, and the corresponding propene radical-cation at m/e 42, produced by loss of propane, gives the base peak.

Similarly, compare the electron impact mass spectra of 2-methylheptane and 2,2,3-trimethyl-pentane (Fig. 6.5). The base peak (m/e = 43) in the top spectrum corresponds to $(CH_3)_2CH^+$ and for the bottom spectrum, $(CH_3)_3C^+$ (m/e 57). Note the presence of $M^{+\bullet}$-$CH_3^\bullet$ (M–15) peaks in the above two spectra: These are generally not observed in the mass spectra of both the compounds. The molecular ion peak is either absent or very weak in branched hydrocarbon.

Cyclic saturated hydrocarbon

Cyclohexane shows molecular ion at m/e 84 (Fig. 6.4C) which is much stronger than the corresponding ions in acyclic compounds. The fragmentation patterns of cycloalkanes may

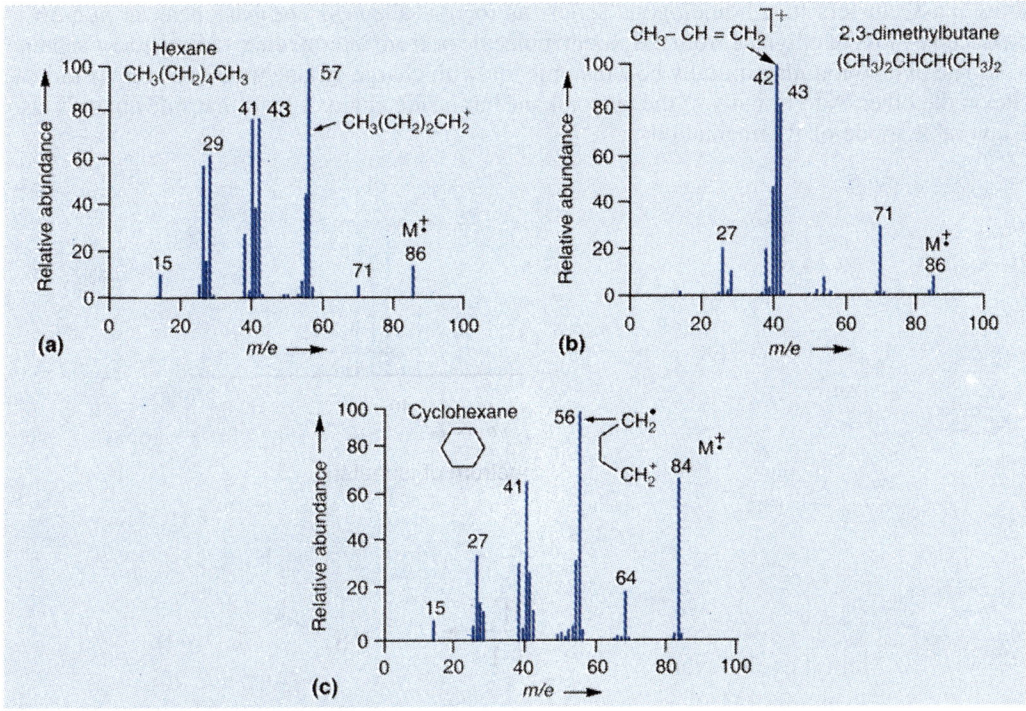

Fig. 6.4: Mass spectrum of isomeric hexane. (a) Hexane, (b) 2,3-Dimethylbutane, (c) Cyclohexane

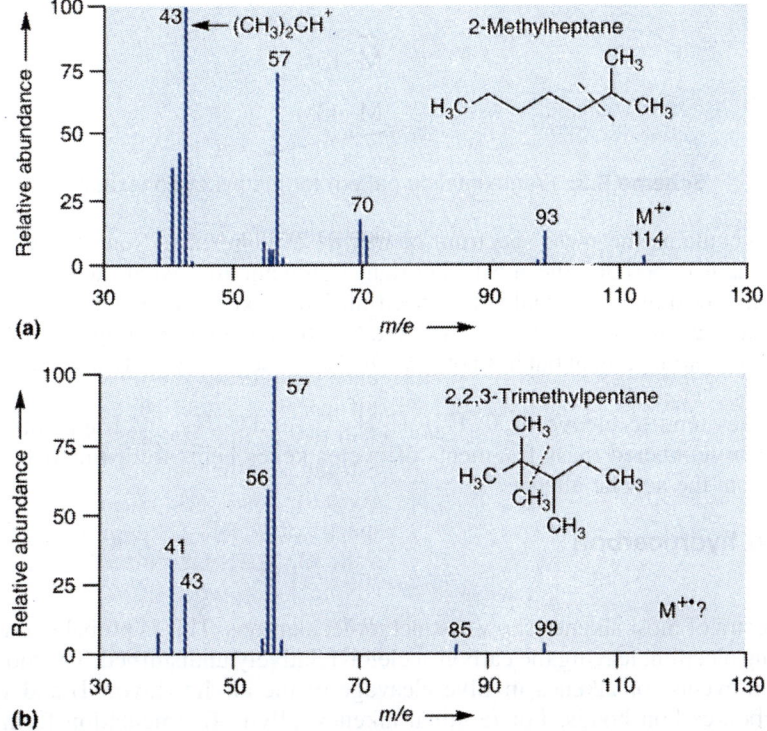

Fig. 6.5: Mass spectrum of (a) 2-methylheptane, and (b) 2,2,3-trimethylpentane

show mass clusters in a homologous series, as for the alkanes. The base peak at m/e 56 is produced by loss of ethylene from the parent molecule or from intermediate radical-ions (Scheme 6.3). The m/e 56 ion may initially be a distonic ion with charge on one terminus and the radical site on the other. Additionally, if the cycloalkane has a side chain, loss of that side chain is also a favorable mode of fragmentation.

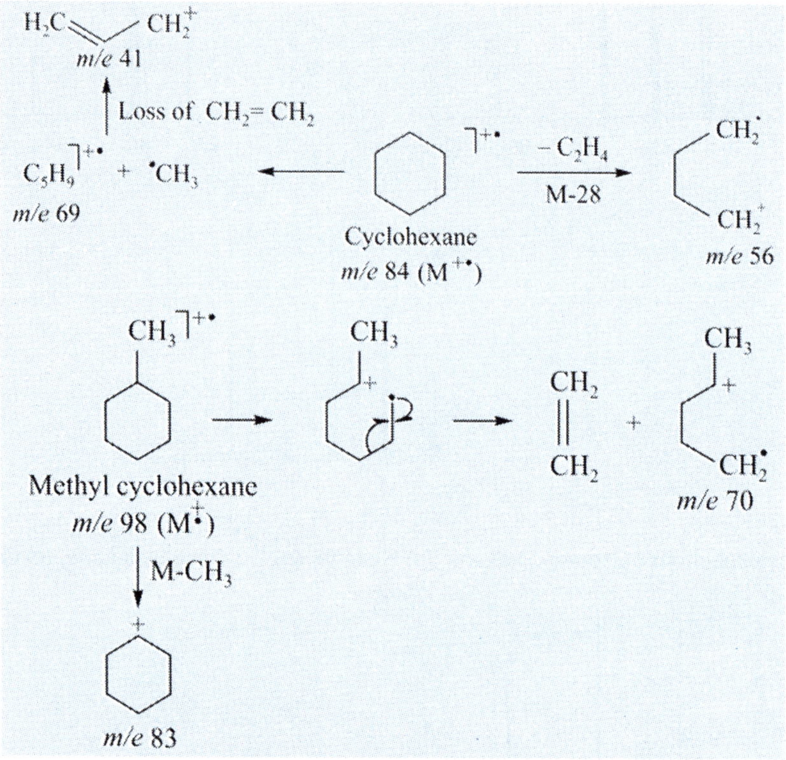

Scheme 6.3: Fragmentation pattern for methyl cyclohexane

Figure 6.6 contains the mass spectrum of methyl cyclohexane. Note that the loss of the methyl side chain is perhaps the most important fragmentation event, and the M-15 ion of m/e 83 gives the most intense signal in the spectrum. The next most abundant ion (m/e 55) may arise by the subsequent loss of ethylene (Scheme 6.3) from the m/e 83 fragment or by loss of a propyl group via rearrangement but it is not clear. It also possesses a significant intense molecular ion peak.

Cycloalkanes tend to cleave in $C_nH_{2n}^+$, $C_nH_{2n-1}^+$, and $C_nH_{2n-2}^+$ fragments. The larger number of even numbered mass fragments of cycloalkenes helps to distinguish this class of compounds from the acyclic alkanes.

Unsaturated hydrocarbon

Alkene

The mass spectra of most alkenes show distinct molecular ions. This is probably due to the loss of a π-bonding electron, leaving the carbon skeleton relatively undisturbed. The most important fragmentation events for alkenes involve cleavage of the allylic (favored) and vinylic (less favored) carbon-carbon bonds. For terminal alkenes, allylic fragmentation forms an allylic carbocation of m/e 41. The fragmentation mechanism for 1-butene shown in Scheme 6.4 illustrates these points. The complete mass spectrum of 1-butene is given in Fig. 6.7.

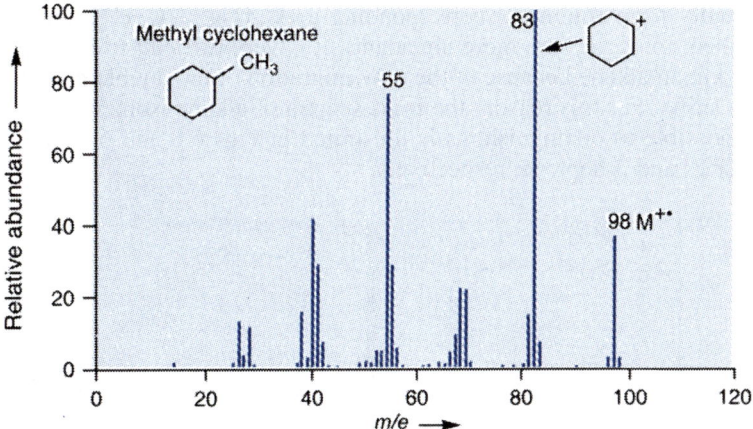

Fig. 6.6: Mass spectrum of methyl cyclohexane

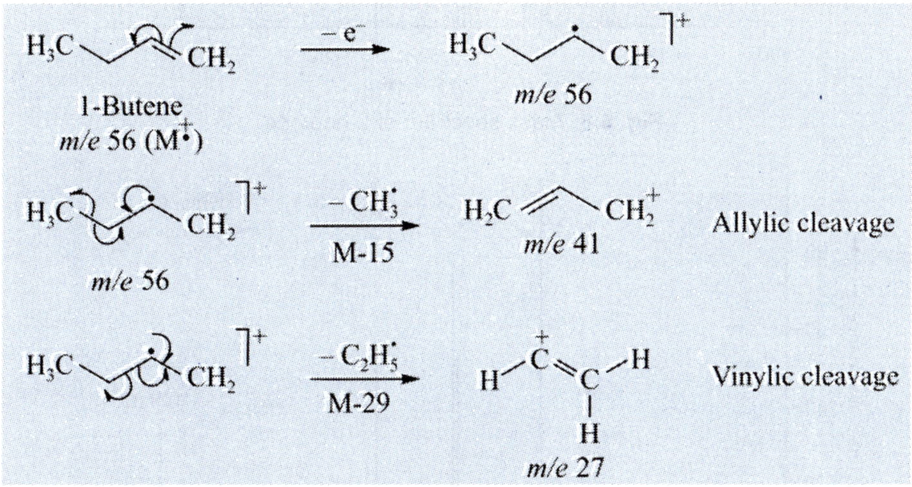

Scheme 6.4: Fragmentation pattern for 1-butene

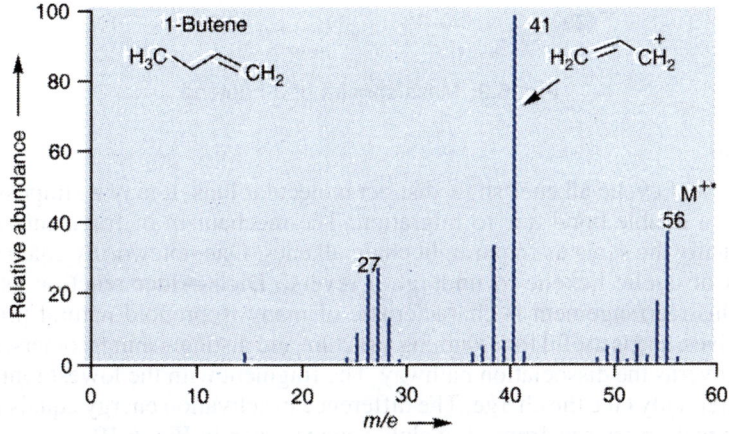

Fig. 6.7: Mass spectrum of 1-butene

Alkenes usually form fragments corresponding to $C_nH_{2n+1}^+$, $C_nH_{2n}^+$, and $C_nH_{2n-1}^+$ (the latter two fragment ion series are more abundant). It is very difficult to locate the position of the double bond in an alkene because of the easy migration of the double bond by hydride and hydrogen atom shifts. For this reason, the mass spectra of alkene isomers are nearly identical and almost impossible to distinguish, as is illustrated in Figs 6.8 and 6.9, which contain the mass spectra for 2- and 3-heptene, respectively.

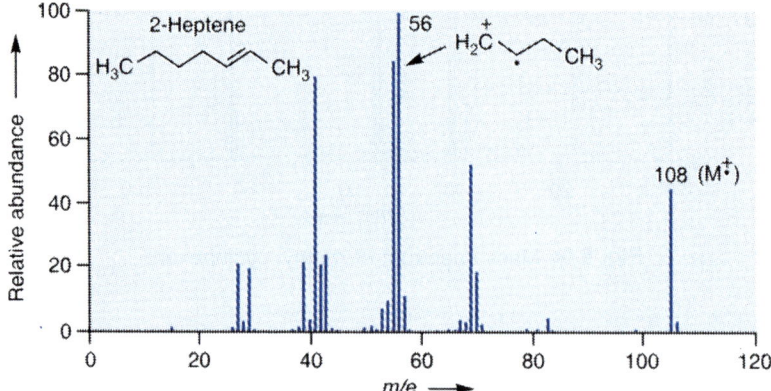

Fig. 6.8: Mass spectrum of 2-heptene

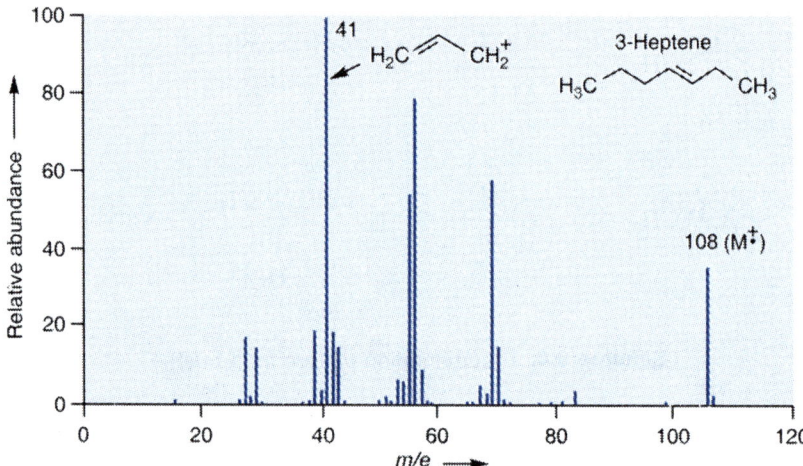

Fig. 6.9: Mass spectra of 3-heptene

Cyclic alkene

The mass spectra of cyclic alkenes show distinct molecular ions. It may be impossible to locate the position of a double bond due to migration. The mechanism of fragmentation for cyclic alkenes is virtually the same as for straight chain alkenes. One noteworthy characteristic is the fragmentation of cyclic hexene to undergo a reverse Diels-Alder reaction as indicated in Scheme 6.5 This rearrangement is characteristic of many isoprenoid natural products and of tetralin derivatives, and is useful for assigning structure and distinguishing isomers. Here, product ion enthalpy governs the dissociation pathway. The fragment with the lowest ionization energy (IE) will preferentially take the charge. The difference in activation energy equals the difference in IE. The complete mass spectrum of cyclohexene is given in Fig. 6.10.

Similarly, methyl cyclohexene shows the fragmentation as per Scheme 6.6.

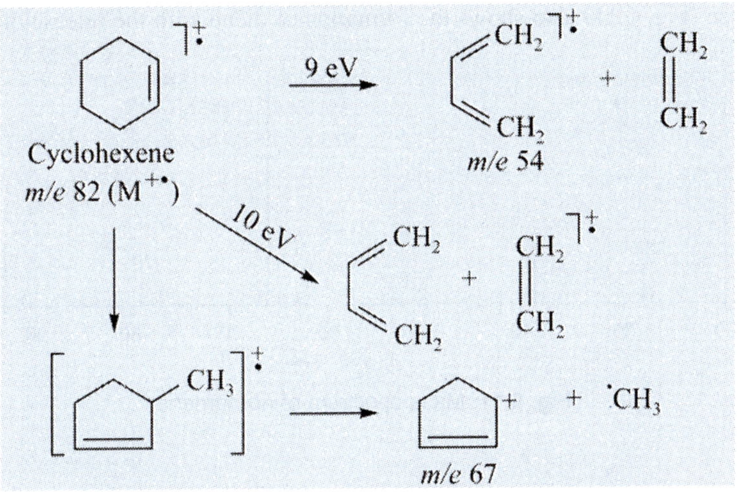

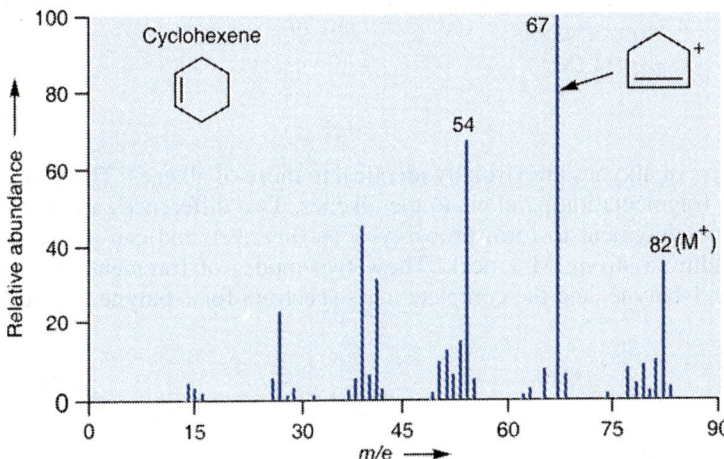

Scheme 6.5: Mass fragmentation pattern for cyclohexene

Fig. 6.10: Mass spectrum of cyclohexene

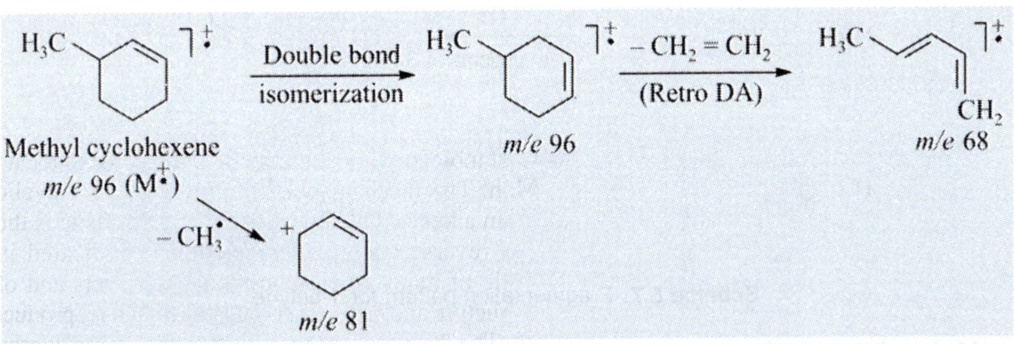

Scheme 6.6: Fragmentation pattern for methyl cyclohexene

Norbornene (Fig. 6.11) also shows the formation of diene with the release of olefin.

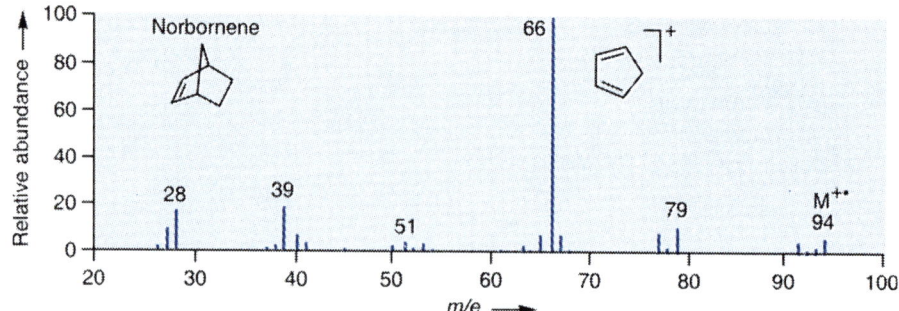

Fig. 6.11: Mass spectrum of norbornene

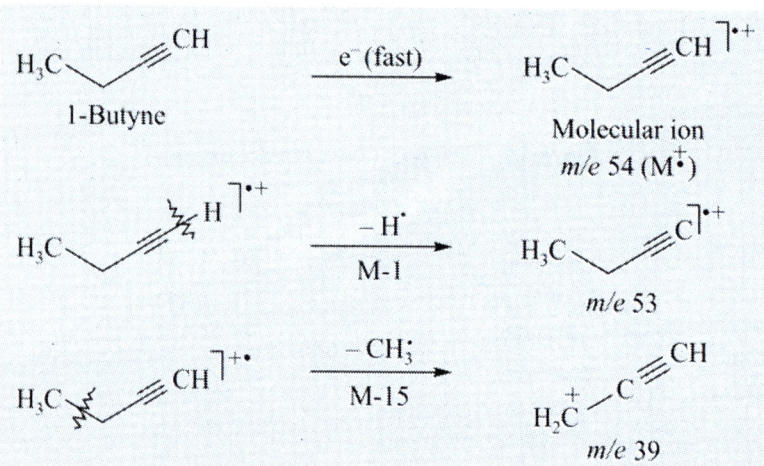

Alkynes

The mass spectra of alkynes are virtually identical to those of alkenes. The molecular ion peak is intense, and fragmentation parallels to the alkenes. Two differences are worth mentioning: Terminal alkynes fragment to form propargyl ions (m/e 39), and can also lose the terminal hydrogen, yielding a strong M-1 peak. These two modes of fragmentation are outlined in Scheme 6.7 for 1-butyne, and the complete mass spectrum for 1-butyne is given in Fig. 6.12.

Scheme 6.7: Fragmentation pattern for 1-butyne

Aromatic hydrocarbons

Unlike other hydrocarbons, the mass spectra of aromatic hydrocarbons are characterized by the presence of comparatively intense molecular ions and by low m/e peaks of low intensity, as illustrated by the EI spectra of benzene, toluene and n-propylbenzene below.

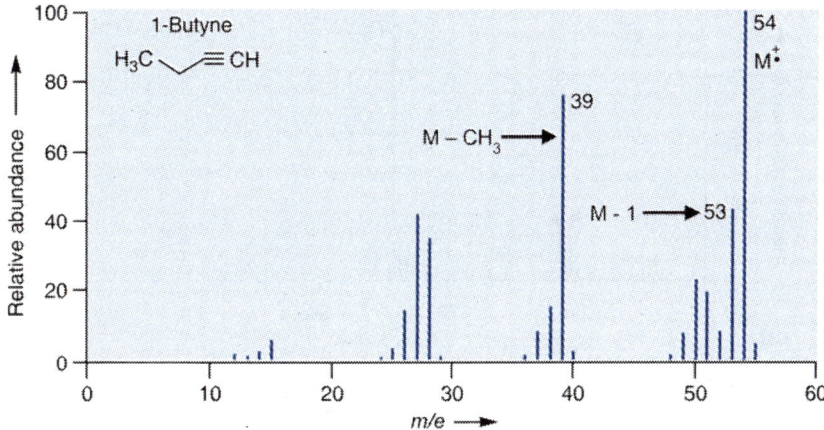

Fig. 6.12: Mass spectrum for 1-butyne

The mass spectrum of benzene shows a strong molecular ion, which is the base peak, and little additional fragmentation is evident. This is because the fragmentation of a benzene ring is a high energy process and simple pathways for the formation of stable ions do not exist. However, some detectable peaks (Fig. 6.13a) are evident as shown in Scheme 6.8.

When an alkyl group is attached to a benzene ring, the preferred site of cleavage is at the benzyl carbon to give a peak at m/e 91. This peak corresponds to the benzyl cation, or its rearrangement product, the **tropylium cation**. The tropylium cation is exceptionally stable because it is aromatic and the positive charge is delocalized over all seven ring carbons. The base peak in toluene (Fig. 6.13b) is due to loss of a hydrogen atom to form relatively stable benzyl cation, which undergoes rearrangement to form the very stable tropylium cation. The frequently observed peak at m/e 65 is due to the loss of neutral acetylene from the tropylium ion and the minor peaks below this arise from more complex fragmentation (Scheme 6.9).

The mass spectrum of propyl benzene is shown in Fig. 6.13c. A decent molecular ion is evident at m/e 120, with the major cleavage being loss of an ethyl radical (M - 29) to give the benzyl (or tropylium) cation at m/e 91. It also undergoes McLafferty rearrangement to exhibit the peak at m/e 92 (Scheme 6.10).

Bicyclic aromatic hydrocarbon

Molecular ion peaks are strong due to the stable structure as shown in Fig. 6.14 for naphthalene. It does not show any appreciable fragmentation. Molecular ion peaks are strong due to the stable structure.

HYDROXY COMPOUNDS

Alcohols

The molecular ion peak is usually small for primary and secondary alcohol, and usually undetectable for tertiary alcohols. Just as with carbonyl compounds, cleavage on either side of the alcohol carbon (α-cleavage) is the most important feature in alcohol fragmentation. This will typically involve the loss of an alkyl group, and often, it is the largest alkyl group that is preferentially lost. If the alkyl chain attached to the alcohol carbon is at least of three carbons in length, then a process similar to McLafferty rearrangements seen for carbonyl compounds can take place. Transfer of a γ-hydrogen to the alcohol oxygen leads to the loss of water from the molecule. This dehydration can be a very important indication for the presence of alcohol functionality. The 1-pentanol has hydroxyl group at the end of five carbon chain. In mass spectrum, 1-pentanol shows a peak at m/e 70 (Fig. 6.15). There are two fragment ions, one at

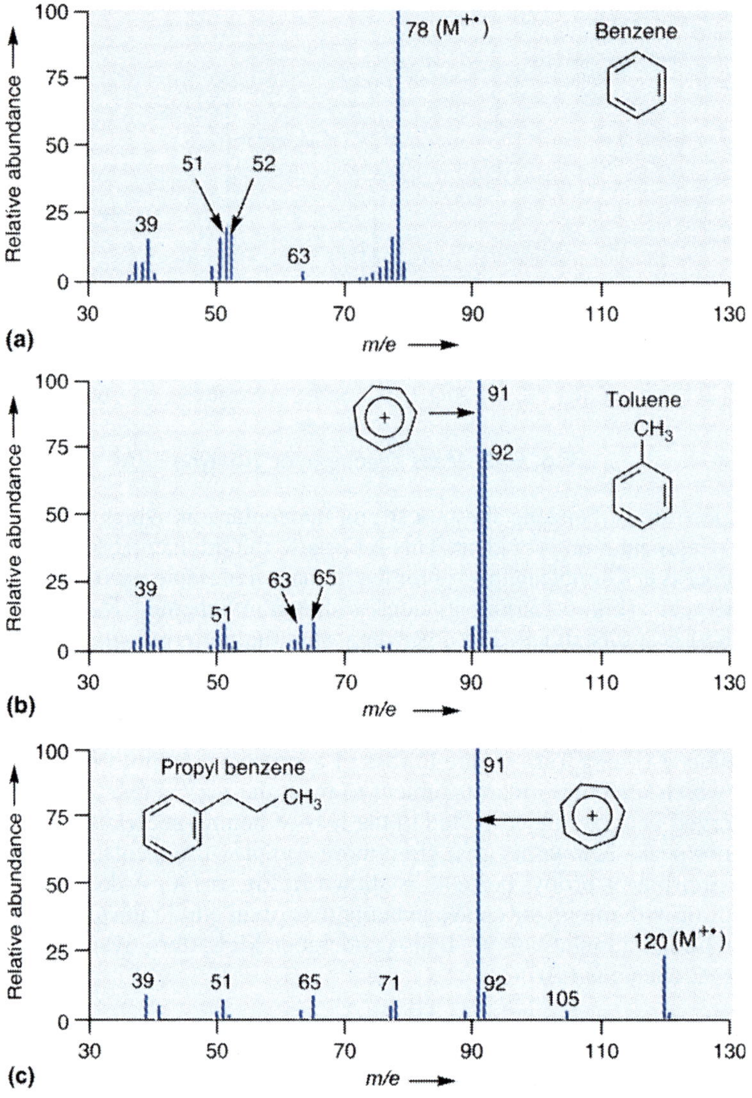

(a)

Relative abundance

100 — 78 (M⁺•)   Benzene

75 —

50 —

51   52

39

63

30   50   70   90   110   130

m/e →

(b)

Relative abundance

100 —

91   Toluene

92   CH₃

75 —

50 —

39

51

63   65

30   50   70   90   110   130

m/e →

(c)

Relative abundance

100 — Propyl benzene   91

CH₃

75 —

50 —

120 (M⁺•)

39   51   65   71   92   105

30   50   70   90   110   130

m/e →

Fig. 6.13: Mass spectrum of (a) benzene, (b) toluene, and (c) propyl benzene

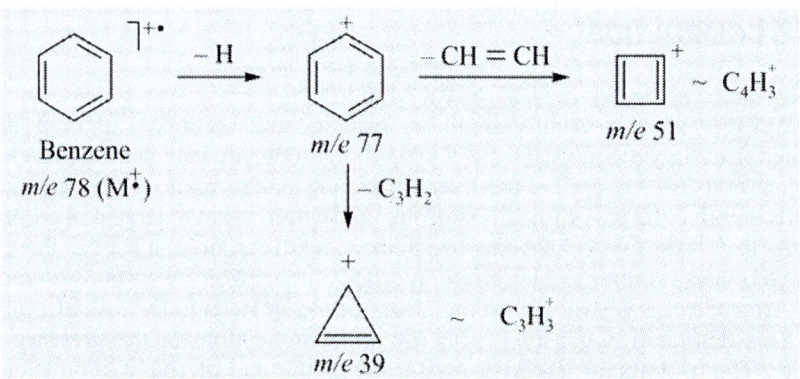

Benzene m/e 78 (M⁺•) →(– H)→ m/e 77 →(–CH=CH)→ m/e 51 ~ $C_4H_3^+$

m/e 77 →(– C_3H_2)→ m/e 39 ~ $C_3H_3^+$

Scheme 6.8: Fragmentation pattern for benzene

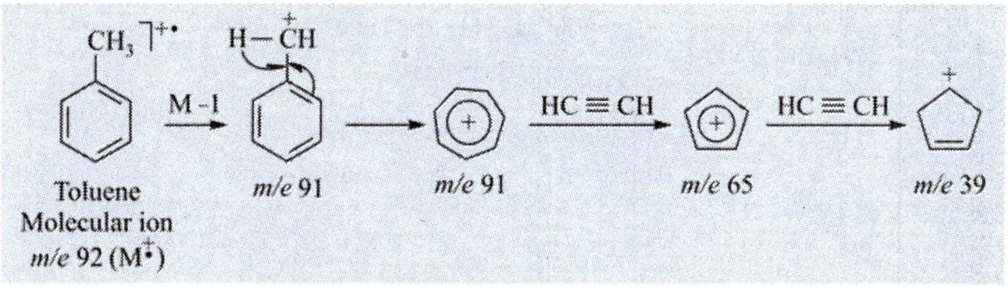

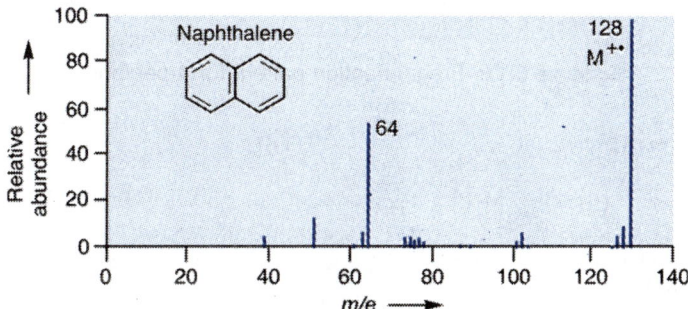

Scheme 6.9: Fragmentation pattern for toluene

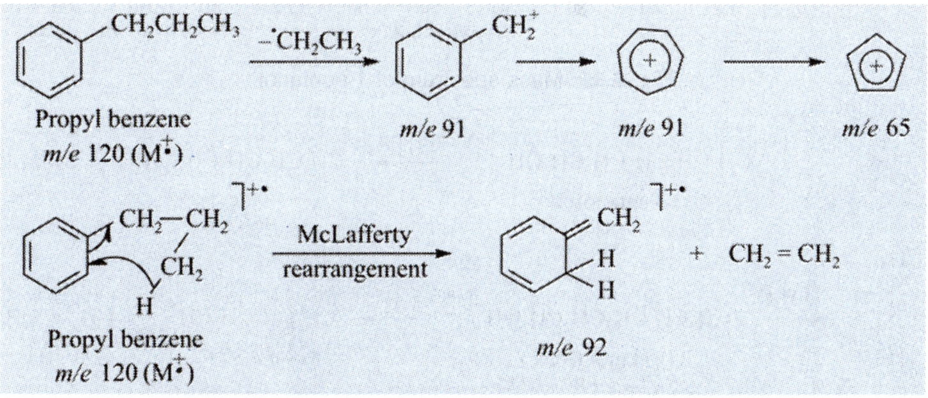

Scheme 6.10: Fragmentation pattern for propyl benzene

Fig. 6.14: Mass spectrum of naphthalene

m/e 70 (loss of water), and the other fragment at m/e 42 (loss of H_2O and ethene). The fragment at m/e 55 is probably due to loss of methyl radical from the m/e 70 ions. The m/e 31 is due to $CH_2=OH$ (Scheme 6.11). The mechanism for alcohol fragmentation is given in Scheme 6.12 for 2-pentanol.

The complete mass spectrum of 2-pentanol is given in Fig. 6.16, which illustrates most of these points. It shows the peaks at m/e 73 and 45 due to α-cleavage.

t-Pentanol shows three significant fragment ions. Fragment due to α-cleavage (loss of an ethyl radical) forms the m/e 59 base peak. Loss of water from this gives a $m/e = 41$ fragment, and loss of ethene from m/e 59 gives a m/e 31 fragment (Scheme 6.13).

Cyclic alcohol

Cyclic alcohol undergoes fragmentation by at least three ways and these are exemplified with the spectrum (Fig. 6.17) of cyclohexanol. The first fragmentation is the α-cleavage and loss of

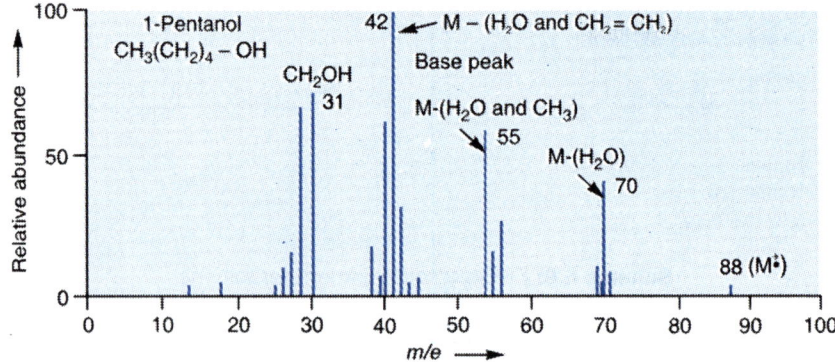

Fig. 6.15: Mass spectrum of 1-pentanol

$$CH_3CH_2CH_2CH_2CH_2OH \rceil^{+\bullet} \xrightarrow{-H^\bullet} CH_3CH_2CH_2CH_2CH_2O^\bullet + H^\bullet$$

1-Pentanol
m/e 88 (M$^{+\bullet}$)   m/e 87

CH$_3$

⬜ $\xleftarrow{-H_2O}$ $CH_3CH_2CH_2CH_2CH_2OH \rceil^{+\bullet} \xrightarrow{-H^\bullet} CH_3CH_2CH_2CH_2^+ + CH_2 = \overset{+}{O}H$

m/e 70

1-Pentanol (M$^{+\bullet}$)   m/e 87   m/e 31

↓ $-\bullet CH_3$ m/e 88

⬜$^+$

m/e 55

Scheme 6.11: Fragmentation pattern for 1-pentanol

$\overset{+\bullet\ddot{O}H}{}$ $\overset{+\ddot{O}H}{}$

H$_3$C⟍⟋⟍⟋⟍CH$_3$ $\xrightarrow[M-15]{-CH_3^\bullet}$ H$_3$C⟍⟋⟍⟋   α-Cleavage

2-Pentanol   m/e 73
m/e 88 (M^{+})
(absent)

$\overset{H\ddot{O}\bullet+}{}$ $\overset{+\ddot{O}H}{}$

H$_3$C⟍⟋⟍⟋CH$_3$ $\xrightarrow[M-43]{-C_3H_7^\bullet}$ ⟍⟋CH$_3$ α-Cleavage

m/e 45

H⟍$\ddot{O}H$ $\xrightarrow[\substack{M-18 \\ \text{Dehydration}}]{-H_2O}$ ⬜$\rceil^{\bullet+}$ CH$_3$ $\xrightarrow[(M-70)-15]{-CH_3^\bullet}$ ⬜$\rceil^+$

‡$\ddot{O}H$⟋CH$_3$ m/e 70   m/e 55

H⟍$\ddot{O}H$ $\xrightarrow[M-(28+18)]{\substack{-H_2O \\ -C_2H_4}}$ H$_2$C$^\bullet$⟋$\overset{+}{}$⟍CH$_3$ Dehydration and elimination

‡$\ddot{O}H$⟋CH$_3$ m/e 42

Scheme 6.12: Fragmentation pattern for 2-pentanol

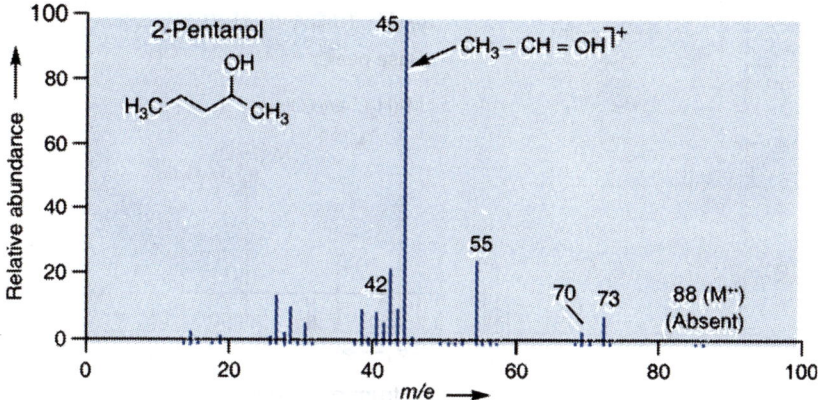

Fig. 6.16: Mass spectrum of 2-pentanol

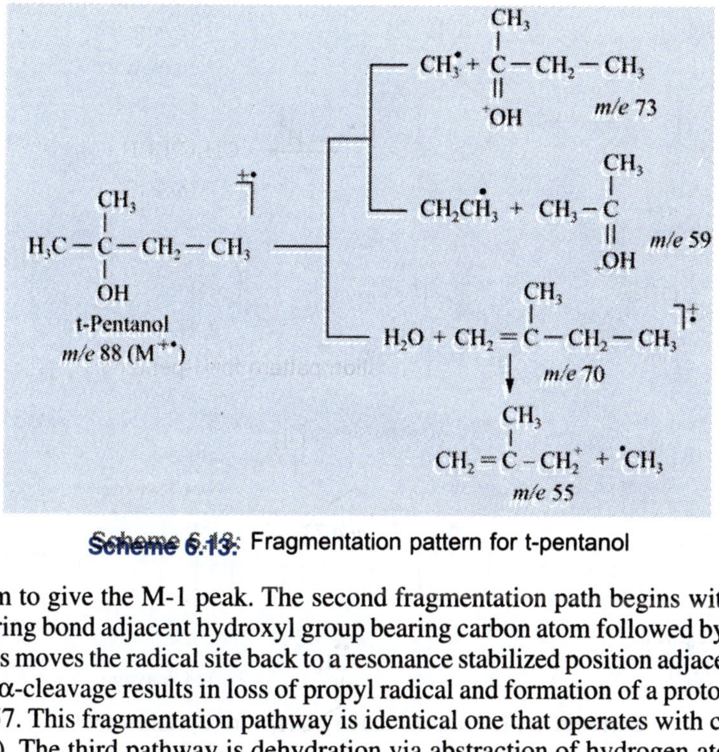

Scheme 6.13: Fragmentation pattern for t-pentanol

hydrogen atom to give the M-1 peak. The second fragmentation path begins with initial alpha cleavage of a ring bond adjacent hydroxyl group bearing carbon atom followed by 1,5 hydrogen migration. This moves the radical site back to a resonance stabilized position adjacent to oxonium ion. A second α-cleavage results in loss of propyl radical and formation of a protonated acrolein ion with m/e 57. This fragmentation pathway is identical one that operates with cyclohexanone (Scheme 6.14). The third pathway is dehydration via abstraction of hydrogen atom from three or four carbon atoms away to produce bicyclic radical cations with m/e 82. The molecular ion peak is usually undetectable.

Similarly, cyclopentyl alcohol shows the H-rearrangement for the prominent peak at m/e 57 (Scheme 6.15).

Aromatic alcohol

The mass spectrum of benzyl alcohol (Fig. 6.18) shows the **molecular ion** at m/e 108. Fragmentation via loss of 17 (–OH) gives a common fragment seen for alkyl benzenes at m/e 91. Loss of 31 (–CH$_2$OH) from the molecular ion gives 77 corresponding to the phenyl cation (Scheme 6.16).

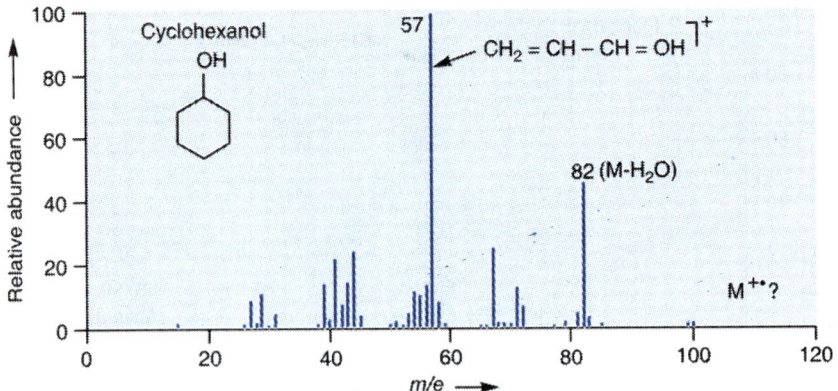

Fig. 6.17: Mass spectrum of cyclohexanol

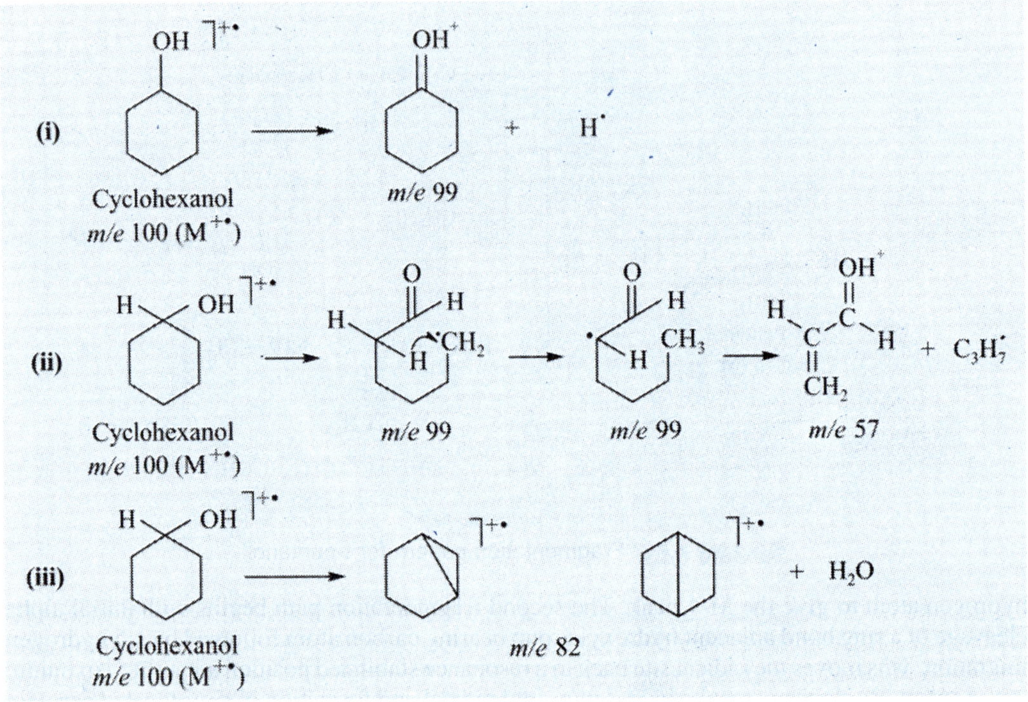

Scheme 6.14: Fragmentation pattern for cyclohexanol

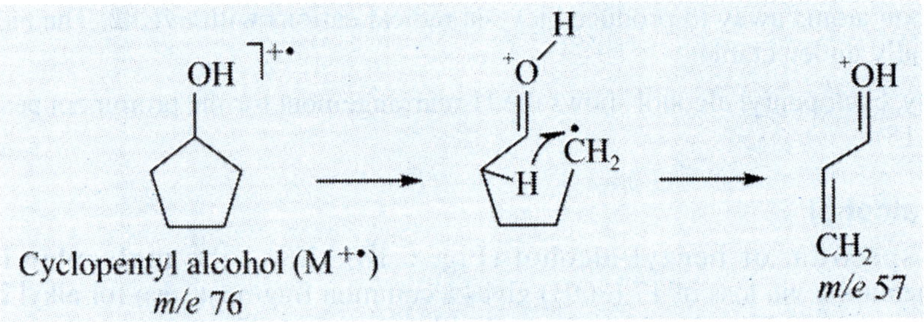

Scheme 6.15: Fragmentation pattern for cyclopentyl alcohol

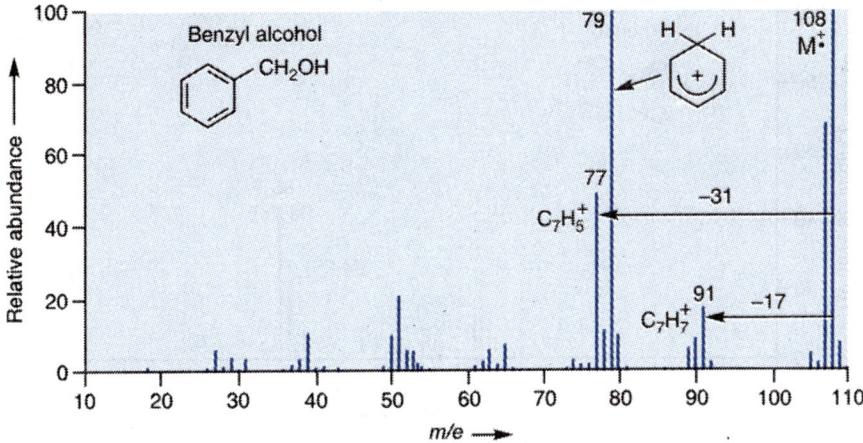

Fig. 6.18: Mass spectrum of benzyl alcohol

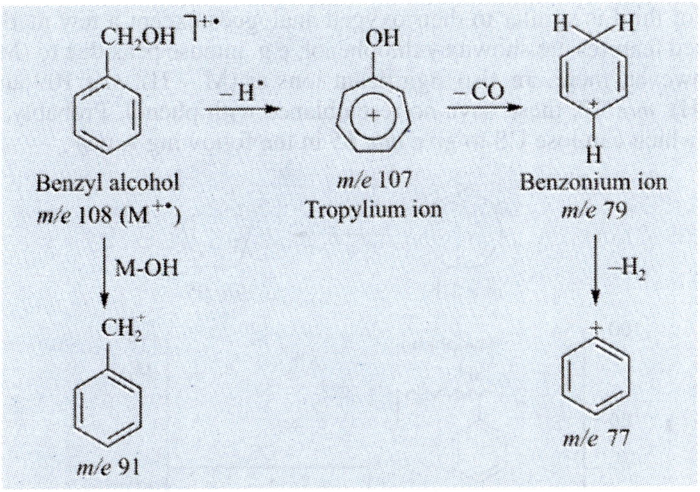

Scheme 6.16: Fragmentation pattern for benzyl alcohol

Phenol

Unlike for aliphatic alcohols, the molecular ion peak for phenols are quite intense. Phenols can lose the stable elements of carbon monoxide to give abundant fragment ions at M-28, and can also lose the elements of the formyl radical (HCO•) to give abundant fragment ions at M-29. The mechanism of this fragmentation is still not clear. However, Fig. 6.19 contains the mass spectrum of phenol, which highlights the production of the fragment ions.

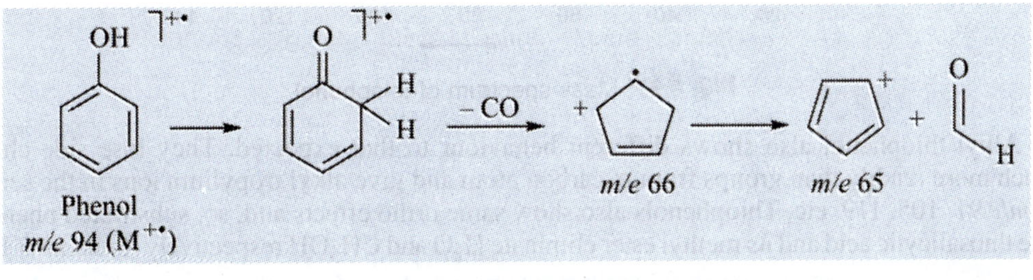

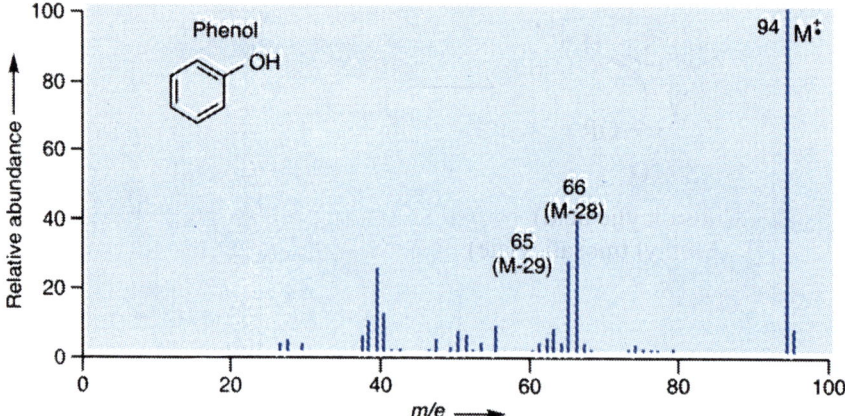

Fig. 6.19: Mass spectrum of phenol

Thiols

The behaviour of thiol is similar to their oxygen analogues except a few marked differences. The two expected features are shown by thiophenol, e.g. intense peak due to (M-CS), at *m/e* 66 (Fig. 6.20). However, there are also significant ions at (M – H), *m/e* 109 and (M – C_2H_2), *m/e* 84 (M – SH), *m/e* 77; these have no resemblance with phenol. Probably, (M – H) is the tropylium ion, which can lose CS to give *m/e* 65 in the following ways:

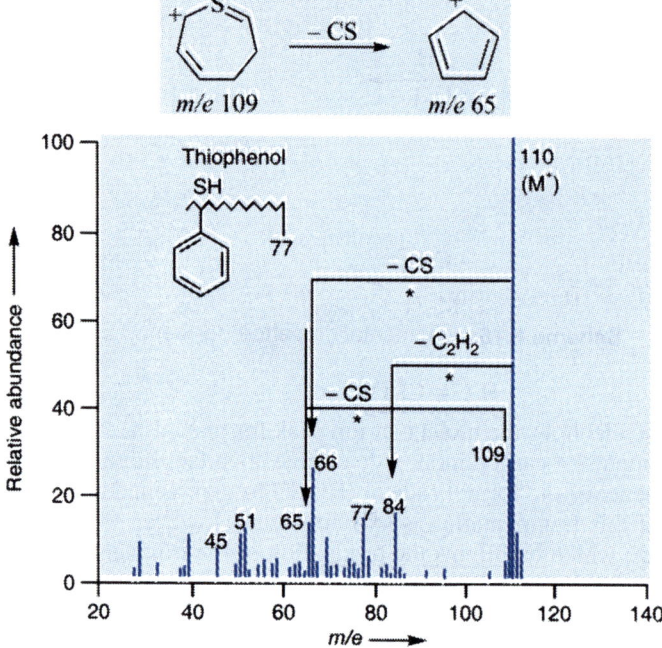

Fig. 6.20: Mass spectrum of thiophenol

Alkyl thiophenol also shows different behaviour to that expected. They lose side chain much more readily than groups from α-carbon atom and give alkyl tropylium ions in the series of *m/e* 91, 105, 119, etc. Thiophenols also show same ortho effects and, so, substituted phenols like thiosalicylic acid and its methyl ester eliminate H_2O and CH_3OH respectively (Scheme 6.17).

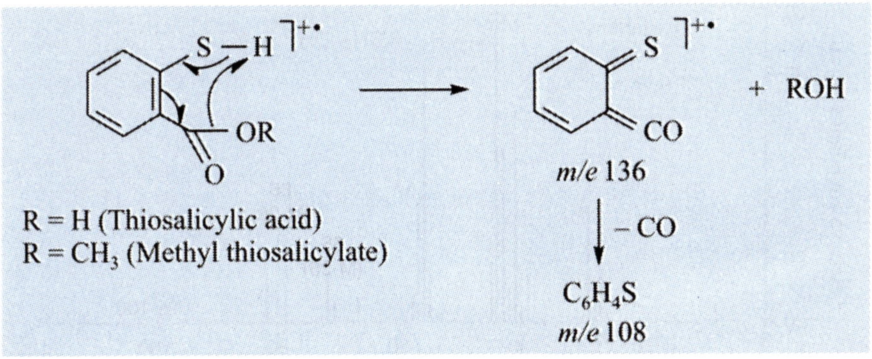

Scheme 6.17: The 'ortho effect' fragmentation of thiosalicylic acid and methyl thiosalicylate

ALDEHYDE

Aliphatic aldehyde

In simple aliphatic aldehyde like hexanal, the molecular ion at m/e 100 is very weak but detectable, as is the M-1 ion at m/e 99. The predominant fragmentation involves cleavage α to the carbonyl group with loss of the larger alkyl radical. This means that a peak with a mass-to-charge ratio (m/e) of 29, corresponding to the formation of CHO^+, is expected. This is illustrated for n-hexanal in Scheme 6.18.

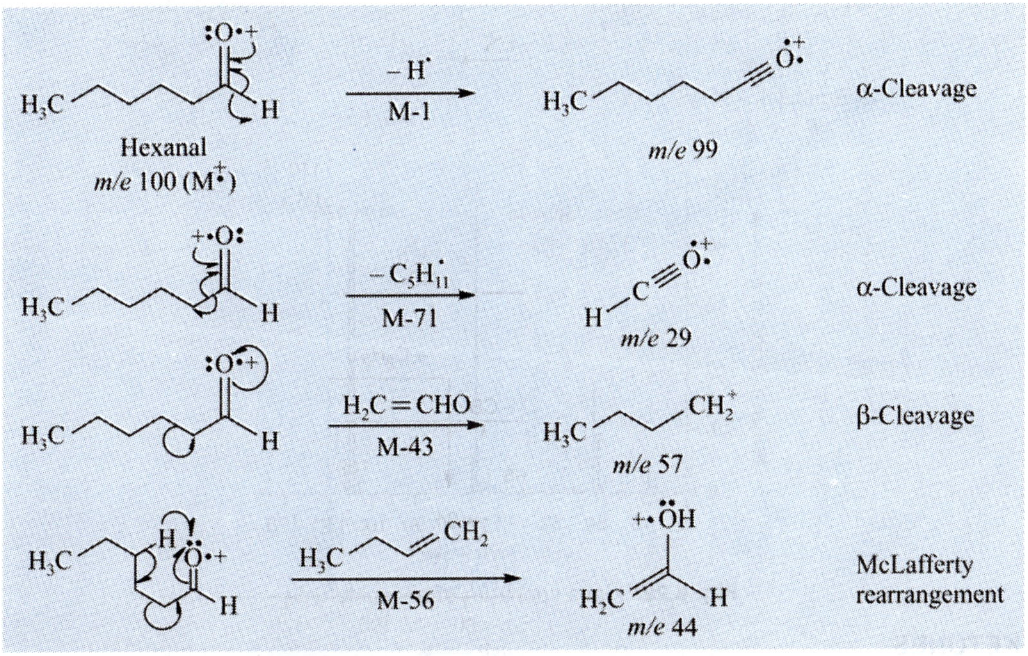

Scheme 6.18: Fragmentation pattern for hexanal

If a γ-hydrogen is present, a peak at m/e 44 can arise from a process known as the McLafferty rearrangement. This involves γ-hydrogen transfer to the carbonyl oxygen atom with the subsequent formation of $C_2H_5O^+$. The complete mass spectrum of hexanal, given in Fig. 6.21, illustrates these points.

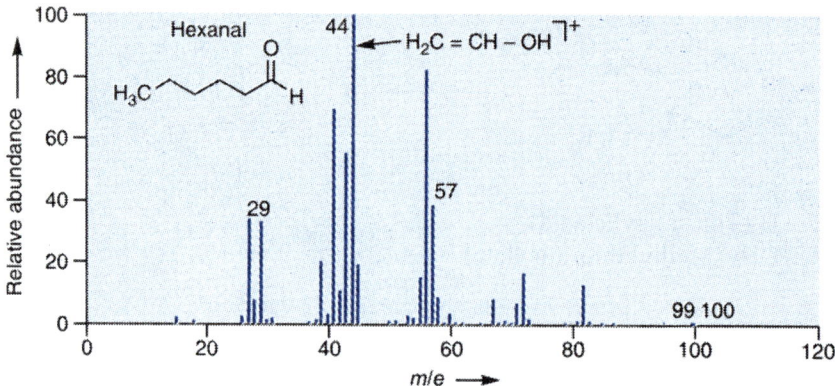

Fig. 6.21: Mass spectrum of hexanal

Aromatic aldehydes

Aromatic aldehydes, such as benzaldehyde (Fig. 6.22), typically display intense molecular ions and M-1 peaks. In the mass spectrum of benzaldehyde, the oxocarbocation formed by loss of a hydrogen atom, expels CO to give the phenyl cation, $C_6H_5^+$ at m/e 77.

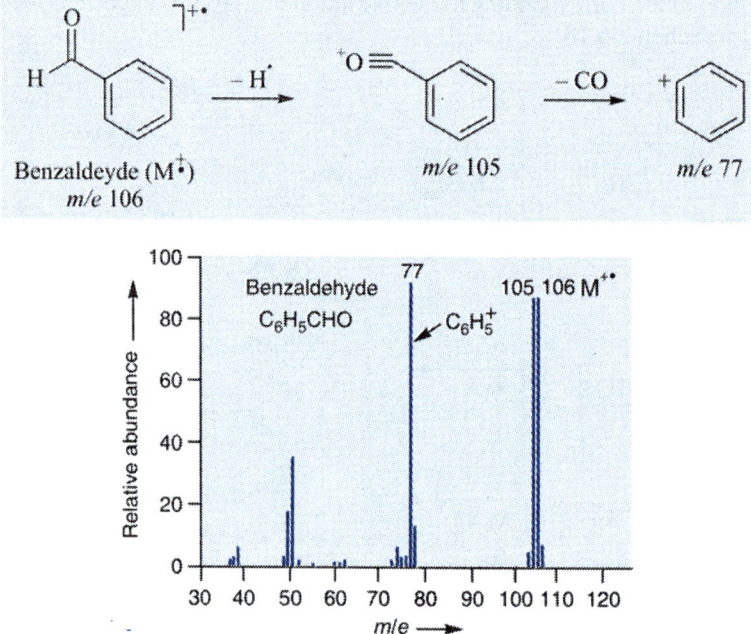

Fig. 6.22: Mass spectrum of benzaldehyde

KETONES

Aliphatic ketones

α-Cleavage is also common for ketones and may occur at either of the two bonds next to the carbonyl group. In general, loss of the larger alkyl fragment is preferred. Thus, for methyl ketones ($CH_3–CO–R$) a prominent peak at m/e 43 ($CH_3–CO^+$) is expected. Gamma (γ) hydrogen transfer due to McLafferty rearrangement may also occur for ketones. This is illustrated with the example of 2-pentanone in Scheme 6.19.

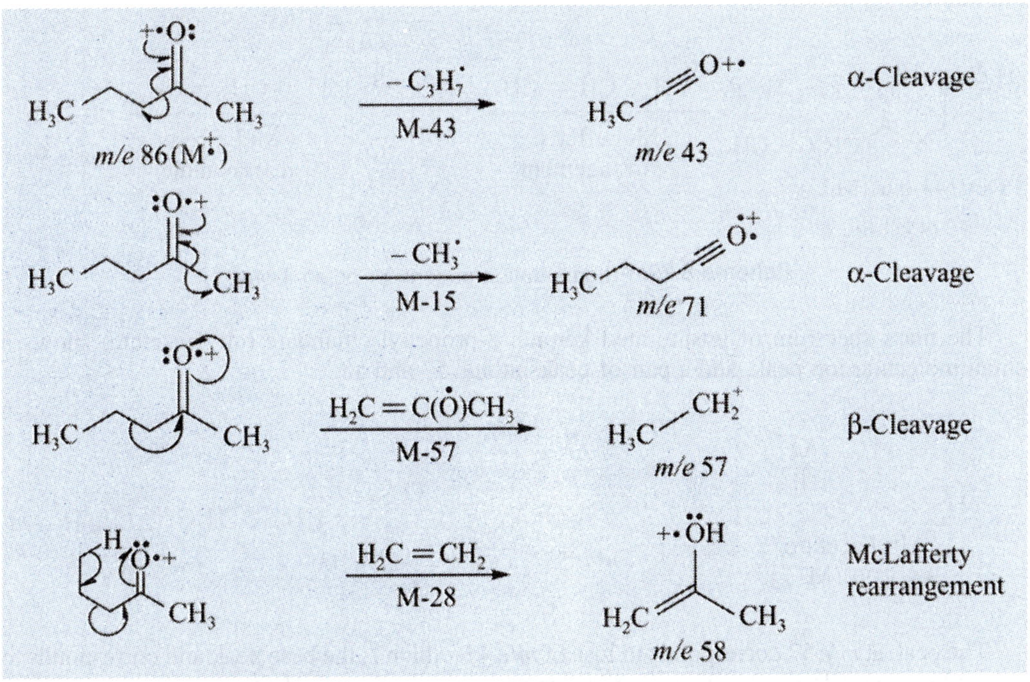

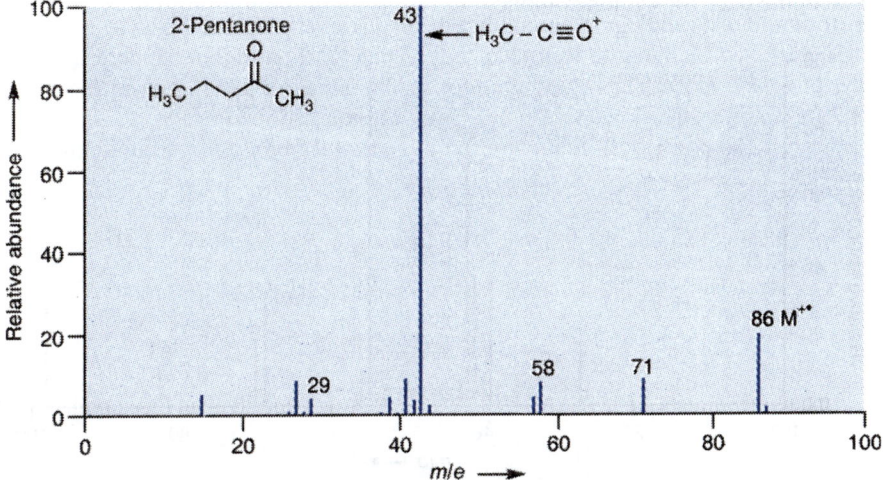

Scheme 6.19: Fragmentation pattern for 2-pentanone

The two ions formed by α-cleavage (*m/e* 43 and 71), then lose CO (decarbonylate) to give carbocations (with masses of 15 and 43, respectively) (Fig. 6.23). When the alkyl chains are long, the resulting alkene will fragment by losses of alkenes and other neutrals (e.g. H_2), in accord with EE ion rule.

Fig. 6.23: Mass spectrum of 2-pentanone

If there is a chain of three or more carbons atoms in each chain of the ketone, then a second McLafferty rearrangement is possible. This is illustrated below for octan-4-one in Scheme 6.20.

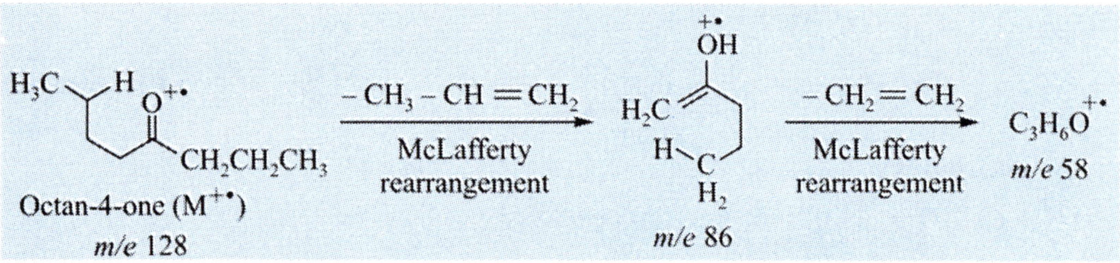

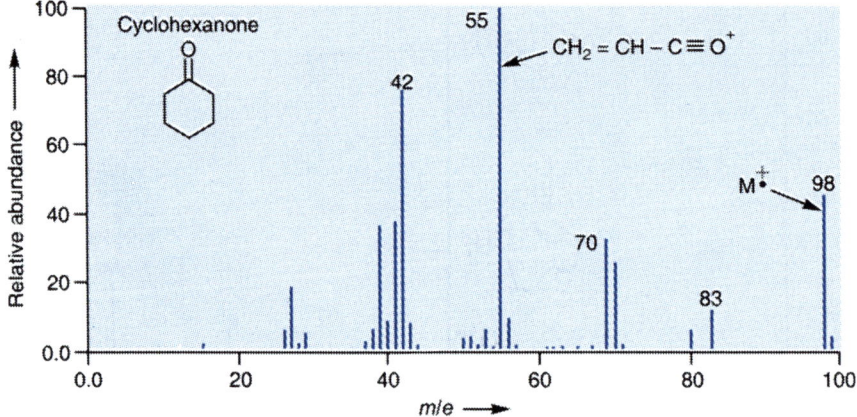

Scheme 6.20: Fragmentation pattern for octan-4-one

The mass spectrum of unsaturated ketone, 2-propenyl ethanoate (allyl acetate) shows a small molecular ion peak, and a pair of peaks at m/e 57 and 58.

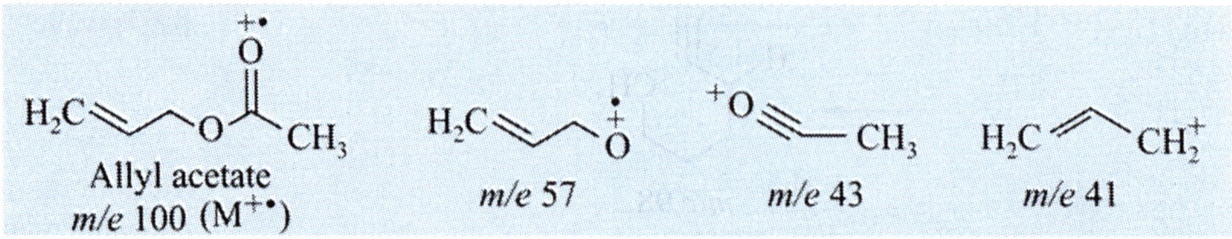

The peak at m/e 57 corresponds to loss of m/e 43, which is the base peak and corresponds to the acylium ion ($CH_3C\equiv O^+$). The m/e 57 fragment corresponds to C_3H_5O, suggesting the original compound was an allyl ($-O-CH_2CH=CH$) or methylvinyl ($-O-CH-CHCH_3$) ester. Either of these would generate the peak observed at m/e 41. The peak at m/e 58 corresponds to the protonated m/e 57 cation radical.

Cyclic ketones

Cyclic ketones (Fig. 6.24) undergo a variety of fragmentation and rearrangement processes as shown in Scheme 6.21 for cyclohexanone.

Fig. 6.24: Mass spectrum of cyclohexanone

Similarly, cyclopentanone reveals the presence of α-cleavage and other fragmentation pathways as per Scheme 6.22.

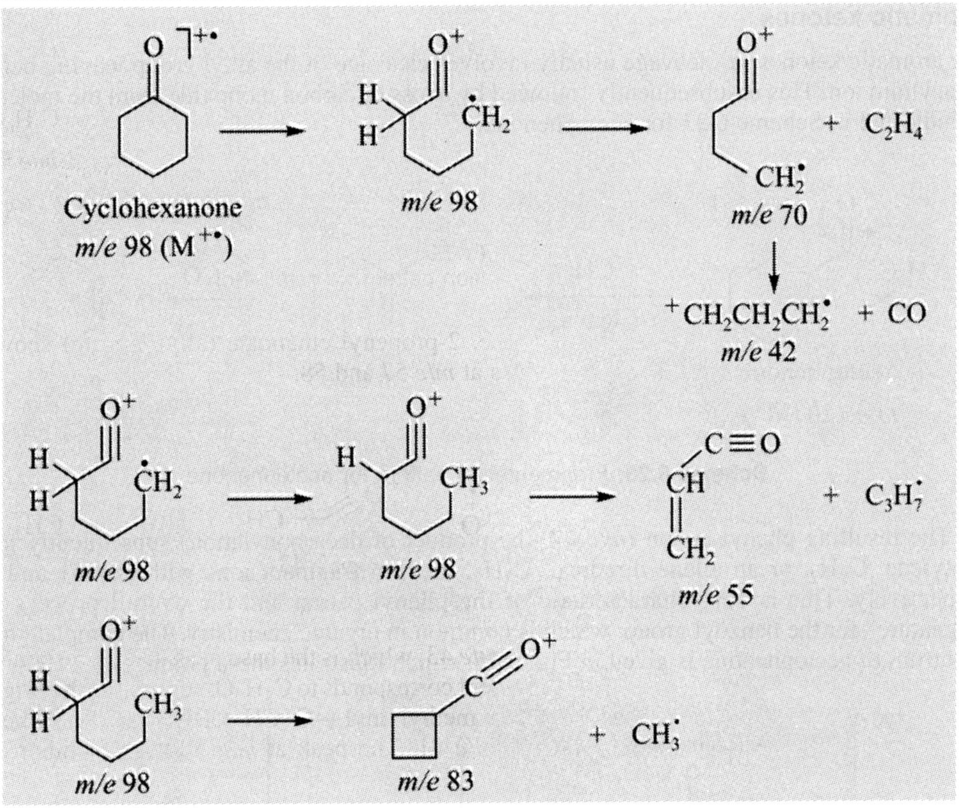

Scheme 6.21: Fragmentation pattern for cyclohexanone

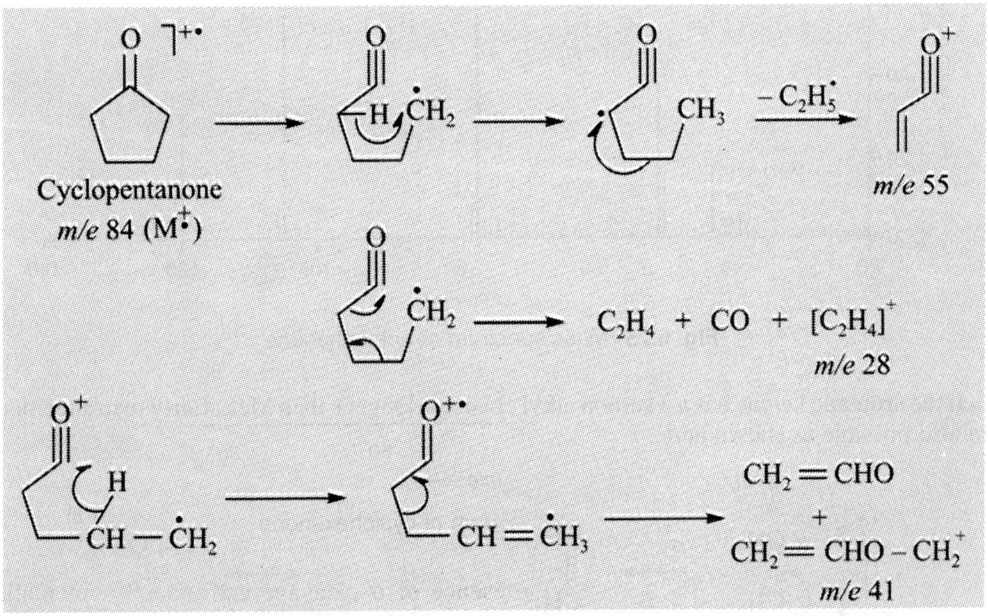

Scheme 6.22: Fragmentation pattern for cyclopentanone

Aromatic ketones

For aromatic ketones, α-cleavage usually involves cleavage of the alkyl group leaving behind an acylium ion. This is subsequently followed by a loss of carbon monoxide from the molecule as indicated in Scheme 6.23 for acetophenone.

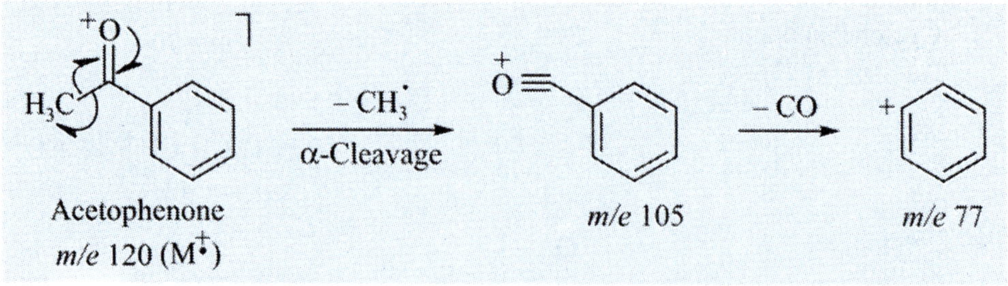

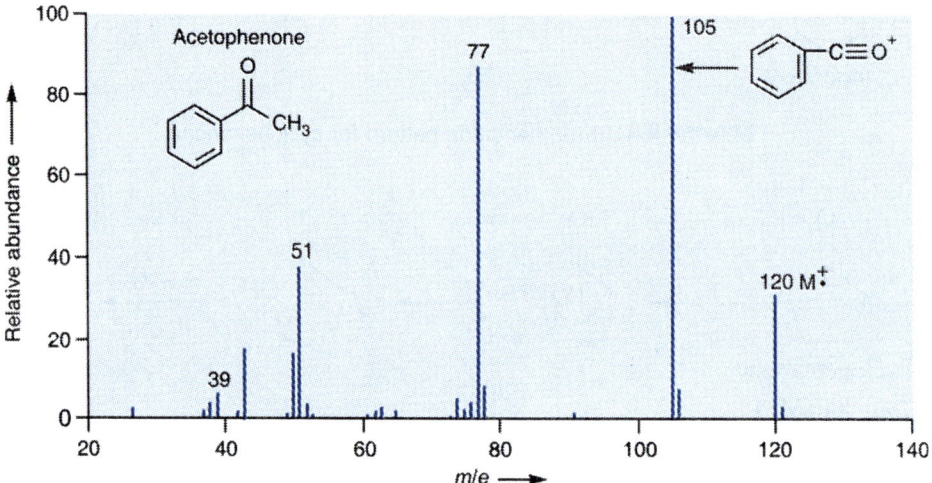

Scheme 6.23: Fragmentation pattern for acetophenone

    The resulting phenyl cation (m/e 77, the product of decarbonylation) subsequently loses acetylene, C_2H_2, or an allene diradical, C_3H_2, to form fragment ions with m/e 51 and 39, respectively. This is very characteristic of the phenyl cation and the overall process is a "signature" for the benzoyl group, which is common in organic chemistry. The complete mass spectrum of acetophenone is given in Fig. 6.25, which illustrates these points.

Fig. 6.25: Mass spectrum of acetophenone

    If the aromatic ketone has a 3 carbon alkyl chain (or longer), then McLafferty rearrangements are also possible as shown here.

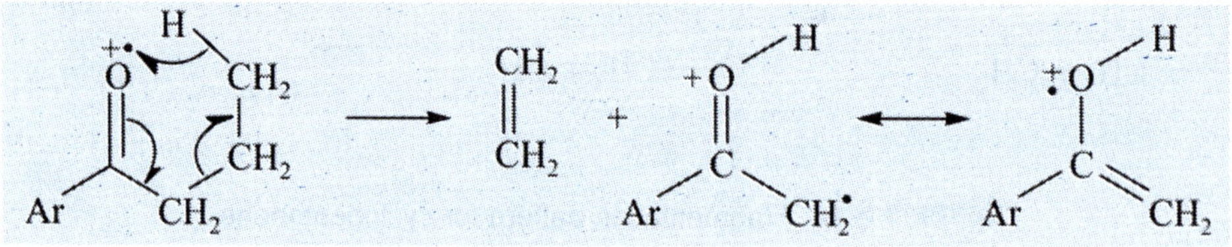

CARBOXYLIC ACID

Aliphatic carboxylic acid

The molecular ion peak is weak but usually observable. An important fragmentation pattern involves α-cleavage (breaking either bond to the carbonyl carbon) as shown in Scheme 6.24 with the example of pentanoic acid.

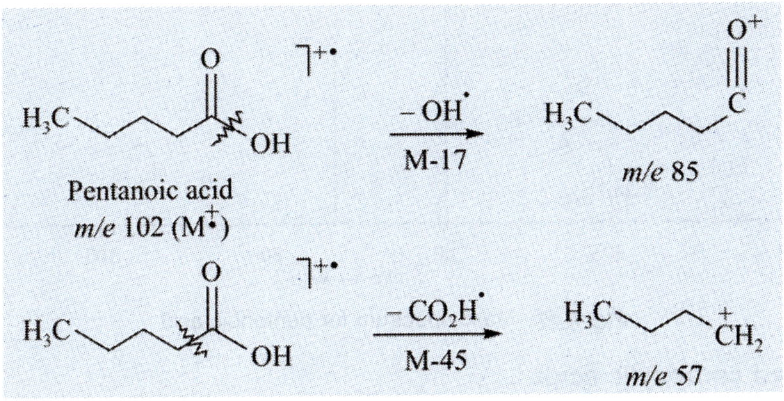

Scheme 6.24: α-Cleavage of pentanoic acid

Loss of the alkyl group as a free radical, leaving CO_2H^+, also occurs as shown in Scheme 6.25.

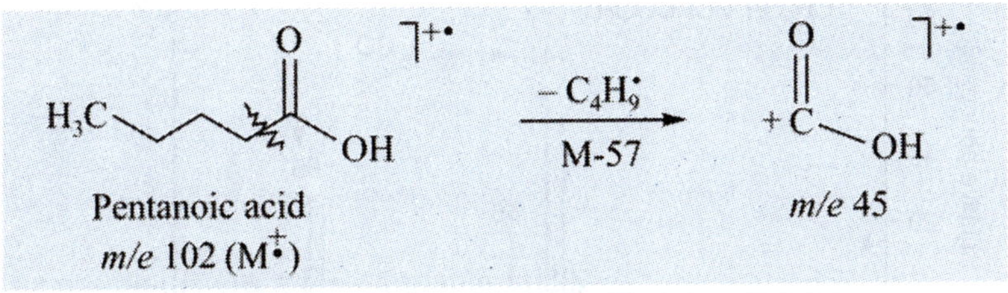

Scheme 6.25: Loss of alkyl radical in pentanoic acid

With acids having γ hydrogens, the principal pathway for rearrangement is the McLafferty rearrangement (Scheme 6.26).

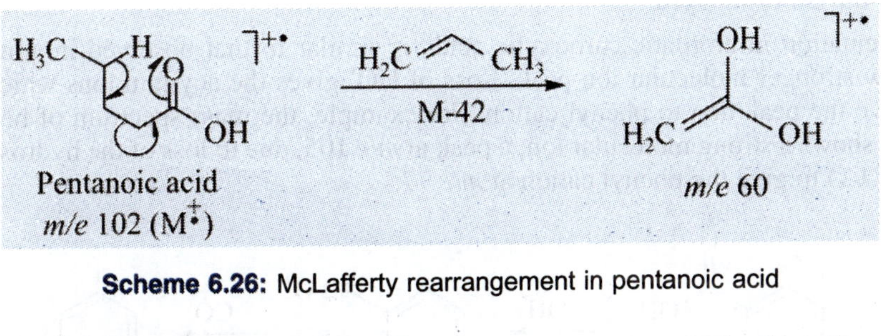

Scheme 6.26: McLafferty rearrangement in pentanoic acid

The complete mass spectrum for pentanoic acid is given in Fig. 6.26, which illustrates most of the above points.

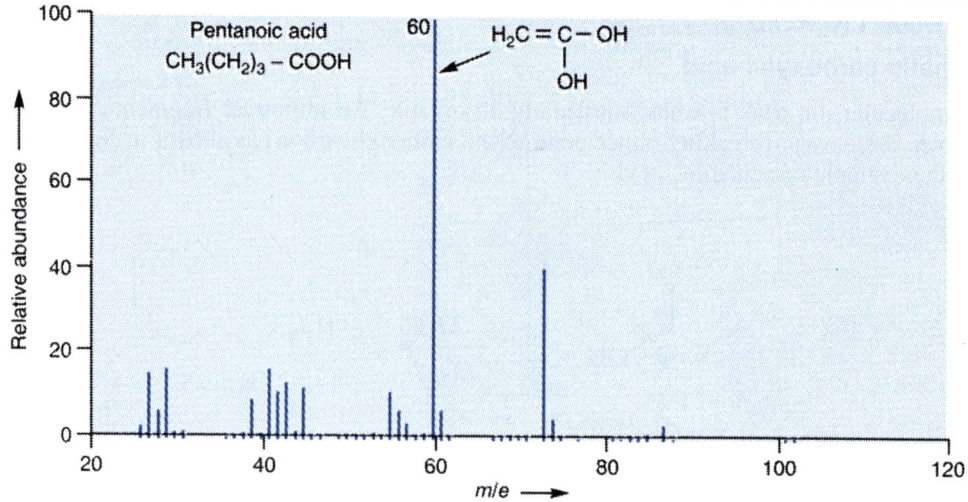

Fig. 6.26: Mass spectrum for pentanoic acid

Unsaturated carboxylic acids

The unsaturated carboxylic acids, e.g. methyl acrylic acid (Fig. 6.27), also show M–OH (m/e 69), M–COOH (m/e 41) ion in the spectrum. The unsaturated acids show intense molecular ion peak.

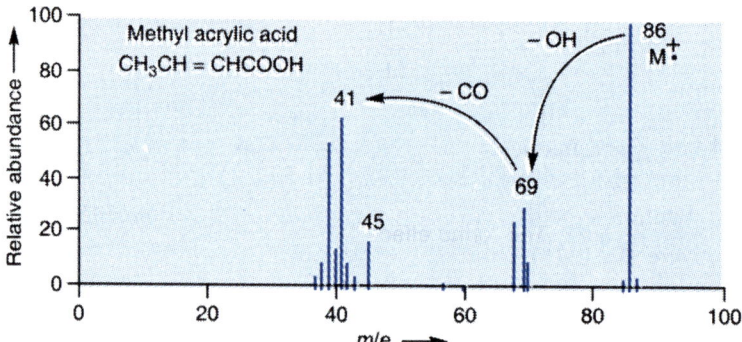

Fig. 6.27: Mass spectrum of methyl acrylic acid

Aromatic carboxylic acid

The fragmentation in aromatic carboxylic acids is similar to that observed in simple esters. These show stronger molecular ion peak. Loss of HO$^{\bullet}$ gives the acylium ions which can lose C≡O to give the peak due to phenyl cation. For example, the mass spectrum of benzoic acid (Fig. 6.28) shows a strong molecular ion, a peak at m/e 105, due to loss of the hydroxyl radical, and loss of CO to give the phenyl cation at m/e 77.

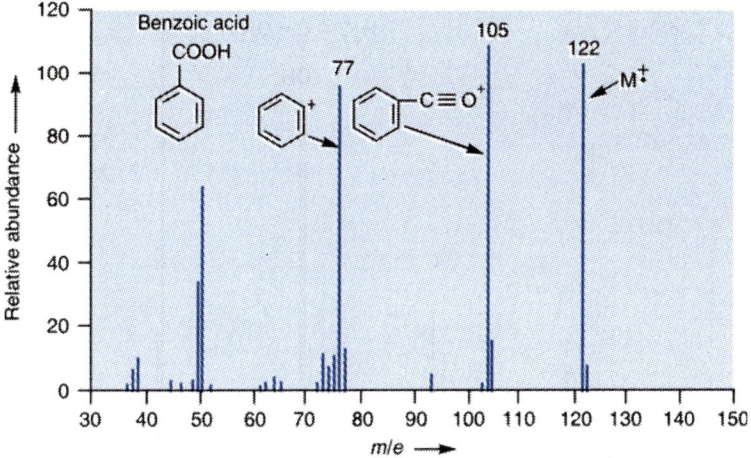

Fig. 6.28: Mass spectrum of benzoic acid

The benzoic acids substituted with alkyl, amino, or hydroxy substituents at the *ortho* position readily dehydrate via proton transfer from the *ortho* substituent to the hydroxyl group (ortho effect). Water is lost, resulting in a major M-18 ion in the mass spectrum. Scheme 6.27 outlines this process for *o*-toluic acid.

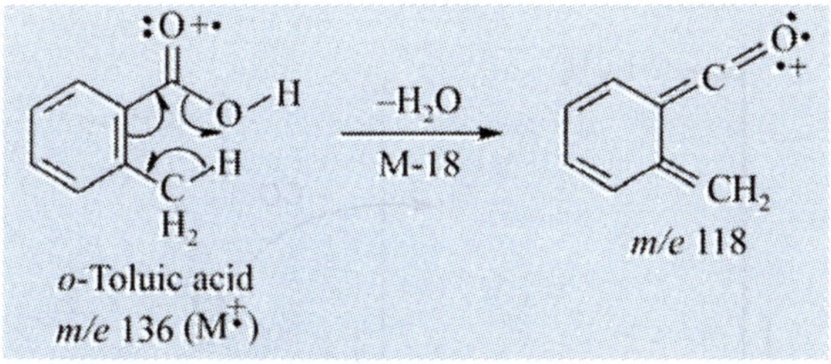

Scheme 6.27: The "ortho effect" fragmentation of *o*-toluic acid

The mass spectrum of *o*-toluic acid (Fig. 6.29) indicates the facile nature of this dehydration event (the signal for the product ion is the base peak).

ESTER

Aliphatic esters

The molecular ion is usually of low abundance but generally observable for esters. As in all carbonyl compounds, α-cleavage in esters also is an important fragmentation process. The most important cleavage reaction of carboxylate esters is loss of the alkoxy group (summarize in Table 6.1) to form the oxocarbocation. In the mass spectrum of methyl propanoate (Fig. 6.30), the molecular ion at *m/e* 88 fragments with loss of a methoxy radical to give the oxocarbocation at *m/e* 57, which can expel CO to give the ethyl carbocation at *m/e* 29. In this spectrum, the peak from the cleavage on the opposite side of the carbonyl is also evident; the fragment CH_3-O-CO^+ at *m/e* 59 (Scheme 6.28).

Methyl esters with alkyl side-chains readily undergo McLafferty rearrangement reactions producing cation radicals with *m/e* 74 (Scheme 6.29).

Since methyl esters are very common and their formation is used in the derivatization of fatty acids for GC/MS, the example below is presented to indicate the most common fragmentation patterns observed for a simple methyl ester, methyl butyrate. Loss of the alkoxy group is the most important of these fragmentations as shown in Scheme 6.30.

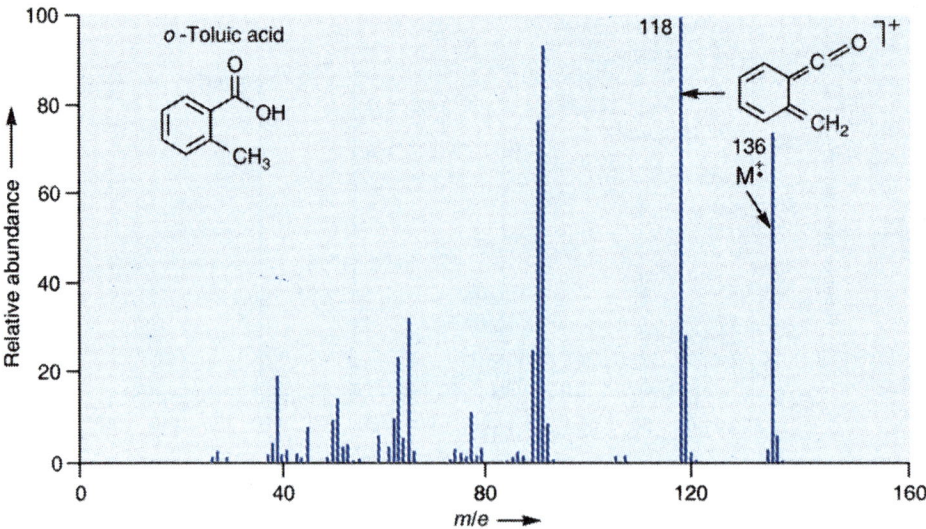

Fig. 6.29: Mass spectrum of *o*-toluic acid

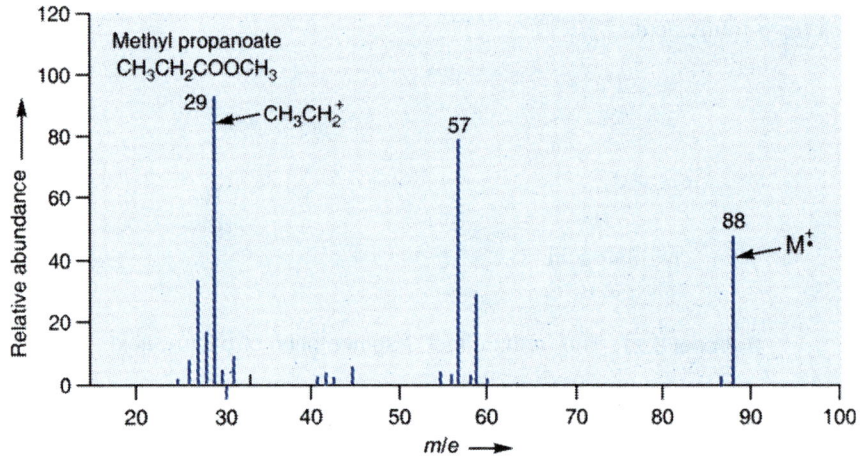

Fig. 6.30: Mass spectrum of methyl propanoate

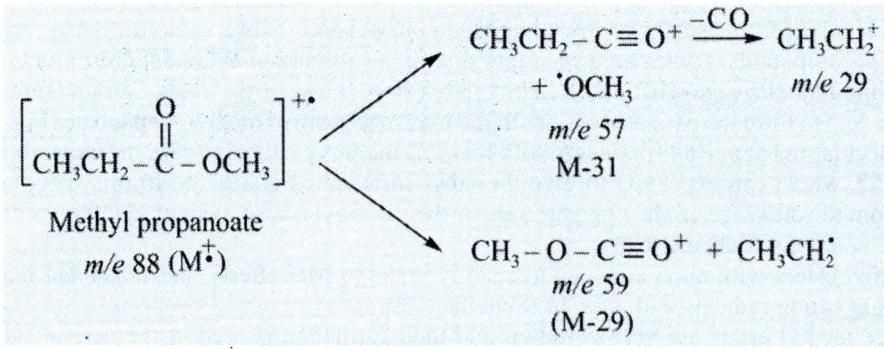

Scheme 6.28: Fragmentation pattern for methyl propanoate

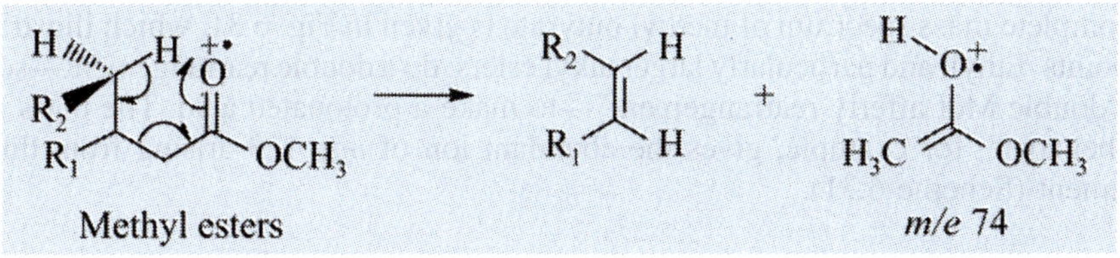

Methyl esters   m/e 74

Scheme 6.29: McLafferty rearrangement in methyl esters

Table 6.1: Alkoxy radicals formed from the most common esters

Ester	Alkoxy radical formed	Ion to observe
Methyl	$CH_3O^{\bullet}$	M-31
Ethyl	$CH_3CH_2O^{\bullet}$	M-45
Propyl (and isopropyl)	$CH_3CH_2CH_2O^{\bullet}$	M-59
Phenyl	$C_6H_5O^{\bullet}$ (PhO$^{\bullet}$)	M-93
Benzyl	$C_6H_5CH_2O^{\bullet}$ (BzO$^{\bullet}$)	M-105

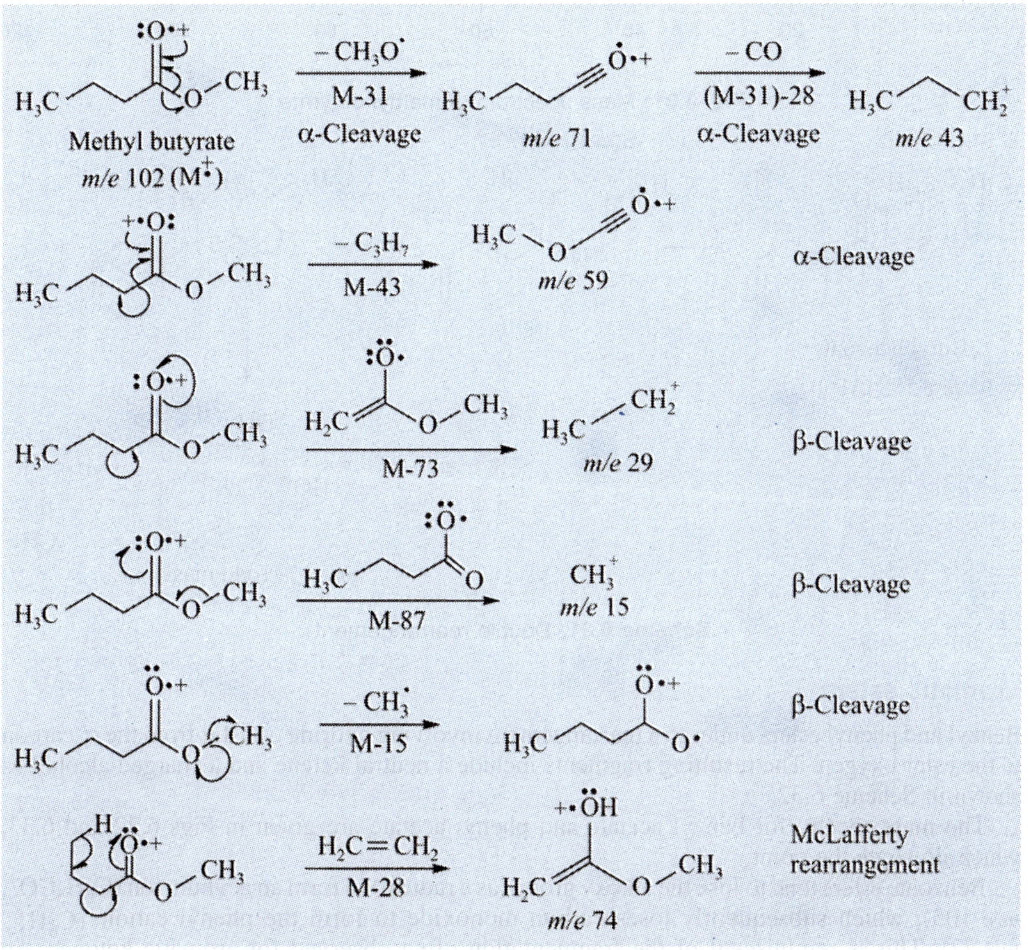

Scheme 6.30: Fragmentation pattern for methyl butyrate

The complete mass spectrum of methyl butyrate is given in Fig. 6.31, which illustrates most of these points. Ethyl, and particularly larger alkyl esters, do a double rearrangement—sometimes called a "double McLafferty rearrangement"—to make a protonated acid. The mass spectrum of butyl benzoate, for example, gives the abundant ion of m/e 123 arising from this double rearrangement (Scheme 6.31).

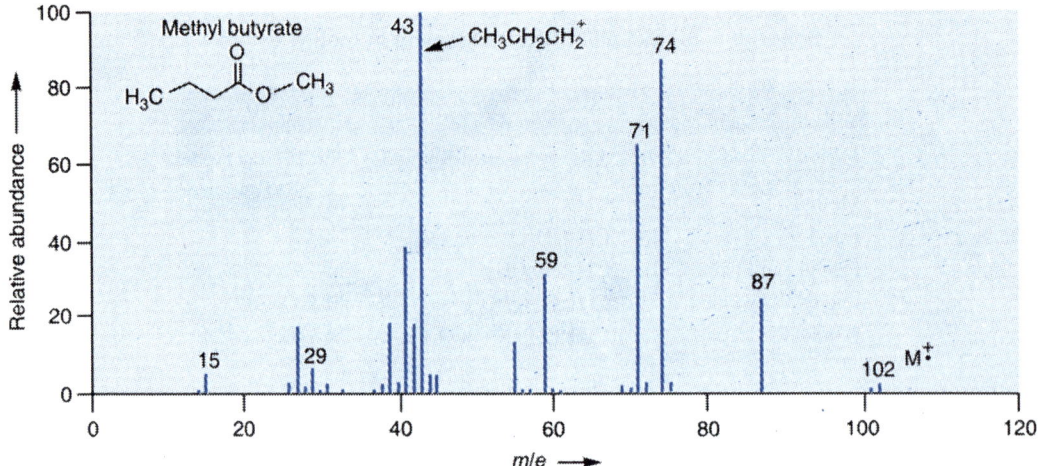

Fig. 6.31: Mass spectrum of methyl butyrate

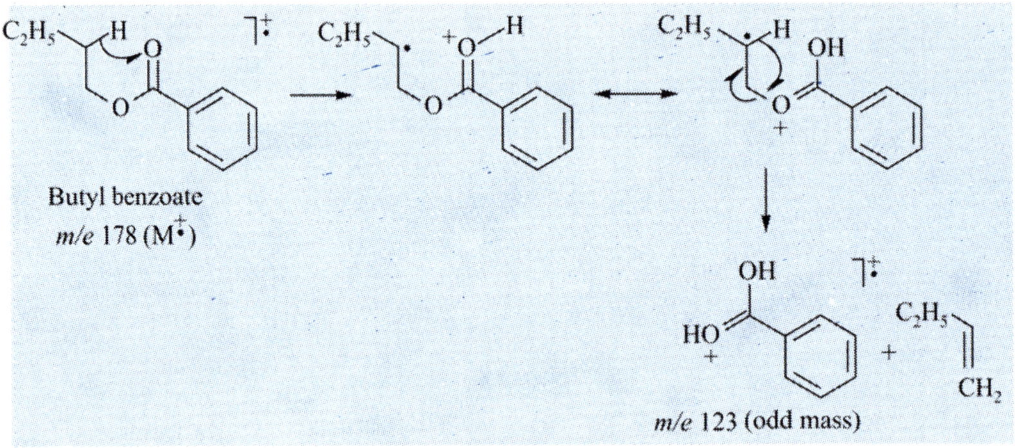

Scheme 6.31: Double rearrangement

romatic esters

Benzyl and phenyl esters undergo a rearrangement involving hydride transfer from the α-carbon to the ester oxygen. The resulting fragments include a neutral ketene and a charged alcohol as shown in Scheme 6.32.

The mass spectra for benzyl acetate and phenyl acetate are given in Figs 6.32 and 6.33, which illustrate the point.

Benzoate esters tend to lose the alkoxy group (as a radical) to form an acylium ion ($C_6H_5CO^+$, m/e 105), which subsequently loses carbon monoxide to form the phenyl cation ($C_6H_5^+$, m/e 77). This is reminiscent of the fragmentation often observed for aromatic ketones. An important exception to this pattern involves benzoate esters bearing alkyl, amino, or hydroxy

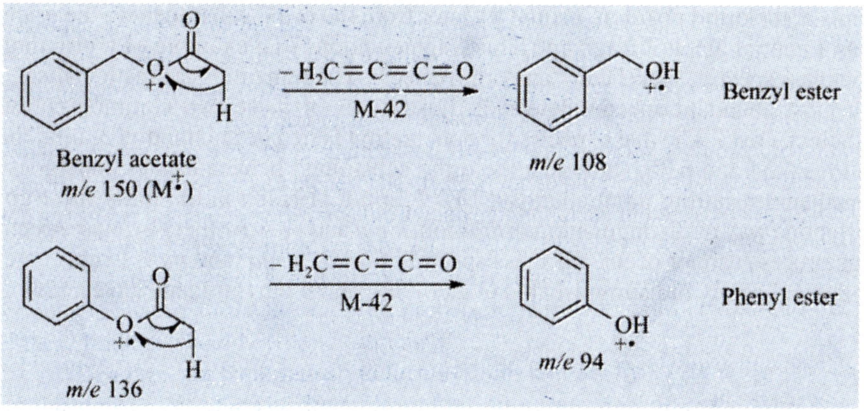

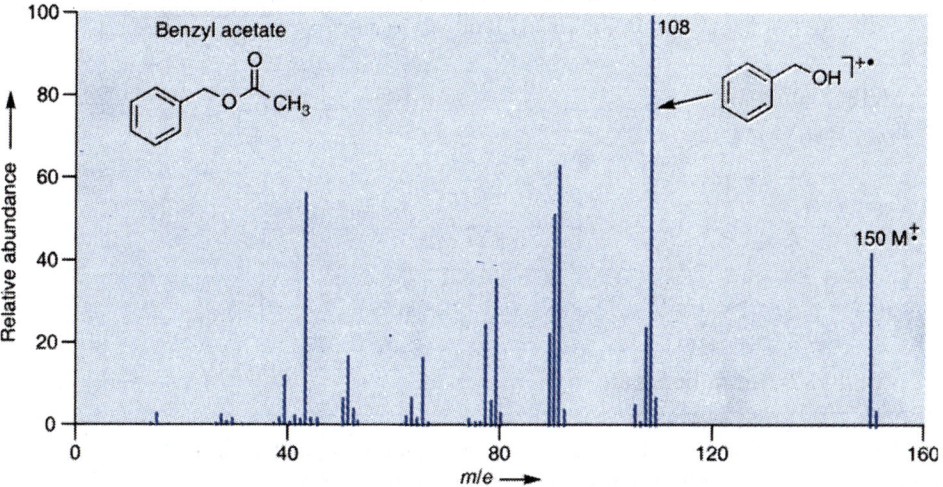

Scheme 6.32: Most common fragmentation involving benzyl and phenyl esters

Fig. 6.32: Mass spectrum of benzyl acetate

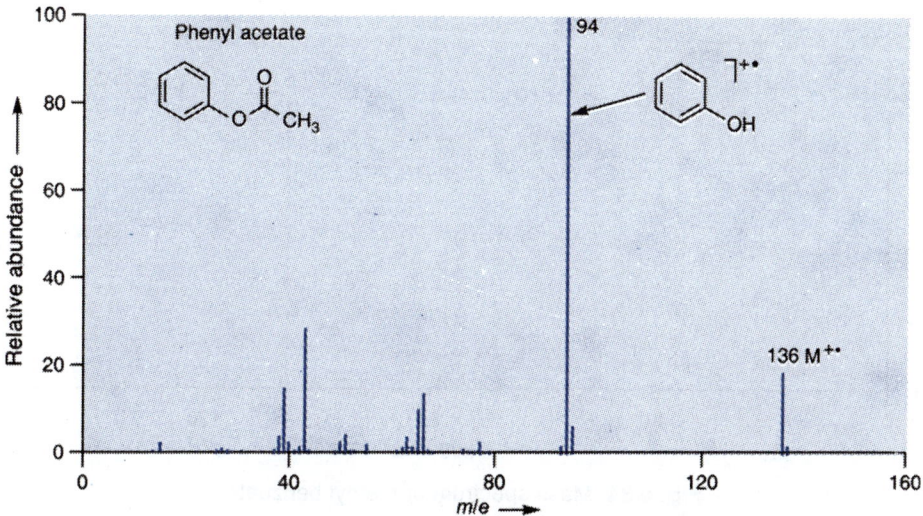

Fig. 6.33: Mass spectrum of phenyl acetate

substituents at the ortho position. Proton transfer from the ortho substituent to the ester oxygen eliminates a neutral alcohol fragment. This fragmentation is an example of a proximity effect in organic mass spectra, specifically an "ortho effect". The remaining aromatic radical cation is often the most abundant species detected. Illustrations of these two common pathways are given in Scheme 6.33. The mass spectra for both methyl benzoate and methyl 2-aminobenzoate, given in Figs 6.34 and 6.35, show that the ortho substituted benzoate ester undergoes a facile loss of methanol resulting in an abundant M − 32 (*m/e* 119) ion in the mass spectrum. Ortho effects offer one way of distinguishing *ortho* from *meta* and *para* isomers by mass spectrometry. When this process cannot occur, the mass spectra of *ortho*, *meta*, and *para* isomers are usually very similar and nearly indistinguishable (e.g., *o*-, *m*-, and *p*-ethyl toluene have nearly identical spectra).

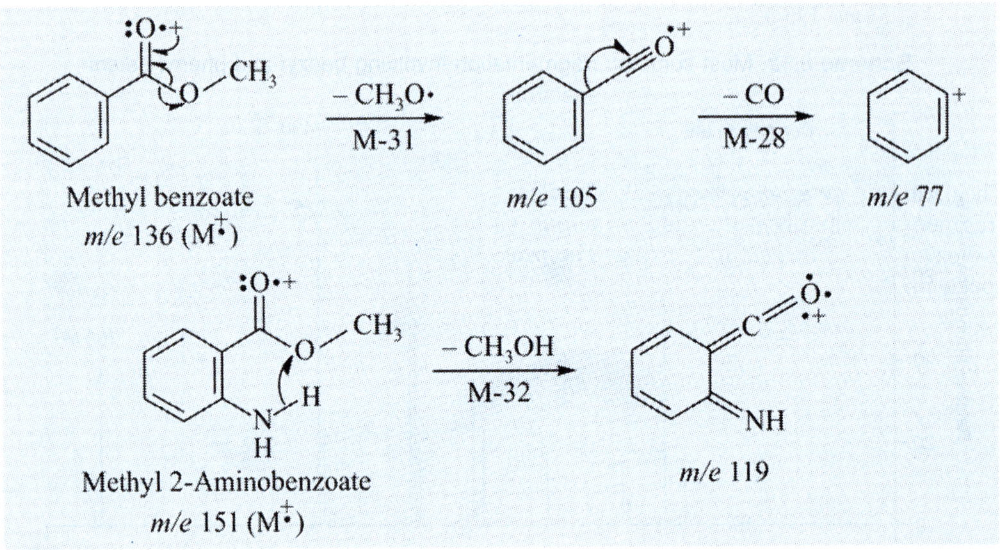

Scheme 6.33: Most common fragmentation involving benzoate and *ortho* substituted benzoate esters

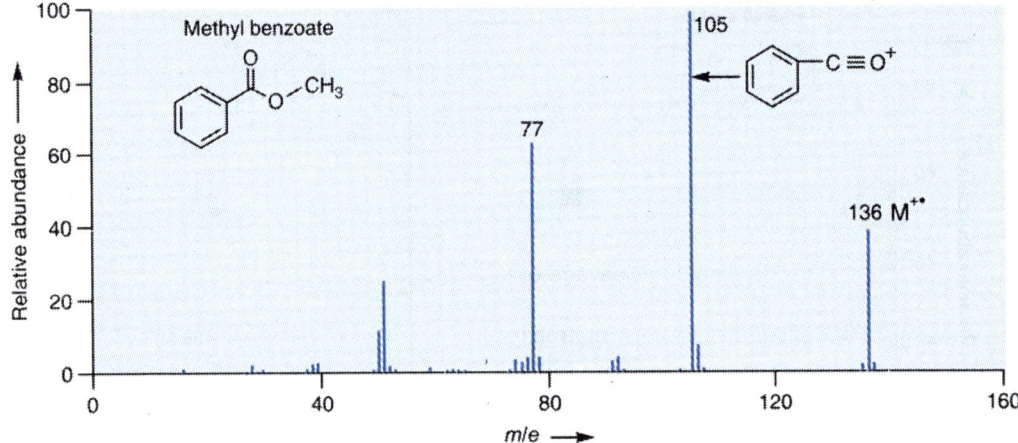

Fig. 6.34: Mass spectrum of methyl benzoate

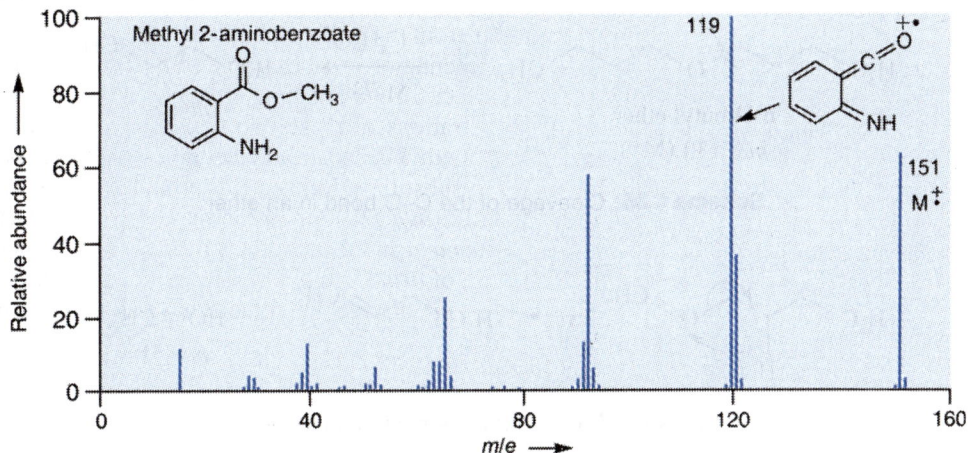

Fig. 6.35: Mass spectrum of methyl 2-aminobenzoate

LACTONES

The molecular ion peak is very weak. The cleavage adjacent to the oxygen atom is a characteristic fragmentation pattern of γ and δ lactone spectrum. There is loss of an alkyl radical by α-cleavage with respect to the ring. The peaks at m/e 42, 70 and 71 are typical of δ-lactone spectrum.

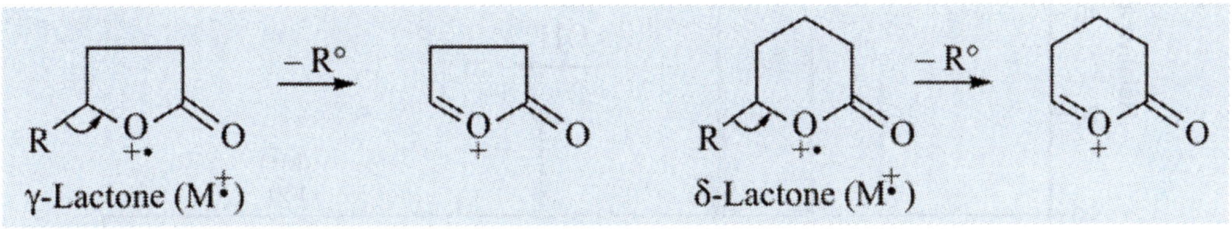

γ-Lactone ($M^{\dot{+}}$) δ-Lactone ($M^{\dot{+}}$)

ETHERS

Aliphatic ethers

The molecular ion peak is usually weak, however, larger than for alcohols. An important fragmentation process for ethers involves the cleavage of the C–C bond to the α-carbon as shown in Scheme 6.34.

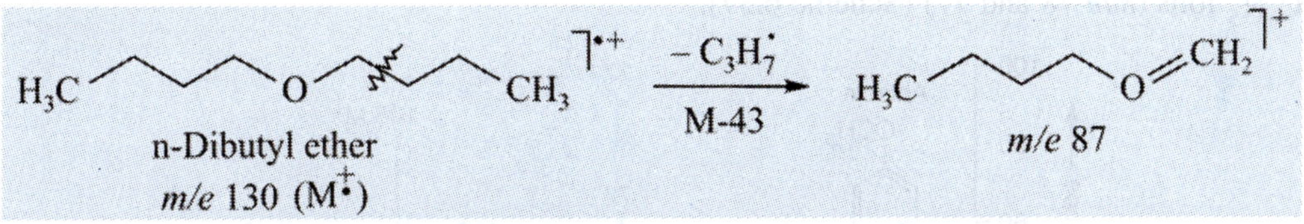

n-Dibutyl ether
m/e 130 ($M^{\dot{+}}$)

$-C_3H_7^{\cdot}$
$M-43$

m/e 87

Scheme 6.34: Cleavage of the C–C bond to the α-carbon

A second common mode of fragmentation involves cleavage of the C–O bond (Scheme 6.35). Hydride transfer from a β-carbon is an important rearrangement process in ethers as shown in Scheme 6.36.

Peaks usually occur in $C_nH_{2n}OH^+$ increments for ethers. The complete mass spectrum for n-butyl ether is given in Fig. 6.36, which illustrates most of these points.

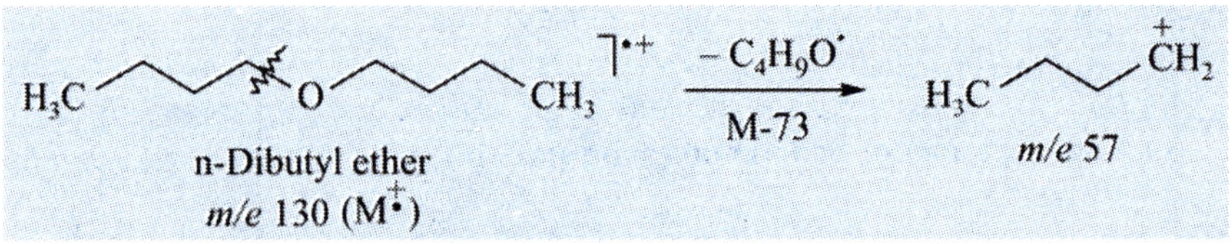

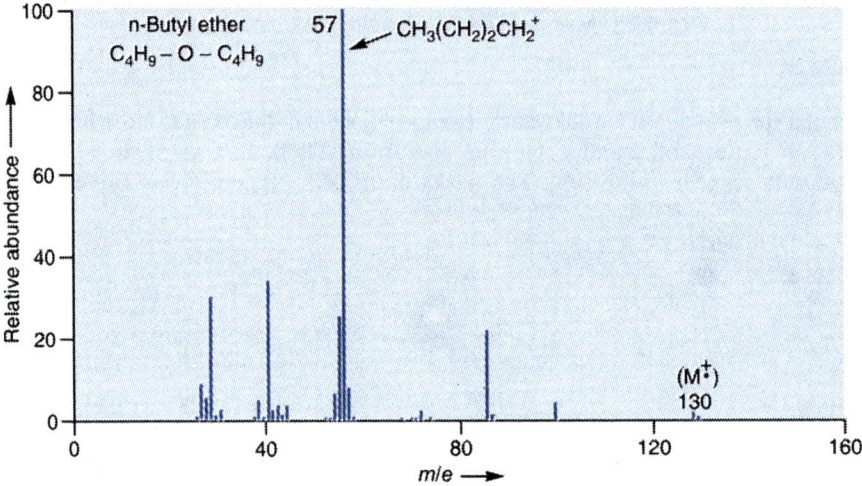

Scheme 6.35: Cleavage of the C–O bond in an ether

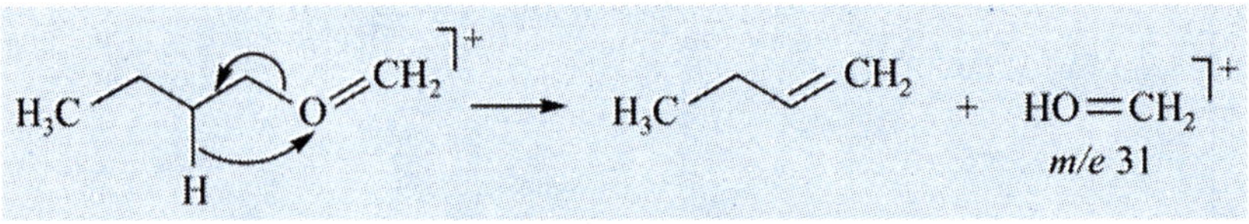

Scheme 6.36: Rearrangement in an ether

Fig. 6.36: Mass spectrum for *n*-butyl ether

Aromatic ethers

Aromatic ethers (phenyl methyl ether; anisole, Fig. 6.37) can fragment to lose the alkyl group, giving the $C_6H_5O^+$ ion (*m/e* 93). Additionally, the alkoxy group can be lost to give $C_6H_6^+$ and $C_6H_5^+$ ions (*m/e* 78 and 77) (Scheme 6.37).

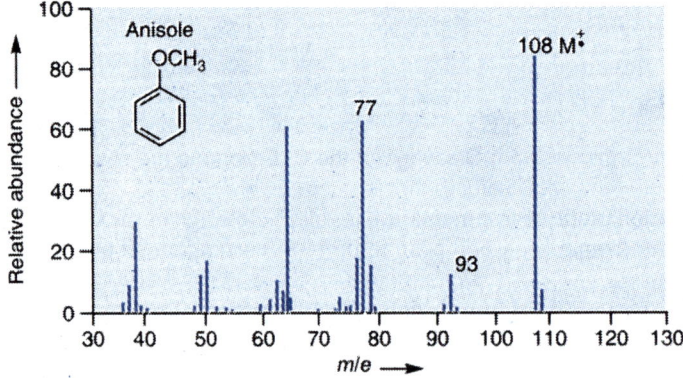

Fig. 6.37: Mass spectrum of anisole

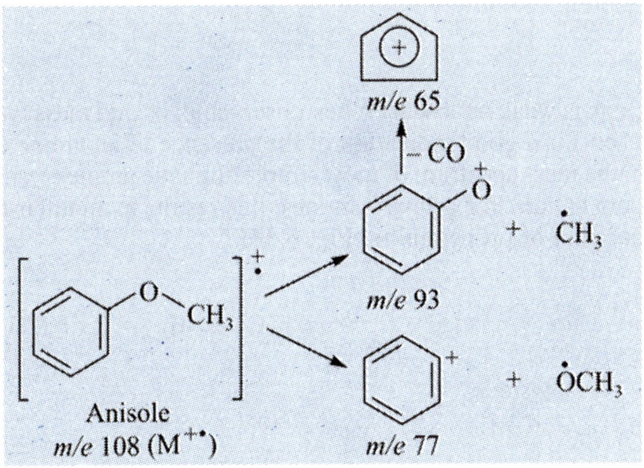

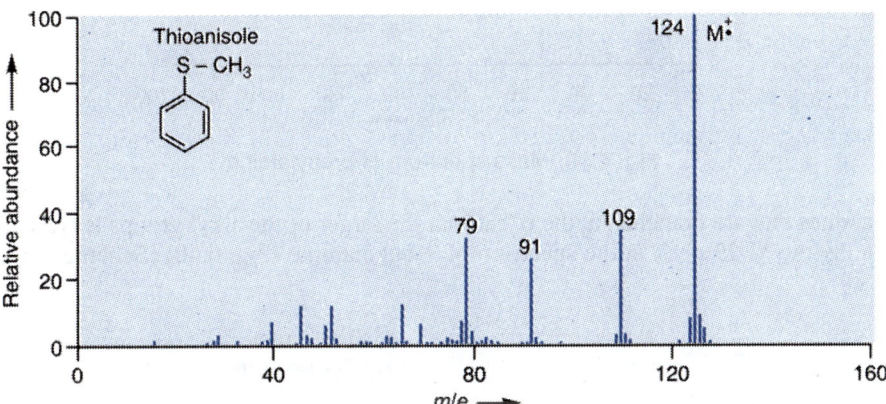

Scheme 6.37: Fragmentation pattern for anisole

Thioether

The molecular ions of thioether are stable in comparison to aliphatic ether. For example, thioanisole (Fig. 6.38) fragments with the loss of SH radical to give intense peak at *m/e* 91 (Scheme 6.38).

Fig. 6.38: Mass spectrum of thioanisole

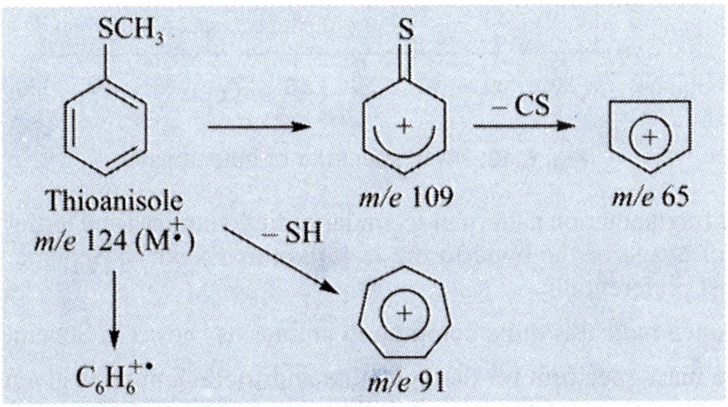

Scheme 6.38: Fragmentation pattern for thioanisole

AMINES

Aliphatic amines

The molecular ion peak is weak or absent. When observable, its odd mass (when an odd number of nitrogens are present) is a good indication of the presence of an amine (nitrogen rule). The most intense peak in the mass spectrum of most simple aliphatic amines results from α-cleavage to give the ammonium cation. For primary amines, this results in an intense peak at m/e 30, as seen in the mass spectrum of propylamine (Fig. 6.39).

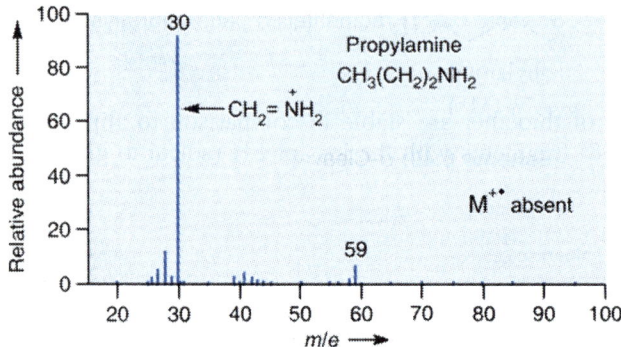

Fig. 6.39: Mass spectrum of propylamine

For amines that are branched at the α-carbon, the larger of the alkyl groups is typically lost, as shown by the M-29 peak in the spectrum of 2-butanamine (Fig. 6.40) (Scheme 6.39).

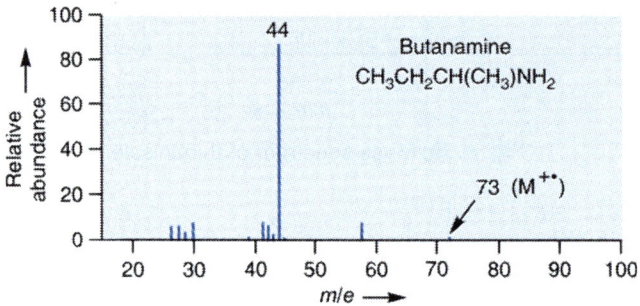

Fig. 6.40: Mass spectrum of butanamine

An important fragmentation pattern in secondary (diethylamine) and tertiary amine (triethyl-amine) involve cleavage of the bond to the α-carbon as shown in Scheme 6.40. The largest alkyl group is lost preferentially.

Loss of hydrogen radical is quite common in amines as shown in Scheme 6.41.

The complete mass spectrum for diethylamine and triethylamine is given in Figs 6.41 and 6.42, which illustrate most of these points. The secondary and tertiary amines also show rearrangement as per Scheme 6.42.

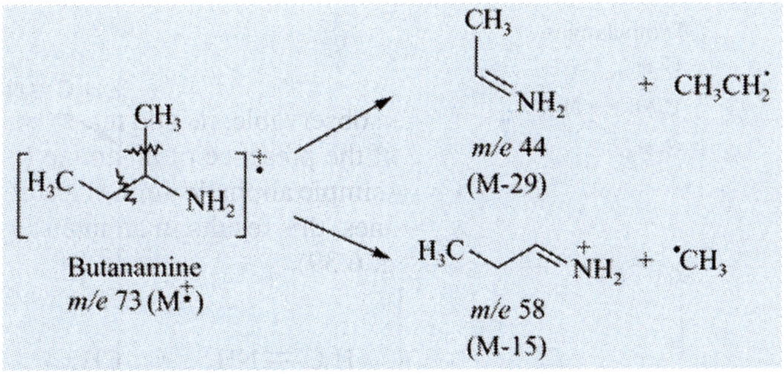

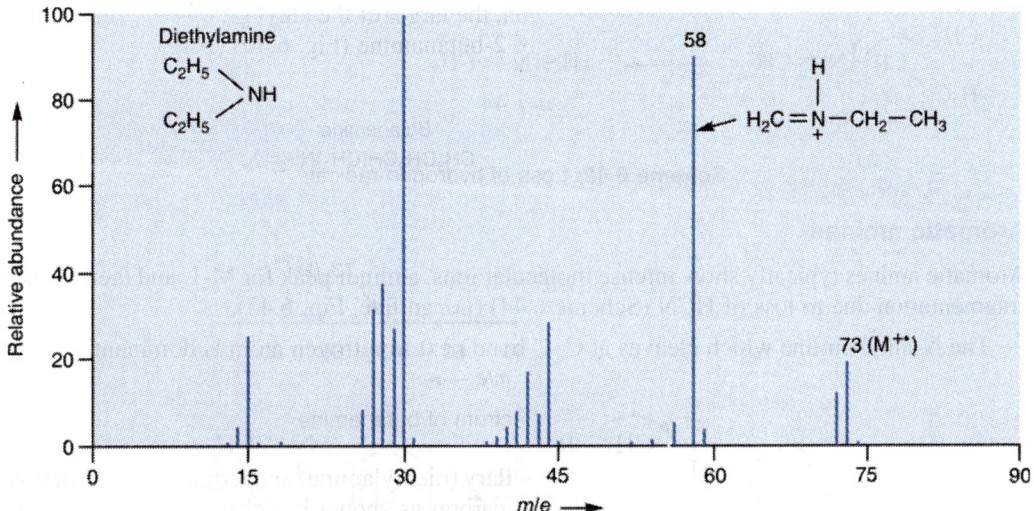

Scheme 6.39: Mass fragmentation pattern for butanamine

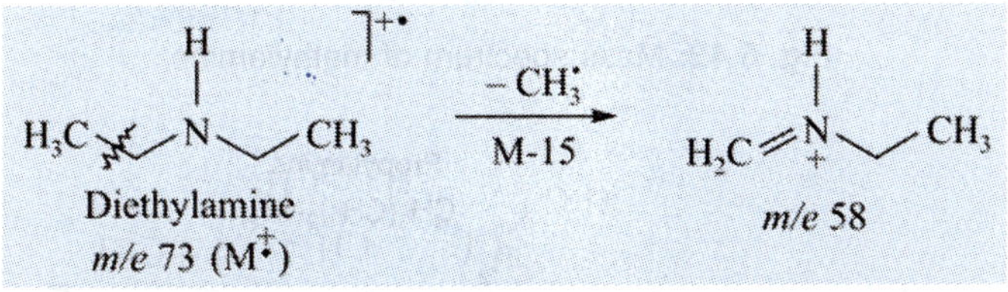

Scheme 6.40: α-Cleavage in secondary amine

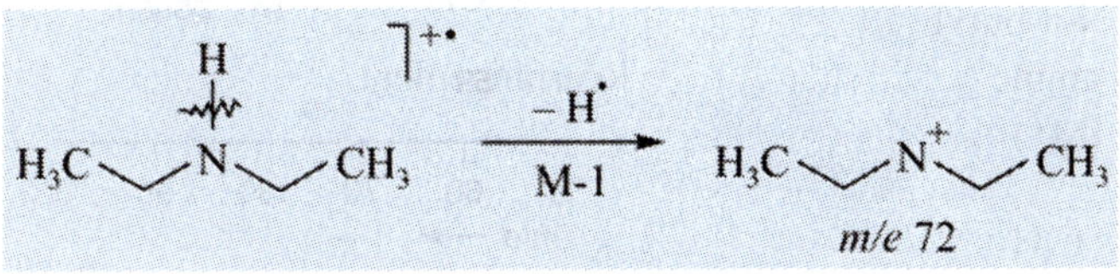

Scheme 6.41: Loss of hydrogen radical

Fig. 6.41: Mass spectrum for diethylamine

Cyclic aliphatic amines produce molecular ion peaks. The loss of hydrogen from the α-carbon is also a prominent peak. The ring is cleaved when the α-bond is broken and subsequent alkene molecules fragment from the remaining ring structure. Their principal modes of fragmentation are shown in Scheme 6.43.

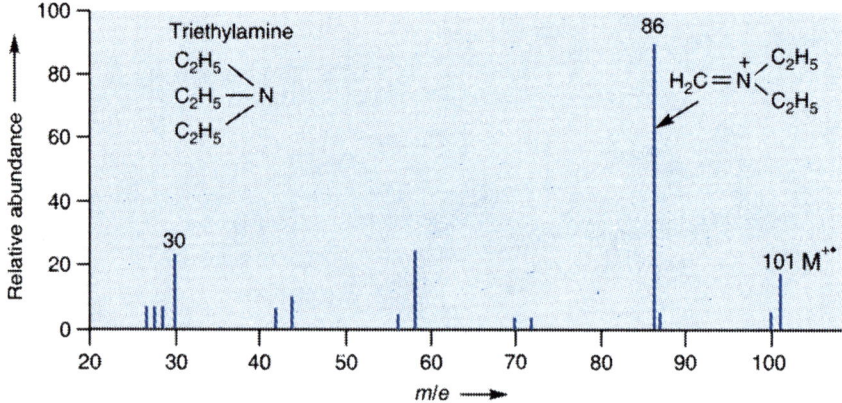

Fig. 6.42: Mass spectrum of triethylamine

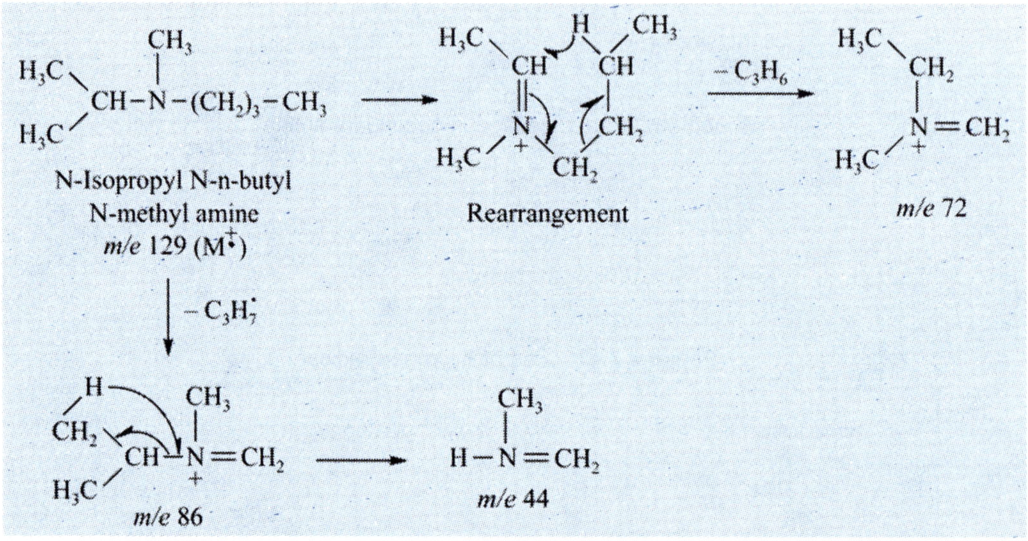

Scheme 6.42: Loss of hydrogen radical

Aromatic amines

Aromatic amines typically show intense molecular ions, a minor peak for M-1, and then further fragmentation due to loss of HCN (Scheme 6.44) (i.e. aniline, Fig. 6.43).

The N-alkyl aniline which cleaves at C–C bond next to nitrogen atom is dominant.

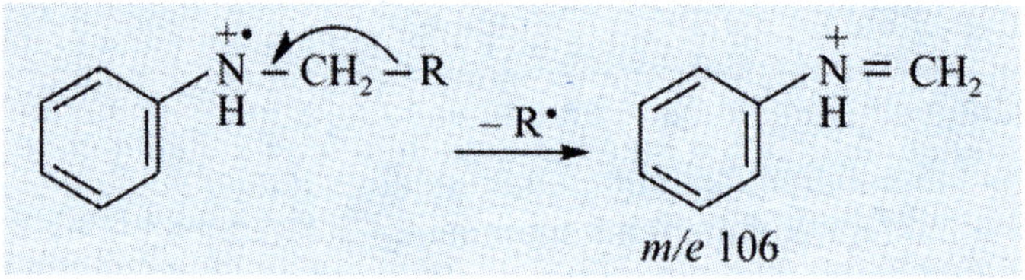

m/e 106

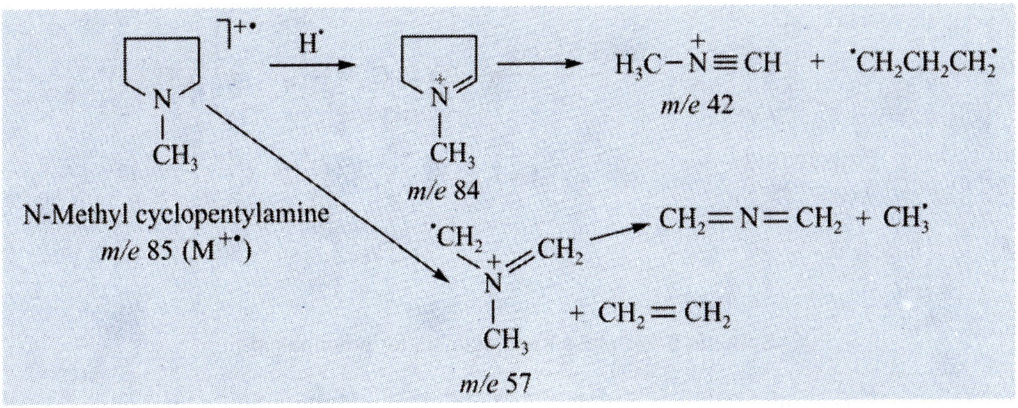

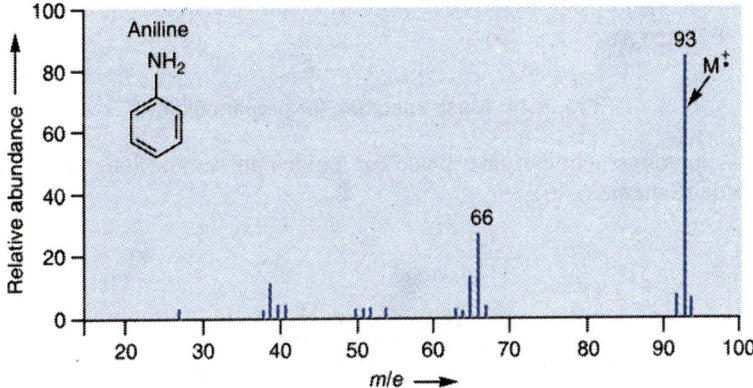

Scheme 6.43: Fragmentation pattern for N-methyl cyclopentylamine

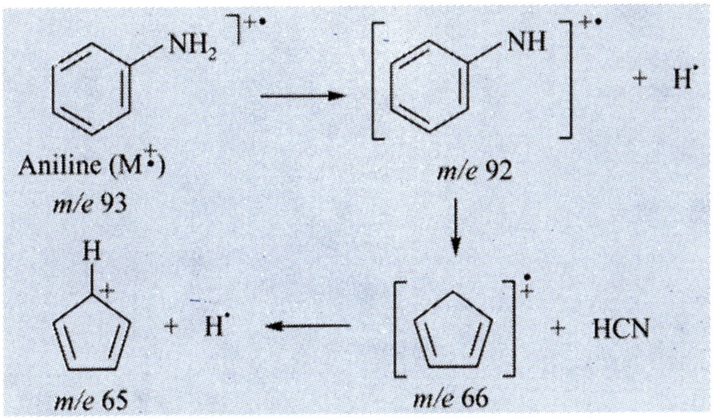

Scheme 6.44: Mass fragmentation for aniline

Fig. 6.43: Mass spectrum of aniline

AMIDE

Aliphatic amides

The molecular ion peak is usually observable, and is a good indication of the presence of an amide (nitrogen rule). An important fragmentation pattern involves α-cleavage (breaking either bond to the carbonyl carbon) as shown in Scheme 6.45.

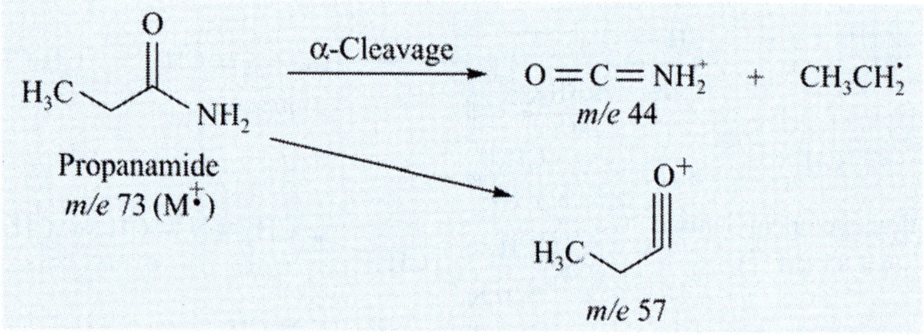

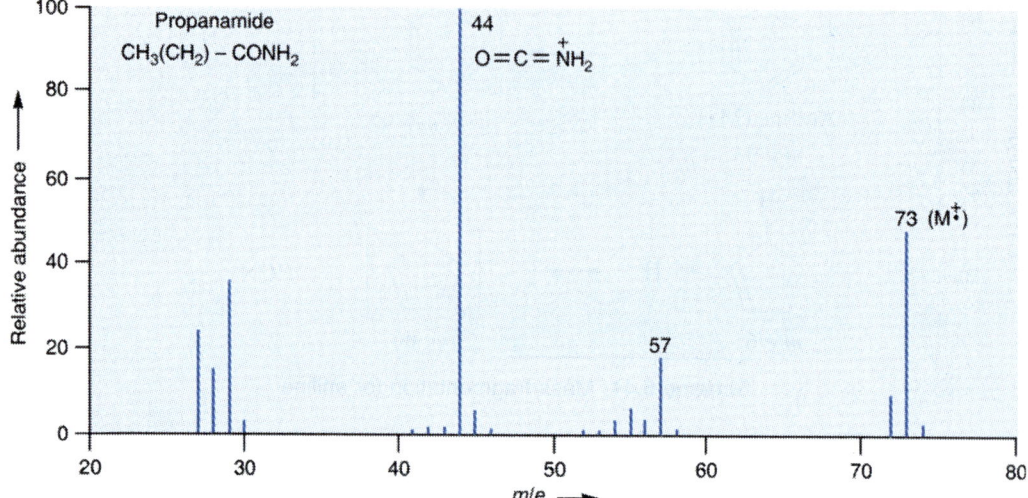

Scheme 6.45: Mass fragmentation for propanamide

The mass spectrum of simple primary amides typically shows a moderate molecular ion and an intense peak at *m/e* 44 due to loss of an alkyl radical (propanamide, Fig. 6.44). It also exhibits loss of amino group to give peak at *m/e* 57.

Fig. 6.44: Mass spectrum for propanamide

McLafferty rearrangement can take place for amides possessing long chain alkyl portion, e.g. pentenamide (Scheme 6.46).

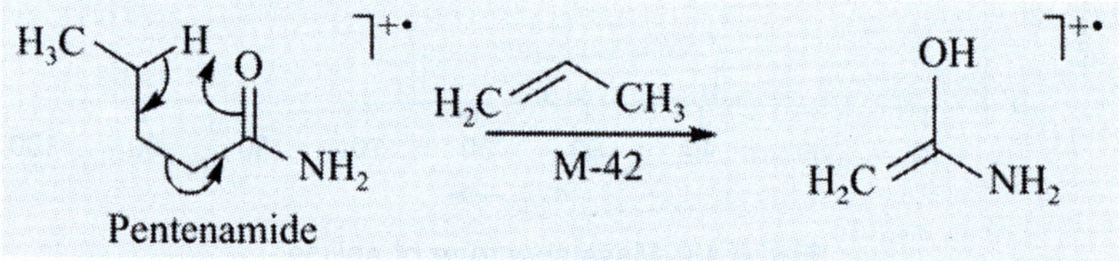

Scheme. 6.46: McLafferty rearrangement of an amide

Aromatic amides

The aromatic amides show intense molecular ion peak. Like aliphatic amides, all show important fragmentation pattern involves α-cleavage (breaking either bond to the carbonyl carbon) as shown in Scheme 6.47 with the example of benzamide (Fig. 6.45).

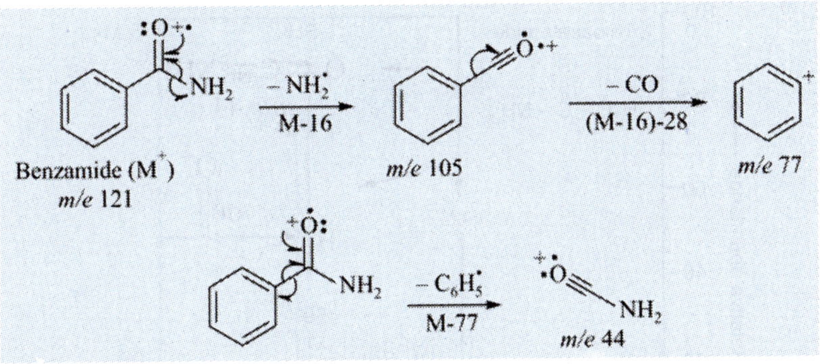

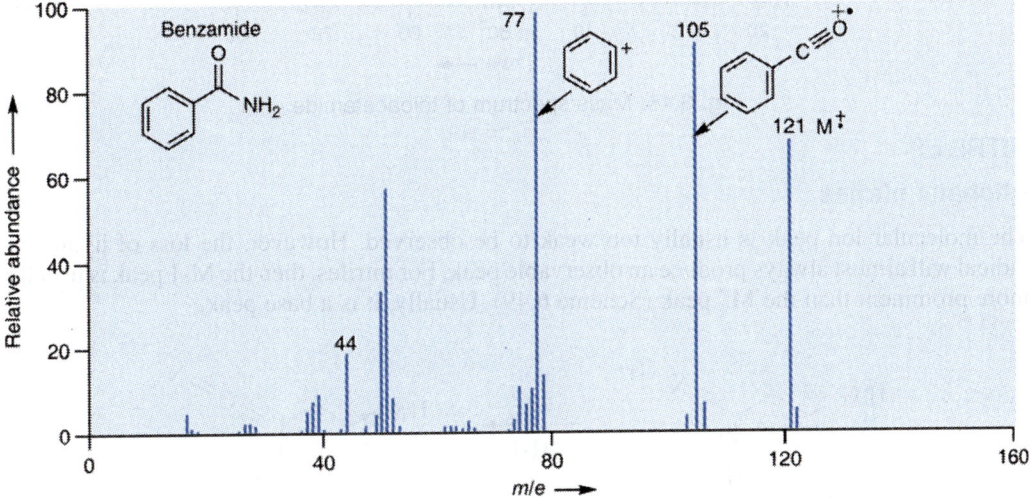

Scheme 6.47: Fragmentation pattern for benzamide

Fig. 6.45: Mass spectrum of benzamide

Thioamides

The mass spectrum of thioamides, e.g. thioacetamide shows strong molecular ion peak. It exhibits the peak at m/e 60 and m/e 59, corresponding to loss of methyl radical and amino radical from the molecular ion respectively. But in addition to thioacetamide contains peak at m/e 42 due to loss an M–$^{\bullet}$SH ion (Fig. 6.46 and Scheme 6.48).

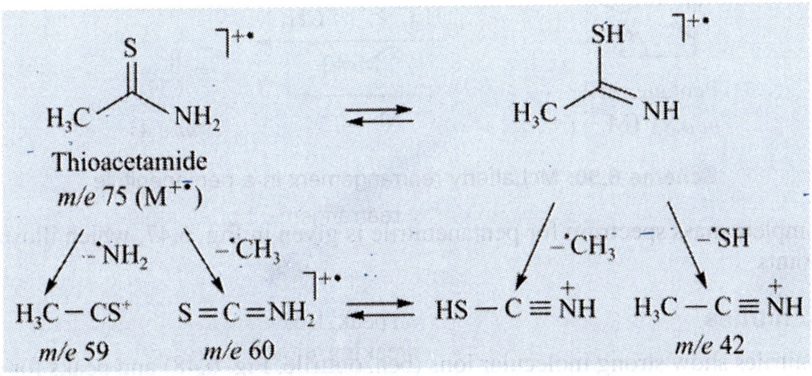

Scheme 6.48: Mass fragmentation for thioacetamide

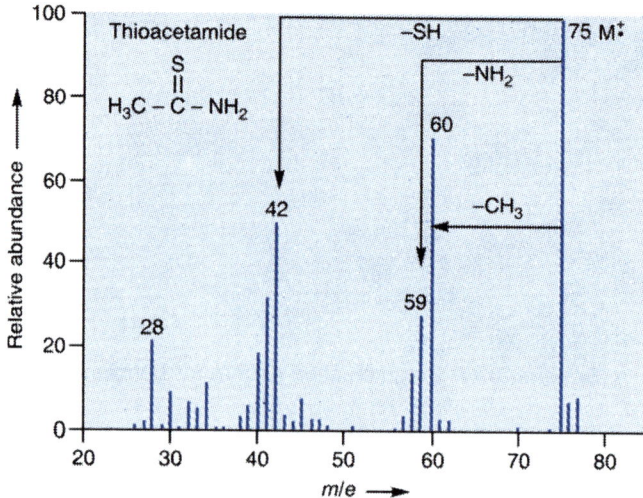

Fig. 6.46: Mass spectrum of thioacetamide

NITRILES

Aliphatic nitriles

The molecular ion peak is usually too weak to be observed. However, the loss of hydrogen radical will almost always produce an observable peak. For nitriles, then the M-1 peak is usually more prominent than the M^+ peak (Scheme 6.49). Usually, it is a base peak.

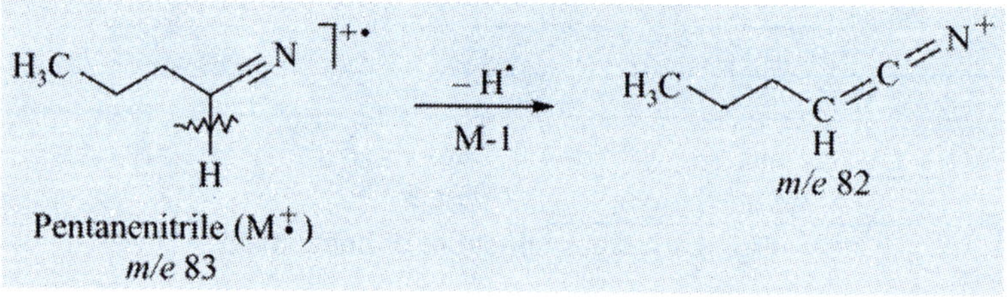

Scheme 6.49: Loss of hydrogen radical

McLafferty rearrangement can take place for nitriles (Scheme 6.50).

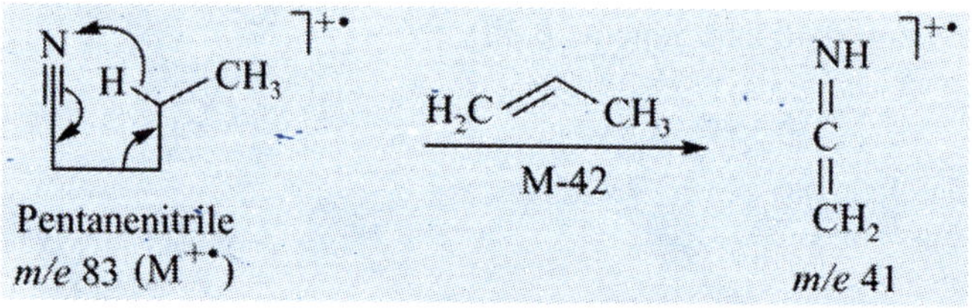

Scheme 6.50: McLafferty rearrangement in a pentanenitrile

The complete mass spectrum for pentanenitrile is given in Fig. 6.47, which illustrates most of these points.

Aromatic nitriles

Aromatic nitriles show strong molecular ions (benzonitrile, Fig. 6.48) and peaks for the loss of cyanide and the elements of HCN (*m/e* 77 and 76, respectively).

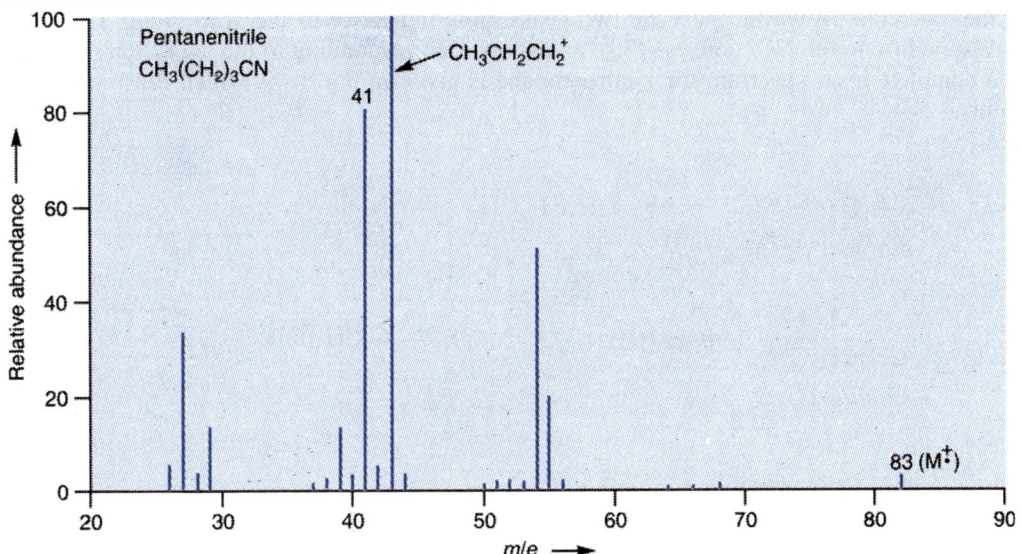

Fig. 6.47: Mass spectrum for pentanenitrile

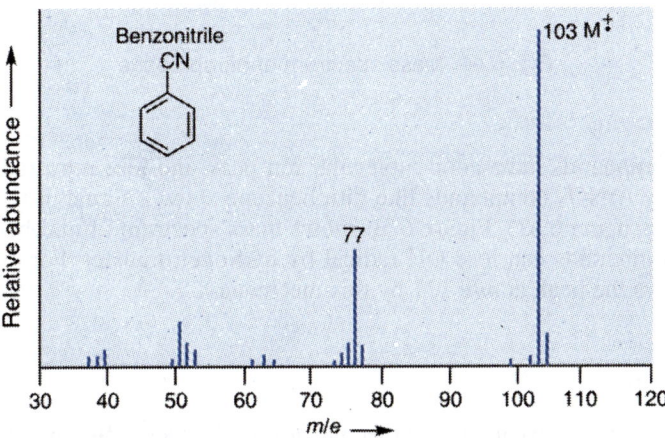

Fig. 6.48: Mass spectrum of benzonitrile

NITRO COMPOUNDS

Aliphatic nitro compounds

The molecular ion peak for aliphatic nitro compounds is seldom observed. The mass spectrum observed for aliphatic nitro compounds is usually due to the fragmentation of the alkyl portion of the molecule. However, there are two peaks quite indicative of the nitro group: One peak corresponding to the NO⁺ ion (*m/e* 30), and another corresponding to the NO_2^+ ion (*m/e* 46). The complete mass spectrum for 1-nitropropane is given in Fig. 6.49, which illustrates these points.

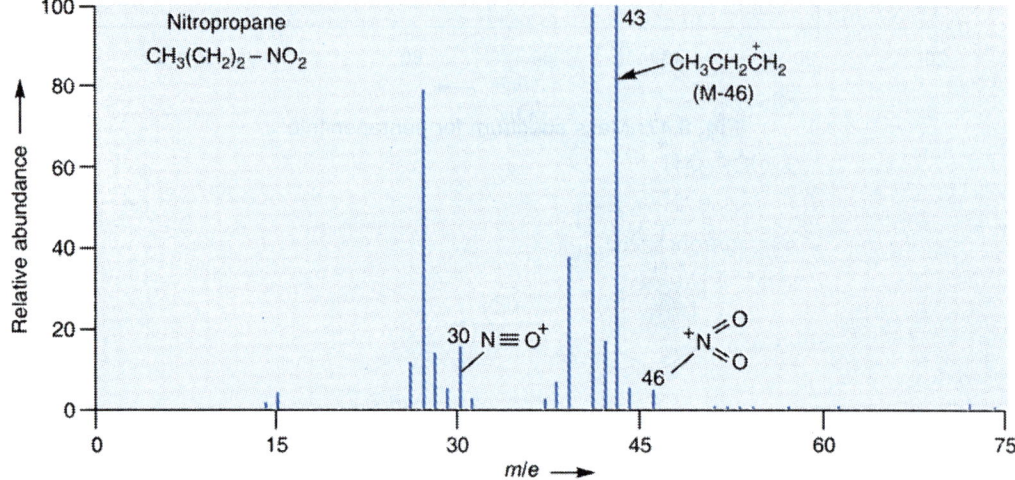

Fig. 6.49: Mass spectrum of nitropropane

Aromatic nitro compounds

Aromatic nitro compounds have good molecular ion peak and lose nitro radical to show the peak of Ar⁺. Many $ArNO_2$ compounds like nitrobenzene show a signal of $(M – O)^+$ which is quite distinctive (Scheme 6.51). Figure 6.50 shows mass spectrum of nitrobenzene.

Ortho nitro compounds can lose OH radical by hydrogen transfer. For example, 2-amino nitrobenzene shows the peak at *m/e* 121 by this mechanism.

EPOXIDES

Aliphatic epoxides

The molecular ion peak is weak. The molecular ion undergoes γ fission with respect to the epoxide ring giving an intense peak.

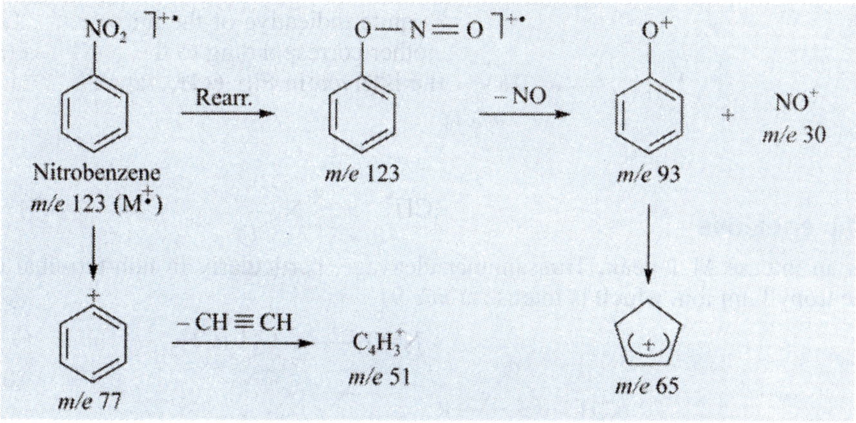

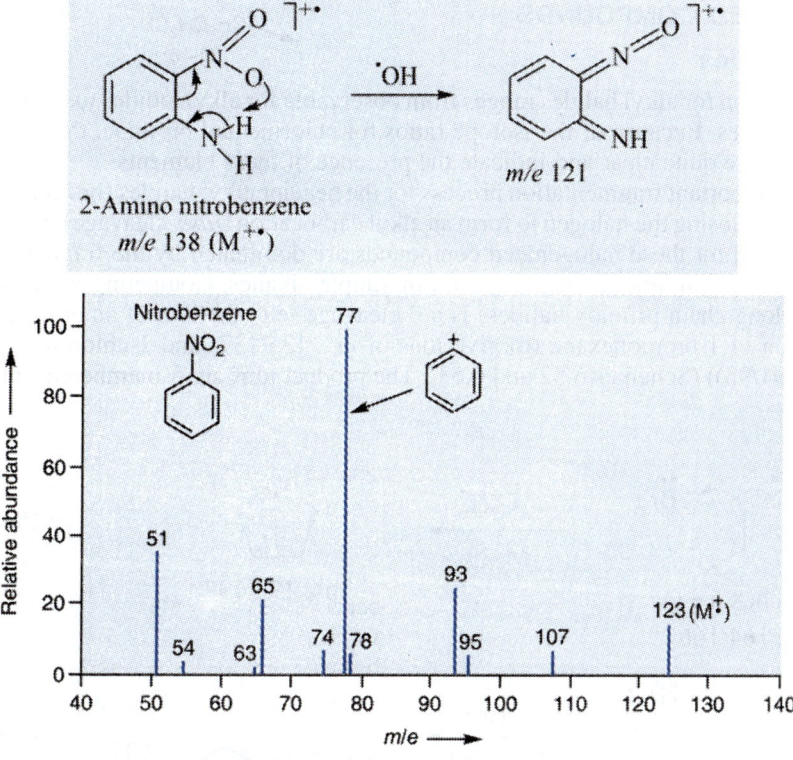

Scheme 6.51: Fragmentation pattern for nitrobenzene

Fig. 6.50: Mass spectrum of nitrobenzene

There can be two types of McLafferty rearrangement, with respect to suitably substituted epoxides.

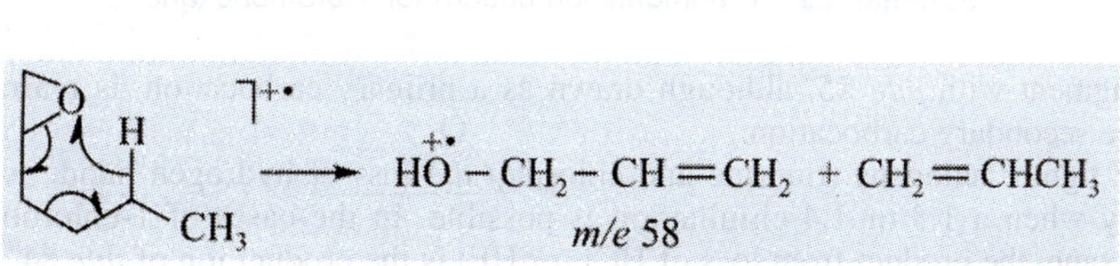

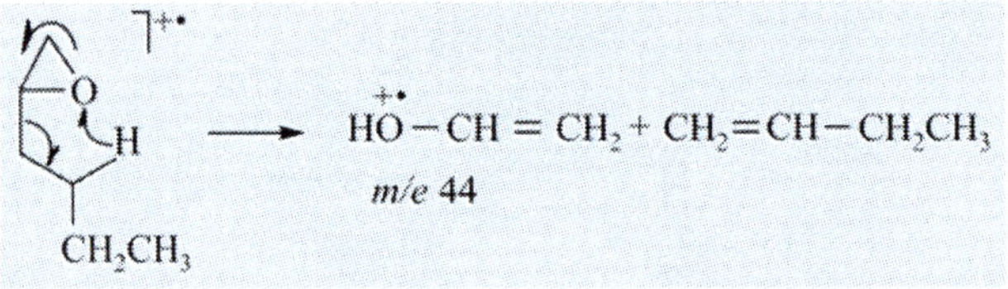

$$\text{[epoxide]}^{+\bullet} \longrightarrow \overset{+\bullet}{HO}-CH=CH_2 + CH_2=CH-CH_2CH_3$$

$$m/e \ 44$$

Aromatic epoxides

There is an intense M-1 peak. Transannular cleavage, particularly in non-terminal epoxide, gives the tropyllium ion, which is intense at m/e 91.

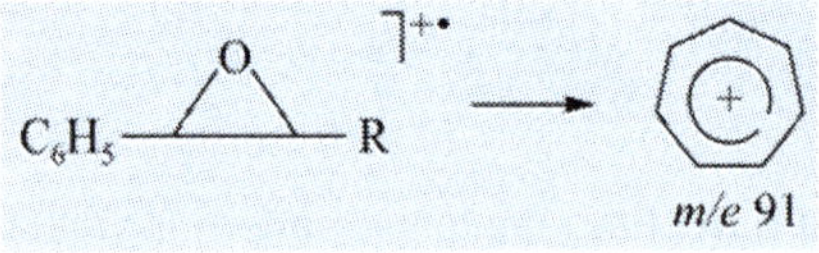

$$[C_6H_5\text{-epoxide-}R]^{+\bullet} \longrightarrow \text{[tropyllium]}^{+}$$

$$m/e \ 91$$

HALOGENATED COMPOUNDS

Aliphatic halides

The molecular ion for alkyl halides ranges from observable for alkyl iodides to barely detectable for alkyl fluorides. Because of the isotope ratios for chlorine and bromine, the M^+ and $M + 2$ observed ions are quite clear and indicate the presence of these elements.

The most important fragmentation process for the heavier alkyl halides (bromine and iodine) involves simply losing the halogen to form an alkyl carbocation (*ipso*-cleavage). For this reason, the mass spectra for these halogenated compounds are dominated by the fragmentation of an alkyl ion and, thus, mimic the mass spectra of simple alkanes. Competing with this loss (for sufficiently long chain primary halides) is a d cleavage (e.g. the loss of an ethyl group in the fragmentation of 1-bromohexane (to give ions of m/e 135/137) and 1-chlorohexane (to give ions of m/e 91/93)) (Schemes 6.52 and 6.53). The product ions are 5-membered ring halonium ions.

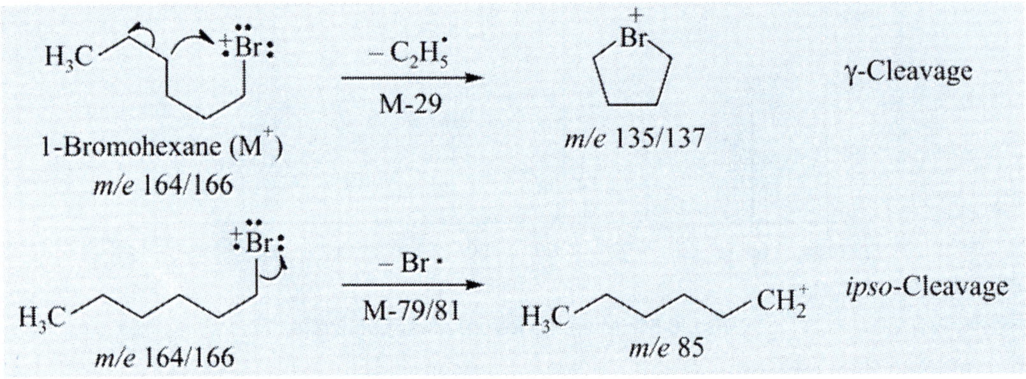

1-Bromohexane (M^+)
m/e 164/166

$-C_2H_5$
M-29

m/e 135/137

γ-Cleavage

m/e 164/166

$-Br\bullet$
M-79/81

m/e 85

ipso-Cleavage

Scheme 6.52: Fragmentation pattern for 1-bromohexane

The fragment with m/e 85, although drawn as a primary carbocation, is rearranged to a more stable secondary carbocation.

For the lighter halogens (fluorine and chlorine) the loss of hydrogen halide is important particularly when a 1,3 or 1,4-elimination is possible. In the cases of 1-chlorohexane and 1-fluorohexane, the product from loss of HCl (or HF) is the product ion of m/e 84.

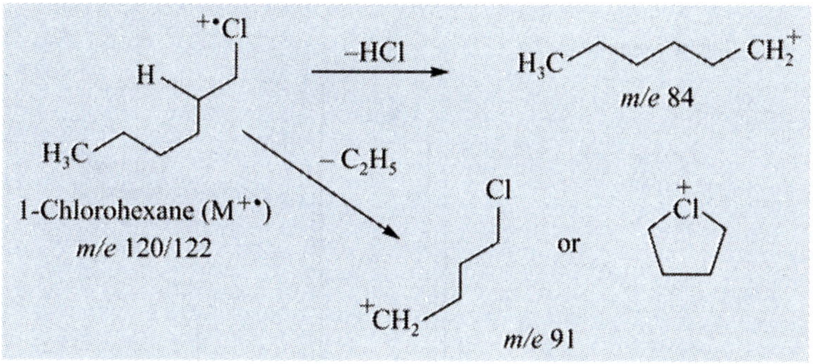

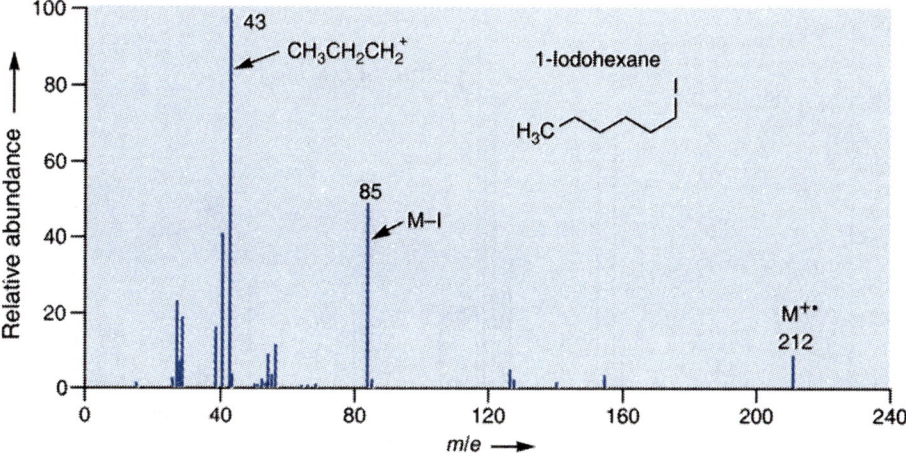

Scheme 6.53: Fragmentation pattern for 1-chlorohexane

The complete mass spectra of 1-iodohexane, 1-bromohexane, 1-chlorohexane, and 1-fluorohexane are given in Figs 6.51 to 6.54, respectively.

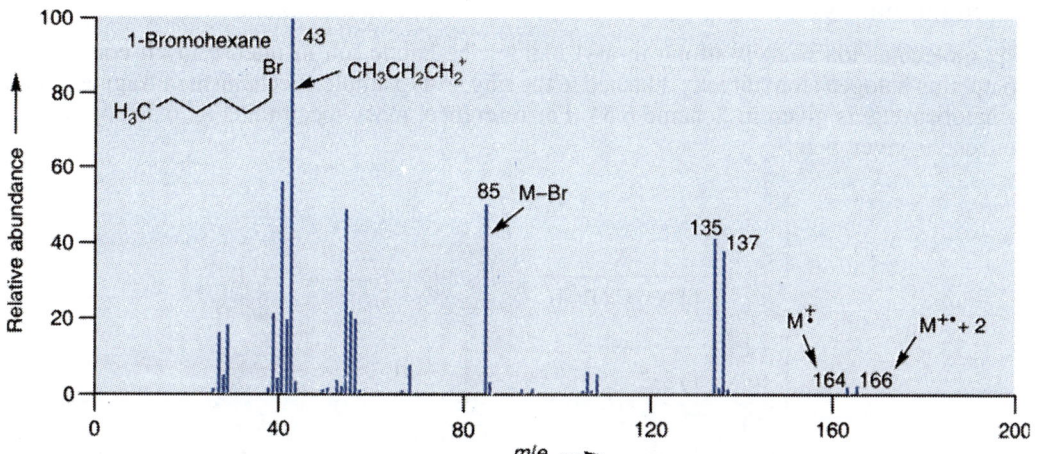

Fig. 6.51: Mass spectrum of 1-iodohexane

Fig. 6.52: Mass spectrum of 1-bromohexane

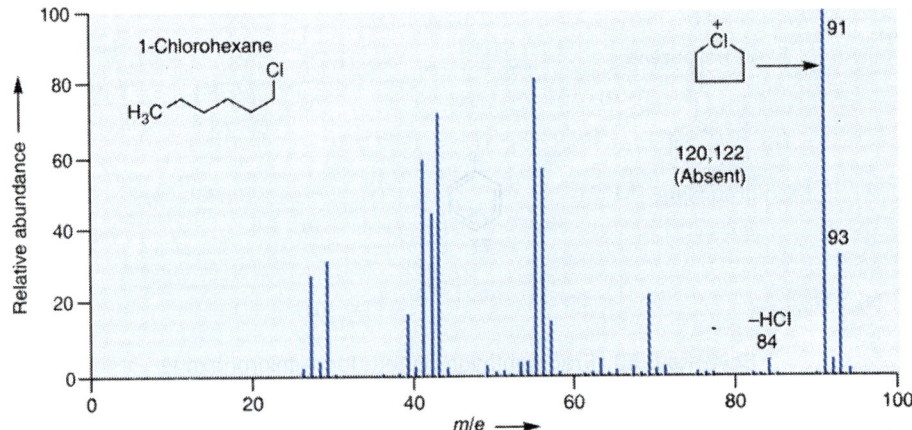

Fig. 6.53: Mass spectrum of 1-chlorohexane

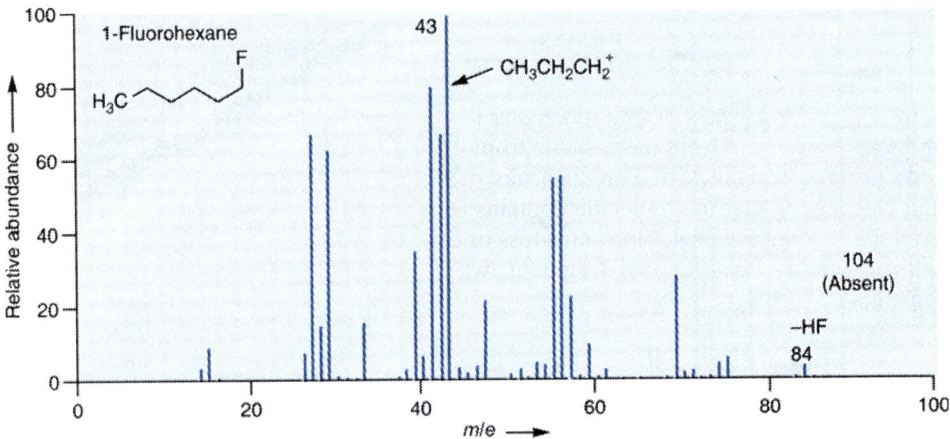

Fig. 6.54: Mass spectrum of 1-fluorohexane

Aromatic halides

The molecular ion peak is strong in aryl halides. M-halide ion in intense in all compounds containing halogen atom directly attached to the ring. For example, mechanism of fragmentation of halobenzene is given in Scheme 6.54. For reference, mass spectrum (Fig. 6.55) of chlorobenzene is given here.

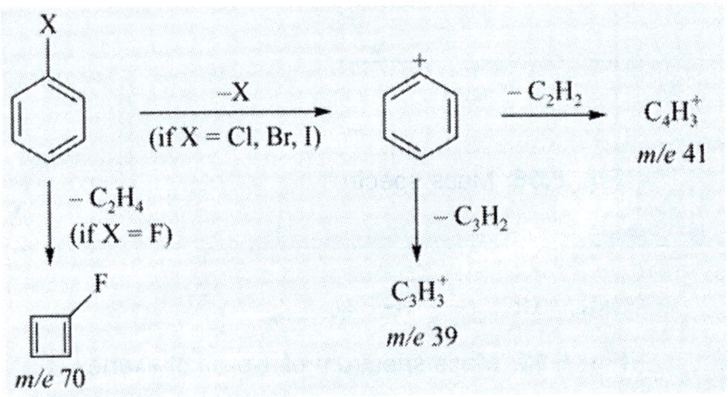

Scheme 6.54: Fragmentation pattern for aromatic halides

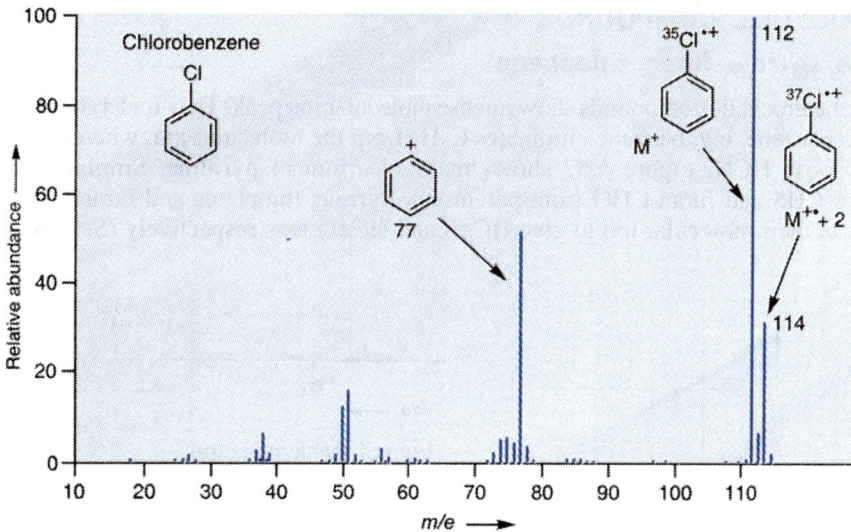

Fig. 6.55: Mass spectrum of chlorobenzene

Benzyl halide ions show strong molecular ion peak and also lose halide ion to form benzyl cation which changes to form more stable tropyllium ion. For example, the spectrum of benzyl bromide (Fig. 6.56) shows two small peaks of equal intensity in the molecular ion region, strongly suggesting that the molecule contains bromine (equal concentrations of the ^{79}Br and ^{81}Br isotopes). The base peak represents loss of this bromine to give the peak at *m/e* 91, which is highly suggestive of a benzyl fragment, which rearranges to form the tropylium cation (Scheme 6.55).

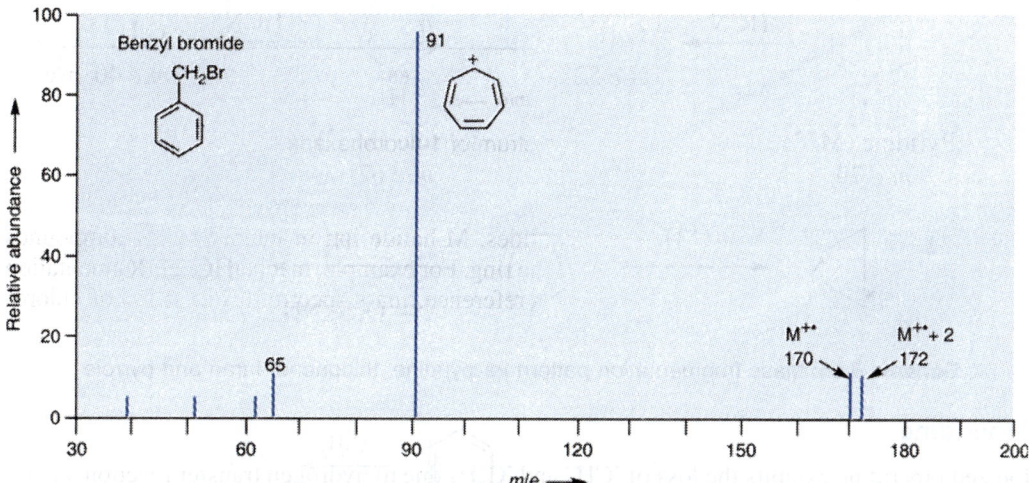

Fig. 6.56: Mass spectrum of benzyl bromide

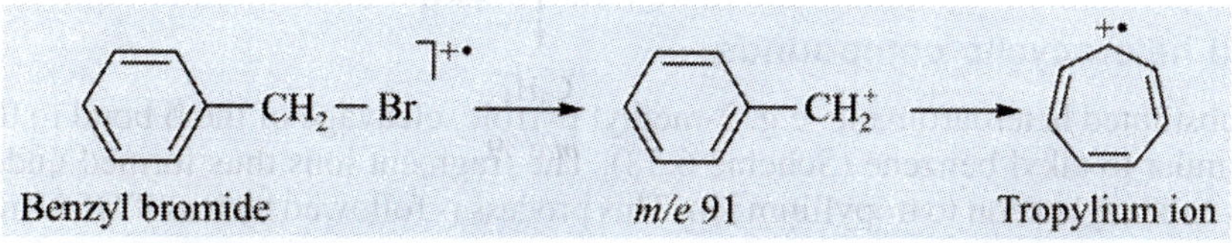

Scheme 6.55: Mass fragmentation pattern for benzyl bromide

HETEROCYCLIC COMPOUNDS

Pyridine, pyrrole, furan, thiophene

Aromatic heterocyclic compounds show intense molecular ion peak. They undergo fragmentation similar to benzene, e.g. benzene eliminates C_2H_2 from the molecular ion, whereas pyrrole and pyridine loose HCN. Figure 6.57 shows mass spectrum of pyridine. Similarly, thiophene eliminates CHS and furan CHO from parent ion. Pyrrole, thiophene and furan also eliminate $C_3H_3^•$ from their molecular ion to give $HC\equiv S$ and $HC\equiv O$ ions respectively (Scheme 6.56).

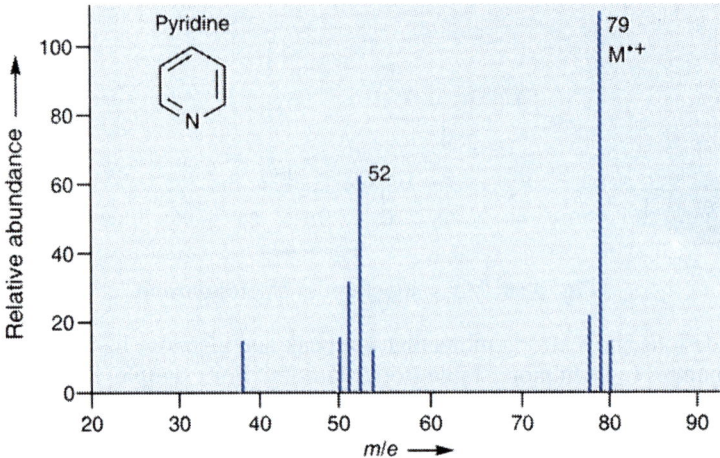

Fig. 6.57: Mass spectrum of pyridine

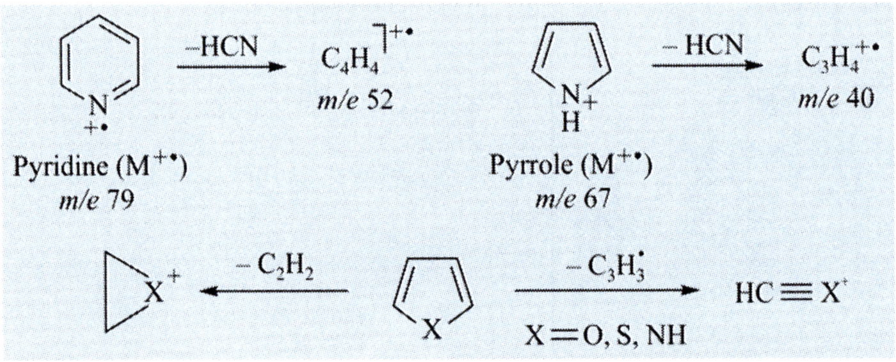

Scheme 6.56: Mass fragmentation pattern for pyridine, thiophene, furan and pyrrole

Piperidine

Ionized piperidine exhibits the loss of $^•CH_3$ and $^•C_2H_5$ due to hydrogen transfer reaction and of C_2H_5 due to double α-cleavage to produce ions at m/e 70, 56 and 57 respectively (Scheme 6.57). Abundant M-1 ion (m/e 84) also results due to loss of alpha H.

Alkylated heterocyclic compounds

In alkyl substituted heteroaromatic, e.g. 2-methyl pyrrole, cleavage of the β bond to the ring is favored similar to alkyl benzene (Scheme 6.58). The fragment ions thus formed undergo ring expansion as benzyl cation to tropyllium ion. This process is followed by loss if HCN in nitrogen heterocycles.

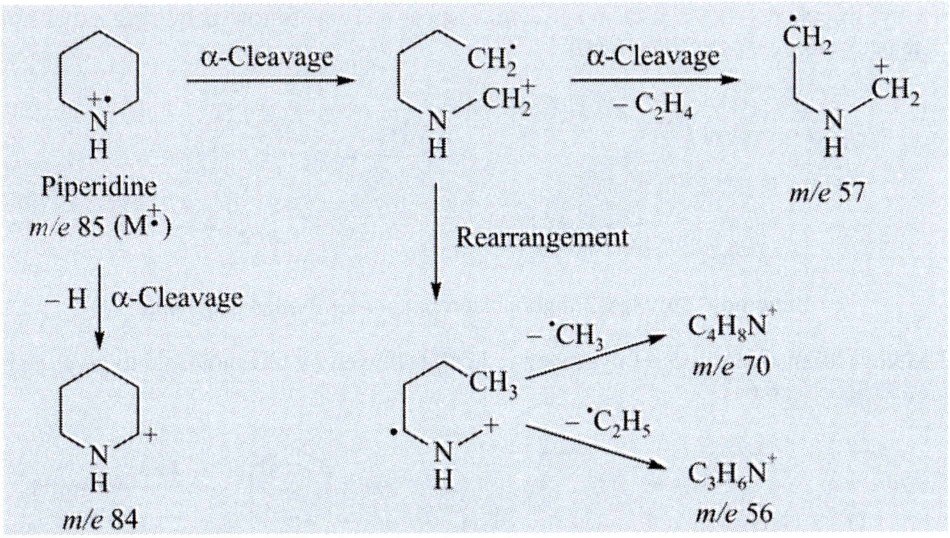

Scheme 6.57: Mass fragmentation pattern for piperidine

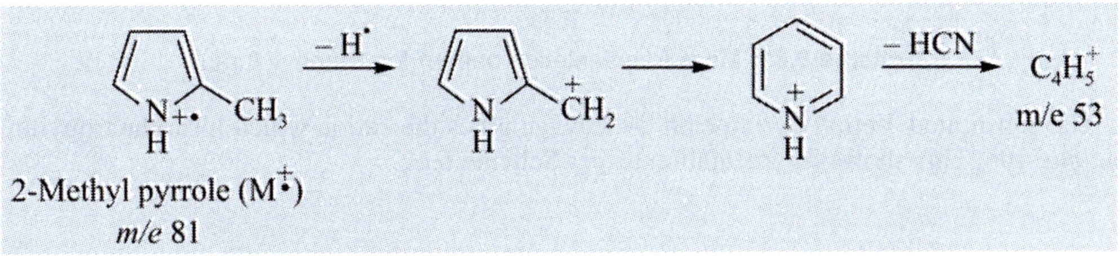

Scheme 6.58: Mass fragmentation pattern for 2-methyl pyrrole

Two main processes occur for fragmentation of N-methyl pyrolidine as shown in Scheme 6.59. The first one of these is loss of alpha hydrogen from the ring, followed by the formation of $CH_3-N^+=CH$ ion while the second is alpha cleavage within the ring and expulsion of the alkene to form $^+N=CH_2CH_2$ radical.

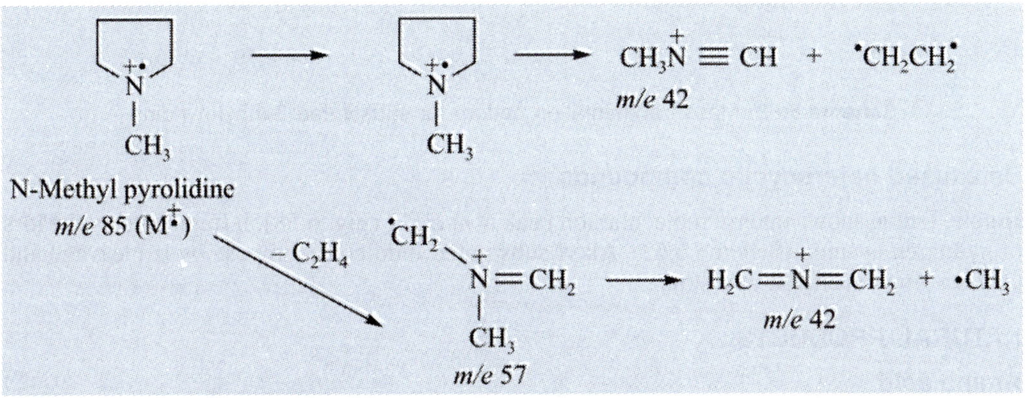

Scheme 6.59: Mass fragmentation pattern for N-methyl pyrolidine

3-Ethyl thiophene shows β-cleavage with respect to ring, followed by ring expansion to give the peak at *m/e* 97 (Scheme 6.60).

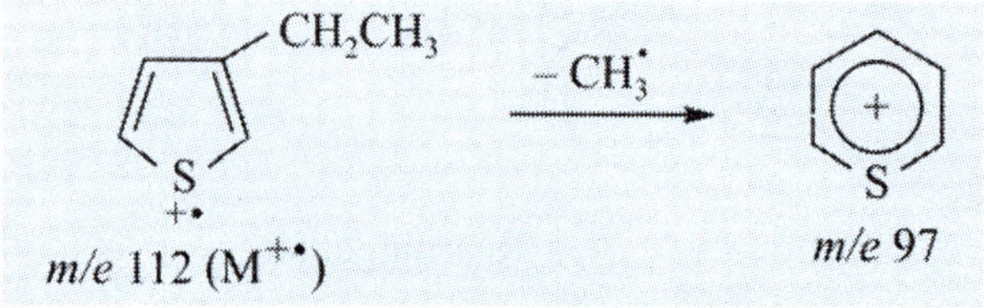

Scheme 6.60: Mass fragmentation pattern for 3-ethyl thiophene

2-Methyl furan shows loss of hydrogen radical, followed by CO molecule to give the peak at *m/e* 53 (Scheme 6.61).

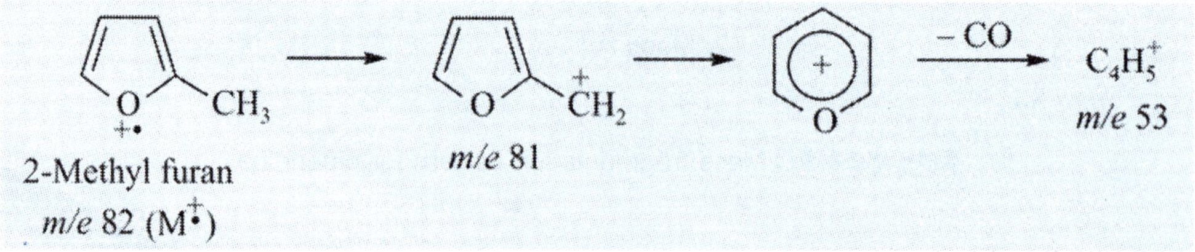

Scheme 6.61: Mass fragmentation pattern for 2-methyl furan

The substituted 3-ethyl pyridine on β-cleavage gives the cation which form the tropyllium ion, *m/e* 106. This shows fragmentation as per Scheme 6.62.

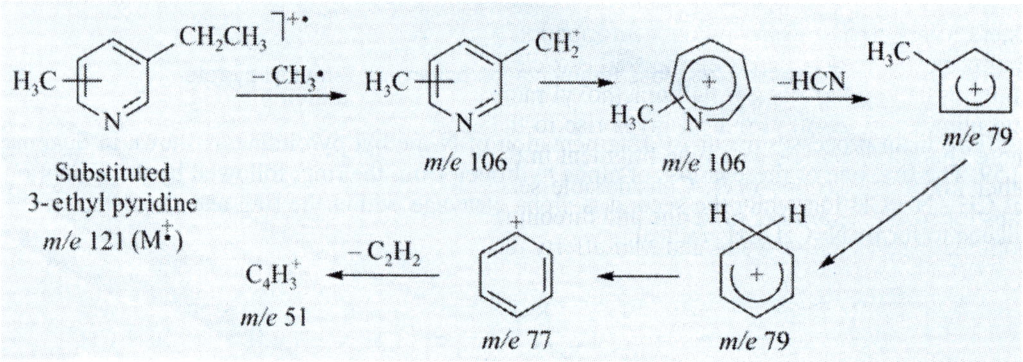

Scheme 6.62: Mass fragmentation pattern for substituted 3-ethyl pyridine

Benzfused heterocyclic compounds

Indole. Indole shows intense molecular ion peak at *m/e* 117 (Fig. 6.58). It fragments by the loss of hydrogen cyanide (Scheme 6.63). Alkyl substituted indoles decompose by α-cleavage and undergo ring expansion reactions.

NATURAL PRODUCTS

Amino acid

The amino acids are non-volatile in nature and so they are derivatized for mass analysis. Some of these derivatives include ethyl ester, N-acetyl, N-trimethylsilyl, etc. The mass spectra of

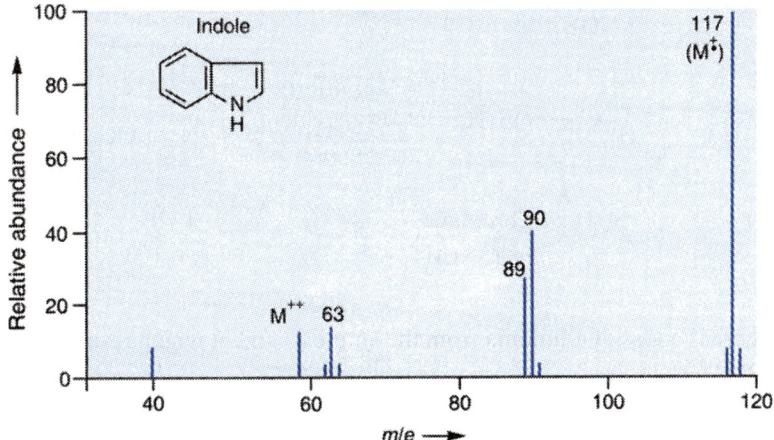

Fig. 6.58: Mass spectrum of indole

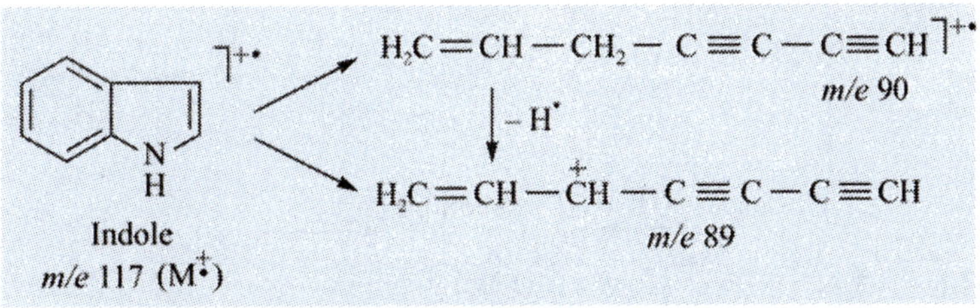

Scheme 6.63: Mass fragmentation pattern for indole

esters furnish more clear structural information than those of all other derivatives. The major fragmentation mode of an α-amino acid is α-cleavage (cleavage of carbon-carbon bond next to amino group). Thus, loss of carbomethoxyl radical leads to the formation of amine fragment and elimination of alkyl radical gives rise to the aldamine fragment. The amine fragment is more abundant than the aldamine fragment in the spectra of amino acid which do not contain other functional groups in R. Considerable secondary fragmentation is observed when R is substituted. Thus, in ester of serine and threonine, the presence of the hydroxyl group triggers both the elimination of water and McLafferty rearrangement (Scheme 6.64).

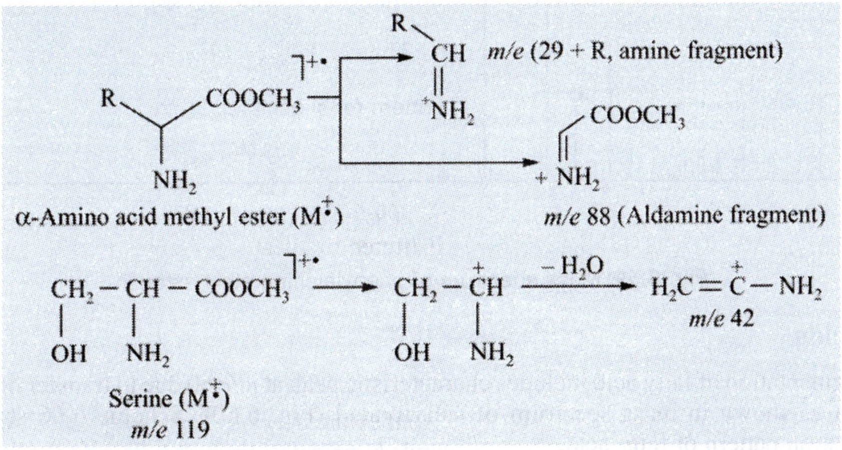

Scheme 6.64: Fragmentation pattern for serine and threonine (*contd.*)

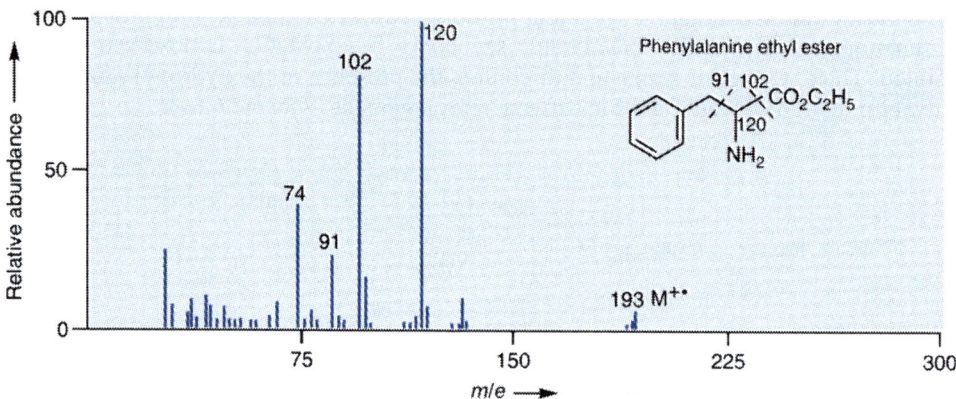

In lysine, there is a loss of ammonia from the amine fragment which results in the formation of base peak (m/e 84).

Phenylalanine ethyl ester undergoes benzylic cleavage. Figure 6.59 shows the mass spectrum of phenylalanine ethyl ester. The charge is retained by either of fragment. The aldamine fragments (Scheme 6.65) at m/e 102 are in abundance due to stability if benzyl radical which is lost.

Fig. 6.59: Mass spectrum of phenylalanine ethyl ester

Fatty acids

Mass fragmentation of fatty acid includes characteristic peak at m/e 60 due to transfer of gamma hydrogen as shown in mass spectrum of stearic acid (Fig. 6.60). Scheme 6.66 shows the fragmentation pattern of fatty acid.

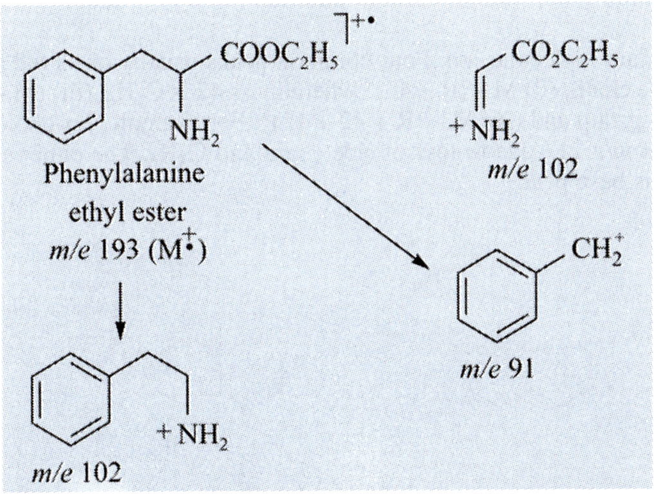

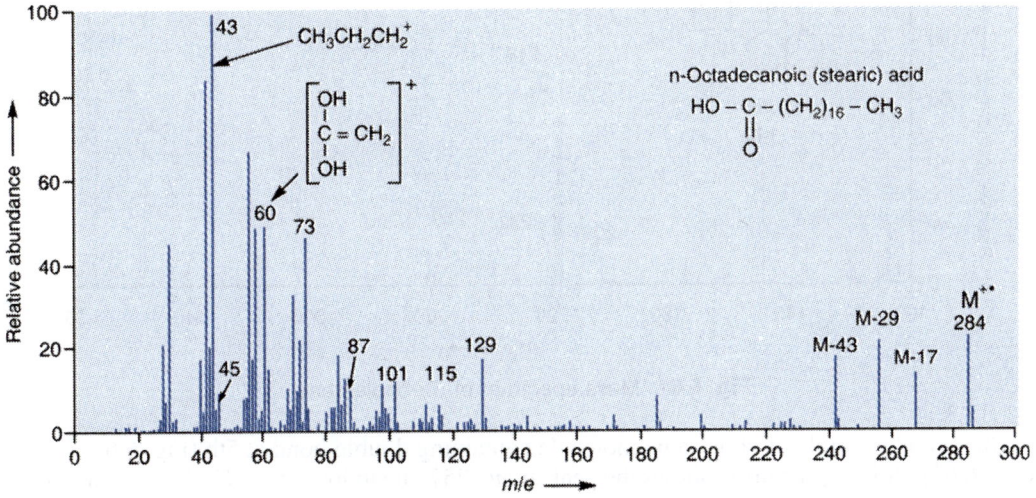

Scheme 6.65: Fragmentation pattern for phenylalanine ethyl ester

Fig. 6.60: Mass spectrum of stearic acid

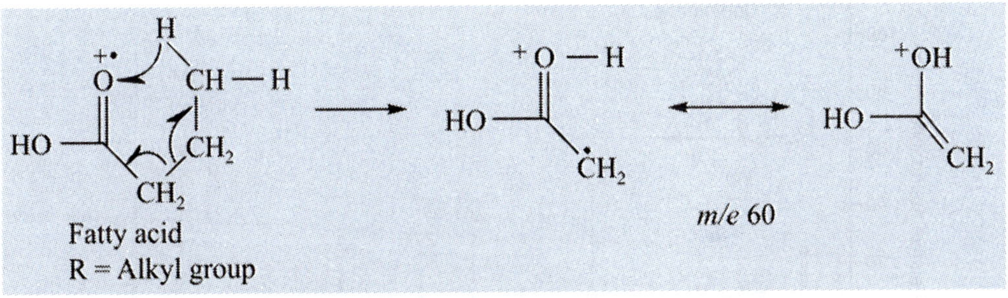

Scheme 6.66: Fragmentation pattern for fatty acid

Steroids

These show abundant molecular ion. Four common peaks usually observed are: (i) $(M - R)^+$, where R is the side chain, (ii) $M - (R + 42)^+$ where mass 42 is C_3H_6, (iii) $(M - 15)^+$ due to loss of angular methyl group and (iv) $M - (R + 42 + 15)^+$. For example, 5α-cholestene (Fig. 6.61) gives base peak at m/e 217 due to loss of side chain and C_3H_6. The either structure I or II is responsible for this base peak.

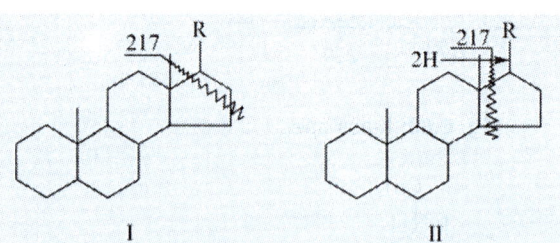

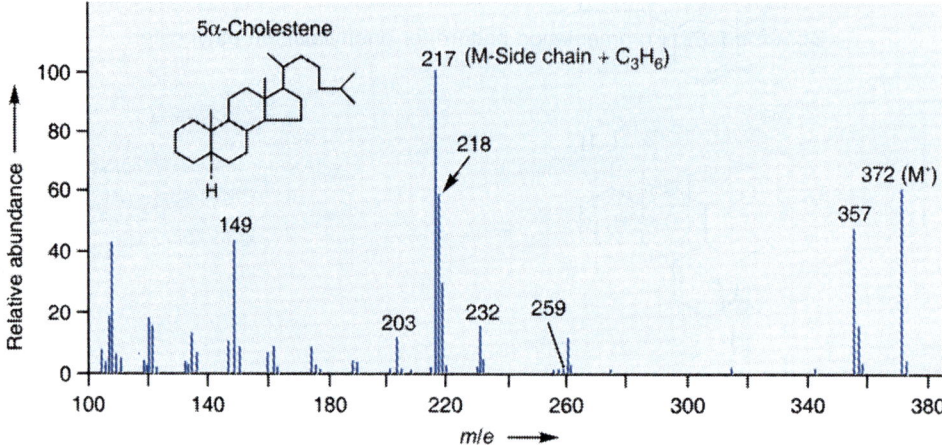

Fig. 6.61: Mass spectrum of 5α-cholestene

In second example of cholestene molecule containing double bond at 5th (Fig. 6.62) or 7th (Fig. 6.63) position, spectrum shows the peak at m/e 257 due to loss of the C-17 side chain with allylic fission of the nuclear double bonds leading to species e and f respectively (Scheme 6.67).

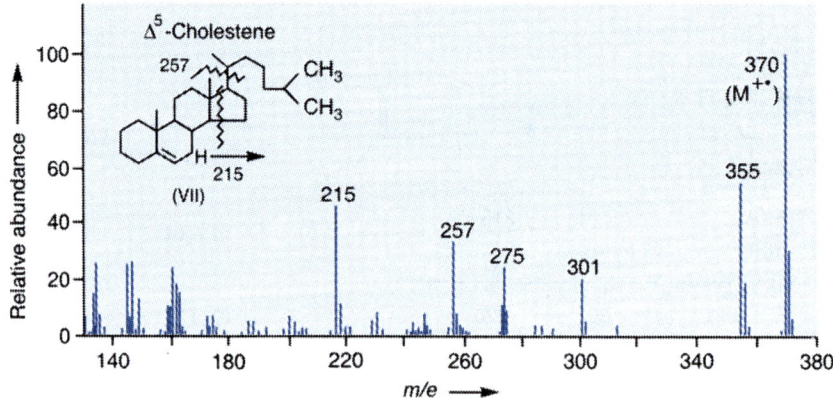

Fig. 6.62: Mass spectrum of Δ^5-cholestene

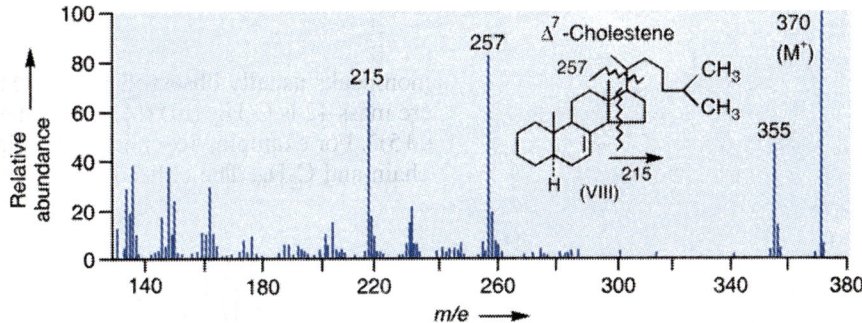

Fig. 6.63: Mass spectrum of Δ^7-cholestene

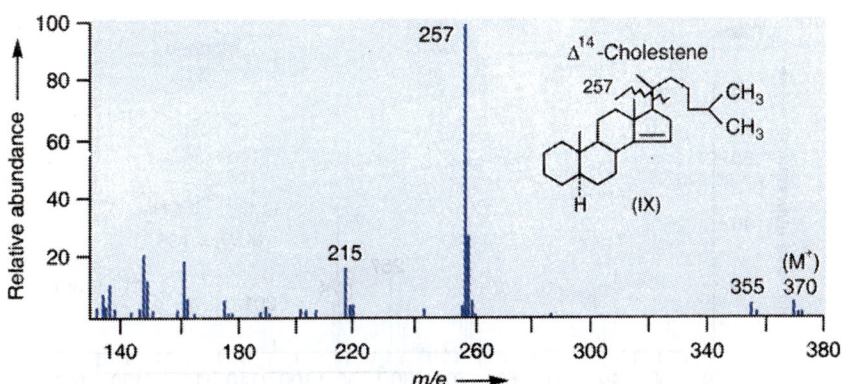

Scheme 6.67: Mass fragmentation pattern for Δ^5-cholestene and Δ^7-cholestene

Such a supposition is clearly not feasible in Δ^{14}-cholestene and yet this is the substance where such side chain loss occurs to the greatest extent, virtually overshadowing all of the other fragment peaks and especially the molecular ion peak (Fig. 6.64). In this, peak at m/e 215 is very weak (Scheme 6.68).

Fig. 6.64: Mass spectrum of Δ^{14}-cholestene

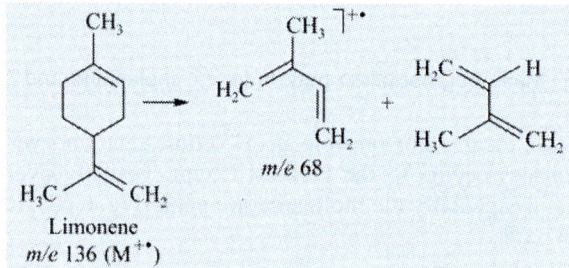

Scheme 6.68: Mass fragmentation pattern for Δ^{14}-cholestene

Terpenes

Monoterpenes

Hydrocarbon

The mass spectrum of limonene (Fig. 6.65) shows the major fragment at *m/e* 68 by a reverse Diels-Alder reaction.

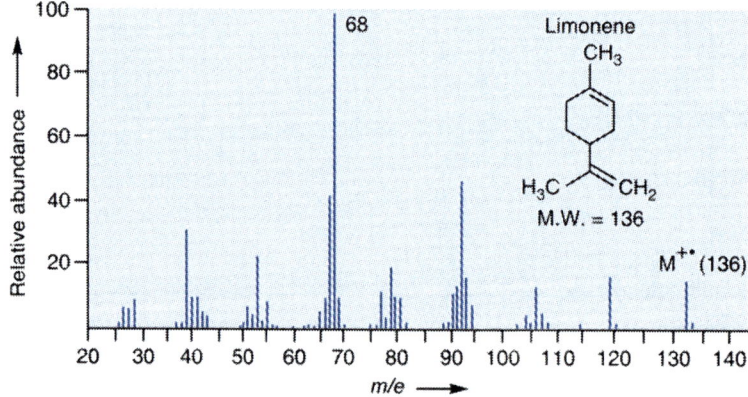

Fig. 6.65: Mass spectrum of limonene

Aldehydes

The mass spectra of citronellal, geranial and cuminic aldehyde are presented in Figs 6.66 to 6.68 respectively. The first two spectra (citronellal and geranial) show the base peak at m/e 69, due to the fragment (Schemes 6.69 and 6.70) which formed by splitting the bond in allylic position to the double bond.

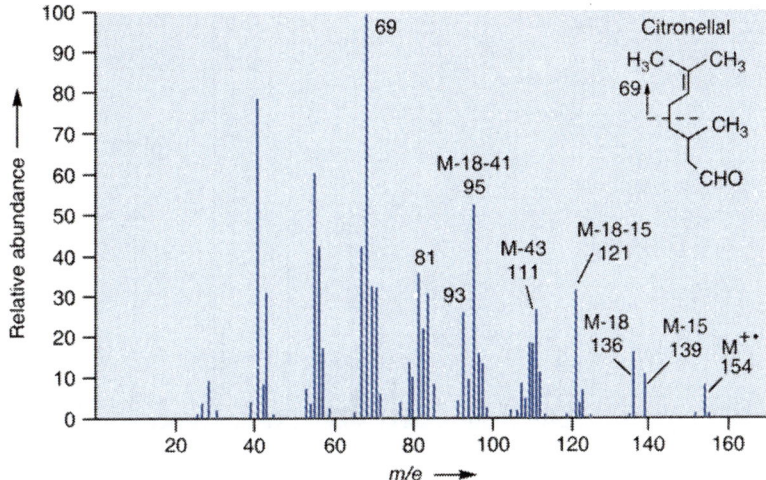

Fig. 6.66: Mass spectrum of citronellal

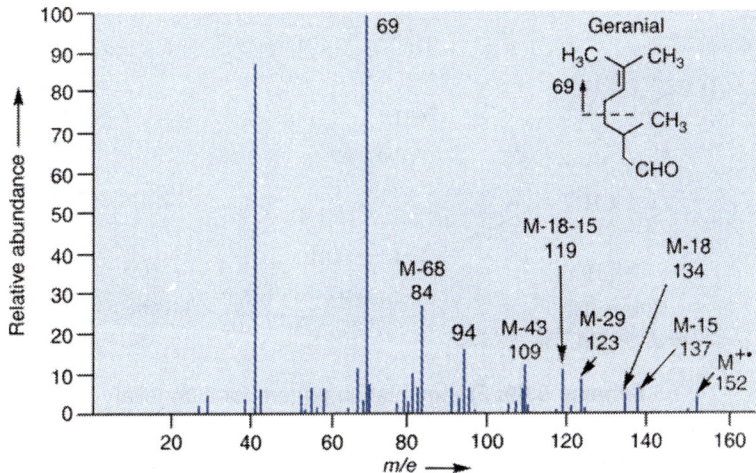

Fig. 6.67: Mass spectrum of geranial

The other peaks M–CH$_3$, M–H$_2$O, etc. are visible in the spectra. In case of cuminic aldehyde, there seems to be no extra hydrogen atom available for comparative loss of water for M-18 peak. The base peak at m/e 133 is due to loss of methyl group. The peak at m/e 105 (M-43) is due to loss of isopropyl group.

Ketones

The mass spectra of piperitinone and menthone are presented in Figs 6.69 and 6.70, respectively. They differ from aldehyde in that their mass spectra show a very small peak at M-18 position, corresponding to loss of one water molecule from the parent molecule. The piperitinone shows the peak at m/e 82 due to allylic cleavage (Scheme 6.71).

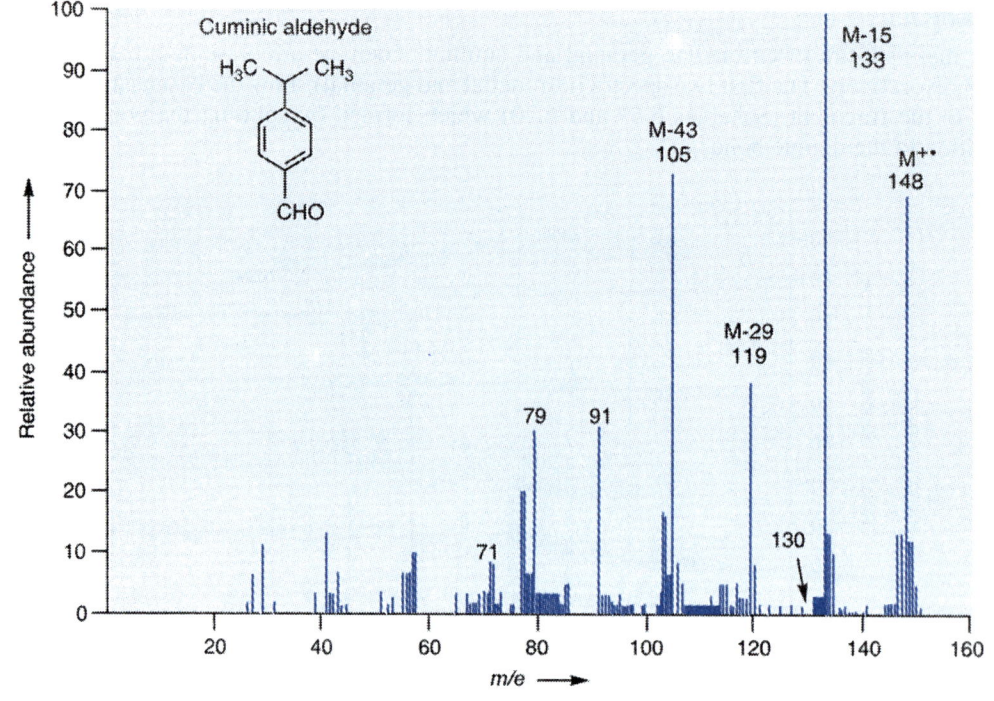

Fig. 6.68: Mass spectrum of cuminic aldehyde

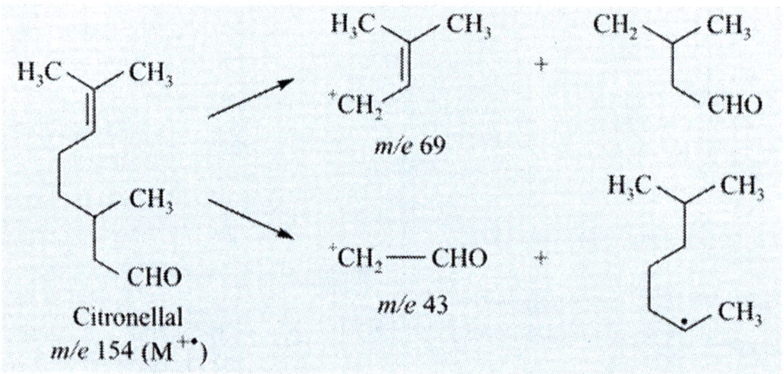

Scheme 6.69: Fragmentation pattern for citronellal

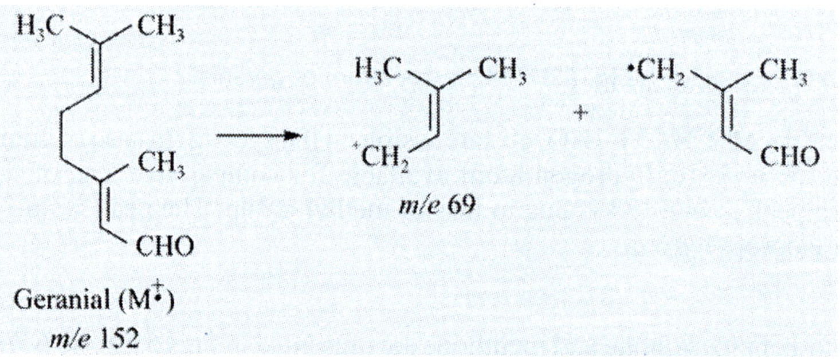

Scheme 6.70: Fragmentation pattern for geranial

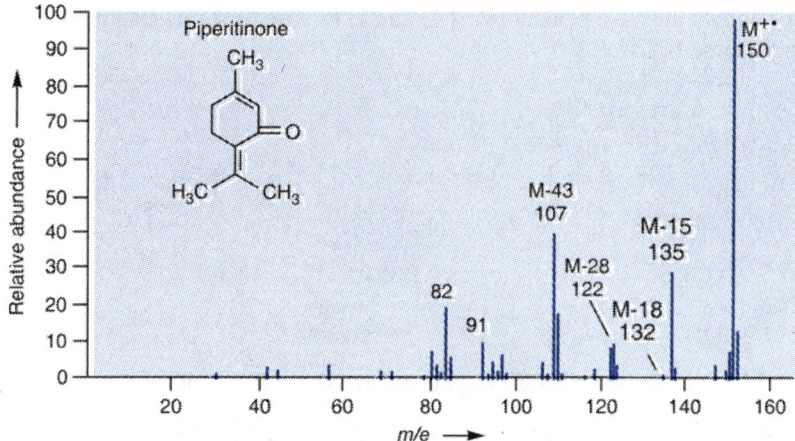

Fig. 6.69: Mass spectrum of piperitinone

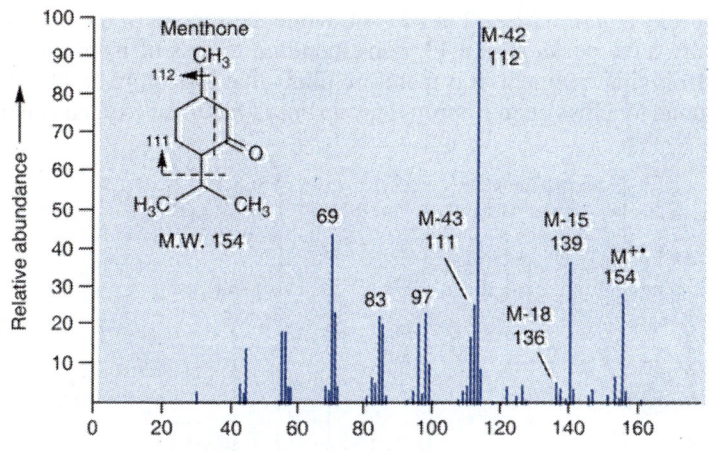

Fig. 6.70: Mass spectrum of menthone

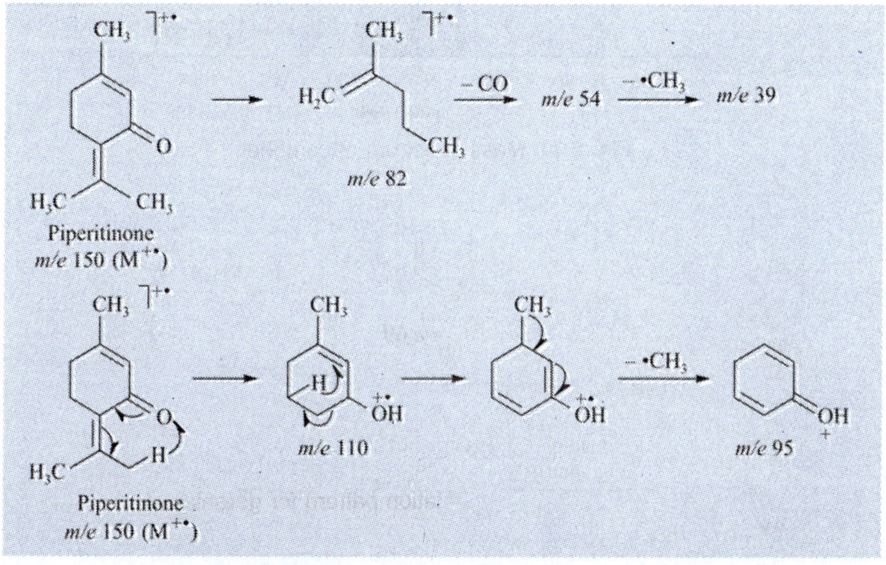

Scheme 6.71: Fragmentation pattern for piperitinone

Menthone shows the base peak at *m/e* 112 due to release of stable fragment, CH_3–$CH=CH_2$ as shown in Scheme 6.72.

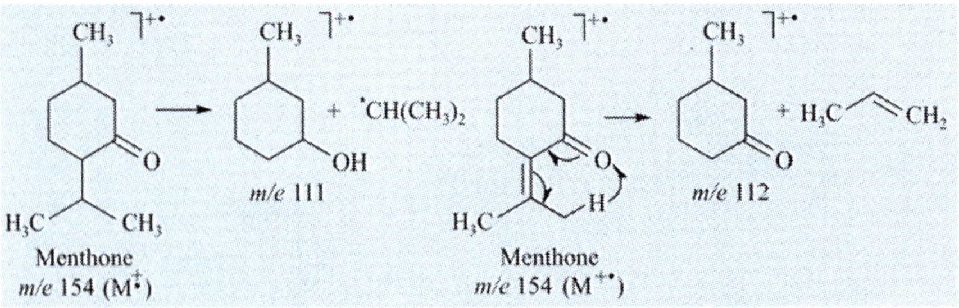

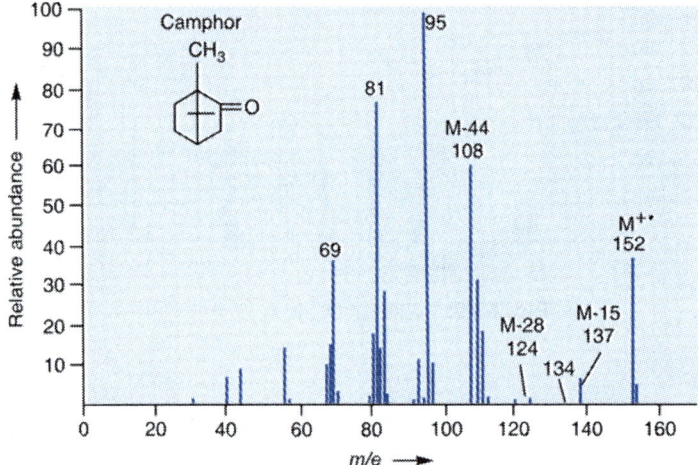

Scheme 6.72: Fragmentation pattern for menthone

Bicyclic ketones. The mass spectrum of camphor is given in Fig. 6.71. The spectrum of camphor revealed one diffuse peak at 76.8 corresponding to the breakdown reaction of $152^+ \rightarrow 108^+ \rightarrow {}^+ 44$ (Scheme 6.73). The mass spectra of the monoterpene ketone show the presence generally have peaks at *m/e* = M – 42 or M – 44, while the spectra of the hydrocarbons and the alcohols generally have peaks at *m/e* 43, corresponding to loss of isopropyl group, or at *m/e* values derived from this fragment. It is therefore likely that loss of 42 or 44 units from ketones does not correspond to a loss of an isopropyl group but rather to an oxygen containing fragment (Scheme 6.73).

Fig. 6.71: Mass spectrum of camphor

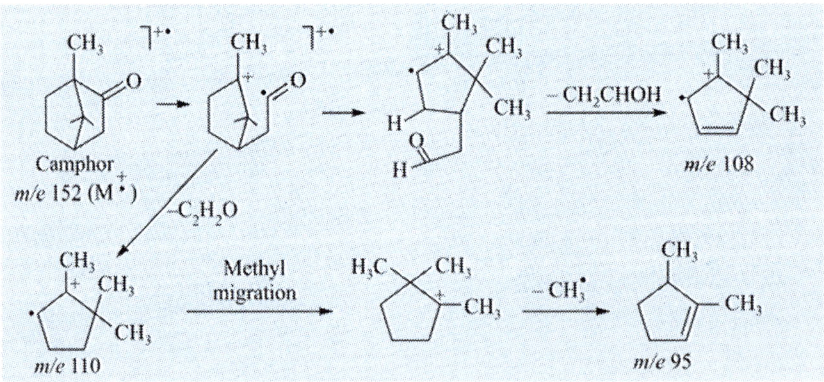

Scheme 6.73: Fragmentation pattern for camphor

Alcohol

(a) Acyclic

The mass spectra of β-citronellol, geraniol, and linalool are given in Figs 6.72, 6.73 and 6.74 respectively. With the exception of alcohol, linalol, they have m/e 69 as the base peak. This fragment is formed by the splitting of bond in allylic position to the double bond. Linalol has $m/e = M + 1$ as the molecular ion peak and $m/e = M - 18 - 43 = 93$ as the base peak.

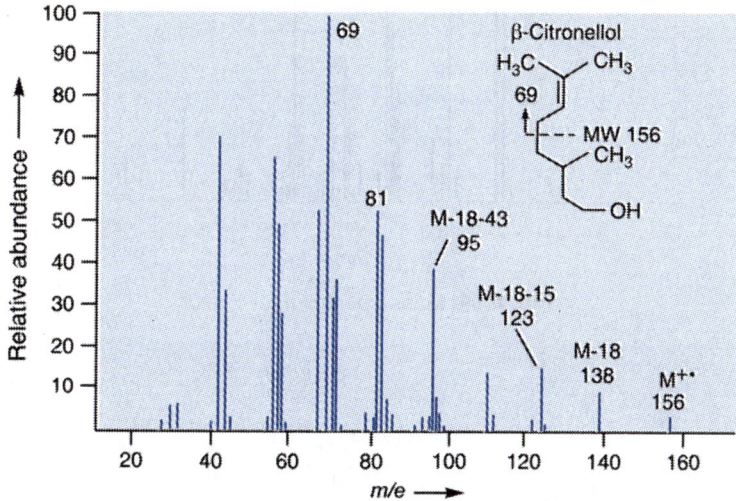

Fig. 6.72: Mass spectrum for β-citronellol

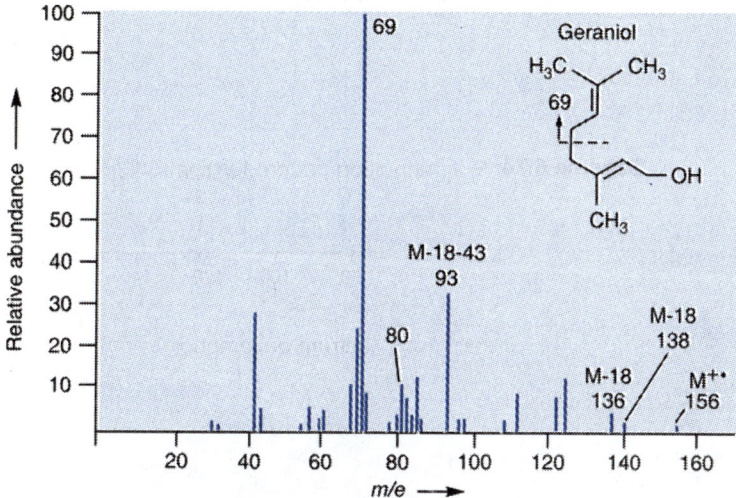

Fig. 6.73: Mass spectrum for geraniol

(b) Aromatic

The mass spectra of thymol and carvacrol are given in Figs 6.75 and 6.76, respectively. These compounds are much more stable than the acyclic compounds as revealed by the presence of high concentration of ions with large m/e values. The strongest peaks in each case, i.e. $m/e = M$ and M^+-15, indicate the firmness with which the hydroxyl group is bound (Scheme 6.77).

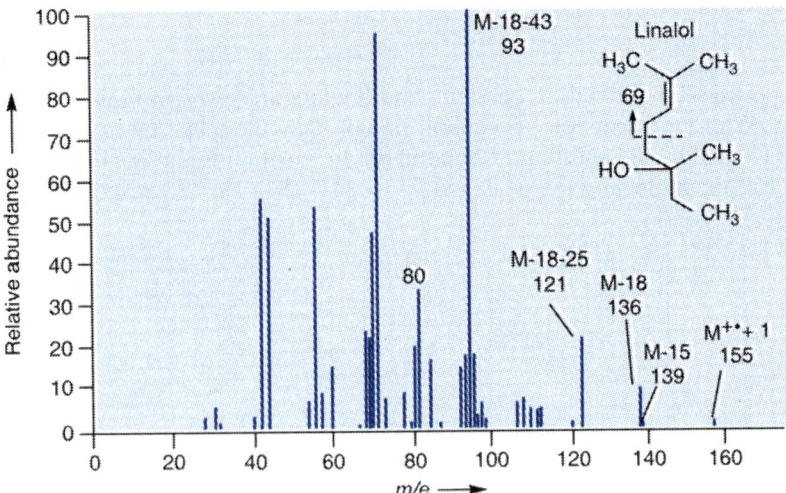

Fig. 6.74: Mass spectrum for linalol

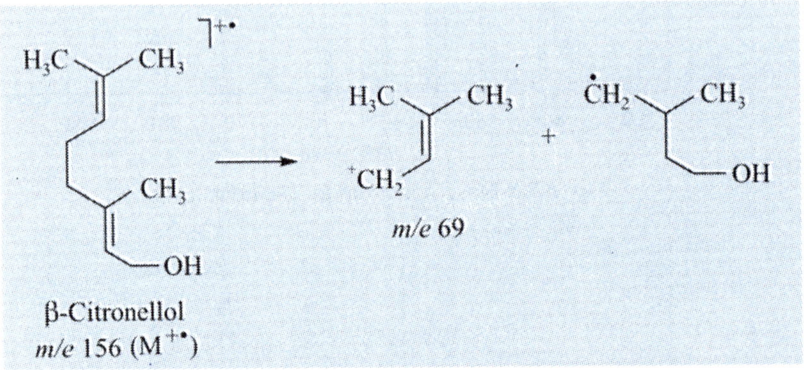

Scheme 6.74: Fragmentation pattern for β-citronellol

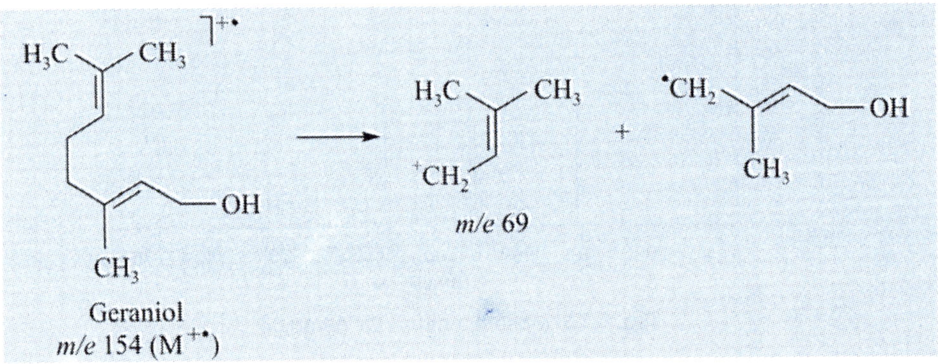

β-Citronellol
m/e 156 (M⁺•)

m/e 69

Scheme 6.75: Fragmentation pattern for geraniol

Geraniol
m/e 154 (M⁺•)

m/e 69

(c) Non-aromatic

The mass spectra of alpha-terpineol and menthol are given in Figs 6.77 and 6.78 respectively. The α-terpineol is tertiary alcohol and so base peak at *m/e* 59 is due to tertiary cation (Schemes 6.78 and 6.79).

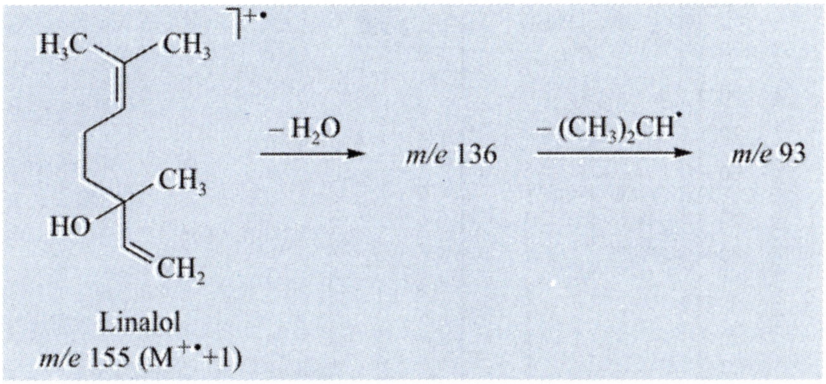

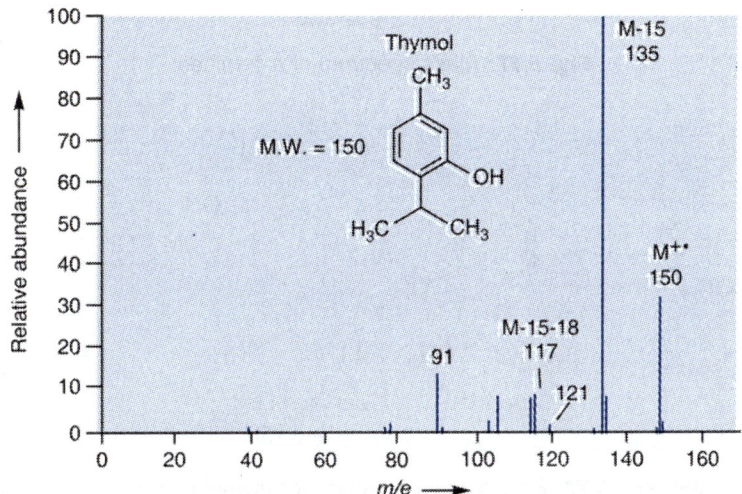

$$\text{Linalol}$$
$$m/e\ 155\ (M^{+\bullet}+1)$$

Scheme 6.76: Fragmentation pattern for linalol

Fig. 6.75: Mass spectrum for thymol

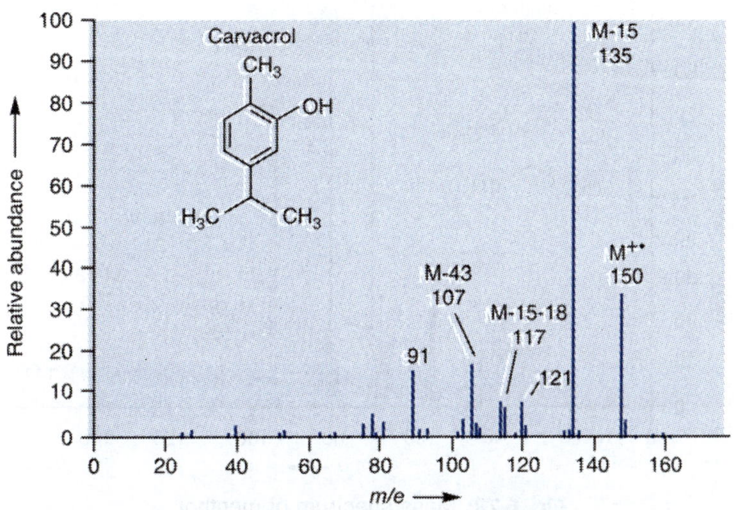

Fig. 6.76: Mass spectrum for carvacrol

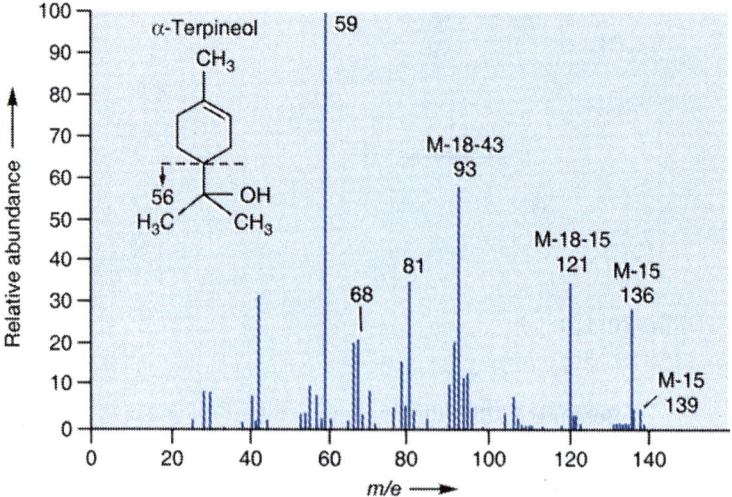

Fig. 6.77: Mass spectrum of α-terpineol

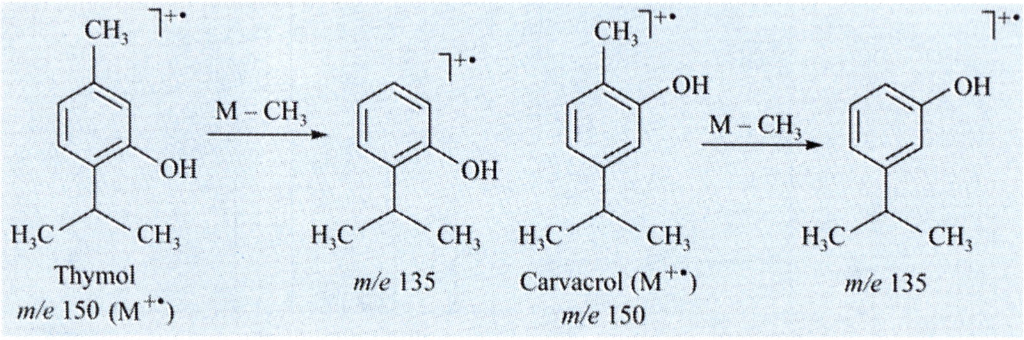

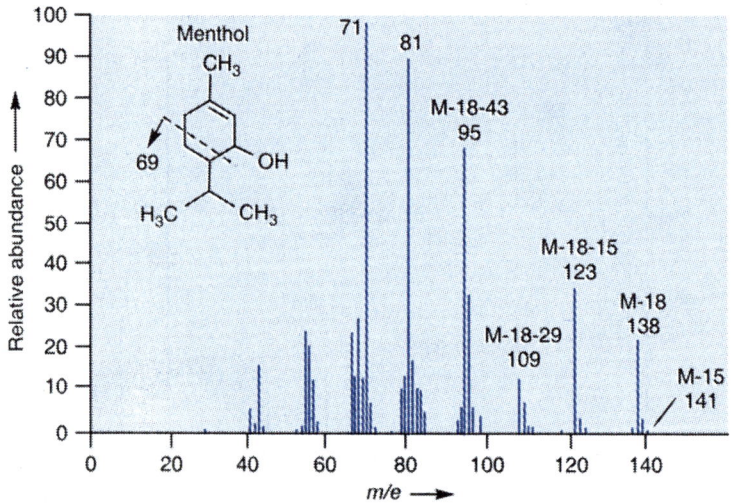

Scheme 6.77: Fragmentation pattern for thymol and carvacrol

Fig. 6.78: Mass spectrum of menthol

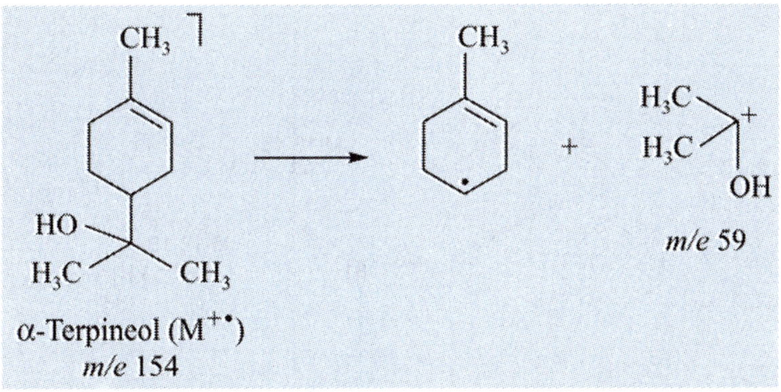

Scheme 6.78: Fragmentation pattern for α-terpineol

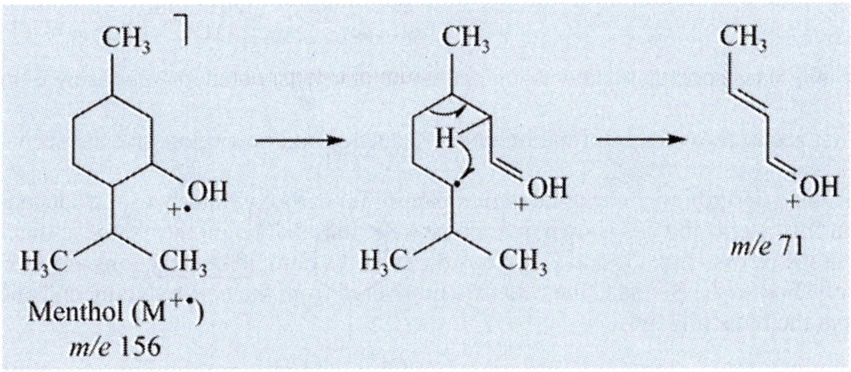

Scheme 6.79: Fragmentation pattern for menthol

Menthol shows the base peak at m/e 71 due to formation of fragment, $CH_3–CH=CH–OH^+$. Both the compounds show M-18 peak due to loss of H_2O molecule and M-18-43 due to loss of H_2O and isopropyl group.

Triterpenoids

Triterpenes that contain Δ^{12}-double bond (1) undergo a retro Diels-Alder reaction to form fragments containing the ABC*-rings and the C*DE-rings (Scheme 6.80).

Compound	R^1	R^2	R^3	R^4	R^5	R^6
			Substituent group			
β-Amyrin	CH_3	CH_3	H	CH_3	CH_3	CH_3
α-Amyrin	CH_3	CH_3	CH_3	CH_3	H	CH_3
Oleanolic acid	CH_3	CH_3	H	COOTMSi	CH_3	CH_3
Ursolic acid	CH_3	CH_3	CH_3	COOTMSi	H	CH_3

Carbohydrates

Under different ionization conditions the fragmentation patterns are similar and depending on determining factors can be classified into two groups:

1. Glycosidic cleavages of bond linking two sugar rings
2. Cross-ring cleavages—breaking of two bonds

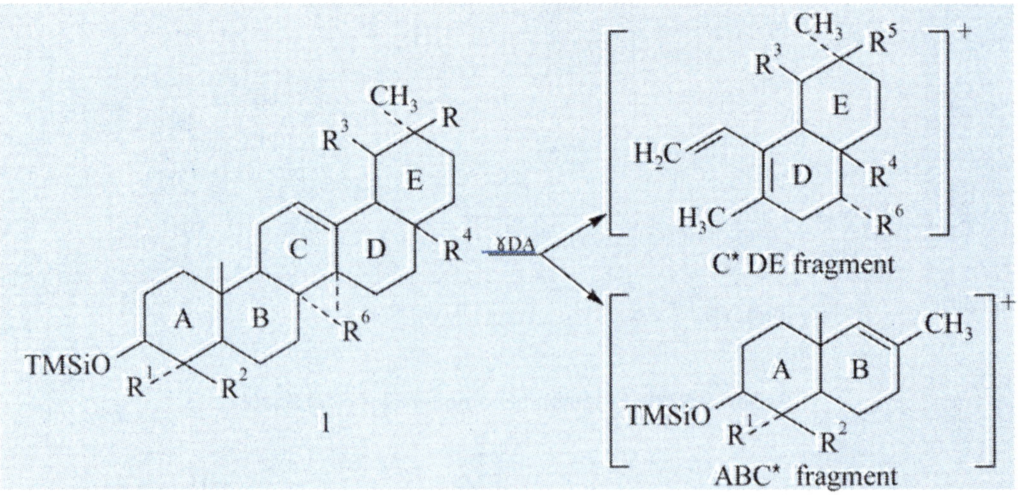

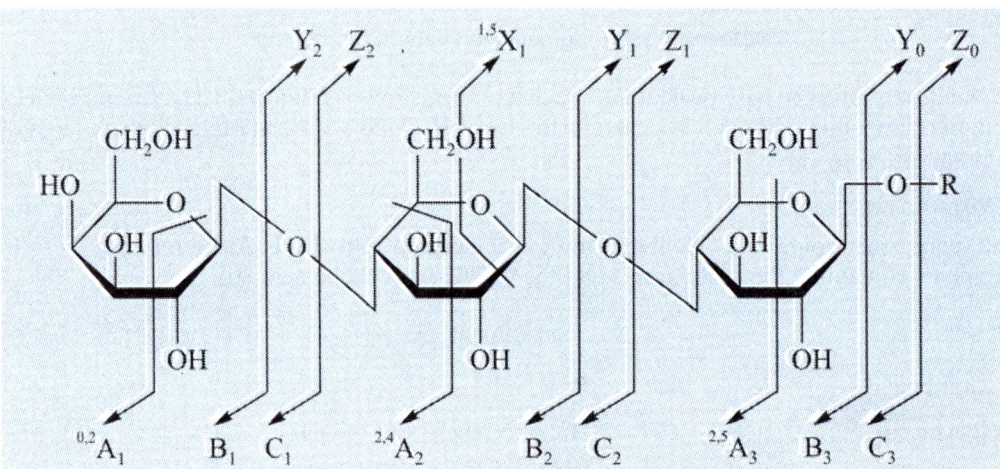

Scheme 6.80: Mass spectral fragmentation pattern of trimethylsilylated triterpenes by EI ionization

The first group provides information about sequence and branching and the second group about linkage.

A schematic describing the fragmentation pattern for carbohydrates was introduced in 1988 by Domon & Costello and it is shown in Scheme 6.81. Ions that retain the charge at the reducing terminus are X (cross-ring), Y, and Z (glycosidic) and the complementary ions are A, B, and C respectively. For the A, B, and C the rings are numbered from the non-reducing end and for the others from the reducing end.

Scheme 6.81: Schematic representation of the major fragment ions from carbohydrates

Flavonoids

The mass fragmentation pattern of some flavonols, flavanones, dihydroflavonol are given in Scheme 6.82. Each subgroup exhibits a characteristic fragmentation behavior which includes dehydration, loss of CO and fission of C-ring. The relative abundance of these fragments is highly associated with the applied collision energy. If the collision energy is low, the main fragment in the spectrum produced is $[M + H]^+$. By increasing the collision energy, a complete fragmentation of protonated molecule is obtained.

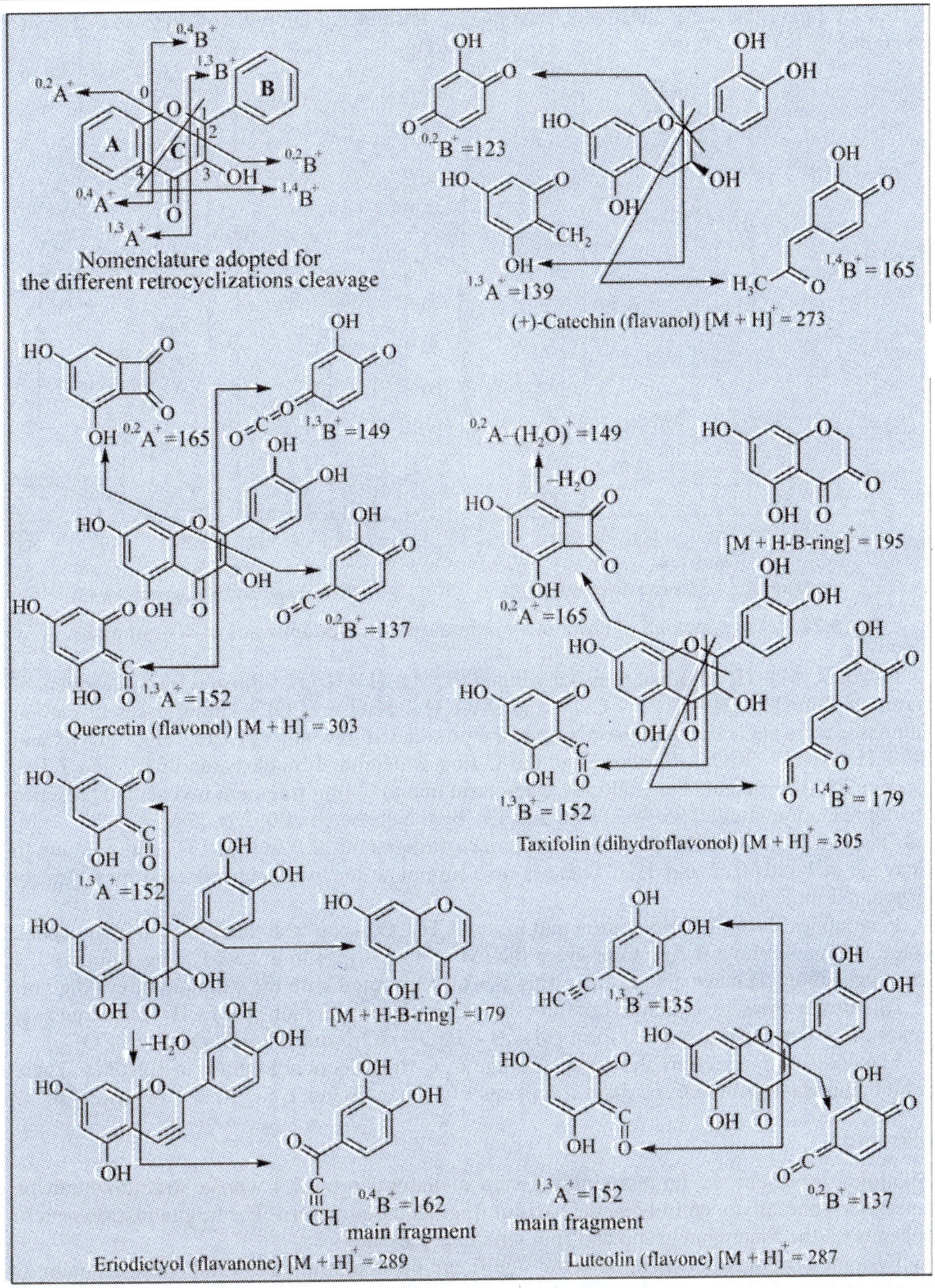

Nomenclature adopted for the different retrocyclizations cleavage

$^{0,2}B^+ = 123$

$^{1,3}A^+ = 139$

(+)-Catechin (flavanol) $[M + H]^+ = 273$

$^{1,4}B^+ = 165$

$^{0,2}A^+ = 165$ $^{1,3}B^+ = 149$

$^{0,2}B^+ = 137$

$^{1,3}A^+ = 152$

Quercetin (flavonol) $[M + H]^+ = 303$

$^{0,2}A-(H_2O)^+ = 149$

$^{0,2}A^+ = 165$

$^{1,3}B^+ = 152$

Taxifolin (dihydroflavonol) $[M + H]^+ = 305$

$[M + H-B-ring]^+ = 195$

$^{1,4}B^+ = 179$

$^{1,3}A^+ = 152$

$[M + H-B-ring]^+ = 179$

$^{0,4}B^+ = 162$
main fragment

Eriodictyol (flavanone) $[M + H]^+ = 289$

$^{1,3}B^+ = 135$

$^{1,3}A^+ = 152$
main fragment

$^{0,2}B^+ = 137$

Luteolin (flavone) $[M + H]^+ = 287$

Scheme 6.82: Fragmentation pattern for flavonoids (5,7,3′,4′-hydroxy substituted compounds).
Source: Molecules, 2007, 12, 593.

For example, the mass spectra of flavonol for normalized energies of 40% and 45% are given in Fig. 6.79.

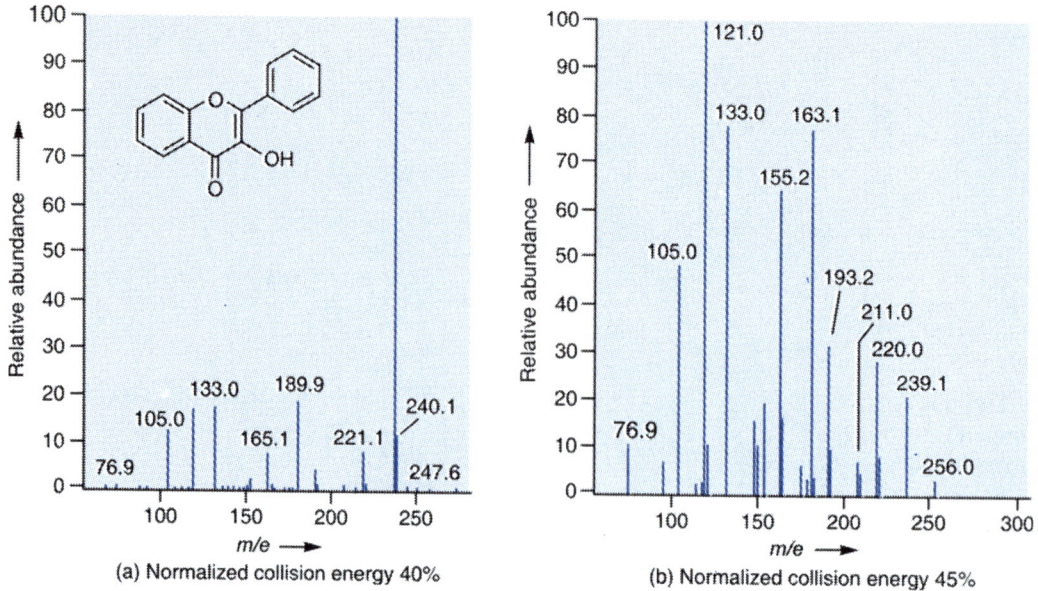

(a) Normalized collision energy 40%

(b) Normalized collision energy 45%

Fig. 6.79: Mass spectrum of flavonol for normalized collision energies of 40% and 45%

Flavonol $[M + H]^+$ product ions dehydrated to $[M + H - H_2O]$, followed by two sequential losses of CO: $[M + H - H_2O - CO]^+$ and $[M + H - H_2O - 2CO]^+$. These losses of carbon monoxide were also observed directly from the protonated flavonol: $[M + H - H_2O - CO]^+$ and $[M + H - H_2O - 2CO]^+$. Furthermore, the C-ring is degraded by cleavage of 0,2 and 1,3 to produce RDA fragment. For example, Quercetrin due to C-ring fragment reveals the presence of fragmentation at m/e 165, 153, 149, and 137 by the cleavage of 0, 2 & 1, 3.

Flavones (e.g. luteolin) exhibit the presence of fragments at m/e 153, 137, and 135 due to cleavage of bonds 0, 2 and 1, 3. There is also loss of water molecule followed by CO from protonated molecule.

Flavonones show the dehydration and loss of CH_2CO group from the protonated molecule. There is loss of whole B-ring to produce the $[M + H - B\text{-ring}]$ fragment. C-ring 1, 3 and 0, 4 retrocyclization cleavage also occurs. This is well-illustrated with the example of eriodictyol.

Dihydrofavonols, for example, taxifolin show an initial dehydration of $[M+H]^+$ and sequential losses of two carbonyl groups to form $[M + H - H_2O - CO]^+$ and $[M + H - H_2O - 2CO]^+$.

Flavanols (e.g. catechin) show the cleavage of C-ring identical to those of flavones. There is high abundance of the three main fragments of C-ring fission, e.g. $^{1,2}B^+$, $^{1,3}A^+$ and $^{1,4}B^+$.

Alkaloids

Alkaloids represent the largest single group of natural products whose structures can be determined partially or entirely on the basis of fragmentation pattern. The fragmentation pattern depends on the functional group and ring structure present.

Pyridine alkaloids like nicotine show significant molecular ion and (M-1) peaks. The mass fragmentation of nicotine is shown in Scheme 6.83. The spectra is characterized by abundant molecular ions. The other peak results from splitting out CH_3N neutral fragment and rupture of the bond between the pyridyl and the cyclic aminyl nuclei (parent minus seventy eight). The most abundant peak (base peak) in the mass spectrum appears at the m/e 84.

Cocaine shows many identifying ions at m/e 272, 198, 182, 105, 96, 82, 77, etc. The drug abuse of cocaine is identified on the basis of some prominent peaks.

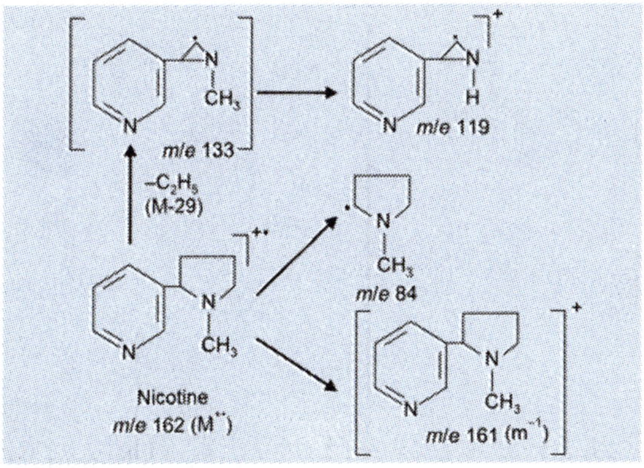

Scheme 6.83: Mass fragmentation pattern for nicotine

The quinoline alkaloids like quinine show major fragment at m/e 136 and 159 due to quinuclidine and quinoline moiety. There is also fragment at m/e 160 due to H-rearrangement during the rupture of C_8-C_9 bond.

The mass spectrum of morphine is presented in Fig. 6.81. The spectrum shows a molecular ion peak at m/e 285 with a relative abundance 100%. Its fragmentation pattern is given in Scheme 6.84.

Papaverine ($C_{20}H_{21}NO_4$, MW:339), a benzylisoquinoline alkaloid, revealed an apparent protonated molecular $(M + H)^+$ at m/e 340. The tandem MS analysis of MH^+ provided the MS/MS product ions at m/e 202, 187, 171, 151 and 135 (Scheme 6.85).

Ephedrine gives an EI spectrum (Fig. 6.82a & b) dominated by the m/e 58 fragment ion. Methane CI gives an MH^+ at m/e 166 and fragments at m/e 148, 135 and 58 due to protonation on the OH and $NHCH_3$ groups.

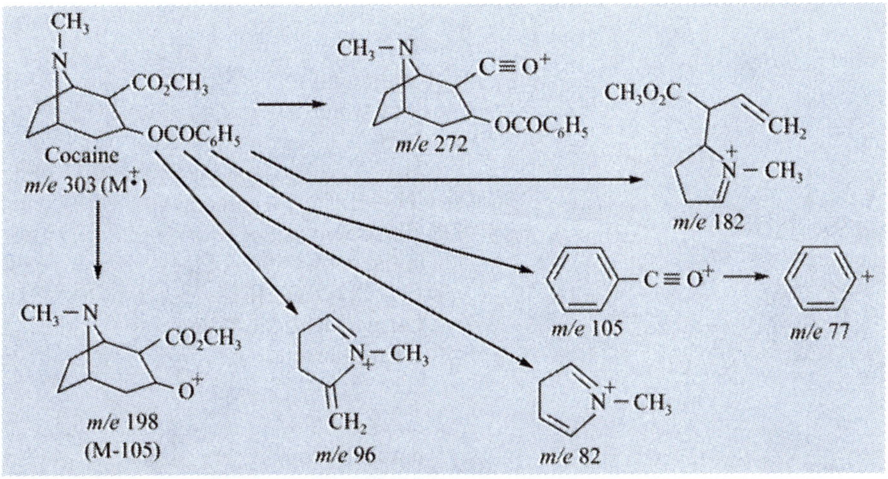

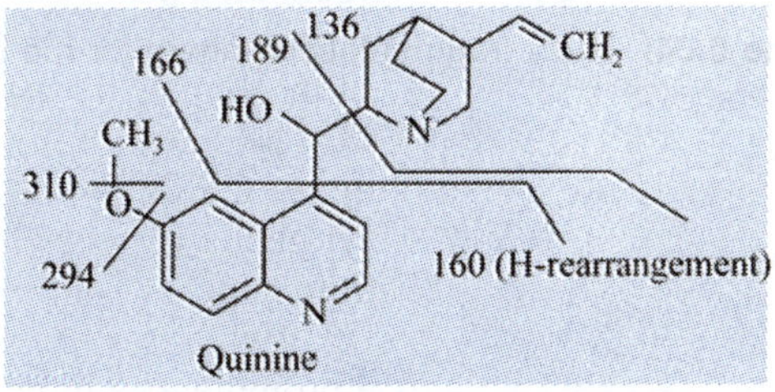

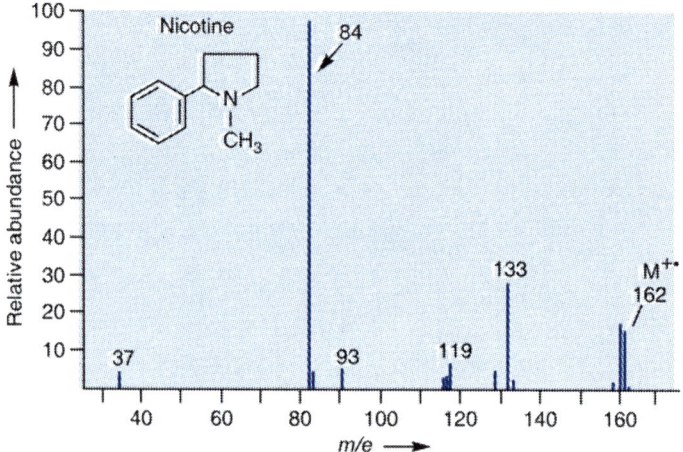

Fig. 6.80 : Mass spectrum for nicotine

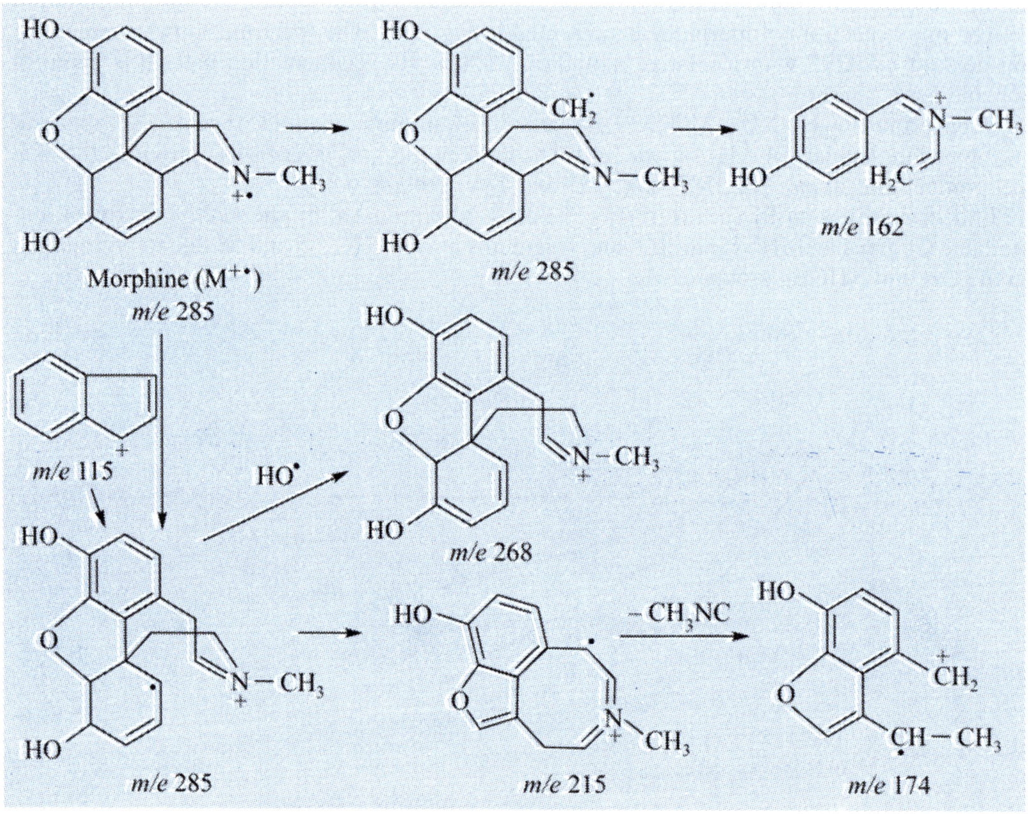

Scheme 6.84 : Mass fragmentation pattern for morphine

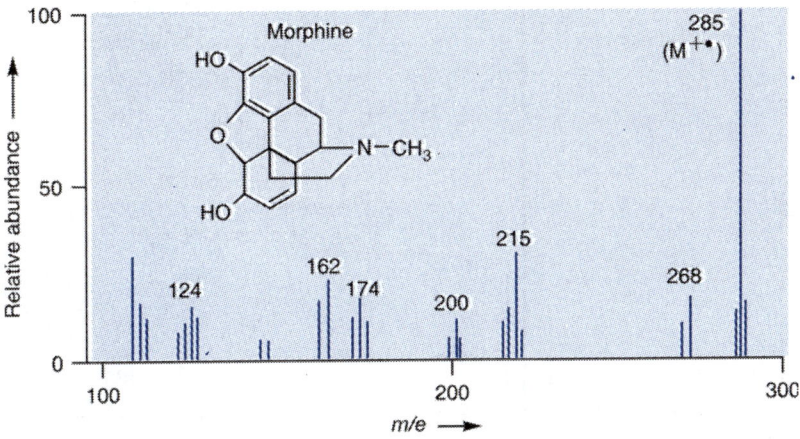

Fig. 6.81: Mass spectrum for morphine

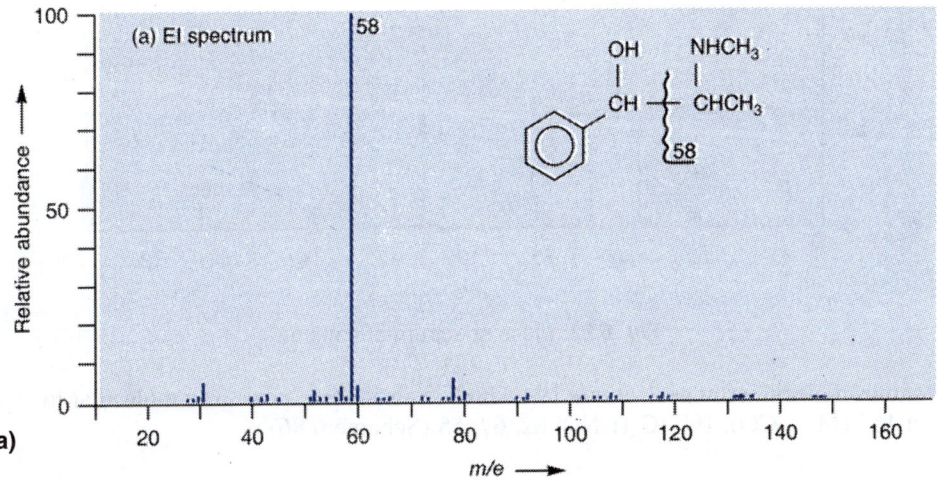

Scheme 6.85: Mass fragmentation pattern for papaverine

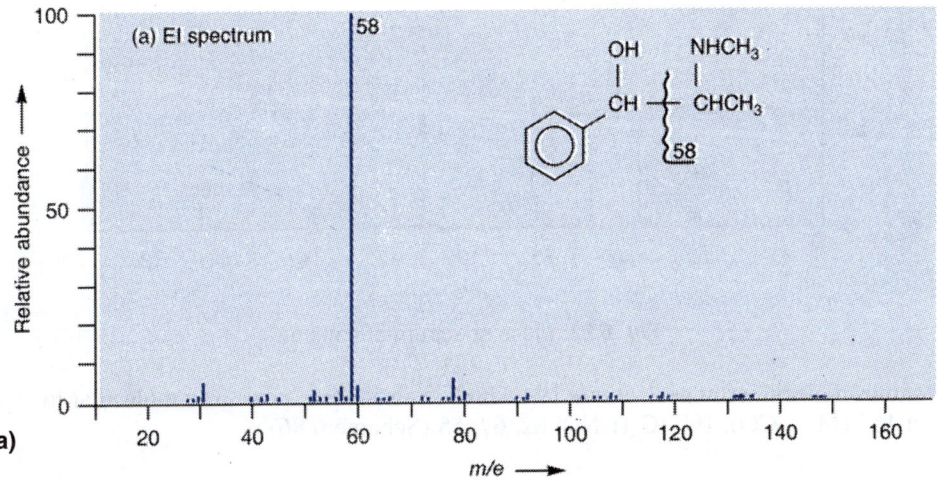

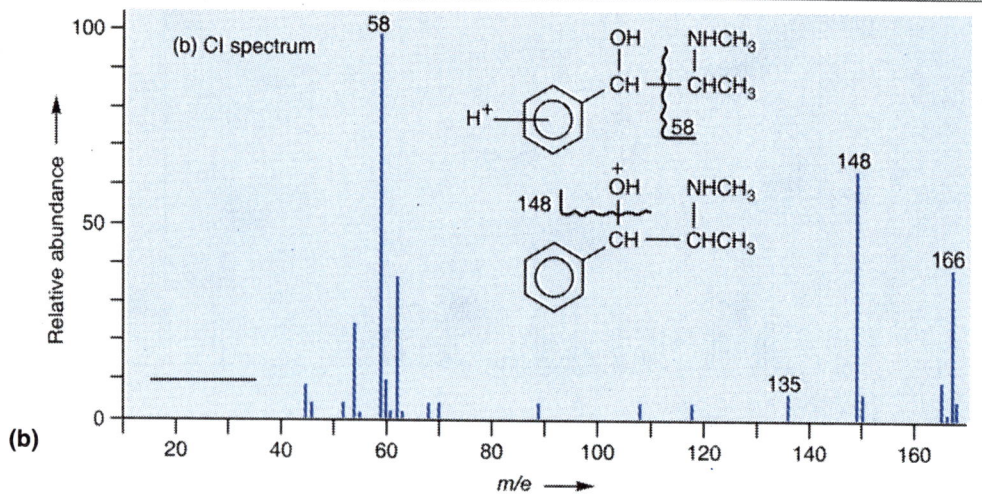

Fig. 6.82: (a) El spectrum of ephedrine, (b) Cl spectrum of ephedrine

Xanthines

The fragmentation pattern of xanthines is illustrated with the example of caffeine (Fig. 6.83).

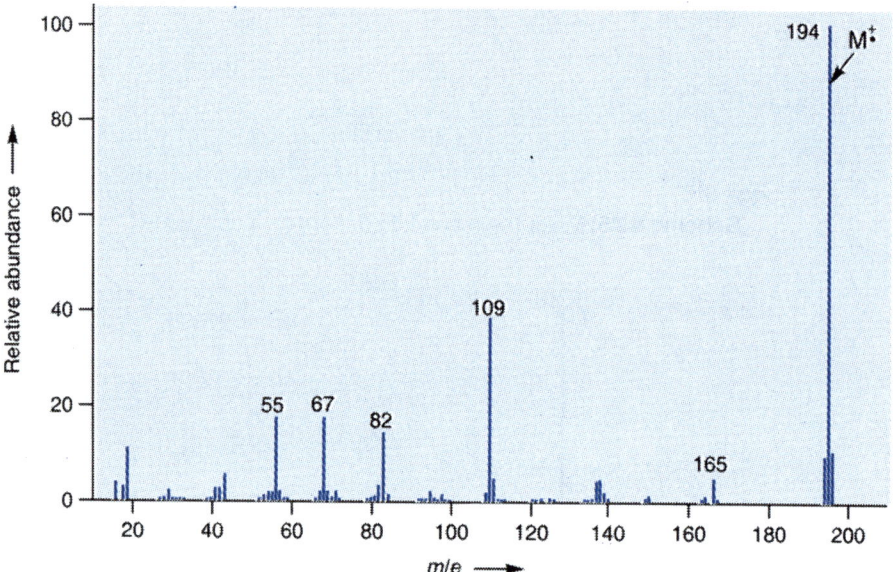

Fig. 6.83: Mass spectrum of caffeine

It shows molecular ion peak at m/e 194 which is also a base peak. The molecular ion forms peak at 165 ($M^+ - CO$), 109 ($C_5H_7N_3$), 82, 67, 55 (Scheme 6.86).

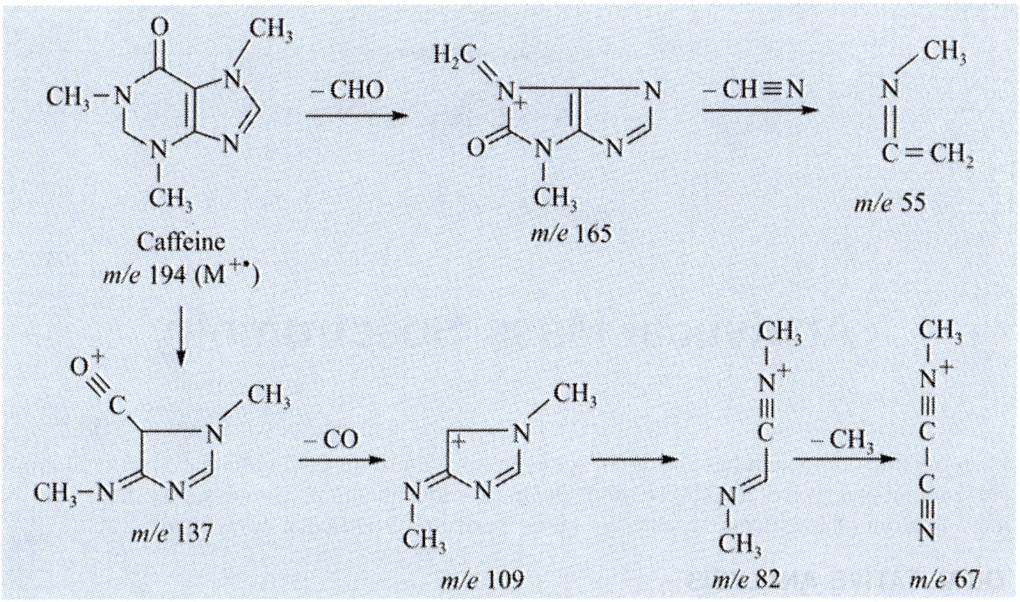

Scheme 6.86: Mass fragmentation pattern for caffeine

7

Analytical Mass Spectrometry

Each organic compound has specific fragmentation pattern depending on the functional group present in it and type of ion source used. Qualitative and quantitative analysis are done with the identification of fragments appeared in mass spectrum of particular compound.

QUALITATIVE ANALYSIS

The fundamental equation of a mass spectrometer is:

$$\frac{m}{e} = \frac{H^2 R^2}{2V}$$

This equation expresses the relationship between the mass to charge ratio (m/e) of an ion, the magnetic field (H), the radius of the trajectories of the ions (R) and the applied accelerating voltage (V). This equation can also be modified as:

$$m = \frac{k}{V}$$

where $k = \frac{1}{2} H^2 R^2$

The mass of the ion focussed on the collector of the mass spectrometer is, therefore, inversely proportional to the applied accelerating voltage and this relationship suggests that the different ions can be focussed by varying the acceleration potential at constant magnetic field. By releasing appropriate values of the voltage, the specific ion (single ion detection) can be focussed for analysis. By suitably varying the accelerating voltage between certain fixed mass numbers, the ion corresponding to these prescribed values can be focussed, successively (multiple ion detection).

Some of the compounds like hallucinogens are active in the order of micrograms, quantities and therefore, their identification from tissues or biological liquids requires the use of a very sensitive method, such as mass spectrometry. It can be detected by either single or multiple ions detection technique. Similarly, the identification of performance enhancing compound in the blood or urine of an athlete prior to competition may alone be sufficient to ban the athlete from competing.

Single ion detection

Only one ion is focussed (generally an intense one). This technique can be used either by coupling with gas chromatograph or by using high resolution mass spectrometer.

In the first method, mass spectrometer is combined with gas chromatograph. Here, mass spectrometer is used as a selective gas chromatograph detector. This technique can increase the sensitivity of detection by a factor of between 1,000 and 10,000 with respect to normal gas chromatographic detectors like electron capture, flame ionization or thermal conductivity detector. An example of this application is the determination of 2,5-dimethoxy-4-methyl amphetamine. Its mass spectrum shows prominent peak at m/e 166 (Fig. 7.1). Hence, acceleration potential corresponding to this mass is focussed and mass spectrum is recorded. Mass fragmentogram is a graph of total ion current versus time.

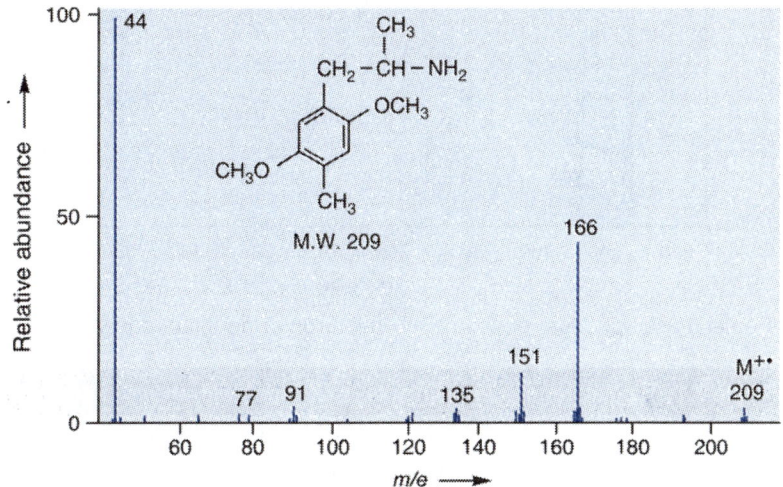

Fig. 7.1: Mass spectrum of 2,5-dimethoxy-4-methyl amphetamine

Using mass spectrometric method, it can be detected at picogram level of DDT, means in trace amounts as low as 10 picogram level. Similarly, testosterone is derivatized with N-methyl-N-trimethylsilyl-trifluoroacetamide (MSTFA), N-trimethylsilylimidazole (TMSI) and dithio-erythritol, and subjected to MS analysis for qualitative identification.

In second method (high resolution mass spectrometry), it is possible to differentiate the ions that have the same nominal m/e ratio but have different elemental composition. This method can be applied for determination of p-tyramine, a precursor of catecholamines, isolated from brain extract. The sample is introduced directly into the ion source and a doublet is observed at the nominal mass 108. The high resolution mass spectrometer is focussed on m/e 108.0575, an ionic fragment corresponding to lipid-hydrocarbon that has a peak at m/e 108.0939.

Multiple ion detection

When the mass spectrometer is employed as a multiple ion detector, it focuses on two or three ionic fragments within a 10–30% range of the magnetic scan of the instrument or on up to eight fragments over the whole field using a quadrupole mass spectrometer. For example, the mass spectrum of desmethyl chloropromazine-trifluoroacetate shows three prominent peaks at m/e 232, 234 and 246. The component separated by gas chromatograph (Fig. 7.2) is focussed on the three peaks at m/e 232, 234 and 246.

Desmethyl chloropromazine trifluoroacetate

Similarly, drugs of abuse like morphine, 6-acetyl morphine, heroin, codeine, etc. can be determined in the urine sample in the forensic analysis of drug abused persons. To the urine sample sodium carbonate (10%) is added to adjust the pH 9.5 and then extraction is done with mixture of $CHCl_3$-isopropyl alcohol (3 : 1). Then sample is subjected to GC-MS analysis. The presence of suspected drugs of abuse can be detected by typical multiple ion detection, given in Table 7.1 and Fig. 7.3.

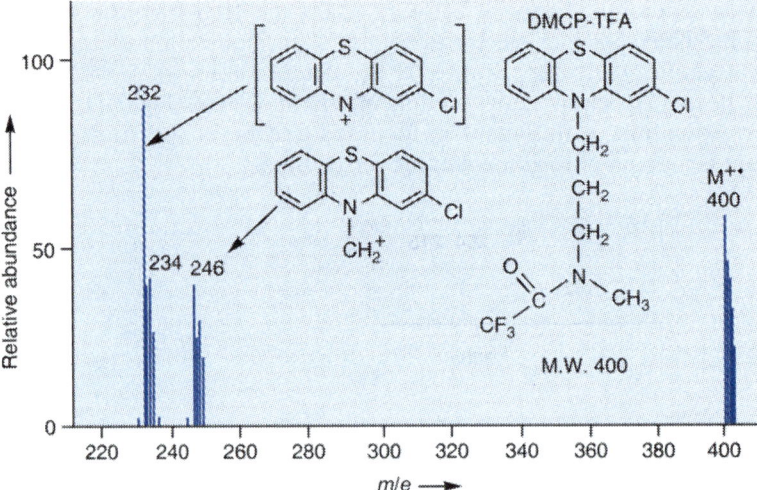

Fig. 7.2: Partial mass spectrum of desmethyl chlorpromazine trifluoroacetate (DMCP-TFA)

Table 7.1: Typical fragment ions peaks (*m/e*) along with their % intensity	
1. Codeine	299(100), 229(24), 162(46), 124(23)
2. Morphine	285(100), 268(16), 162(58), 124(20)
3. Thebaine	311(100), 296(30), 242(11), 211(9)
4. 6-Acetyl morphine	327(100), 268(53), 162(11), 124(13)
5. Heroin	369(59), 327(100), 310(36), 268(47), 162(11)
6. Desmethyl papaverine	325(77), 324(100), 310(30), 266(13)

Quantitative estimation

This method is analogous to those illustrated above using the single ion detection method but has the additional advantage of greater specificity. This result is obtained by focussing on more than one characteristic ionic fragment of a molecule and this ensures the identity of compound displayed. For quantitative estimation, calibration curve is plotted with the help of internal standard, and then from this calibration curve, quantity of unknown sample is determined. The internal standard has the same characteristics and is of a similar (though not the same) size to the compounds of interest. This ensures that the ionization and detection efficiencies of the internal standard and the compound under investigation are essentially identical. It is desirable to have the peak area under the ion signal rather than the height of an ion signal for quantitation because the latter is affected by mass resolution of measurement.

Example 1

This method is useful in the quantitative determination of imipramine. The fragmentation pattern of imipramine shown here. It shows $[M + H]^+$ peak at *m/e* 281. The spectrometer is either focussed at *m/e* 281 or *m/e* 86 to perform quantitative estimation. The quantitative estimation is given by the relative peak areas of the ion fragment. The quantity is directly read out from calibration curve, obtained from standard. This method is useful to estimate the compound in the presence of complex mixtures as in extracts from biological fluids without interference.

M-86(194) - - - -

$CH_2 \dagger CH_2 \dagger CH_2 \dagger N(CH_3)_2$

M-58   M-44(236)

M-72(208)

Imipramine

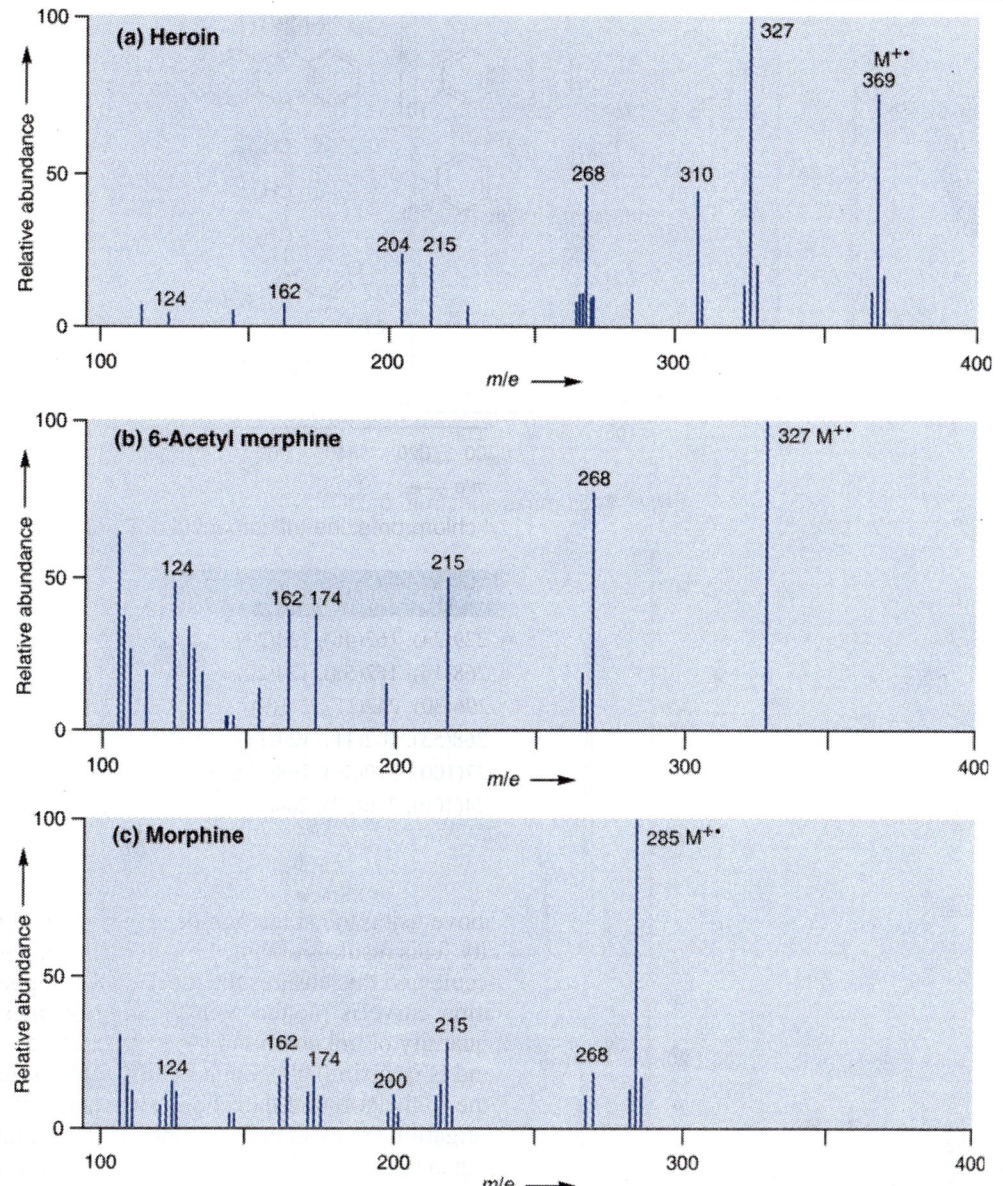

Fig. 7.3: Mass spectra of (a) Heroin, (b) 6-Acetyl morphine; and (c) Morphine

Example 2

DDT [$C_{14}H_9Cl_5$, 352 ($M^{+\bullet}$)] can be quantified by mass spectrometric technique. It shows the fragments at m/e 352 ($M^{+\bullet}$), 282 ($M - 2Cl$)$^+$, 237 ($M + 2 - CCl_3$)$^+$, 235 ($M - CCl_3$)$^+$, 212 ($M - 2Cl - 2Cl$)$^+$, 199 ($M - CCl_3 - Cl$)$^+$ and 165 ($M + 2 - CCl_3 - 2Cl$)$^+$. It can be quantified by calculating the peak area of prominent peak at m/e 235 (base peak) and 165 (Fig. 7.4.).

Example 3

A quantitative mass spectral analysis of the diethyl derivative of 5-fluorouracil in blood plasma can also be done as illustrated in Fig. 7.5. In this analysis, the response from the substance of interest is measured relative to that from an internal standard added to the sample. The top trace

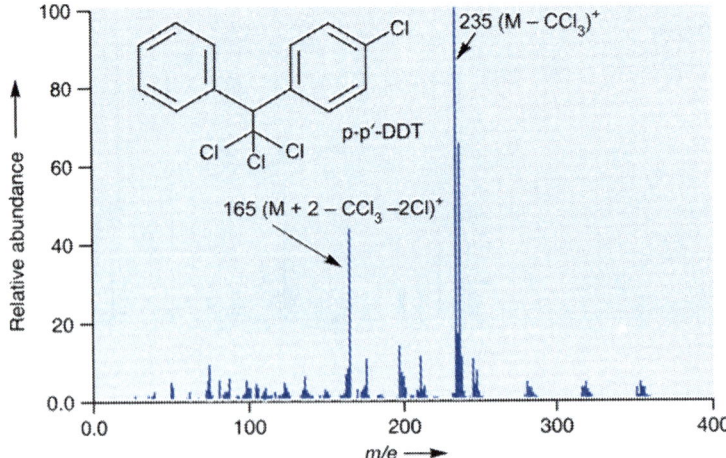

Fig. 7.4: EI mass spectrum of DDT

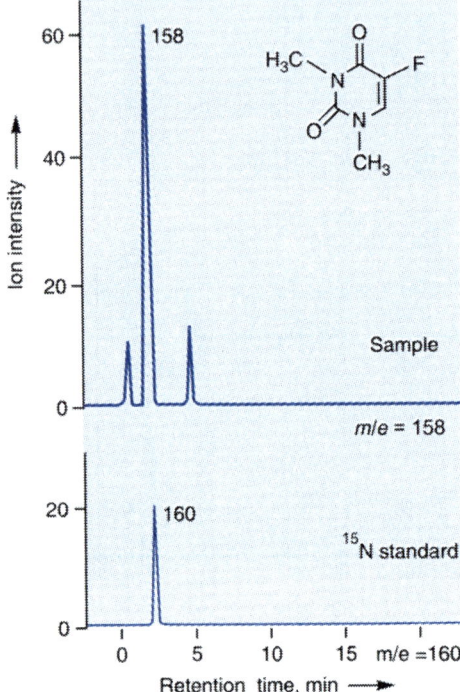

Fig. 7.5: Quantitative MS analysis of the dimethyl derivative of 5-flurouracil in blood plasma

shows the SIM profile of the molecular ion at m/e 158. The compound of interest produces the center peak. Two minor peaks arise from other sample components that also produces ions at m/e 158. The lower trace is the SIM profile of the same molecule having 15N substituted for the normal 14N at both nitrogen positions. Since both nitrogen's are substituted, the signal from its molecular ion is observed at m/e 160 (2 mass unit higher). Although both the sample and standard have the same retention time, they are detected separately by virtue of their different masses. Standards may be a closely related substance or may be chemically identical but synthesized by substituting an isotope of one of the elements as in this example.

Example 4

The dopamine level can be determined by focusing the peak at *m/e* 124 (Fig. 7.6) in tandem mass spectrometry.

Dopamine is a neurotransmitter which plays pivotal roles in neuronal communications in CNS. The detemination of dopamine is necessary for diagnosis of psychiatric diseases.

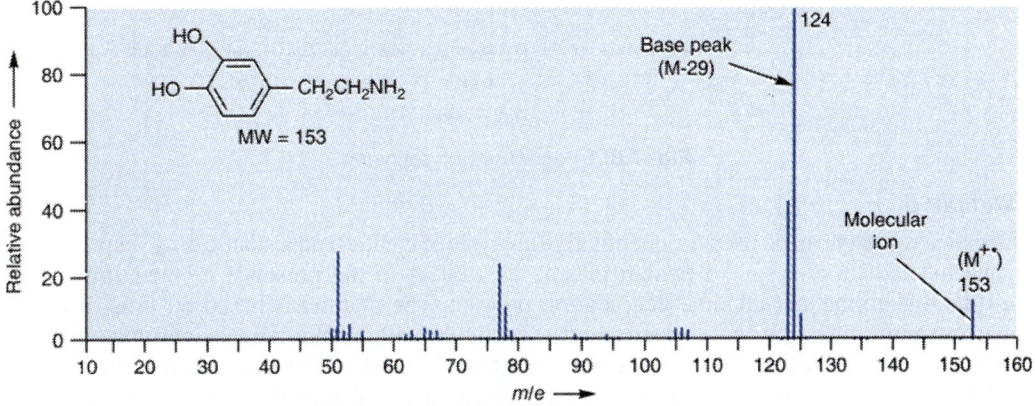

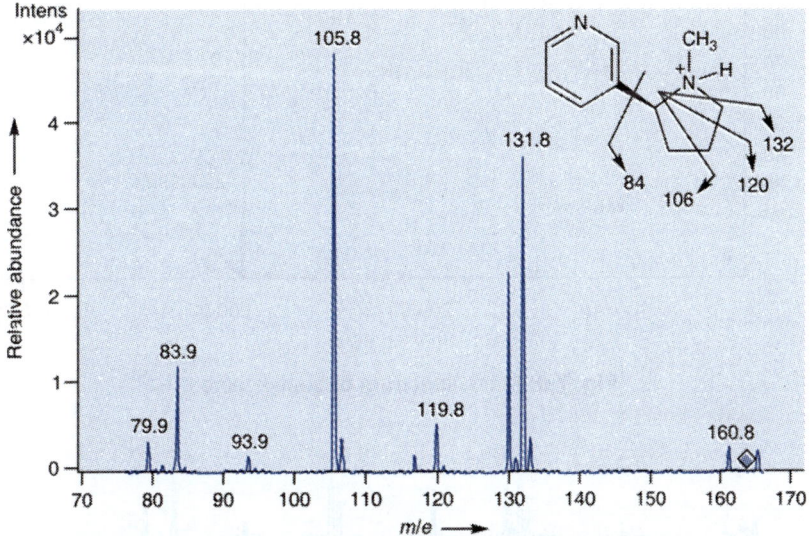

Fig. 7.6: Mass spectrum of dopamine

Example 5

Nicotine can be quantified in tobacco sample by mass spectrometric. The MS spectrum of nicotine obtained after isolation and fragmentation of *m/e* 162.9 is shown in Fig. 7.7. The most abundant ions, 105.8 and 131.8, are chosen for quantification.

Fig. 7.7: Mass spectrum of nicotine and proposed fragmentation pattern

Nicotine is determined in biological fluids on the basis of its metabolite by mass spectrometric technique. Cotinine is a metabolite of nicotine that has been detected in both smokers and non-smokers by single (selected) ion technique in the blood, saliva and urine sample. The EI mass spectrum of cotinine exhibits a fragment ion at *m/e* 98 and molecular ion at *m/e* 176 (Fig. 7.8). Selected ion monitoring of the base peak at *m/e* 98 affords optimal sensitivity for its quantitative estimation. The internal standard, d^3-deutrocotinine is used for making calibration curve. Detection limit is 5–50 ng ml^{-1}.

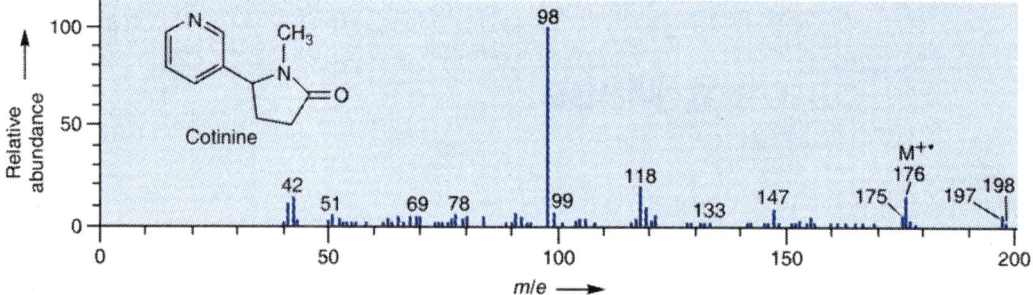

Fig. 7.8: EI spectrum of cotinine

Example 6

Mass spectrometry can be used for quantitative analysis of the local anaesthetic drug, bupivacaine in pharmaceutical and spiked human plasma as well as in the presence of their impurities 2,6-dimethylaniline and alkaline degradation product. The method is based on time of flight electron spray ionization mass spectrometry technique without preliminary chromatographic separation and makes use of ropivacaine as internal standard. The quantification limit is 23.8 ng ml^{-1} and the method is not only highly sensitive and selective, but also simple and effective for determination or identification of drug in authentic and biological fluids. Under the conditions of TOF ES-MS in positive mode, the spectra displays intense peaks of $[M + H]^+$ with ions of the highest mass to charge, 289.2226 for bupivacaine (Fig. 7.9).

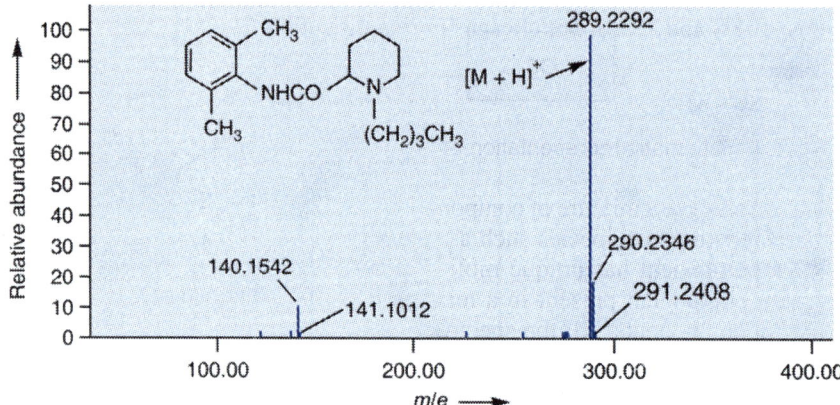

Fig. 7.9: Mass spectrum of bupivacaine

8

Tandem Mass Spectrometry

Introduction

Mass spectrometry is commonly combined with separation devices such as gas chromatograph (GC) and liquid chromatograph (LC) in hyphenated technique. The GC or LC separates the components in a mixture and the each component is introduced, one by one into the mass spectrometer (Fig. 8.1). The tandem mass spectrometry (MS/MS) is an analogous technique where first stage device is another mass spectrometer.

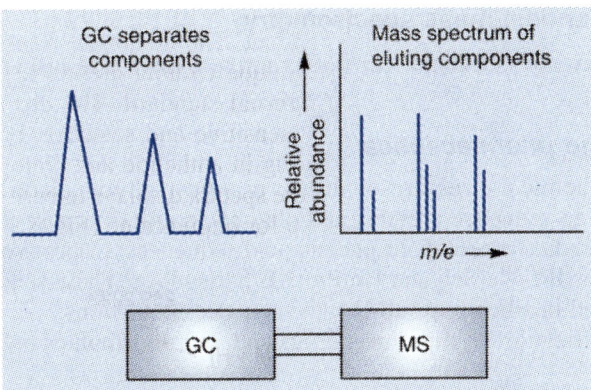

Fig. 8.1: Schematic representation of mass spectrometer coupled to a GC

Suppose that we analyze a mixture of components by soft ionization method. Each component produces a characteristic ionic species such as $[M + H]^+$. To keep the discussion simple, let's assume that each component has unique molecular weight. The mass spectrum of a mixture contains peaks for component present in a mixture. Now, suppose we would like to identify one of the mixture components. All the spectrums tell us the molecular weight but we would like to see the fragment ions that provide the structural information of component of interest (Fig. 8.2).

The simplest form of tandem mass spectrometer combines two mass spectrometers. The first mass spectrometer is used to select single (precursor) mass that is characteristic to the given analyte in a mixture. The mass selected ions pass through the region where they are activated in some way that causes them to fall apart to produce fragment (product) ions. This is usually done by colliding the ions with a neutral gas in a process called collision activation (CA) or collision-induced dissociation (CID). The second mass spectrometer is used to separate the fragment ion according to mass. The resulting MS/MS spectrum consists of only of product

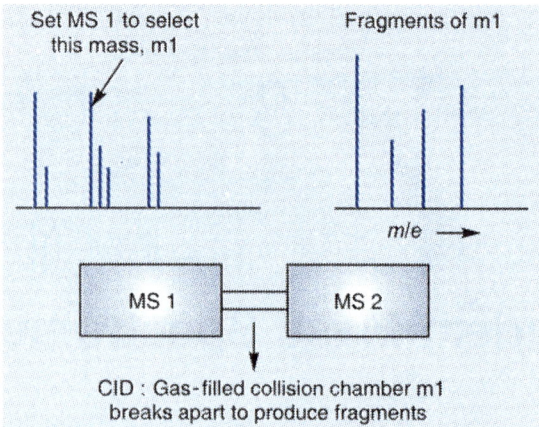

Fig. 8.2: Schematic representation of tandem mass spectrometry (MS/MS)

ions from the selected precursor. Chemical background and other mixture components are absent.

Basically, tandem mass spectrometry uses 2 (or more) mass analyzers in a single instrument. One purifies the analyte ion from a mixture using a magnetic field. The other analyzes fragments of the analyte ion for identification and quantification.

Applications of Tandem mass spectrometry

It is used in industry and academia for both routine assays and research purposes. Some applications are:

Biotechnology and pharmaceutical

1. To determine chemical structure of drugs, proteins, oligonucleotides, etc. MS/MS is commonly used to sequence proteins and oligonucleotides because the fragments can be used to match predicted peptide or nucleic acid sequences, respectively, that are found in databases such as IPI, RefSeq and UniProtKB/Swiss-Prot. These sequence fragments can then be organized in silico into full-length sequence predictions.
2. Detection/quantification of impurities, drugs and their metabolites in biological fluids and tissues.
3. High throughput drug screening
4. Analysis of liquid mixtures
5. Nutraceuticals/herbal drugs/tracing source of natural products or drugs

Clinical testing and toxicology

It is helpful to detect the inborn disorders of metabolism. For detecting inborn disorder, MS/MS works by separating and measuring substances according to their weight. When a drop of a newborn's blood is placed in the MS/MS machine, a special chamber in the machine causes all the substances in the blood to separate and move. The smaller, lighter substances move faster than the larger, heavier substances. The substances move across a special filter and a computer records which substances are present. In addition to recording which substances are present, the computer also measures the quantity of a particular substance. For example, in Fig. 8.3, a blood spot is put through the MS/MS machine and the results show that there are three substances—A, B, and C—and that there is twice as much of substance B as there is of A and C.

Some newborn disorders detected by tandem mass spectrometry are listed in Table 8.1.

Apart from detection of inborn disorders, it is also helpful in diagnosis of cancer, diabetes, various poisons, drugs of abuse, etc.

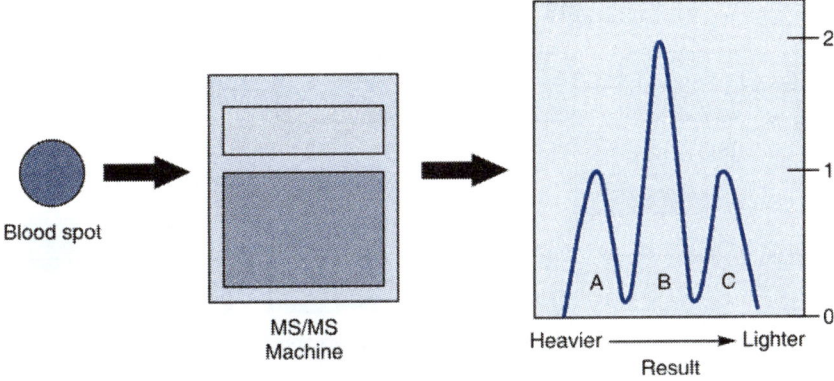

Fig. 8.3: Use of tandem mass spectrometry in clinical testing

Table 8.1: Some disorders detectable by tandem mass spectrometry	
Disorder	**Diagnostic metabolite**
Amino acidemias	
Phenylketonuria	Phenylalanine and tyrosine
Maple syrup urine disease	Leucine + isoleucine
Homocystinuria (CBS deficiency)*	Methionine
Organic acidemias	
Propionic acidemia	C3 acylcarnitine
Methylmalonic acidemia(s)	C3 acylcarnitine
Isovaleric acidemia	Isovalerylcarnitine
Fatty acid oxidation disorders	
SCAD deficiency*	C4, 6 acylcarnitines
MCAD deficiency*	C8, 10:1 acylcarnitines
VLCAD deficiency*	C14, 14:1, 16, 18 acylcarnitines

* CBS (cystathione beta synthase), SCAD (short chain acyl-CoA dehydrogenase), MCAD (median chain acyl-CoA dehydrogenase), and VLCAD (very long chain acyl-CoA dehydrogenase)

MS/MS used with or without chromatographic separation offers the advantage of (i) analytical sensitivity, and (ii) selectivity for drug and drug metabolite analysis. These advantages are achieved through reduction in interferences from other sample components. This allows the development of analytical methods for complex mixtures that are fast, require less stringent sample preparation, less chromatographic separation and therefore consumes much less solvent, allowing higher sample throughput.

MS-MS instrumentation

Single stage mass spectrometry measures the molecular mass of a compound and/or its fragments. **MS/MS consists of two or more** mass spectrometer analyzers all in a single instrument. Figure 8.4 illustrates the comparison of single mass spectrometer (MS) with MS-MS instrumentation.

Tandem mass spectrometers like other mass spectrometers have 4 main parts:

1. Ionization source
2. Sample inlet system
3. Analyzer
4. Detector

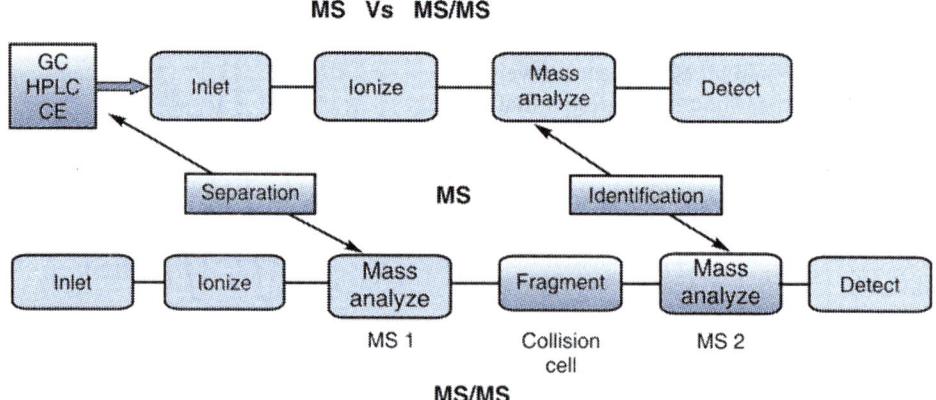

Fig. 8.4: Comparison of single mass spectrometer with MS-MS instrumentation

1. Ionization source

A sample introduced into the ionization source (typically by HPLC) becomes ionized. Ionization process makes the sample components easier to manipulate. The resulting ions are entered into the analyzer where they are separated according to their **mass-to-charge ratios** (*m/e*). The separated ions are then fragmented. The fragments are detected and the signal created by the detected fragment ions is sent to a data system where the *m/e* ratios are stored together with their relative abundance. This data is presented in the format of *m/e* **spectrum**.

The analyzer and detector of the mass spectrometer, and sometimes the ionization source, are maintained under high vacuum to give the ions a reasonable chance of traveling from one end of the instrument to the other without any hindrance from air molecules. Most MS/MS the ionization sources use an **atmospheric pressure ionization** (API) technique.

Methods of sample ionization

The choice of ionization method depends on the type of sample under investigation and the mass spectrometer. Ionization methods most commonly used in tandem mass spectrometry include:

1. Electrospray Ionization (ESI) is an **API** technique (Chapter 2) suited for analysis of polar molecules ranging from less than 100 Da to more than 1,000,000 Da in molecular mass.

    During electrospray ionization the sample is dissolved in a volatile polar solvent and pumped through a narrow stainless steel capillary (~100 µm diameter) at a flow rate between 1 µL/min and 1 mL/min. The "electrospray" is generated by applying a high voltage (~3000 V) to the tip of the capillary. This creates a strong electric field that causes the solvent and sample to elute from the tip of the capillary as **an aerosol of highly charged droplets**. This is assisted by a nebulizing gas (usually nitrogen) flowing around the outside of the capillary and directs the spray towards the analyzer part of the instrument. The charged droplets decrease in size as the solvent is evaporated by the flow of heated nitrogen gas that blows across the front of the ionization source. The charged ions eventually become freed from the solvent then pass through an orifice (called a sampling cone) into a chamber that is under vacuum from which it is then directed through a small aperture into the analyzer section where molecules are separated according to molecular mass.

    Nanospray ionization is a special type of ESI that uses **very low flow rates (30 to 1000 nL/min)** and small sample volumes (1–4 µL) at concentrations of 1–10 µmol/L are used. This technique bypasses the need of a HPLC.

    ESI is operated in 2 modes: In **positive ionization** mode, an organic acid like formic acid is often added in order to add a H^+ ion to the sample molecules; **protonated molecular**

ions have molecular weight of the formula $(M + H)^+$. In **negative ionization** mode a trace of ammonia solution or a volatile amine is added to aid in removal of an H^+ ion from the sample molecules to give a molecular weight determined by the formula $(M - H)^-$.

If the sample has functional groups that readily accept a proton (H^+), then positive ion detection is used, **e.g. amines $R-NH_2 + H^+ = R-NH_3^+$ as in proteins or peptides**. If the sample has functional groups that readily lose a proton, then negative ion detection is used, **e.g. carboxylic acids $R-CO_2H = R-CO_2^-$ and alcohols $R-OH = R-O-$ as in saccharides or oligonucleotides**.

ESI is known as a "soft" ionization method as the sample is ionized by the addition or removal of a proton (H^+), with very little extra energy remaining to cause fragmentation of the sample ions. This is an advantage when analyzing drug metabolites and conjugates.

2. Atmospheric Pressure Chemical Ionization (APCI) uses heat to vaporize the column eluate and a corona discharge to ionize the solvent molecules and produce analyte ions by a variety of chemical ionization methods.

3. Atmospheric Pressure Photoionization (APPI) like APCI uses heat to vaporize the column eluate but uses UV light and chemicals added to the solvent to produce analyte ions through charge and proton transfer reactions.

4. Matrix assisted-laser desorption/ionization (MALDI) uses **laser** light to ionize the sample. Typically a sample is pre-mixed with a special matrix compound for consistent and reliable results. The matrix compound allows the laser light to excite the sample causing it to ionize and move from the surface of the mixture and into the mass spectrometer. This technique does not require a chromatography step.

MALDI is also a "soft" ionization method and is mainly adapted to high molecular weight compounds like proteins.

The first 3 ionization techniques are commonly adapted for small molecular weight compounds.

2. Sample inlet system

Sample can be inserted directly into the ionization source, or can undergo some type of chromatography en route to the ionization source. This latter method of sample introduction usually involves: High pressure liquid chromatography (HPLC), gas chromatography (GC) or capillary electrophoresis (CE) separation column.

3. Mass analyzers

The main function of the **mass analyzer** is to **separate** ions, formed in the ionization source, according to their m/e ratios. More popular tandem mass spectrometers (based on analyzers) include those of the (i) **quadrupole-quadrupole type (also known as triple quadrupole instruments)**, or (ii) **the hybrid types** (which include magnetic sector/quadrupole, and more recently, the quadrupole/time-of-flight (Q-TOF) geometries).

(i) Triple Quadrupole Mass spectrometer

Figure 8.5 is a scheme graph of triple quadrupole mass analyzer. As can be seen from Fig. 8.5 the analyzer consists of three single quadrupole mass analyzers in series. The first quadrupole acts like a mass filter, only those ions with specific m/e value can pass Q1; the second quadrupole Q2 is not used as mass separation device. It acts as a collision cell where, the ions coming out from first quadrupole Q1 collides with some gas atoms and breaks into small fragments and focuses any product ions into the third quadrupole Q3. Both sets of quadrupole may be controlled to allow the transmission of single m/e or range of m/e to give the desired analytical information. In the quadrupole Q2 or collision cell, an inert gas (typically argon or xenon) is admitted to collide with the selected sample ions to bring out their fragmentation. The process is known as collision-induced dissociation (CID). The actual change in mass from a precursor molecule ion

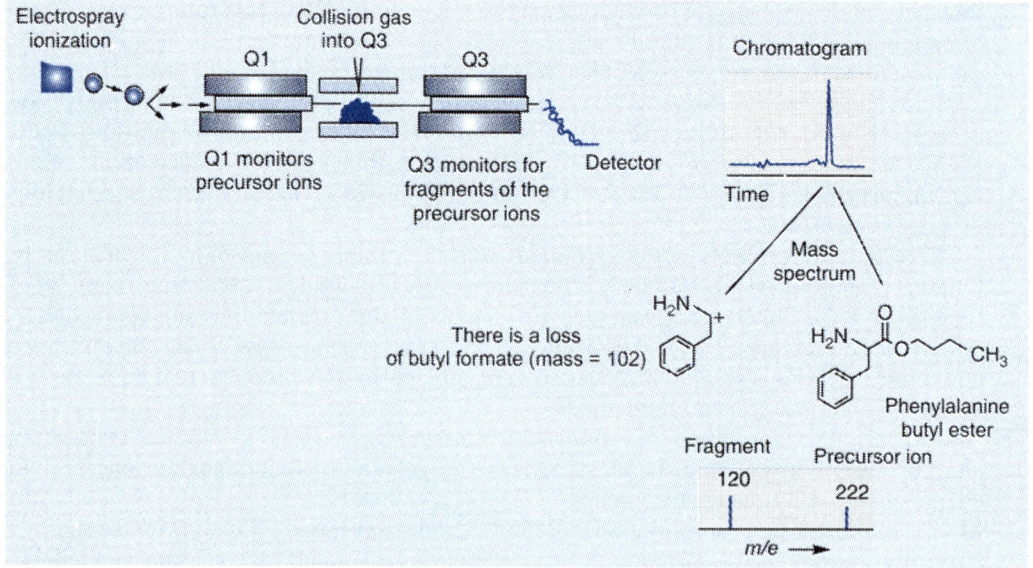

Fig. 8.5: Schematic diagram of tandem mass spectrometry (MS/MS) with triple quadrupole

to a product molecule ion is referred to as a transition and is written as 315.1 > 109 for a molecule of mass M + 1 = 315.1 producing a molecule of mass 109 following reaction in the collision cell.

(ii) Hybrid mass spectrometer

A hybrid mass spectrometer is a device for tandem mass spectrometry that consists of a combination of two or more *m/e* separation devices of different types.

(a) Magnetic sector quadrupole mass spectrometer

When the first quadrupole of the triple quadrupole is replaced by double focusing mass analyzers, the instrument is termed hybrid (hybrid of magnetic sector and quadrupole technologies). The advantage of this configuration is that the MS instrument can be used as high resolution condition to select the ion of interest. Despite the efficiency of modern chromatography, complete resolution of components, especially if mixtures are involved, is not possible and background signal may be observed. Low resolution mass spectrometers allow us to differentiate between components with similar retention properties but whose mass spectra contains ions at different *m/e* ratios while high resolution mass spectrometers allow us to do this but also to differentiate between components whose mass spectra contain ions with the same nominal mass but which have different atomic compositions. In this way, further specificity is conferred on the analysis.

(b) Quadrupole:Time of flight mass spectrometer

In this instrument, the final stage quadrupole of triple quadrupole is replaced by the orthogonal time of flight (TOF) mass analyzer as shown in Fig. 8.6. The configuration is typical of the latest generation of TOF instrument in which a number of reflectrons, in this case two, are used to increase the flight path of the ions and thus increase the resolution that may be achieved. The difference between this and other MS-MS instruments is the way in which the MS2 unit operates. To reiterate, in contrast to others mass analyzers which are scanned sequentially through the *m/e* interest and provide MS-MS spectra of user-selected masses, the TOF analyzer detects all of the ions that enter it at a specific time. It is therefore possible, particularly in view of the high-scan-speed capability of this instrument, to provide, continuously a full MS-MS product ion spectrum of each ion produced in the mass spectrometer.

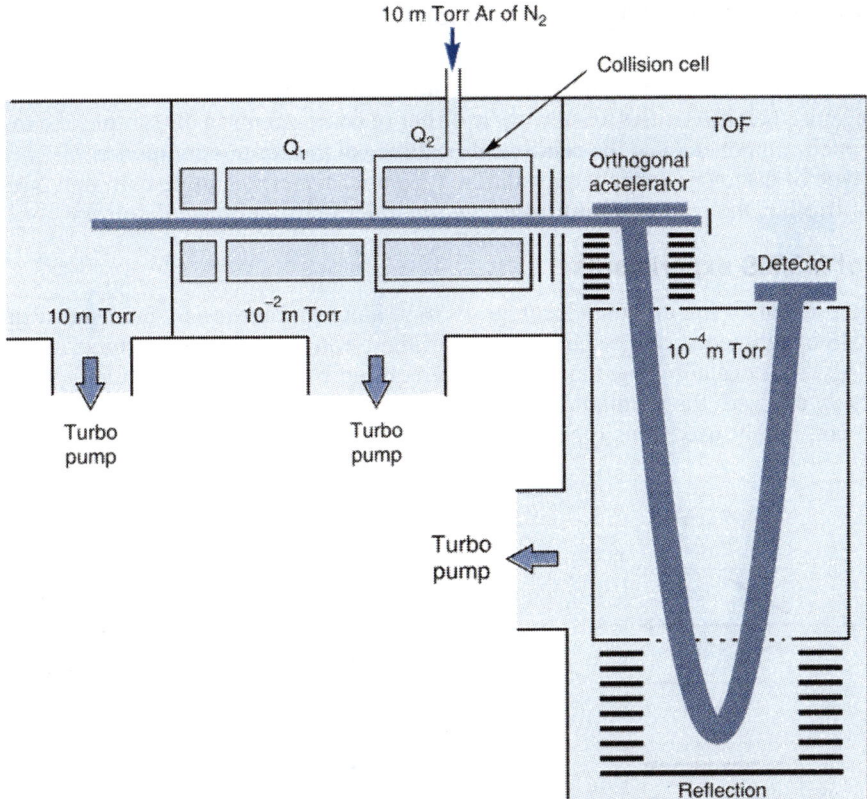

10 m Torr Ar of N_2

Collision cell

Q_1

Q_2

TOF

Orthogonal accelerator

Detector

10 m Torr

10^{-2} m Torr

10^{-4} m Torr

Turbo pump

Turbo pump

Turbo pump

Reflection

Fig. 8.6: Schematic diagram of a hybrid mass spectrometer including a quadrupole analyzer, a quadrupolar collision cell and an orthogonal acceleration time-of-flight analyzer

The disadvantage of this mode of operation is that it renders the Q-TOF system unable to carry out the precursor and constant neutral loss scan.

Note: In an instrument without a reflectron, both the precursor and product ions reach the detector at the same time and are not separated. The reflectrons act as the energy analyzer and product ions with different energies, after passage through the reflectron, will have the different flight time to the detector and thus separated.

(c) Tandem mass spectrometer on the ion-trap

In fact, the principle of ion-trap mass analyzer is the same as the triple quadrupole mass analyzer. First, the trap selects some specific proteins with certain *m/e* values, all the other proteins will fly out of the trap; then the voltage is suddenly increased, which leads to the energy increase of the remaining ions, so the ions collide with the gas atoms and produce many fragments; like in triple quadrupole mass analyzer, the fragments are caught to produce the final mass spectrum.

Triple quadrupole vs ion-trap

The similarity between triple quadrupole and ion-trap is the principle. Both types, first select some specific ion, then break this specific ion into small pieces and catch the fragments to yield the final mass spectrum.

One of the differences between these two analyzers is that they produce different fragments for the same protein ion. Trap-ion trends to fragment the ions more thoroughly. The net result is higher efficiency in fragmenting of molecule.

4. Detection and recording of sample ions.

The **detector** monitors and amplifies the ion current and then transmits the signal to the data system where it is recorded in the form of a **mass spectrum**. The **m/e** values of the ions are plotted against their **intensities** to show the **number of components** in the sample, the **molecular mass** of each component, and the **relative abundance** of the various components in the sample.

     The type of detector is supplied to suit the type of analyzer; the more common ones are the **photomultiplier**, the **electron multiplier** and the micro-channel plate detectors.

Types of MS/MS experiments

Consider a precursor ion m^{1+} that decomposes to produce product ion m^{2+} and a neutral loss N.

     MS/MS experiments can be classified according to which these species (precursor, product, neutral loss) are constant. Note that there is no requirement that the product and precursor ion is positively charged; the terminology is identical to negative charged ions.

     Here commonly used four types of experiments (using triple quadrupole) are shown in Fig. 8.7.

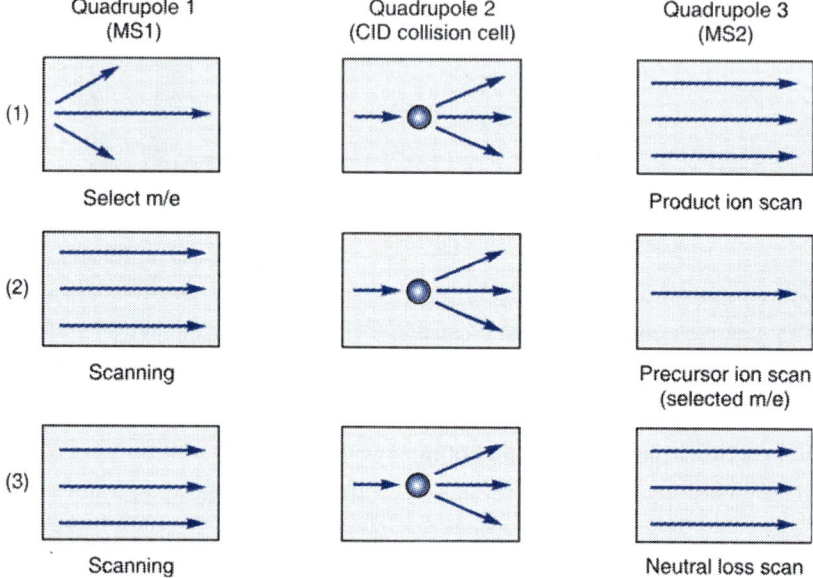

Fig. 8.7: Four types of experiments used in tandem mass spectrometry

1. Product (daughter) ion scanning

The first mass analyzer selects a specified sample ion and so allows only that ion to pass. This ion is usually the molecular-related, i.e. $(M + H)^+$ or $(M – H)^-$ ions. These chosen ions pass into the collision cell, and are bombarded by gas molecules. This breaks up the molecule into fragments. These fragment ions are then separated according to their *m/e* ratios, by the second mass analyzer. All the fragment ions arise directly from the precursor ions specified. The result is a fingerprint pattern specific to the compound under investigation. This type of experiment is particularly useful for providing structural information. For example, the product ion spectrum of the $[M + H]^+$ of buspirone in Fig. 8.8 displays two key fragment ions at *m/e* 122 and *m/e* 265. These ions are attributed to structurally informative product ions on each side of the buspirone molecule as highlighted in fragmentation pathway.

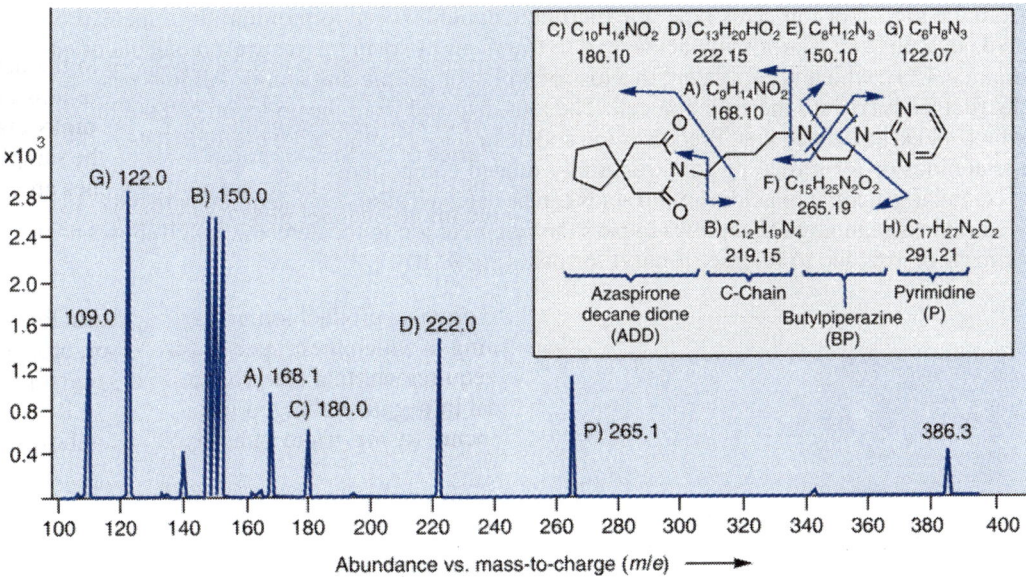

Fig. 8.8: MS/MS spectrum of buspirone by QQQ product ion scan and interpretation of the fragmentation pattern

2. Precursor (parent) ion scanning

The first mass analyzer allows all sample ions to pass through it. The second analyzer is set to monitor specific fragment ion that is generated in the collision cell. All precursor molecules that give rise to the specific fragment will be detected. This type of experiment is useful for monitoring groups of compounds contained within a mixture which fragment to produce common fragment ions. For example, all the acylcarnitine methyl ester cation exhibits a signal at m/e 99 (usually the base peak on CID). As shown in Fig. 8.9, this ion (m/e 99) is derived from the loss of triethylamino group and acyl group from acylcarnitine. In the precursor ion scan mode, quadrupole Q3 of the triple quadrupole transmits ion of m/e 99 while Q1 is scanned from m/e 200 to 450 generates a profile consisting of molecular cation of acylcarnitine with little chemical interferences.

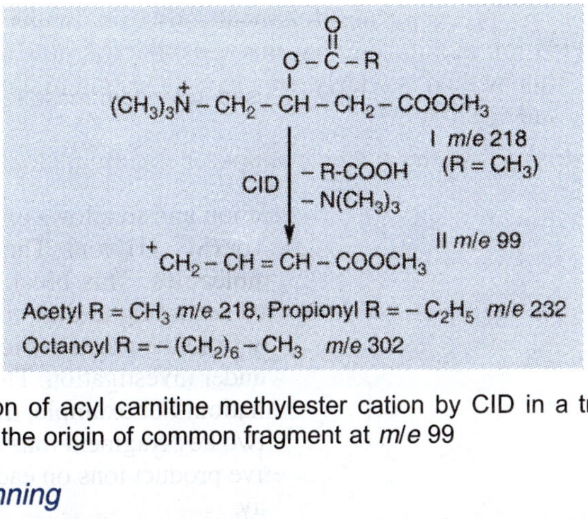

$$O-\overset{\displaystyle \overset{O}{\|}}{C}-R$$
$$(CH_3)_3\overset{+}{N}-CH_2-\overset{|}{CH}-CH_2-COOCH_3$$

I m/e 218
(R = CH₃)

CID $-$ R-COOH
$-$ N(CH₃)₃

II m/e 99

$$\overset{+}{C}H_2-CH=CH-COOCH_3$$

Acetyl R = CH₃ m/e 218, Propionyl R = $-$ C₂H₅ m/e 232
Octanoyl R = $-$ (CH₂)₆ $-$ CH₃ m/e 302

Fig. 8.9: Fragmentation of acyl carnitine methylester cation by CID in a triple quadrupole mass spectrometer showing the origin of common fragment at m/e 99

3. Neutral loss scanning

It involves both analyzers collecting data, across the whole m/e range. But the second analyzer is set that it will only detect products fragments that differ by a specific mass (neutral fragment) from the precursor ion. This type of experiment could be used to monitor all of the carboxylic acids in a mixture. Carboxylic acids tend to fragment by losing a (neutral) molecule of carbon dioxide, CO_2, which is equivalent to a loss of 44 Da or atomic mass units. All ions pass through the first analyzer into the collision cell. The ions detected from the collision cell are those from which 44 Da have been lost. Precursor ion and neutral loss methods are useful to detect unknown metabolites or for screening for structurally related compounds.

Similarly, neutral or acidic amino acids can be detected after derivatization with butyl alcohol. Neutral loss scanning only allows to pass the fragment ion to the detector which shows the loss of mass of 102 due to release of butyl formate (Fig. 8.10).

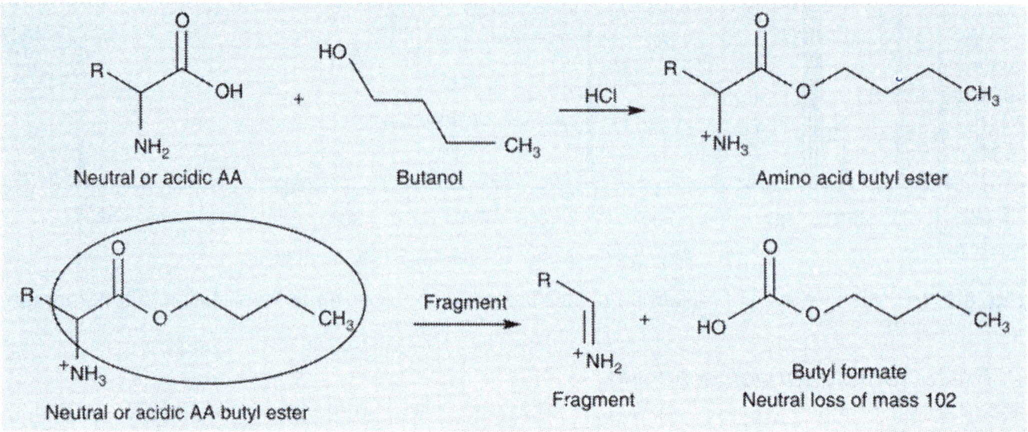

Fig. 8.10: Derivatization and neutral loss scanning

4. Selected reaction monitoring (SRM)

Involves setting both analyzers such that only ions of selected molecular weight are allowed to pass through the first analyzer and only specifically selected fragments arising from these are measured by the second analyzer. It means, for this scan, both the first and second analyzers are focussed on selected masses. The method is quite analogous to selected ion monitoring in standard mass spectrometry. But here the ions selected by the first mass analyzer are only detected if they produce a given fragment, by a selected reaction. The absence ôf scanning allows one to focus on the precursor and fragment ions over longer times, increasing the sensitivity as for selected ion monitoring, but this sensitivity is now associated with a high increase in selectivity. This method is widely used in drug testing in blood and urine samples. It is also very sensitive and specific.

The process is illustrated with the example antipsychotic drug, clozapine.

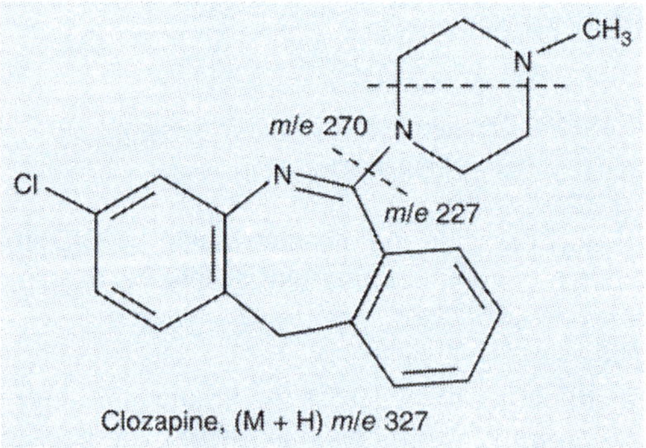

Clozapine, (M + H) m/e 327

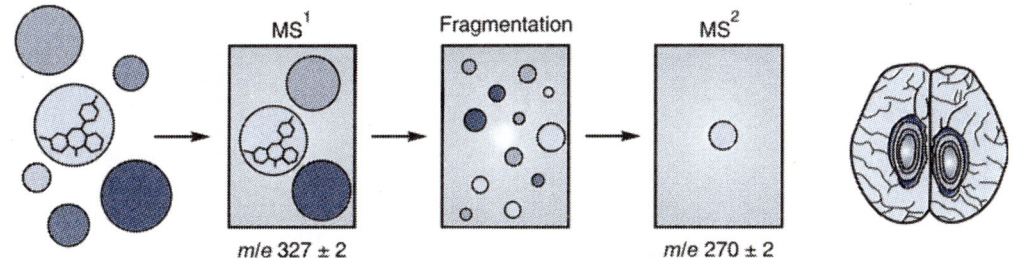

Fig. 8.11: Selected reaction monitoring

After tissue compounds enter the mass spectrometer, the ions with an *m/e* of (327 ± 2) are isolated during the first round of initial mass analysis. Clozapine and its isobaric species are then fragmented, and a second mass analysis isolates the signature fragment of clozapine (*m/z* 270.08) and uses its intensity to create an image. Clozapine and other compounds have been desorbed and ionized prior to mass analysis. The selected reaction monitoring (SRM) is illustrated in Fig. 8.11, the first round of mass analysis removes all ions outside the mass window of 325 Da–329 Da that allows isolation of precursor ions for subsequent fragmentation. The remaining ions are then fragmented. For the second round of mass analysis, ions outside of the 268 Da–272 Da are removed, leaving only the signature fragment of clozapine (at *m/e* 270) to be imaged.

9

Gas Chromatography-
Mass Spectrometry

Introduction

The hyphenated technique is combination (or) coupling of two different analytical techniques with the help of proper interface. Mainly chromatographic techniques are combined with spectroscopic techniques. In the chromatography, the pure or nearly pure fraction of chemical components in a mixture is separated out and spectroscopy produces selective information for identification using standards or library spectra. "The coupling of the separation technique and an on-line spectroscopic detection technology leads to a hyphenated technique." The term hyphenated techniques ranges from the combination of separation-separation, separation-identification and identification-identification techniques.

The term "hyphenation" was first adapted by Hirschfeld in 1980 to describe a possible combination of two or more instrumental analytical methods in a single run. The aim of the coupling is to obtain an information-rich detection for both identification and quantification compared to that with a single analytical technique.

Advantages
1. Hyphenated techniques give fast and accurate analysis
2. There is a higher degree of automation.
3. There is a higher sample throughput.
4. It gives better reproducibility.
5. There is a reduction of contamination due to its closed system.
6. There is a separation and quantification at the same time.

Types of hyphenated techniques
1. Double hyphenated techniques.
2. Triple hyphenated techniques.

1. Double hyphenated techniques
- LC-MS
- LC-NMR
- LC-IR
- CE-MS
- GC-IR
- GC-MS
- GC-FTIR

2. Triple hyphenated techniques

- LC-API-MS
- APCI-MS-MS
- ESI-MS-MS
- LC-ESI-MS
- LC-NMR-MS

GAS CHROMATOGRAPHY-MASS SPECTROMETRY (GC-MS)

The GC-MS is one of the hyphenated analytical techniques. As the name implies, it is actually two techniques that are combined to form a single method of analyzing mixtures of chemicals. Gas chromatography is a technique capable of separating, detecting and partially characterizing the organic compounds particularly when present in small quantity. For example, Fig. 9.1 shows the separation of two components present in a mixture in gas chromatogram. Each component goes into the mass spectrometer which provides some definite information regarding the original compound, may be present in small quantity by providing mass spectrum.

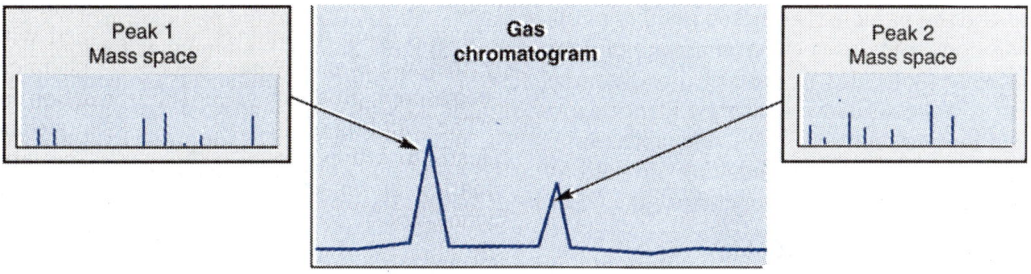

Fig. 9.1: Concept of GC-MS

The use of a mass spectrometer as the detector in gas chromatography was adopted during the 1950s by Roland Gohlke and Fred McLafferty. These sensitive devices were bulky, fragile, and originally limited to laboratory settings. The development of affordable and miniaturized computers has helped in the simplification of the use of this instrument, as well as allowed great improvements in the amount of time it takes to analyze a sample.

Combination of GC-MS provides extremely powerful tool because it permits direct and effectively continuous correlation of chromatographic and mass spectrometric properties (Fig. 9.2). The separation and identification of the components of complex natural and synthetic mixture are achieved more quickly than any other technique with less amount of sample.

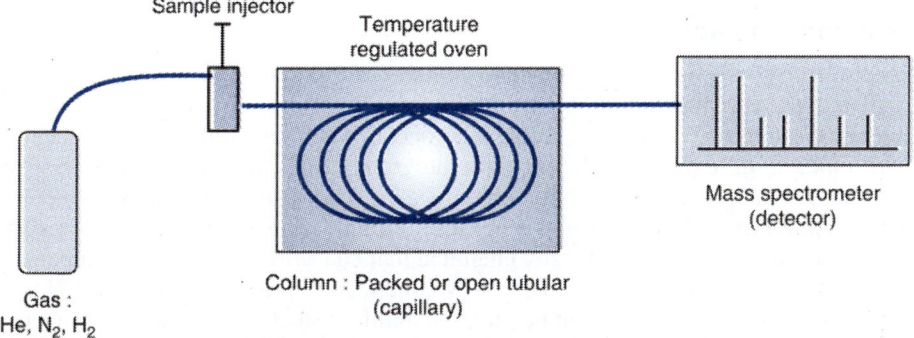

Fig. 9.2: Schematic diagram of GC-MS

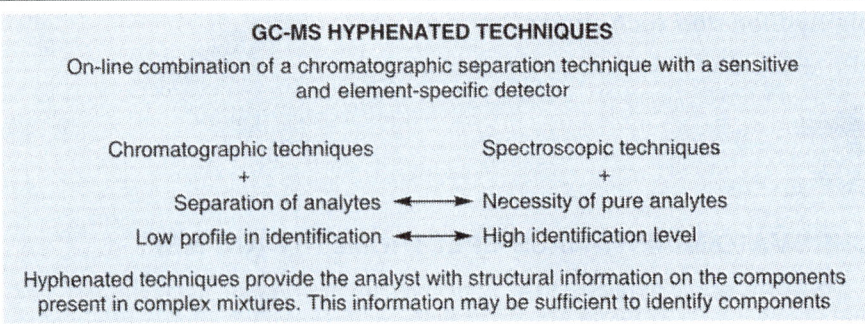

Fig. 9.3: Advantages of GC-MS

Advantages of GC-MS

GC requires the analyte to have significant vapor pressure between 30° and 300°C. GC presents an insufficient proof of the nature of the detected compounds. The identification is based on retention time matching that may be inaccurate or misleading. GC-MS represents the mass (m) of a given particle (Da) to the number of electrostatic charges (e) that the particle carries. GC-MS commonly uses electron impact (EI) and chemical ionization (CI) techniques for ionizing organic molecules to produce positive ions. The main features of enhanced molecular ion, improved confidence in sample identification, significantly increased range of thermally labile and low volatility samples amenable for analysis, much faster analysis, improved sensitivity particularly for compounds that are hard to analyze and the many other features and options provide compelling reasons to use the GC-MS in broad range of areas.

Disadvantages of GC-MS

In GC only volatile samples or the sample which can be made volatile, are separated by this method. During injection of the gaseous sample, proper attention is required.

Components of GC-MS instrumentation

It consists of the following components:

1. Gas chromatograph
2. Mass spectrometer
   - Ionizations source: For example, electron impact (EI) or chemical ionization (CI)
   - Mass analyzer: For example, magnetic sector, quadrupole, ion-trap, time-of-flight
   - Mass detector
3. GC-MS interface
4. Software/data display

1. Gas chromatograph

In general, chromatography is used to separate mixtures of chemicals into individual components. Once isolated, the components can be evaluated individually.

In all chromatographic techniques, separation occurs when the sample mixture is introduced (injected) into a **mobile phase**. In liquid chromatography (LC), the mobile phase is a solvent. In gas chromatography (GC), the mobile phase is an inert gas such as helium.

The mobile phase carries the sample mixture through what is referred to as a stationary phase. The **stationary phase** is a usually chemical that can selectively attract components in a sample mixture. The stationary phase is usually contained in a tube of some sort. This tube is referred to as a column. Columns can be glass or stainless steel of various dimensions.

The mixture of compounds in the mobile phase interacts with the stationary phase. Each compound in the mixture interacts at a different rate. Those that interact the fastest will exit (elute from) the column first. Those that interact slowest will exit the column last. By changing characteristics of the mobile phase and the stationary phase, different mixtures of chemicals can be separated. Further refinements to this separation process can be made by changing the temperature of the stationary phase or the pressure of the mobile phase.

Our GC has a long, thin column containing a thin interior coating of a solid stationary phase (5% phenyl, 95% dimethylsiloxane polymer). The length of column generally varies from 30 to 60 m. The internal diameter is 0.25 mm. This column is referred to as a capillary column. By covering the outside surface of these capillary columns with a polymeric coating, these flexible fused silica GC columns are made more durable. This particular column is used for semi-volatile, non-polar organic compounds. The compounds must be in an organic solvent.

The capillary column is held in an oven that can be programmed to increase the temperature gradually (or in GC terms, ramped). This helps in separation of different components present in a mixture. As the temperature increases, those compounds that have low boiling points elute from the column sooner than those that have higher boiling points. Therefore, there are actually two distinct separating forces, temperature and stationary phase interactions mentioned previously.

The organic compounds are introduced into the GC column by injecting a few microlitres (μL) of the concentrated solvent extract into an injection port (non-volatile organics) or by heating the sorbent trap (volatile organics). An inert carrier gas (He, N_2, H_2), is used to sweep the extracted organic compounds, which are now in the vapor state, through the GC column. Compounds due to different solubility/affinity in the liquid phase of the GC column will take different times to traverse the length of the column. For a specific set of experimental conditions, the time taken by particular compound to travel from injection port (time zero) through a GC column to detector is a physical property of that compound and is referred to as its **retention time**. Generally, higher molecular weight compounds will have greater retention times than lower molecular weight compounds. Also, compounds that have a similar polarity to that of the liquid phase (stationary phase adsorbed on solid support) will be more soluble in the stationary phase and will have greater retention times than compounds less soluble in the liquid phase. Therefore, organic compounds in a mixture can be separated from each other by using gas chromatography, and the retention times of these compounds can be used to assist in their identification. Some environmental samples are so complex that there are hundreds of compounds present in their concentrated organic extracts. There are currently no GC columns available that can completely separate all components of such complex mixtures from each other.

As the compounds are separated, they elute from the column and enter a detector. The detector is capable of creating an electronic signal whenever the presence of a compound is detected. The greater is the concentration in the sample, the bigger the signal. The signal is then processed by a computer. The time from when the injection is made (time zero) to when elution occurs is referred to as the **retention time** (RT).

While the instrument runs, the computer generates a graph from the signal (Fig. 9.4). This graph is called a **gas chromatogram**. Each of the **peaks** in the chromatogram represents the signal created when a compound elutes from the GC column and goes into the detector. The x-axis shows the RT, and the y-axis shows the intensity (abundance) of the signal. In Fig. 9.4, there are several peaks labeled with their RTs. Each peak represents an individual compound that was separated from a sample mixture. The peak at 4.97 minutes is from dodecaine, the peak at 6.36 minutes is from biphenyl, the peak at 7.64 minutes is from chlorobiphenyl, and the peak at 9.41 minutes is from hexadecanoic acid methyl ester.

If the GC conditions (oven temperature ramp, column type, etc.) are the same, a given compound will always exit (elute) from the column at nearly the same RT. By knowing the RT for a given compound, we can make some assumptions about the identity of the compound. However, compounds that have similar properties often have the same retention times. Therefore, more information is usually required before an analytical chemist can make an identification of a compound in a sample containing unknown components.

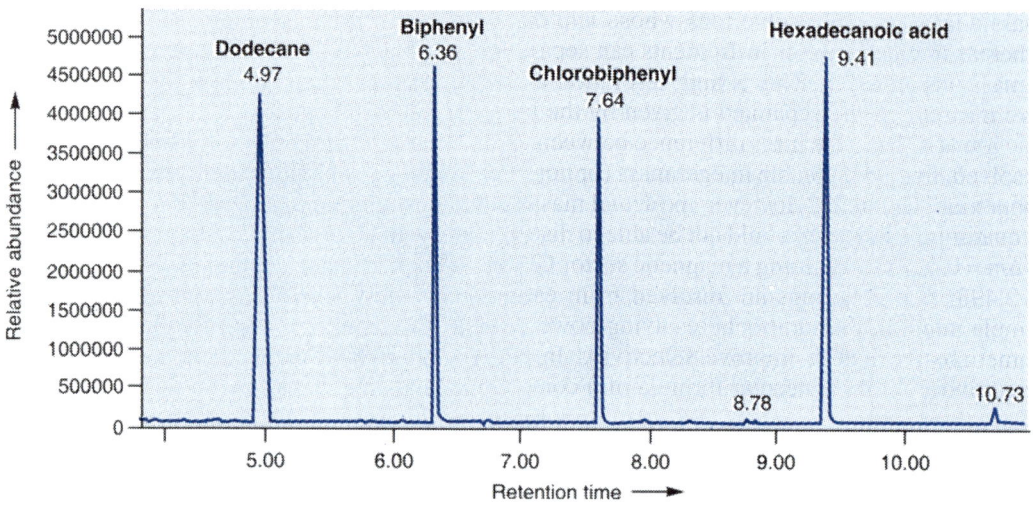

Fig. 9.4: Chromatogram generated by a GC

2. Mass spectrometer

As the separated sample components elute from the GC column, they are monitored using any of a large number of detectors developed for this purpose. The most versatile of these detectors is the mass spectrometer (MS). When an MS detector is used to detect the compounds that elute from a GC column, the combined technique is called gas chromatography-mass spectrometry (GC-MS). A schematic drawing of a GC-MS instrument is given in Fig. 9.2. Initially, molecules enter the source chamber of the mass spectrometer maintained under high vacuum, where they are bombarded by electrons. The energy transferred to molecules in this process causes them to ionize and dissociate into various fragment ions. Ions may be singly or multiply-charged. The positive ions formed are made to traverse an analyzer section, maintained at 10^{-5} to 10^{-7} torr. After the ions traverse the analyzer section where they are separated according to their mass-to-charge ratio (m/e), they are detected by an extremely sensitive device called an electron multiplier. By plotting the abundance of ions detected versus their m/e, a mass spectrum is obtained. The mass spectrum of a compound is like a fingerprint that can be used to identify the original organic structure. It consists of a bar graph representation of the m/e of the ions and their abundances normalized to the most abundant ion (base peak). By matching the GC retention time of a sample component and its mass spectrum with those of a standard reference compound analyzed under the same conditions, a positive identification of the sample component is obtained. Several different mass analyzers have been developed. One of the most common designs consists of a square array of four parallel metal rods. By controlling radio-frequency (RP and DC voltages) to these rods, an oscillating electric field is generated and this allows ions to be filtered according to their m/e. At a specific setting of voltages, only ions of the desired m/e will have a stable trajectory and will be able to reach the electron multiplier. By changing the applied voltages in a specified manner, the mass spectrum of a compound can be generated as the ions of various m/e are scanned. The entire process is performed in about one second. This design is called a quadrupole mass analyzer.

In another design, ions travel through a magnetic field where their momentum is affected by the magnetic field strength. Conditions can be controlled to allow the analyzer to scan across a range of m/e to form a mass spectrum. This design is called a magnetic sector mass analyzer. Other designs are described in detail in references given previously. An important concept in GC-MS is resolution. In GC, resolution refers to the ability of the GC column to separate components in a mixture from each other. In mass spectrometry, mass resolution refers to the ability of an analyzer to separate ions that have similar m/e. For example, quadrupole mass analyzers can resolve ions whose m/e differ by one unit (i.e. mass 16 from mass 17), whereas magnetic sector instruments can separate ions whose m/e differ by one thousandth of a mass unit or better. This is high resolution mass spectrometry. The numerical resolution for two masses that are separated is given by the formula $m/\Delta m$, where m is the nominal mass of one ion and Δm is the mass difference between that ion and the next higher mass ion that is just resolved. For example, the integer mass (or nominal mass) of both nitrogen gas (N_2) and carbon monoxide (CO) is 28. However, the actual mass of CO is 27.99492 while that of N_2 is 28.00615. A quadrupole analyzer would not be able to distinguish between these two ions (mass difference $= \Delta m = 0.01123$). By using a magnetic sector GC-MS analyzer at resolution $m/\Delta m = 28/0.01123 = 2,493$, these two ions are resolved from each other. Since the m/e values of ions are not simple integers, the additional resolving power of high resolution magnetic sector analyzers is sometimes needed to improve selectivity. In addition, it is sometimes possible to establish unequivocally the molecular formula of a compound by accurate mass determinations.

3. GC-MS interface

Gas chromatography and mass spectrometry are, in many ways, highly compatible techniques. In both techniques, the sample is in the vapor phase, and both techniques deal with about the same amount of sample (typically less than 1 ng). Unfortunately, there is a major incompatibility between the two techniques: The compound exiting the gas chromatograph is a trace component in the GC's carrier gas at a pressure of about 760 torr, but the mass spectrometer operates at a vacuum of about 10^{-5} to 10^{-6} torr. This difference in pressure of 8 to 9 orders of magnitude creates a considerable problem. For appropriate performance, the mass spectrometer analyzer housing has to be maintained at a very low pressure, typically about 10^{-6} torr, otherwise analyte ions will not reach the ion detector owing to collisions with other neutral atoms and molecules.

For this reason, the introduction of a sample from the gas chromatograph must be accompanied by some kind of interface/restrictor to keep the total mass (solvent/gas) entering in the mass spectrometer low enough to be compatible with the pumping capacity of the vacuum system. The earliest approach, dating from the late 1950s, simply split a small fraction of the gas chromatographic effluent into the mass spectrometer. Depending on the pumping speed of the mass spectrometer, about 1 to 5% of the GC effluent was split off into the mass spectrometer, venting the remaining 95 to 99% of the analytes into the atmosphere. It was soon recognized that this was not the best way to maintain the high sensitivity of the two techniques, and hence, improved GC-MS interfaces were designed. These interfaces reduced the pressure of the GC effluent from about 760 torr to 10^{-6} to 10^{-5} torr, but at the same time, they passed all (or most) of the analyte molecules from the GC into the mass spectrometer.

Different types of interfacing devices are effusion separator, jet separator and membrane separator.

(a) Effusion separator or Watson-Biemann interface

Effusion separator interface is based on the molecular filtering of the gas effluent by means of porous glass frit. One end of this tube is connected to the GC column and other to the ion source of mass spectrometer. In this, the carrier gas molecules are separated by an effusion chamber which is made of ultra fine-porosity sintered glass (Fig. 9.5) tube with an average pore size of about one micrometer. The porous barrier is surrounded by vacuum chamber. The lighter carrier gas permeates the effusion barrier in preference to the heavier sample organic molecules. Enrichment is typically five-to-six fold and the yield (sample throughput) is an about 27%. Drawbacks: High dead volume added and high surface area shows discrimination effects in the case of smaller molecules.

(b) Jet separator

The classical jet separator or molecular jet interface for GC/MS was developed from the original Becker jet separator. This device works in the basis of the differences in diffusibility between

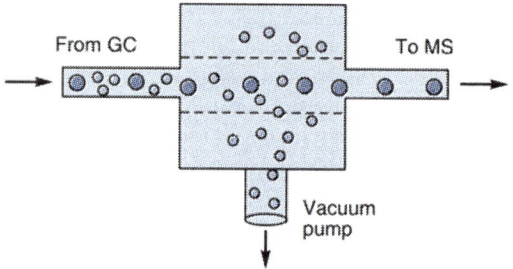

Fig. 9.5: Effusion separator

the carrier gas and the organic compound. The carrier gas is almost always a small molecule such as helium (mass 4) or hydrogen (mass 2) with a high diffusion coefficient, whereas the organic molecules (higher mass) have much lower diffusion coefficients. In operation, the GC effluent (the carrier gas with the organic analytes) is sprayed through a small nozzle, indicated as d_1 in Fig. 9.6, into a partially evacuated chamber (about 10^{-2} torr). Because of its high diffusion coefficient, the helium is sprayed over a wide solid angle (tremendous expansion of gases), whereas the heavier organic molecules are sprayed over a much narrower angle and tend to go straight across the vacuum region. By collecting the middle section of this solid angle with a skimmer (marked d_3 in Fig. 9.6) and passing it to the mass spectrometer, the higher-molecular-weight organic compounds are separated from the carrier gas, which is removed by the vacuum pump. Most jet separators are made from glass by drawing down a glass capillary, sealing it into a vacuum envelope, and cutting out the middle spacing (marked d_3 in Fig. 9.6). It is important that the spray orifice and the skimmer be perfectly aligned. The short path through the interface to the ion source reduces dead volume, which gives better peak separation.

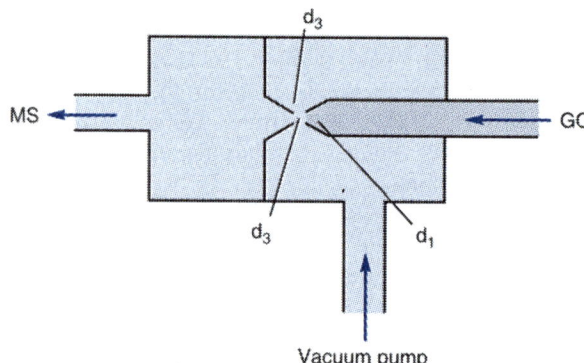

Fig. 9.6: Jet separator, device for interfacing GC with an MS

The jet interface is very versatile, inert and efficient. Despite it, there is a disadvantage of reduced efficiency with more volatile compounds and potential plugging problems at the capillary orifice.

(c) Permselective membrane

In this system dimethyl silicon rubber membrane of thickness 0.025–0.040 mm is used. The membrane is a sandwich (Fig. 9.7) between spiral channels, column effluent on one side under pressure, MS on the other side under vacuum—relies on differential permeability of carrier gas vs analyte molecules. The silicone-rubber membrane transmits organic non-polar molecules and acts as a barrier for (non-organic) carrier gases. Despite being a very effective enrichment procedure, it also suffers from discrimination effects with more polar analytes and produces significant band broadening of their chromatographic peaks

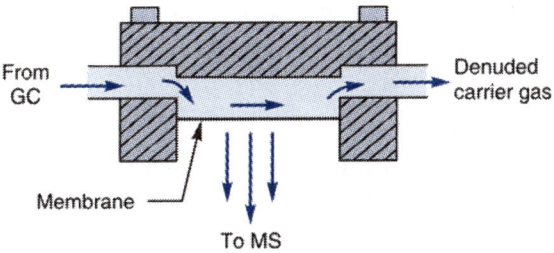

From
GC

Denuded
carrier gas

Membrane

To MS

Fig. 9.7: Permselective membrane

(d) Direct introduction

Capillary columns use optimum flow rates of gas of about 1–2 ml min^{-1}, instead of more than 10–20 ml min^{-1} used with packed columns, allowing all the effluent to be directed to the mass spectrometer. This is usually done through a direct coupling where the column exit is introduced into the ion source without a capillary restriction (Fig. 9.8). It is extensively used in analytical laboratories.

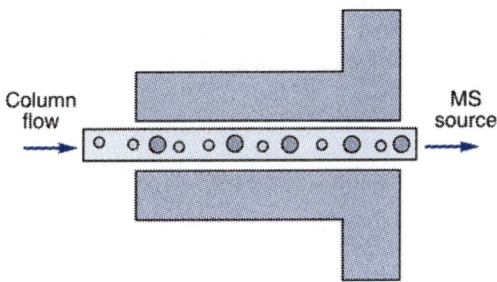

Column
flow

MS
source

Fig. 9.8: Direct introduction

4. Software/data display

The ions which reach to the detector give the signal and displayed on the screen. These data are stored.

Applications for GC-MS

GC-MS provides enhanced sample identification, higher sensitivity, an increased range of analyzable samples, and faster results, which increases the utility of applications for GC-MS technique in several areas.

Medicine

GC-MS is used in screening tests for the detection of several congenital metabolic diseases. It detects trace levels of compounds present in the urine of patients with genetic metabolic disorders. It can also detect the presence of oils in ointments, creams, and lotions.

Environmental monitoring

Monitoring environmental pollutants is a major application of GC-MS. It is widely used in the detection of dibenzofurans, dioxins, herbicides, sulphur, pesticides, phenols, and chlorophenols in air, soil, and water. This method is very sensitive and effective.

Food and fragrance analysis

Aromatic compounds such as fatty acids, esters, aldehydes, alcohols, and terpenes present in food and beverages can be easily analyzed using GC-MS. The technique can also be used to detect spoilage or contamination of food. The analysis of a wide range of oils such as lavender oil, olive oil, spearmint oil, and essential oils, perfumes, fragrances, allergens, menthol, and syrups is also possible using GC-MS.

Pharmaceutical applications

In the pharmaceutical industry, GC-MS is used in research and development, production, and quality control. It is used in identification of impurities in active pharmaceutical ingredients. In medicinal chemistry, GC-MS is used in the synthesis and characterization of compounds and in pharmaceutical biotechnology.

Forensic applications

Using GC-MS, fire debris analysis can be performed as per the American Society for Testing Materials (ASTM) standards. GC-MS is widely used in forensic toxicology to identify poisons and steroids in biological specimens and in anti-doping labs to detect performance enhancing drugs such as anabolic steroids.

Biological analysis

GC-MS can be used for the bioanalysis of body fluids to detect narcotics, barbiturates, alcohols, and drugs such as anticonvulsants, anesthetics, antihistamines, sedative hypnotics, and anti-epileptic drugs. It is also useful in detecting pollutants and metabolites in serum and in fatty acid profiling in microbes.

Chemical warfare

Explosive detection systems in public places use GC-MS technique for the analysis and detection of chemical warfare agents.

Geochemical research

Due to its structurally significant mass spectral peaks, extended range of analyzable low volatility samples, enhanced molecular ions, and valuable isotope ratio information, GC-MS is a powerful tool for geochemical applications.

Industrial applications

GC-MS is ideal for the analysis of inorganic gases and aromatic solvents, detection of impurities and allergens in cosmetics. It is also used in the synthesis of cellulose acetate, polyethylene, polyvinyl, and synthetic fibers.

Therefore, we may conclude that automated GC-MS systems offer rapid and reproducible results in several applications.

10

Liquid Chromatography-
Mass Spectrometry

Liquid chromatography-mass spectrometry (LC-MS) is an analytical chemistry technique that combines the physical separation capabilities of liquid chromatography (or HPLC) with the mass analysis capabilities of mass spectrometry (MS).

Coupled chromatography-MS systems are popular in chemical analysis because the individual capabilities of each technique are enhanced synergistically. While liquid chromatography separates mixtures with multiple components, mass spectrometry provides structural identity of the individual components with high molecular specificity and detection sensitivity.

The LC-MS technique may be applied in a wide range of sectors including biotechnology, environment monitoring, food processing, and pharmaceutical, agrochemical, and cosmetic industries.

LC-MS SYSTEM

Typical LC-MS system is combination of HPLC with MS using interface (ionization source) (Fig. 10.1). The sample is separated by LC, and the separated sample species are subjected to atmospheric pressure ion source, where they are converted into ions in the gas phase. The mass analyzer is then used to sort ions according to their mass-to-charge ratio and detector counts the ions emerging from the mass analyzer and may also amplify the signal generated from each ion. As a result, mass spectrum (a plot of the ion signal as a function of the mass-to-charge ratio) is created, which is used to determine the elemental or isotopic nature of a sample, the masses of particles and of molecules, and to elucidate the chemical structures of molecules.

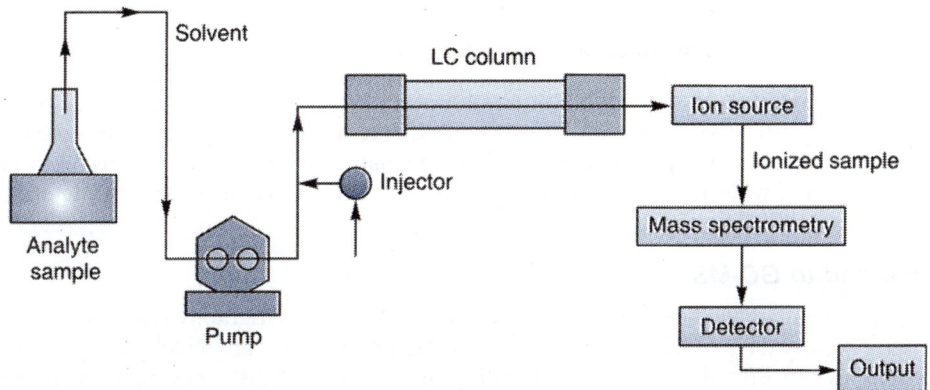

Fig. 10.1: Hyphenated liquid chromatography–mass spectrometry technique

Advantages of LC-MS

In general, liquid chromatography (LC) separates the components of a sample based on differences in their affinity (or retention strength) for the stationary phase or mobile phase, then detects the separated components using UV, fluorescence, or electrical conductivity based on their properties. Such detectors primarily qualify substances based on retention time and quantitate substances based on peak intensity and peak area. Chromatography offers great resolution, but accurately qualifying and quantitating substances can be difficult if multiple components elute approximately at the same time, such as during simultaneous multianalyte analysis. In contrast, mass spectrometry (MS) offers a highly sensitive detection technique that ionizes the sample components using various methods, then separates the resulting ions in vacuum based on their mass-to-charge ratios and measures the intensity of each ion. Since the mass spectra provided by MS can indicate the concentration level of ions that have a given mass, it is extremely helpful for qualitative analysis. That is because mass is information particular to specific molecules and MS enables obtaining that information directly. However, that only applies when measuring a single component. If multiple components are injected simultaneously, it becomes extremely difficult to analyze the spectra. Therefore, LC-MS systems combine the outstanding separation resolution of liquid chromatography with the outstanding qualitative capabilities of mass spectrometry. The mass spectra obtained from these scan measurements provide molecular mass and structural information for eluted components, which supplements the qualitative information based on retention times obtained using other LC detectors (Fig. 10.2).

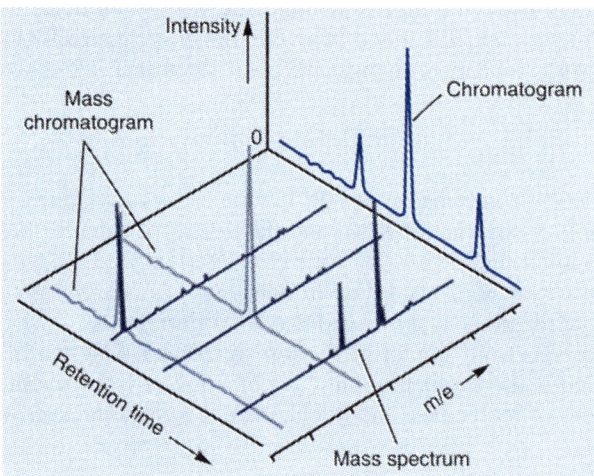

Fig. 10.2: Chromatogram and mass spectrum

In addition, SIM (selected ion monitoring) measurements detect substances based on mass, which is a highly-selective parameter. This enables quantitative analysis that avoids the effects of impurities even when separation by LC is inadequate. In terms of providing both broad applicability for a wide range of substances and high selectivity, mass spectrometers offer excellent characteristics as an LC detector.

Comparison to GC-MS

Mass spectrometers have been used as a detector for gas chromatography (GC) systems since the relatively early days of mass spectrometry and their advantages are widely recognized. GC-MS provides a very effective means for separating and qualifying substances, but its applicability is limited to gases or volatile compounds with relatively low molecular mass, and samples with high thermal stability. In contrast, if the substance is dissolved in a mobile phase (liquid), then

LC is able to analyze even least volatile or thermally-unstable compounds that are difficult to analyze using GC-MS. In other words, one of the advantages of LC-MS is its broad applicability.

Interfacing LC and MS

Mainly the LC-MS contains liquid chromatography assembly, ion generation unit/ ionization source, mass analyzer and mass spectrometric data acquisition. The effluent mobile phase with separated compound from the liquid chromatography is interfaced with the ionization source of the mass spectrometer.

There has been a major focus on improving the interface between the LC and the MS. Liquid chromatography uses high pressure to separate a liquid phase and produces a high gas load. Mass spectrometry requires a vacuum and a limited gas load. For example, common flow from an LC is 1 ml/min of liquid which, when converted to the gas phase, is 1 l/min. However, a typical mass spectrometer can accept only about 1 ml/min of gas.

Furthermore, an LC operates at near ambient temperature, whereas an MS requires an elevated temperature. There is no mass range limitation for samples analyzed by the LC but there are limitations for an MS analyzer.

Finally, LC can use inorganic buffers and MS prefers volatile buffers.

Hence, there is need of interface which can produce the compromising situation between two incompatible analytical techniques. These interfaces are as follows.

Interfaces using desorption techniques

The interfaces in this category are based on the selective vaporization of the elution solvent before it enters the spectrometer source.

Moving-belt interface

The first interface to be made available commercially was the moving-belt interface, shown schematically in Fig. 10.3. The operation of the interface may be divided conveniently into four stages, as follows: (i) Application of mobile phase and analyte(s) to a continuously moving belt; (ii) removal of the mobile phase by passage of the belt under an infrared heater and through a number of differentially pumped regions; (iii) After removal of the solvents, the analyte molecules are (thermally) desorbed from the belt into the ion source and mass analyzed; (iv) cleaning of the belt with a heater and/or a wash-bath to remove any in volatile materials or excess sample prior to the application of further mobile phase and analyte(s) and a repeat of steps (ii)–(iv).

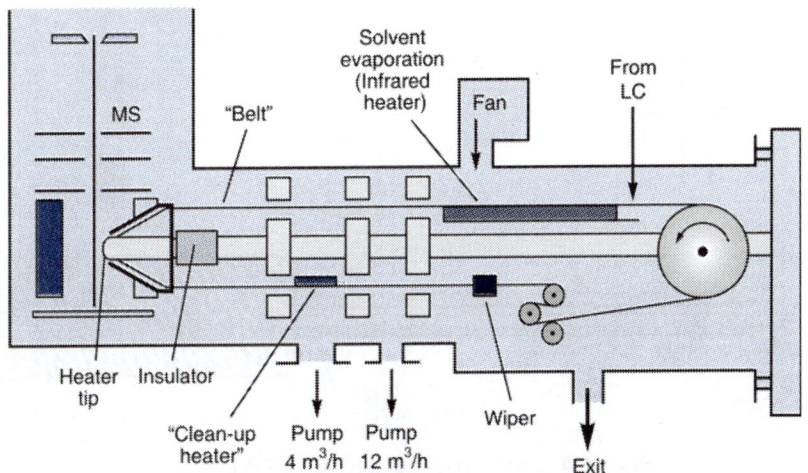

Fig. 10.3: Moving belt interface

Advantages

- The interface can be used with a wide range of HPLC conditions, flow rates and mobile phases, both normal and reverse phase, particularly if spray deposition is employed.
- The analyst does have some choice of the ionization method to be used; EI, CI and FAB are available, subject to certain limitations, and thus both molecular weight and structural information may be obtained from the analyte(s) under investigation.

Disadvantages

- An intense chemical background from the material from which the belt was made is often observed in the mass spectra generated by this type of interface unless adequate conditioning is carried out.
- The belt is prone to break during operation.
- Problems may be encountered in the analysis of thermally labile compounds, as heat is required for mobile-phase removal and for the transfer of analyte from the belt into the source of the mass spectrometer, and highly in volatile compounds which cannot be desorbed from the belt, unless FAB is used for ionization.

Continuous flow fast atom bombardment (CF-FAB) interface

In a continuous flow fast atom bombardment (CF-FAB) interface, typically a 5–15 µl/min liquid stream, mixed with 5% glycerol (or) thioglycerol (or) nitrobenzyl alcohol as FAB matrix, flows through a narrow-bore fused-silica capillary towards either a stainless steel frit or a (gold plated) FAB target (Fig. 10.4). At the target or frit, a uniform liquid film is formed due to a subtle balance between solvent evaporation and solvent delivery. The solvent evaporates but the glycerol (or other matrix forming substances) stays at the nozzle surface. Ions are generated by bombardment of the liquid film by fast atoms or ions, common to FAB. The fast atom like Ar or Xe is used for bombarding the sample and ions are sputtered out of the solution and enter into MS.

Advantages

- Can study thermally labile materials, since the only heat applied to the probe tip is that required to prevent freezing of the mobile phase as it evaporates.

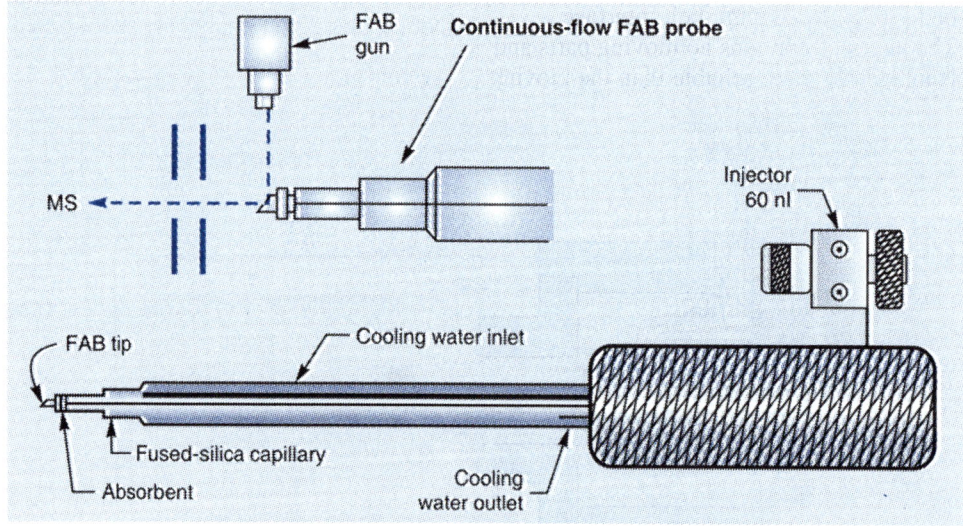

Fig. 10.4: Common designs for CF-FAB interface

- Due to low flow rates used with micro bore capillary, it produces good sensitivity small amounts of sample.
- The interface is simple in design and relatively easy to construct.
- Mobile phases with a high percentage of water may be used.

Disadvantages

The presence of the matrix can cause chromatographic problems if added to the mobile phase before the column, especially if this is of small diameter. The low flow rates that are used require an increased concentration of matrix to be present in the mobile phase to ensure an appropriate amount reaches the probe tip. If conventional HPLC columns are used, with splitting of the eluate to provide the necessary flow rate, an overall decrease in sensitivity usually results.

Interfaces using aerosols and additional ion generation

In these methods for interfacing liquid chromatography and mass spectrometry, there is production of aerosol from the column eluate. The size of droplets is allowed to decrease in size by evaporation until the solid analyte remains and then generate ions. Several methods have been employed to produce an aerosol of as small as possible either in vacuum or under atmospheric pressure. In vacuum, the fine droplets tend to aggregate and heat transfer has to be optimized to minimize this effect.

Direct liquid introduction (DLI) interface

In the DLI interface, the column effluent is split at the entrance of ion source. Only about 1–2% of the effluent is allowed to enter the mass spectrometer through a 5 μm aperture, made in a stainless steel diaphragm at the end of large-bore capillary. The rest of effluent is discarded (Fig. 10.5). The capillary is water cooled to prevent solvent boiling before ejection of the liquid jet through small orifice. The droplets of the jet are vaporized in a heated desolvation chamber, after which the sample enters the ion source. After desolvation of the droplets in a desolvation chamber, the analytes can be analyzed using solvent-mediated CI with the LC solvents as the reagent gas.

Advantages

- No heat is applied to the interface and it is therefore able to deal with thermally labile materials better than the moving-belt interface.
- The interface contains no moving parts and is cheap and simple to construct and operate and is inherently more reliable than the moving-belt interface.

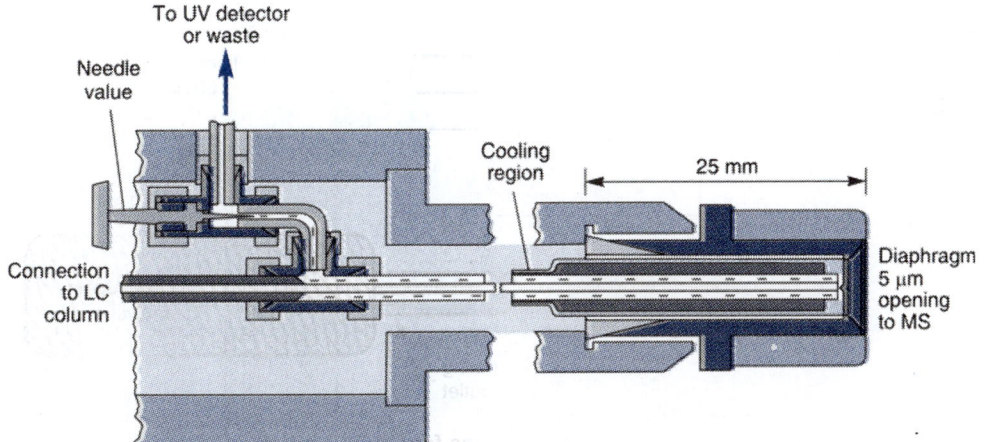

Fig. 10.5: Schematic diagram for DLI interface

- Both positive- and negative-ion CI spectra can be generated and the interface provides molecular weight information, plus it can also be used for sensitive quantitative and semi-quantitative procedures.

Disadvantages

- Non-volatile compounds are not usually ionized with good efficiency.
- The pinhole is prone to blockage and therefore the system must be kept completely free of solid materials.
- Only a small proportion of the flow from a conventional HPLC column is able to enter the source of the mass spectrometer and sensitivity is consequently low.
- Ionization is brought about by CI-like processes and structural information is therefore limited unless a mass spectrometer system capable of MS-MS is used.

Particle beam (PB) interface

The LC eluent is forced through a small nebulizer using a helium (He) gas flow to form a stream of uniform droplets (Fig. 10.6). These droplets move through a desolvation chamber and evaporate to a solid particle. During the desolvation of the droplets, the less volatile analyte molecules coagulate into small particles, typically 50–300 nm. The solvent vapor and nebulization gas are separated from the particles by means of molecular-beam technology. The mixture is expanded at a nozzle into a vacuum chamber. The low mass solvent molecules show a greater tendency to diffuse away from the centre of the expansion, while the heavier analyte particles are kept in the core of the vapor jet. The core of the jet is then sampled by a skimmer. The region between the first set of orifices is maintained at between 2 and 10 torr, and that between the second set between 0.1 and 1 torr. In the momentum separator, sufficient pressure reduction is achieved to generate EI and solvent-independent CI mass spectra.

Advantages

A number of thermally labile and relatively in volatile compounds which do not yield EI spectra when using more conventional inlet methods do so when introduced via the particle-beam interface.

Disadvantages

- The sensitivity of the particle-beam interface is dependent not only on the specific analyte but also on the experimental conditions employed. Detection limits are invariably higher than are desirable.

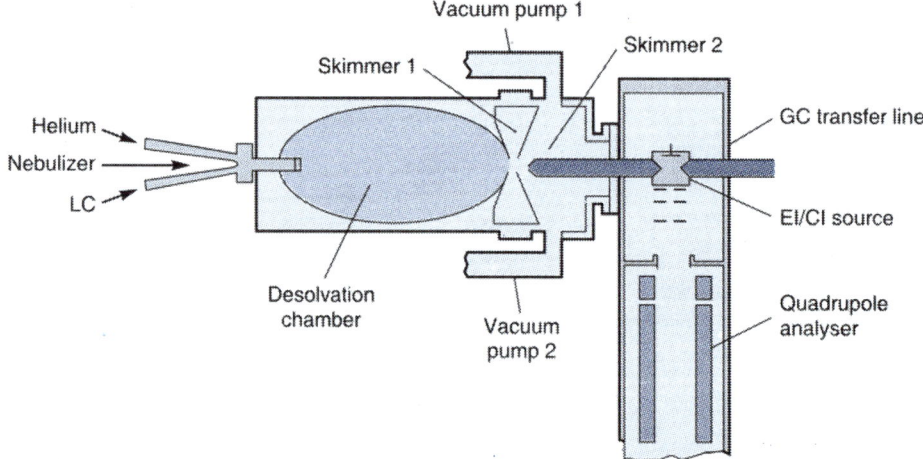

Fig. 10.6: Schematic diagram showing the principal components of particle beam interface

- Neither extremely volatile nor extremely nonvolatile compounds are ideal for investigation using the particle-beam interface.
- The performance of the particle-beam interface deteriorates as the percentage of water in the HPLC mobile phase increases.

Interfaces using aerosols and additional ion generation

In these techniques, the generation of ions is an integrated part of the interface, which therefore dictates the ionization mode.

Thermospray (TSP) interface

The thermospray (TSP) interface, pioneered in 1980 by *Vestal and coworkers* is most widely used LC/MS technique for routine applications in the chemical industry. The technique is sketched in Fig. 10.7. A jet of vapor and small droplets is formed out of a heated vaporizer tube (capillary tube) into a low-pressure region. Nebulization takes place as a result of the disruption of the liquid by the expanding vapor that is formed upon evaporation of part of the liquid in the tube. Before the onset of the partial inside-tube evaporation, a considerable amount of heat is transferred to the solvent, which assists in the desolvation of the droplets in the low pressure region. By applying efficient pumping directly at the ion source up to 2 ml/min of aqueous solvents can be introduced into the MS vacuum system. The ionization of the analytes takes place by means of ion-molecule reactions and ion evaporation processes. The CI reagent gas can be made either in a conventional way using energetic electrons from a filament or a discharge electrode, or in a process called TSP ionization, where the volatile buffer dissolved in the eluent is involved. In the filament-on mode, high-energy electrons (0.4–1.0 keV) emitted from a heated filament are accelerated into the ion source. In the discharge-on mode, a continuous gas discharge is used to generate electrons. Solvent-mediated CI spectra are obtained in both filament-on and discharge-on mode.

Advantages

- The interface was much easier to use at flow rate of 1 ml/min and mobile phases containing high percentages of water
- Can allow unequivocal determination of molecular weight.

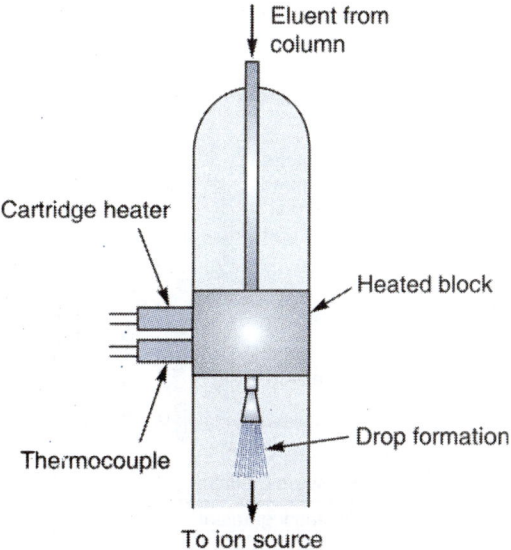

Fig. 10.7: Thermospray interface

- Can be used to study thermolabile compounds.
- High sensitivity.

Disadvantages

- The mobile phase used should be volatile.
- Decomposition of some thermally labile analytes is observed.
- The interface is not suitable for high-molecular-weight (>1000 Da) analytes.
- The reproducibility of analytical results is affected by a number of experimental parameters and is sometimes difficult to control.
- The formation of adducts may confuse the assignment of molecular weight. Difficult to interpret the ionic species generated after repelled-induced fragments.

Electrospray interface

A diagram of electrospray interface is shown in Fig. 10.8. The electrospray interface differs from the thermospray in that it is operated at atmospheric pressure, whereas the thermospray usually functions at a reduced pressure (1–10 torr). The ES source comprises two electrodes; that is, the ES capillary and the atmospheric pressure aperture plate of the mass spectrometer as the counter electrode. A high-voltage supply maintains a potential difference of about 3 kV between both electrodes. The solution is sprayed from the end of a stainless steel capillary situated 1 cm from the atmosphere pressure aperture plate. Under influence of the applied electric field, ions of the same polarity migrate toward the liquid at the capillary tip, where the liquid surface is drawn out of the capillary forming a 'Taylor cone'. When the build-up of an excess of ions of one polarity at the surface of the liquid reaches the point that coulombic forces are sufficient to overcome the surface tension of the liquid, droplets (1 μm) enriched in one ion polarity are emitted from the capillary (Fig. 10.9). These droplets shrink by solvent evaporation and repeated disintegration, leading to very small (3–10 nm) charged droplets capable of producing gas-phase ions, yielding a very soft ionization technique. The ions are sampled through a set of skimmer electrodes and finally analyzed in the MS analyzer.

Advantages

- Ionization occurs directly from solution and consequently allows ionic and thermally labile compounds to be studied.
- Allows to study of molecules with higher molecular weights

Disadvantages

- Electrospray is not applicable to non-polar or low-polarity compounds.

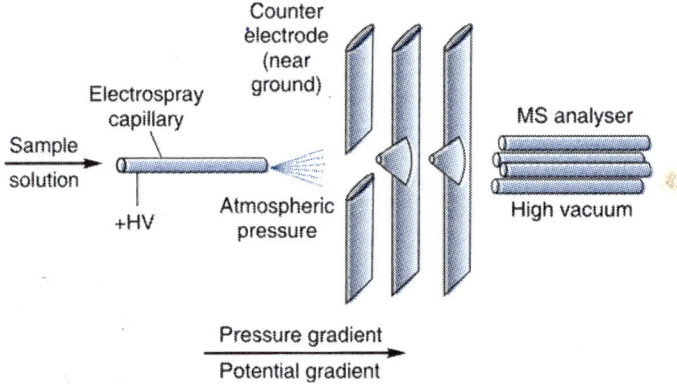

Fig. 10.8: Basic elements of an electrospray interface

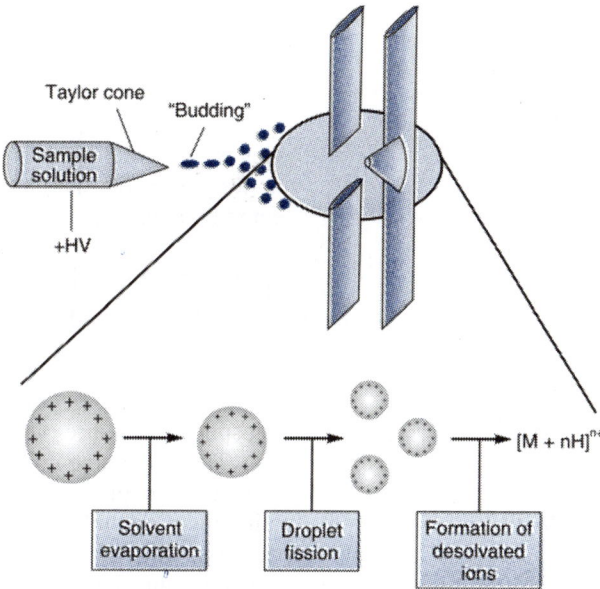

Fig. 10.9: Droplet and ion production under ES conditions

- The mass spectrum produced from an analyte, in terms of the m/z range of the ions observed and their relative intensities, depends upon a number of factors and spectra obtained using different experimental conditions may therefore differ considerably in appearance.
- Suppression effects may be observed and the direct analysis of mixtures is not always possible.
- Electrospray is a soft-ionization method producing intact molecular species and structural information is not usually available.

Atmospheric pressure chemical ionization

Atmospheric pressure ionization is one of the earliest techniques and its potential for an LC/MS interface has already been described in 1960 by Homing and coworkers. But almost 20 years passed before the technique was transformed into a mature method, which holds promise for several important applications. The technical realization is shown in Fig. 10.10. A stable spray is generated by heating an aerosol from the liquid effluent of the LC with a sheath flow of gas at atmospheric pressure. Ions are made by a corona discharge in the spray, forming chemical ionization plasma. The ions are extracted into the mass spectrometer by means of a skimmer system with a curtain of drying gas to reduce the background cluster ions from the solvent.

Advantages

- Applicable to study thermolabile compounds without degradation.
- It is best applied to compounds with low to moderately high polarities.
- APCI is a soft ionization technique which usually enables the molecular weight of the analyte under study to be determined.

Disadvantages

- APCI spectra can contain ions from adducts of the analyte with the HPLC mobile phase or organic modifiers, such as ammonium acetate, that may be present. The presence of ions such as $(M + NH_4)^+$ and $(M + CH_3COO)^+$ may hinder interpretation of the spectra obtained.
- Structural information is not usually available unless cone-voltage fragmentation or MS–MS is used.
- APCI is not able to function effectively at very low flow rates.
- APCI is not suitable for analytes that are charged in solution.

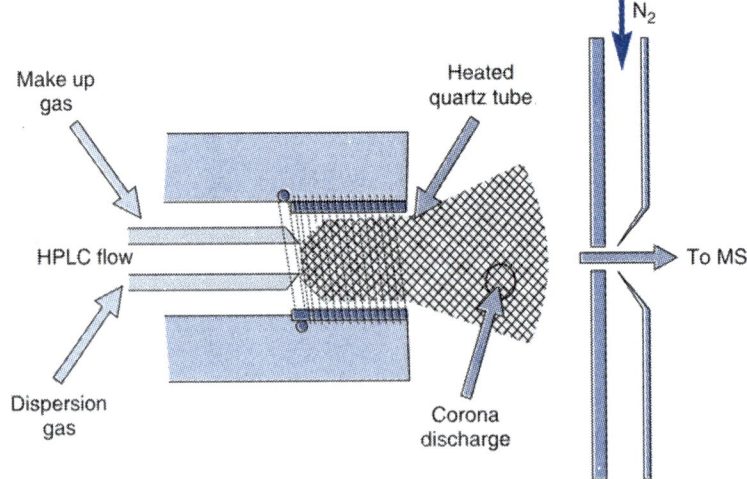

Fig. 10.10: Atmospheric pressure chemical ionization

Applications of LC-MS

LC-MS/LC-MSMS are most widely used in food industries, pharmaceutical and chemical industries for quantitative and qualitative analysis. Some of the applications of LS-MSMS are as follows.

Molecular weight determination

Able to determine the molecule weight of chemical substance, pharmaceutical substances, proteins, etc.

Structural determination/elucidation

Tandem mass spectrometry used to determine structural information using mass spectral fragmentations.

Pharmaceutical applications

It is used to determine the pharmacokinetic profile of the pharmaceuticals like drug, drug metabolites/degradation product, impurities and chiral impurities. The separation and detection of chiral impurities in pharmaceuticals are of great importance because the D-isomer of a drug can have different pharmacological, metabolic and toxicological activity from the L-isomer.

Clinical and biochemical applications

MALDI-TOF MS is used in SNP genotyping, quantification of DNA, gene expression analysis, DNA and RNA sequencing.

Food and environmental applications

It is used to identify aflatoxins (toxic metabolic product in certain fungi), determine the vitamin D_3 in poultry fed supplements, etc. The technique can also be useful for detection of (a) Pesticides, (b) Polycyclic aromatic compounds, and (c) Organometallic compounds.

Capillary electrophoresis/MS applications

Used for analysis of peptides. The proteomics LC-MS approach generally involves protease digestion (usually trypsin as a protease) and denaturation (usually urea to denature tertiary structure and iodoacetamide to cap cysteine residues) followed by LC-MS with peptide mass fingerprinting or LC-MS/MS (tandem MS) to derive sequence of individual peptides.

Chapter

11

Biological Mass Spectrometry

Introduction

The application of mass spectrometry (MS) to biology began in the 1940s, when heavy stable isotopes were used as tracers to study processes such as CO_2 production in animals. Since that time, advances in technology have increased the range of sample types that can be ionized and the range of masses that can be measured, which has diversified the applications of MS. The biological applications of MS currently encompass such diverse areas as screening newborns for metabolic disorders, comparing protein expression levels between cells grown in different media, determining the bioavailability of minerals in food, and studying how pharmaceutical drugs are metabolized *in vivo*. The wide varieties of sample types that can now be analyzed, and the breadth of information that can be obtained, have helped MS to permeate into an extensive range of research areas. The result is that MS has become an essential analytical tool in biological research. The current importance of MS to biological research is highlighted by the 2002 Nobel Prize in Chemistry, which was awarded to *John Fenn and Koichi Tanaka* "for their development of soft desorption ionization methods for mass spectrometric analysis of biological macromolecules".

The goal of this chapter is to familiarize the student with MS and some of its current applications to biological research. The total number of applications is too large to cover in a single chapter, so we provide a general overview with an emphasis on protein characterization, one of the newest and most widespread applications of biological MS.

Applications of MS in biological research

The range of MS applications to biology is extensive. Here, we divide them into three basic categories: Isotope ratio MS, small organic molecule MS, and macromolecular MS. For each of these groups we will discuss common applications.

Isotope ratio MS

Isotope ratio MS (IRMS) is a technique that measures the relative stable isotopic abundance of elements. The elemental isotope ratio can be analyzed for a complex system (bulk IRMS), for a specific compound within a mixture, or for a specific position within a compound. As an example, the $^{13}C/^{12}C$ ratio can be measured for a tissue sample, for a specific fatty acid within the tissue, or for a specific carbon position within the fatty acid. IRMS can reveal information regarding the origin or state of a complex system. For instance, the ratio of $^{13}C/^{12}C$ in apple juice can indicate if the juice has been adulterated with corn syrup. The ratios of $^{13}C/^{12}C$, $^{15}N/^{14}N$, and $^{87}Sr/^{86}Sr$ can pinpoint the geographic origin of ivory, which can help conservationists in preventing future poaching.

In biomedical research, IRMS is frequently used with projects involving stable isotopic tracers. For these experiments, compounds containing elements with abnormal stable isotope ratios are introduced into a system of interest (e.g. by feeding or by injection). The change in the isotopic ratio of the element in the system is then recorded with IRMS. Stable isotope tracers can be used to study the absorption, retention, and utilization of a nutrient *in vivo*. Stable isotope tracers can also be used to study energy expenditure or as a clinical tool for disease diagnosis. IRMS uses magnetic sectors as the mass analyzers and Faraday cups as the ion detectors. The ion source is usually an electron impact or thermal ionization source. An article by Davidsson and coworkers describes an experiment to test the effectiveness of two different iron absorption enhancers (Na_2 EDTA and ascorbic acid) to increase the iron absorbed by children in school meal program. On the first day, children were fed a breakfast "shake" containing one of the enhancers and $^{57}FeSO_4$. The next day, the same children were fed a breakfast "shake" containing either the original enhancer at a different concentration or a second enhancer and $^{58}FeSO_4$. Blood samples were taken before the first meal was given and 2 weeks after the second meal was consumed. The ratio of $^{58}Fe/^{57}Fe$ in the blood was measured to determine which meal allowed more iron absorption. It was found that ascorbic acid at a concentration of 70 mg/meal increased the iron absorbed to 115% of the daily requirement. This double stable isotope technique is useful because two variables can be tested in the same patient at the same time.

Small organic molecules MS

The mass spectrometric analysis of organic compounds gives information on the molecular mass, chemical formula, chemical structure, or quantity of the analyte. Based on the measured *m/e* and their peak intensities, the formula and chemical structure can be determined manually and/or by comparison with a reference database of spectra. As an example, Lavermicocca et al. used MS to identify novel antifungal compounds produced by *Lactobacillus plantarum* strain 21B (a bacterium used in making sourdough bread). Components of a culture filtrate were separated and tested for antifungal activity. Those that showed the highest activity were then characterized using gas chromatography–MS. The spectra were compared with those in a MS spectra library, and the compounds present were identified.

MS is also a valuable tool in the high-throughput analysis of compounds created in combinatorial libraries, such as those created to discover new drugs. If the mass spectrometer is connected to an LC system with appropriate columns, both structural information and information regarding the binding affinity of a molecule can be obtained. MS can also be used to study drug metabolism by identifying the primary metabolites from *in vitro* studies and by measuring them *in vivo* to determine pharmacokinetic parameters. An example of this type of experiment is a study done by Oo et al. In this study, researchers administered oseltamivir (a treatment for influenza) to healthy patients between 1 and 12 years of age. The concentrations of oseltamivir and one of its metabolites were then monitored over time in the urine and plasma using MS. The pharmacokinetic data obtained allowed appropriate dosing schedules for children to be developed. For instrumentation, small organic molecule MS often uses an electron impact or chemical ionization ion source. A wide variety of mass analyzers are used, and tandem MS is often employed to increase the structural information obtained.

Macromolecular MS

MS analysis of macromolecules includes the study of proteins, peptides, oligonucleotides, oligosaccharides, and lipids. We defer a discussion of applications to the study of proteins and peptides to "Protein Characterization." The research on the analysis of oligonucleotides has focussed on studying modified oligonucleotides that may not be compatible with the current enzymatic techniques used to sequence DNA. McLafferty and coworkers were able to sequence oligonucleotides up to 100 nucleotides in length using an ESI source coupled to a Fourier transform-ion cyclotron resonance analyzer. MS can also be used to analyze DNA modifications such as methylation. The characterization of oligosaccharides is more difficult than that of proteins and oligonucleotides because of the isomeric nature of the subunit and its ability to form branched structures. However, the structures of a wide variety of oligosaccharides have been determined by using tandem mass spectrometry.

Many classes of molecules within the "lipid" category, including fatty acids, acylglycerols, and steroids, have been characterized by MS. The characterization of fatty acylcarnitines in blood can be used to screen newborns for hereditary fatty acid oxidation disorders. MS is used to measure the chain length and abundance of fatty acylcarnitines present in the blood. By comparing the resulting spectrum to that of a normal infant, doctors can diagnose if a fatty acid oxidation disorder is present, and in some cases they can determine the exact enzyme in which the newborn is deficient.

Most macromolecular research relies on the use of ESI or MALDI as an ionization source. A variety of mass analyzers, including the ion trap and TOF, are used. Tandem MS is often employed to elucidate structural details or sequence information.

PROTEIN CHARACTERIZATION

MS has become a vital tool in proteomic research. It can give information on the identity of a protein, the amount of the protein that is present, and the modifications in the protein content.

Protein identification

The majority of protein sequence analysis today uses mass spectrometry. There are several steps in analyzing a protein. These are as follows:

1. Digest the protein to peptides (in gel or solution). Mass spectrometry currently gets limited sequence data from whole proteins, but can easily analyze peptides.
2. Trypsin is usually the protease most researchers use in digesting proteins due to a number of reasons. First, it cleaves proteins into its component peptides with an average size of 700 to 1500 Daltons (the ideal size for mass spectrometry). Second, it specifically cleaves the protein at the carboxyl side of arginine and lysine residues. The C-terminals of these peptides are charged and are therefore easily detectable by mass spectrometry. Third, trypsin is highly active and can tolerate a number of additives. And lastly, trypsin can be modified by the methylation of lysine's to prevent self-digestion at these sites.
3. After the proteins have been digested into peptides, they are then separated using reverse phase column (usually 75 µm in diameter) with acetonitrile gradient. Note: Acetic acid should be used with the solvents since trifluoroacetic usually interferes with the ionization process. Separate peptides, usually on reverse phase column with acetonitrile gradient.
4. The ionized peptides are then made to pass through the column eluate which contains peptides and solvent. After the solvent evaporates, their charged surfaces move the ionized peptides into the mass spectrometer. This method, which uses chromatography to introduce molecules into a mass spectrometer, is called high performance liquid chromatography or HPLC.
5. Then, the mass of peptides are measured.
6. The peptides are fragmented by collisions with gas molecules. These peptides fragment (cleaved) at peptide bonds. This is CID (collision induced dissociation) or CAD (collisionally-activated dissociation). The mass of new fragments from peptides are measured. Because there are two steps of mass spectrometry (mass of peptide, mass of fragment of peptide), this is called MS/MS, or MSn because there can be 2 or more fragmentation steps. A two-step process is also called tandem mass spectrometry. The weakest bond in a peptide is the one between the amino acids. Therefore, for low-energy CID, the resulting MS/MS spectrum is a series of peaks that represent peptides that differ only in the number of amino acids they contain. By measuring the mass difference between each peak, one can determine the amino

acid sequence of the original peptide. High-energy CID also causes side-chain fragmentation. This information can be useful to differentiate between amino acids of the same molecular mass (e.g. leucine and isoleucine). In Fig. 11.1 the peptide with an m/e of 2574.30 was chosen for tandem MS analysis. The peptide was isolated and fragmented and the mass spectrum of the resulting fragments is shown. The partial amino acid sequence can be found by calculating the difference in mass between peaks. For example, the mass difference between the peaks with $m/e = 1577.80$ and $m/e = 1449.76$ is 128.04.

  This is within error of the mass of glutamine (Q), which is 128.06 Da. The probability of fragmentation, however, is not the same for all of the amide bonds. This is evident by the wide range of peak intensities in the tandem MS data and the fact that some peaks are missing. For instance, due to the structure of proline, fragmentation at the C-terminal end of proline is rare. This is why in Fig. 11.1, we do not see a peak at $m/e = 2020.94$, which is equal to $m/e = 2117.99$ minus the mass of proline.

7. Use fragment mass data to determine the sequence of the peptide by seeing which combinations of amino acids give the observed masses of peptide fragments.

  The two mass measurements in steps 5 and 7 require a tandem mass spectrometer, or MS/MS. The two measurements can be performed in

- Two different parts of the instrument—tandem in space
- One cell of the instrument that switches modes—tandem in time

  Most data analysis is done by computer, by comparison with known sequences; SEQUEST is the best known program. For new sequences and confirmation of important sequences, data analysis is done by hand.

Reason for not getting complete sequence data for every protein

Seeing enough peptides to show 70% of the sequence of a protein (70% coverage) is a very successful protein analysis. In a project by the Cell Migration Consortium to analyze a number of protein involved in cell migration, 80% coverage of a protein is considered sufficient. There are several reasons why an analysis does not find all amino acids. These may be:

- Protein does not digest well
- Peptides too hydrophilic or small—they pass through the reverse phase column with salt and are not analyzed
- Peptides too large/hydrophobic—they stick in gel, adsorb to tubes, do not elute from column, or are too large for the mass spectrometer to analyze because of poor fragmentation.
- Peptides fragment in ways which cannot be analyzed. Many spectra in an analysis cannot be interpreted. Some spectra only give limited data; proline, histidine, internal lysine and arginine are some reasons peptides do not give complete fragmentation data.

  If more data is needed, another proteinase is used for digestion.

Note:

- Although some work has been done to identify simultaneously all the proteins in a complex mixture, protein separation prior to enzymatic digestion is often performed.
- Proteins are separated using LC or gel electrophoresis (either one- or two-dimensional gel electrophoresis). If separated by gel electrophoresis, the protein band or spot is excised, the proteins within are digested "in-gel", and the resulting peptides are extracted for MS analysis.
- A plug is excised from an electrophoresis gel, and the protein in the gel plug is digested and extracted.

Mass mapping, protein mass fingerprinting

A less complex method for identifying proteins relies on databases of protein sequences. After digesting a protein with trypsin or some other specific proteinase, the masses of the intact

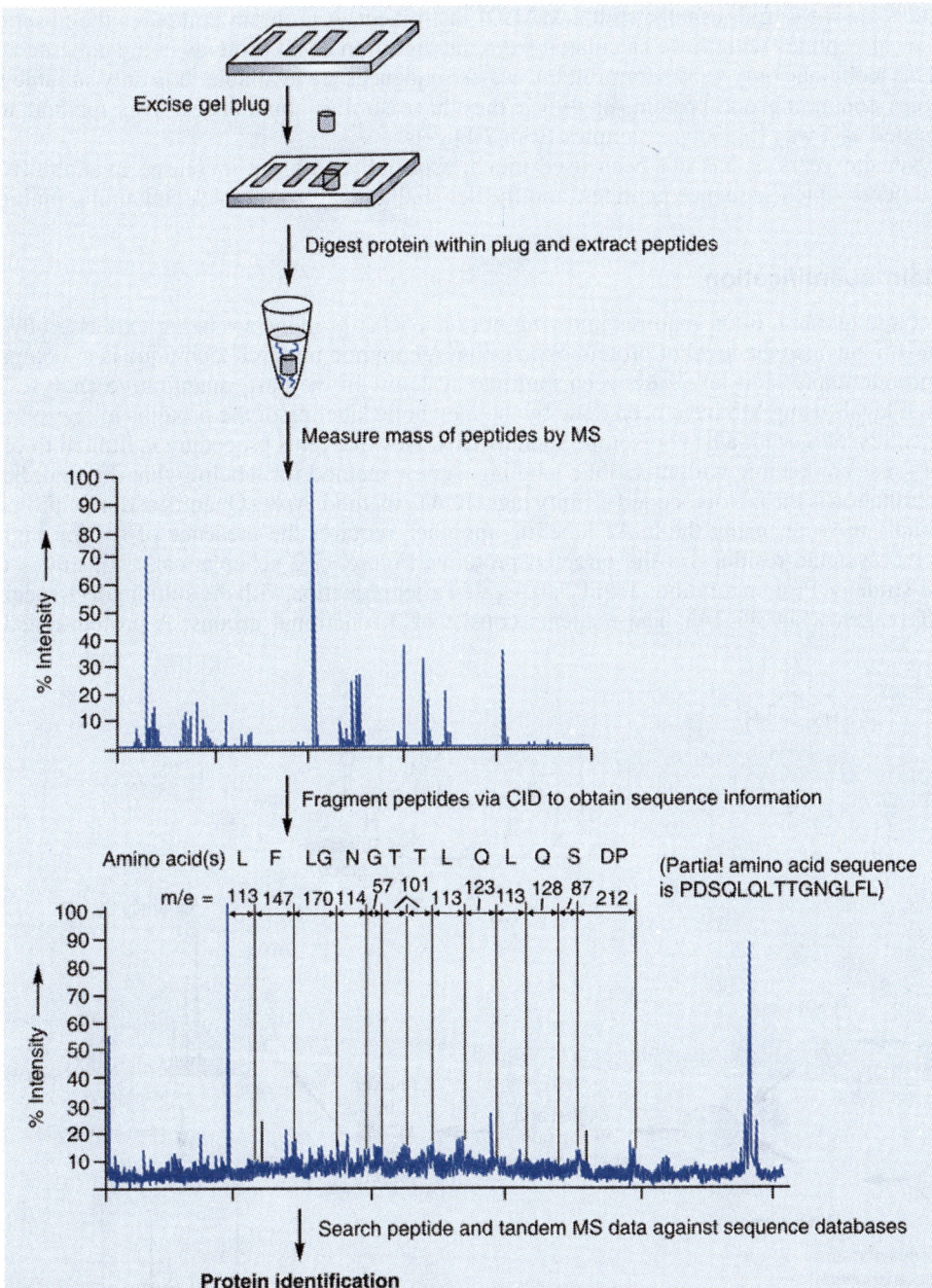

Fig. 11.1: Overview of the process of protein identification. A plug is excised from an electrophoresis gel, and the protein in the gel plug is digested and extracted. The masses of the peptides in the extract are then measured by MS to obtain the peptide mass fingerprint. Next, peptides can be selected to undergo fragmentation via tandem MS. Here, the peptide with *m/e* = 2574 (labelled with an arrow) has been selected and fragmented. Finally, both the MS and tandem MS data are searched against protein sequence databases to determine the protein that was present in the gel spot.

peptides are measured, usually with a MALDI instrument. A program compares the observed masses of peptides with those calculated from the digestion of all proteins in a database.

This technique only works for proteins whose sequences are available. It is only suitable for samples containing one protein, or two if they have similar abundances. This method was suggested as a way to analyze samples from 2-D gels.

Over the years, it has not been used much, especially with the increased availability of instruments which sequence peptides, and the demand to analyze samples containing multiple proteins.

Protein quantification

Proteomic research often requires knowing not only what proteins are being expressed by an organism, but also the level of protein expression. A common research technique is to compare the protein expression levels between multiple systems. In the past, quantitative analyses of protein levels using MS have been done by the metabolic labeling of the proteins in one system (e.g. cell lysates) with a heavy isotope such as 15N. However, this procedure is limited to cells and tissues compatible with metabolic labeling. A new method for labeling that does not have this limitation is the isotope-coded affinity tags (ICAT) method. *Note:* Quantification of proteins in simple mixtures using the ICAT labeling approach requires the presence of several highly reactive cysteine residues in the targeted proteins. Figure 11.2 schematically illustrates the ICAT strategy. Protein mixtures 1 and 2 are treated after reduction with the sulphhydryl-specific ICAT reagent (Fig. 11.2A). The reagents consist of 3 functional groups: A protein-reactive

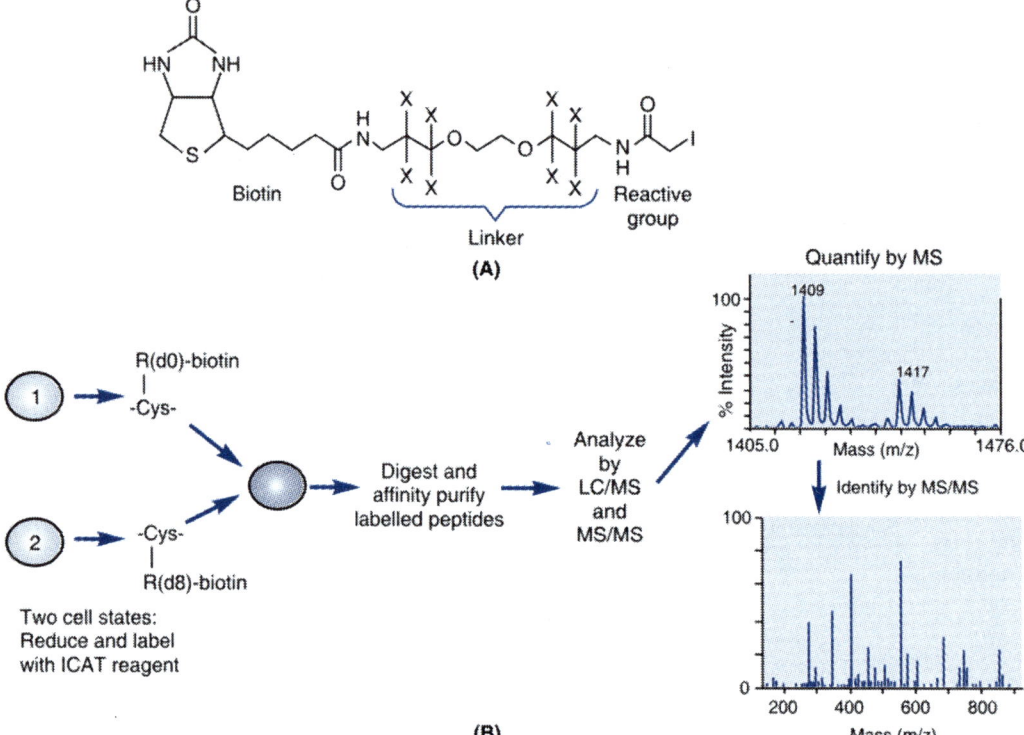

Fig. 11.2: Protein quantification using ICAT method. (A) Structure of isotope-coded affinity tag (ICAT) reagent. The reagent has 3 elements: An affinity tag (biotin) that is used to isolate ICAT-labeled peptides, a linker that can incorporate stable isotopes, and a reactive group with specificity toward thiol groups (cysteines [Cys]). The reagent exists in 2 forms: Heavy (contains 8 deuteriums [d8]) and light (no deuteriums [d0]). (B) ICAT strategy for quantitating differential protein expression.

group (iodoacetate) that is used to covalently attach the reagent at a specific site to the protein; a linker group (ethylene glycol) that exists in 2 forms, isotopically normal and heavy; and an affinity tag (biotin) that is used to selectively extract the reagent-peptide conjugates from the sample mixture. The ICAT reagents exist in 2 forms: Isotopically light (contains no deuteriums [d0]) and isotopically heavy (contains 8 deuteriums [d8]). The heavy and light forms are used to derive the proteins in samples 1 and 2, respectively. After treatment with the ICAT reagents, the samples are mixed. At this point, any optional fractionation technique can be performed to enrich for low-abundance proteins or to reduce the complexity of the mixture, whereas the relative quantities are maintained. The combined protein sample is then proteolyzed by digestion with enzyme-trypsin. The digest is fractionated using cation exchange chromatography, which also removes any neutral species from the tryptic peptides. The tagged peptides (those that contain a cysteine) in each fraction are then separated from the others on an affinity column, and the biotin portion of the tag is cleaved off. The peptides are then separated using LC and analyzed by MS. The ratio of the intensities of the light- and heavy-labeled peptides gives a measurement of the ratio of protein expression in the two systems. The peptide is then analyzed with tandem MS to identify the parent protein.

Proteins with post-translational modifications

The function and activity of a protein is determined in part by any post-translational modifications that may be present. Over 200 distinct types of covalent modifications have been reported; the most common of these include phosphorylation, glycosylation, and ubiquitination. Here, the use of MS to study phosphorylation is reviewed. It is estimated that over 30% of proteins are phosphorylated. Most common sites of phosphorylation include Serine, Threonine and Tyrosine.

The conventional biochemical approach to study phosphopeptides requires the use of radioactive labels and Edman sequencing, but a variety of approaches have been developed to study phosphorylation using MS. If the amino acid sequence of the protein is already known, the sample is digested and the peptides analyzed with MS. The m/e of the peptides measured with MS is compared with the predicted values. An increase in a peptide mass by 80 or 160 Da indicates covalent modification by phosphate (HPO_3) (mono- and diphosphorylated species, respectively). The phosphorylated peptide can then be analyzed using MS/MS to determine the exact position of the modification. If the protein is unknown, the sample is divided after digestion, and one portion is treated with a phosphatase. This step removes any phosphate groups, and a comparison of spectra before and after phosphatase treatment will indicate which peptides are modified. The modified peptides can be analyzed with MS/MS to identify the parent protein and the site of phosphorylation. For example, Fig. 11.3 shows the loss of phosphate group by 80 Da from two site in the mass spectrum after treatment of phosphorylated protein with phosphatase. The presence of phosphorylation can also be determined using different scanning modes on the mass spectrometer. In "neutral loss scanning," the mass spectrometer scans for an ion that when fragmented results in a specified neutral loss (change in mass but not in charge). When searching for phosphorylated peptides, the tandem MS spectra are searched for a fragment that has a mass 98 Da lower than the original peptide mass (with no change in charge).

This neutral loss indicates a loss of H_3PO_4 due to beta elimination. This process cannot detect phosphotyrosines, however, due to stability of the beta protons in the benzene ring. In "precursor ion scanning," the mass spectrometer scans for a specific product ion after fragmentation. For phosphotyrosines, the peptides are fragmented using CID and scanned for a positive fragment ion with 216 m/e (phosphotyrosine immonium ion). For other phosphorylations, the peptides are fragmented using CID and scanned for a negative ion with 79 m/e due to the release of PO_3^- MS has been used to study a multitude of post-translational modifications, often employing the same approaches discussed here for phosphorylation: treatment with an enzyme to remove the modification (with subsequent spectra comparison),

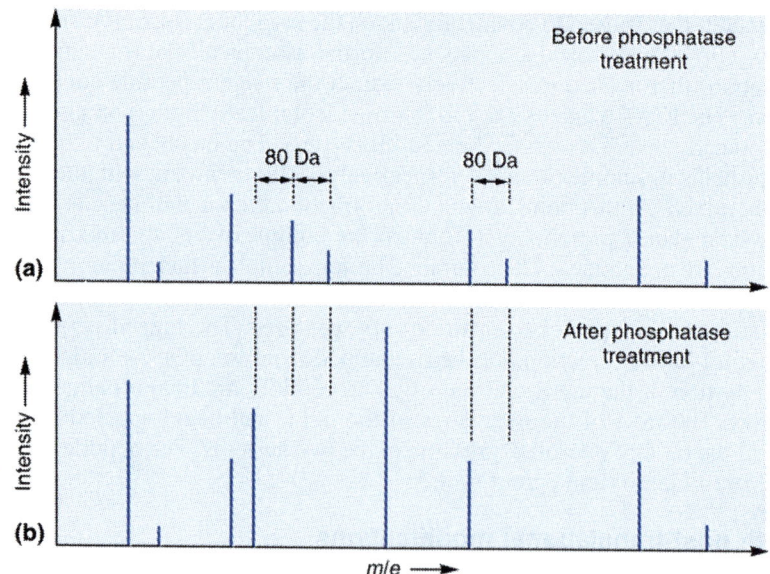

Fig. 11.3: Mass spectrum (a) before phosphatase treatment; (b) after phosphatase treatment

neutral loss scanning, or precursor ion scanning. It should be noted that for modifications by heterogeneous molecules (such as glycans), the structure of the modification could also be analyzed with MS.

Appendices

Resources

- i-mass: *http://www.i-mass.com*
- Ion source: *http://www.ionsource.com*
- Spectroscopy (incorporating Base Peak): *http://www.spectroscopynow.com*

Societies

- American Society for Mass Spectrometry: *http://www.asms.org*
- Australian and New Zealand for Mass Spectrometry: *http://www.anzsms.org/index.htm*
- British Mass Spectrometry Society: *http://www.bmss.org.uk*
- Mass Spectrometry Society of Japan: *http://www.mssj.jp*

Chemical and Biological Online Tools

- Isotope Pattern Calculator: *http://www.shef.ac.ukl~chem/chemputer/isotopes.html*
- FindMod tool: *http://au.expasy.org/tools/findmod/*
- GlycoSuite at Proteome Systems: *http://tmat.proteomesystems.com/glycosuite/*
- Mascot Search-Matrix Science: *www.matrixscience.com/search_form_select.html*
- Peptide Search at EMBL: *http://vsites.unb.br/cbsp/paginiciais/peptsrcseqtag.htm*
- Protein Prospector at UCSF: *http://prospector.ucsf.edu/*
- PROWL at Rockefellar: *http://prowl.rockefellar.com/*
- Web Elements Periodic Table: *http://www.webelements.com/*

Tutorials

- Mass Spectrometric Tutorial: *http://www.astbury.leeds.ac.uk/facil/MStut/mstutorial.htm*
- ASMS-What is Mass Spectrometry: *www.asms.org/whatisms/*
- Mass Spectrometry Tutorial: *www.chem.arizona.edu/massspec/*
- BMSS What is Mass Spectrometry?: *http://www.bmss.org.uk/what_is/whatisframeset.html*
- Cambridge University tutorial: *http://www.methods.ch.cam.ac.uk/meth/ms/theory/*
- i-mass guides: *http://www.i-mass.com/guide/*
- The mass spectrometer - *how it works*: *http://www.chemguide.co.uk/analysis/masspec/howitworks.html*
- Mass spectrometry: *http://www.cem.msu.edu/~reusch/.../Spectrpy/MassSpec/masspec1.htm*

APPENDIX 2
List of Mass Spectrometry Software

Name with link	Type	Description
Andromeda http://www.coxdocs.org/doku.php?id=maxquant:common:download_and_installation	Freeware	Andromeda is a peptide search engine based on probabilistic scoring.
Byonic http://www.proteinmetrics.com/	Proprietary	Database search algorithm released in 2011 by Protein Metrics Inc. It combines peptide identifications to produce protein scores and identification probabilities.
Comet http://comet-ms.sourceforge.net/	Open source	Database search algorithm developed at the University of Washington available for Windows and Linux.
Tide http://crux.ms/download.html	Open source	Tide is a tool for identifying peptides from tandem mass spectra.
Greylag http://greylag.org/	Open source	Database search algorithm developed at the Stowers Institute for Medical Research designed to perform large searches on computational clusters having hundreds of nodes.
InsPecT http://proteomics.used.edu/Software/Inspect/	Open source	A MS-alignment search algorithm available at the Center for Computational Mass Spectrometry at the University of California, San Diego.
Mascot	Proprietary	Performs mass spectrometry data analysis through a statistical evaluation of matches between observed and projected peptide fragments.
MassMatrix http://magneto.case.edu/mm-cgi/downloads.py	Freeware	MassMatrix is database search algorithm for tandem mass spectrometric data. It uses a mass accuracy sensitive probabilistic scoring model to rank peptide and protein matches.
MassWiz https://web.archive.org/web/20101024191936/ http://masswiz.igib.res.in/	Open source	Search algorithm developed at Institute of Genomics and Integrative Biology available as a Windows Command Line Tool.
MS Amanda download (http://ms.imp.ac.at/?goto=msamanda)	Freeware	Search algorithm developed at Institute of Molecular Biotechnology of the Austrian Academy of Sciences (IMBA) available for Windows, Linux and MacOS.
MS-GF + https://github.com/sangtaekim/msgfplus/releases	Open source	MS-GF+ (aka MSGF+ or MSGFPlus) performs peptide identification by scoring MS/MS spectra against peptides derived from a protein sequence database.
MyriMatch http://proteowizard.sourceforge.net/downloads.shtml	Open source	Database search program developed at the Vanderbilt University Medical Center designed to run in a single-computer environment or across an entire cluster of processing nodes.

(Contd.)

Name with link	Type	Description
OMSSA ftp://ftp.ncbi.nlm.nih.gov/pub/lewisg/omssa/CURRENT/	Freeware	The Open Mass Spectrometry Search Algorithm (OMSSA) is an efficient search engine for identifying MS/MS peptide spectra by searching libraries of known protein sequences.
PEAKS DB	Proprietary	Database search engine, run in parallel with *de novo* sequencing to automatically validate search results, allowing for a higher number of found sequences for a given false discovery rate.
pFind http://pfind.ict.ac.cn/	Freeware	pFind Studio is a computational solution for mass spectrometry-based proteomics. It germinated in 2002 in Institute of Computing Technology, Chinese Academy of Sciences, Beijing, China.
Phenyx	Proprietary	Developed by Geneva Bioinformatics (GeneBio) in collaboration with the Swiss Institute of Bioinformatics (SIB). Phenyx incorporates OLAV, a family of statistical scoring models, to generate and optimize scoring schemes that can be tailored for all kinds of instruments, instrumental setups and general sample treatments.
ProbID http://www.bioinfo.org.cn/MSMS/download/index.html	Open source	PI is a powerful suite on analysis of tandem mass spectrum. ProbID seeks to fill the need for the deep analysis of tandem mass spectrum, including the fragmentation rules, preference of cleavage, neutral losses, etc.
Protein Pilot Software http://sciex.com/products/software/proteinpilot-software/	Proprietary	Uses Paragon database search algorithm that combines the generation of short sequence tags ('taglets') for computation of sequence temperature values and estimates of feature probabilities to enable the peptide identification considering hundreds of modifications, non-tryptic cleavages and amino acid substitutions.
Protein Prospector http://prospector.ucsf.edu/ source https://github.com/proteinprospector/	Open source	Protein Prospector is a package of about twenty proteomic analysis tools developed at the University of California, San Francisco. It uses scoring systems tailored to instrument and fragmentation mode to optimize analysis of different types of fragmentation data.
RAId	Lost	Developed at the National Center for Biotechnology Information, Robust Accurate Identification (RAId) is a suite of proteomics tools for analyzing tandem mass spectrometry data with accurate statistics.
SEQUEST	Proprietary	Identifies collections of tandem mass spectra to peptide sequences that have been generated from databases of protein sequences.

(Contd.)

Name with link	Type	Description
SIMS http://emililab.med.utoronto.ca/	Open source	SIMS (Sequential Interval Motif Search) is a software tool design to perform unrestrictive PTM search over tandem mass spectra; users do not have to characterize the potential PTMs. Instead, users only need to specify the range of modification mass for each individual amino acid.
SimTandem http://www.simtandem.org/	Freeware	A database search engine for identification of peptide sequences from LC/MS/MS data; the engine can be used as an external tool in OpenMS/TOPP.
SQID https://research.cbc.osu.edu/ wysocki.11/group-home/bioinformatics/	Open source	SeQuence IDentification (SQID) is an intensity-incorporated protein identification algorithm for tandem mass spectrometry.
X!Tandem http://www.thegpm.org/ TANDEM/index.html	Open source	Matches tandem mass spectra with peptide sequences.

APPENDIX 3
Isotope Masses and Abundances

Isotope	Nominal mass	Mass	Relative abundance	Isotope	Nominal mass	Mass	Relative abundance
^{1}H	1	1.007825	99.985(1)	^{25}Mg	25	24.985837	10.00(1)
^{2}H; D	2	2.014101	0.015(1)	^{26}Mg	26	25.982593	11.01(3)
^{3}H; T	3	3.016049	<0.0001	^{27}Al	27	26.981538	~100
^{4}He	4	4.002603	~100	^{28}Si	28	27.976926	92.22(2)
^{6}Li	6	6.015122	7.5(2)	^{29}Si	29	28.976494	4.69(1)
^{7}Li	7	7.016004	92.5(2)	^{30}Si	30	29.973770	3.09(1)
^{9}Be	9	9.012182	~100	^{31}P	31	30.973761	~100
^{10}B	10	10.012937	19.9(2)	^{32}S	32	31.972070	94.93(31)
^{11}B	11	11.009305	80.1(2)	^{33}S	33	32.971458	0.76(2)
^{12}C	12	12.000000	98.93(8)	^{34}S	34	33.967866	4.29(28)
^{13}C	13	13.003354	1.07(8)	^{36}S	36	35.967080	0.02(1)
^{14}N	14	14.003074	99.632(7)	^{35}Cl	35	34.968852	75.78(4)
^{15}N	15	15.000108	0.368(7)	^{37}Cl	37	36.965902	24.22(4)
^{16}O	16	15.994914	99.757(16)	^{36}Ar	36	35.967546	0.3365(30)
^{17}O	17	16.999131	0.038(1)	^{38}Ar	38	37.962732	0.0632(5)
^{18}O	18	17.999160	0.205(14)	^{40}Ar	40	39.962383	99.6003(30)
^{19}F	19	18.998403	~100	^{39}K	39	38.9637069	93.2581(44)
^{20}Ne	20	19.992440	90.48(3)	^{40}K	40	39.963998	0.0117(1)
^{21}Ne	21	20.993846	0.27(1)	^{41}K	41	40.961826	6.7302(44)
^{22}Ne	22	21.991385	9.25(3)	^{40}Ca	40	39.962591	96.941(156)
^{23}Na	23	22.989769	~100	^{42}Ca	42	41.958618	0.647(23)
^{24}Mg	24	23.985041	78.99(4)	^{43}Ca	43	42.958766	0.135(10)

(Contd.)

Isotope	Nominal mass	Mass	Relative abundance	Isotope	Nominal mass	Mass	Relative abundance
^{44}Ca	44	43.955481	2.086(110)	^{82}Se	82	81.916700	8.73(22)
^{46}Ca	46	45.953693	0.004(3)	^{79}Br	79	78.918338	50.69(7)
^{48}Ca	48	47.952533	0.187(21)	^{81}Br	81	80.916291	49.31(7)
^{45}Sc	45	44.955910	~100	^{78}Kr	78	77.92039	0.35(1)
^{46}Ti	46	45.952630	8.25(3)	^{80}Kr	80	79.916379	2.28(6)
^{47}Ti	47	46.951764	7.44(2)	^{82}Kr	82	81.913485	11.58(14)
^{48}Ti	48	47.947947	73.72(3)	^{83}Kr	83	82.914137	11.49(6)
^{49}Ti	49	48.947871	5.41(2)	^{84}Kr	84	83.911508	57.00(4)
^{50}Ti	50	49.944792	5.18(2)	^{86}Kr	86	85.910615	17.30(22)
^{50}V	50	45.947163	0.250(4)	^{85}Rb	85	84.911792	72.17(2)
^{51}V	51	50.943964	99.750(4)	^{87}Rb	87	86.909186	27.83(2)
^{50}Cr	50	49.946050	4.345(13)	^{84}Sr	84	83.913426	0.56(1)
^{52}Cr	52	51.940512	83.789(18)	^{86}Sr	86	85.909265	9.86(1)
^{53}Cr	53	52.940653	9.501(17)	^{87}Sr	87	86.908882	7.00(1)
^{54}Cr	54	53.938885	2.365(7)	^{88}Sr	88	87.905617	82.58(1)
^{55}Mn	55	54.938049	~100	^{89}Y	89	88.905849	~100
^{54}Fe	54	53.939615	5.845(35)	^{90}Zr	90	89.904702	51.45(40)
^{56}Fe	56	55.934942	91.754(36)	^{91}Zr	91	90.905643	11.22(5)
^{57}Fe	57	56.935398	2.119(10	^{92}Zr	92	91.905039	17.15(8)
^{58}Fe	58	57.933280	0.282(4)	^{94}Zr	94	93.906314	17.38(28)
^{59}Co	59	58.933200	~100	^{96}Zr	96	95.908275	2.80(9)
^{58}Ni	58	57.935348	68.0769(89)	^{93}Nb	93	92.906376	~100
^{60}Ni	60	59.930790	26.2231(77)	^{92}Mo	92	91.906810	14.84(35)
^{61}Ni	61	60.931060	1.1399(6)	^{94}Mo	94	93.905087	9.25(12)
^{62}Ni	62	61.928348	3.6345(17)	^{95}Mo	95	94.905841	15.92(13)
^{64}Ni	64	63.927969	0.9256(9)	^{96}Mo	96	95.904678	16.68(2)
^{63}Cu	63	62.929601	69.17(3)	^{97}Mo	97	96.906020	9.55(8)
^{65}Cu	65	64.927794	30.83(3)	^{98}Mo	98	97.905407	24.13(31)
^{64}Zn	64	63.929146	48.63(60)	^{100}Mo	100	99.90748	9.63(23)
^{66}Zn	66	65.926036	27.90(27)	^{96}Ru	96	95.90760	5.52(20)
^{67}Zn	67	66.927131	4.10(13)	^{98}Ru	98	97.90529	1.88(9)
^{68}Zn	68	67.924847	18.75(51)	^{99}Ru	99	98.905939	12.74(26)
^{70}Zn	70	69.925325	0.62(3)	^{100}Ru	100	99.904219	12.60(19)
^{69}Ga	69	68.925581	60.108(9)	^{101}Ru	101	100.905582	17.05(7)
^{71}Ga	71	70.924707	39.892(9)	^{102}Ru	102	101.904349	31.57(31)
^{70}Ge	70	69.924250	20.84(87)	^{104}Ru	104	103.905430	18.66(44)
^{72}Ge	72	71.922076	27.54(34)	^{103}Rh	103	102.905504	~100
^{73}Ge	73	72.923460	7.73(5)	^{102}Pd	102	101.905607	1.02(1)
^{74}Ge	74	73.921178	36.28(73)	^{104}Pd	104	103.904034	11.14(8)
^{76}Ge	76	75.921403	7.61(38)	^{105}Pd	105	104.905083	2233(8)
^{75}As	75	74.921597	~100	^{106}Pd	106	105.903484	27.33(3)
^{74}Se	74	73.922477	0.89(4)	^{108}Pd	108	107.903895	26.46)9)
^{76}Se	76	75.919214	9.37(29)	^{110}Pd	110	109.905153	11.72(9)
^{77}Se	77	76.919915	7.63(16)	^{107}Ag	107	106.905093	51.839(8)
^{78}Se	78	77.917310	23.77(28)	^{109}Ag	109	108.904756	48.161(8)
^{80}Se	80	79.916522	49.61(41)	^{106}Cd	106	105.90646	1.25(6)

(Contd.)

Isotope	Nominal mass	Mass	Relative abundance	Isotope	Nominal mass	Mass	Relative abundance
^{108}Cd	108	107.90418	0.89(3)	^{138}Ba	138	137.905242	71.698(42)
^{110}Cd	110	109.903006	12.49(18)	^{138}La	138	137.907107	0.090(1)
^{111}Cd	111	110.904182	12.80(12)	^{139}La	139	138.906349	99.910(1)
^{112}Cd	112	111.902758	24.13(21)	^{136}Ce	136	135.90714	0.19(1)
^{113}Cd	113	112.904401	12.22(12)	^{138}Ce	138	137.90599	0.25(1)
^{114}Cd	114	113.903359	28.73(42)	^{140}Ce	140	139.905435	88.48(10)
^{116}Cd	116	115.904756	7.49(18)	^{142}Ce	142	141.909241	11.08(10)
^{113}In	113	112.904062	4.29(5)	^{141}Pr	141	140.907648	~100
^{115}In	115	114.903879	95.71(5)	^{142}Nd	142	141.907719	27.13(12)
^{112}Sn	112	111.904822	0.97(1)	^{143}Nd	143	142.909810	12.18(6)
^{114}Sn	114	113.902783	0.65(1)	^{144}Nd	144	143.910083	23.80(12)
^{115}Sn	115	114.903347	0.34(1)	^{145}Nd	145	144.912569	8.30(6)
^{116}Sn	116	115.901745	14.45(9)	^{146}Nd	146	145.913113	17.19(9)
^{117}Sn	117	116.902955	7.68(7)	^{148}Nd	148	147.916889	5.76(3)
^{118}Sn	118	117.901608	24.22(9)	^{150}Nd	150	149.920887	5.64(3)
^{119}Sn	119	118.903311	8.59(4)	^{144}Sm	144	143.911996	3.1(1)
^{120}Sn	120	119.902199	32.59(9)	^{147}Sm	147	146.914894	15.0(2)
^{122}Sn	122	121.903441	4.63(3)	^{148}Sm	148	147.914818	11.3(1)
^{124}Sn	124	124.905275	5.79(5)	^{149}Sm	149	148.917180	13.8(1)
^{121}Sb	121	120.903822	57.21(5)	^{150}Sm	150	149.917272	7.4(1)
^{123}Sb	123	122.904216	42.79(5)	^{152}Sm	152	151.919729	26.7(2)
^{120}Te	120	119.90403	009(1)	^{154}Sm	154	153.922206	22.7(2)
^{122}Te	122	121.903056	2.55(12)	^{151}Eu	151	150.919846	47.8(1.5)
^{123}Te	123	122.904271	0.89(3)	^{153}Eu	153	152.921227	52.2(15)
^{124}Te	124	123.902819	4.74(14)	^{152}Gd	152	151.919789	0.20(1)
^{125}Te	125	124.904424	7.07(15)	^{154}Gd	154	153.920862	2.18(3)
^{126}Te	126	125.903305	18.84(25)	^{155}Gd	155	154.922619	14.80(5)
^{128}Te	128	127.904462	31.74(8)	^{156}Gd	156	155.922120	20.47(4)
^{130}Te	130	129.906223	34.08(62)	^{157}Gd	157	156.923957	15.65(3)
^{127}I	127	126.904468	~100	^{158}Gd	158	157.924101	24.84(12)
^{124}Xe	124	123.905895	0.09(1)	^{160}Gd	160	159.927051	21.86(4)
^{126}Xe	126	125.90427	0.09(1)	^{159}Td	159	158.925343	~100
^{128}Xe	128	127.903531	1.92(3)	^{156}Dy	156	155.92428	0.06(1)
^{129}Xe	129	128.904780	26.44(24)	^{158}Dy	158	157.924405	0.10(1)
^{130}Xe	130	129.903509	4.08(2)	^{160}Dy	160	159.925194	2.34(6)
^{131}Xe	131	130.905083	21.18(3)	^{161}Dy	161	160.926930	18.9(2)
^{132}Xe	132	131.904155	26.89(6)	^{162}Dy	162	161.926795	25.5(2)
^{134}Xe	134	133.905395	10.44(10)	^{163}Dy	163	162.928728	24.9(2)
^{136}Xe	136	135.90722	8.87(16)	^{164}Dy	164	163.929171	28.2(2)
^{133}Cs	133	132.905447	~100	^{165}Ho	165	164.930319	100
^{130}Ba	130	129.90631	0.106(1)	^{162}Er	162	161.928775	0.14(1)
^{132}Ba	132	131.905056	0.101(1)	^{164}Er	164	163.929197	1.61(2)
^{134}Ba	134	133.904504	2.417(18)	^{166}Er	166	165.930290	33.6(2)
^{135}Ba	135	134.905684	6.592(12)	^{167}Er	167	166.932046	22.95(15)
^{136}Ba	136	135.904571	7.854(24)	^{168}Er	168	167.932368	26.8(2)
^{137}Ba	137	136.905822	11.232(24)	^{170}Er	170	169.935461	14.9(2)

(Contd.)

Isotope	Nominal mass	Mass	Relative abundance	Isotope	Nominal mass	Mass	Relative abundance
^{169}Tm	169	168.934211	~100	^{190}Os	190	189.958445	26.4(4)
^{168}Yb	168	167.933895	0.13(1)	^{192}Os	192	191.961479	41.03(3)
^{170}Yb	170	169.934759	3.05(6)	^{191}Ir	191	190.960591	37.3(5)
^{171}Yb	171	170.936323	14.3(2)	^{193}Ir	193	192.962923	62.7(5)
^{172}Yb	172	171.936378	21.9(3)	^{190}Pt	190	189.95993	0.01(1)
^{173}Yb	173	172.938207	16.12(21)	^{192}Pt	192	191.961035	0.79(6)
^{174}Yb	174	173.938858	31.8(4)	^{194}Pt	194	193.962663	32.9(6)
^{176}Yb	176	175.942569	12.7(2)	^{195}Pt	195	194.964774	33.8(6)
^{175}Lu	175	174.940768	97.41(2)	^{196}Pt	196	195.964934	25.3(6)
^{176}Lu	176	175.942683	2.59(2)	^{198}Pt	198	197.967875	7.2(2)
^{174}Hf	174	173.940042	0.162(3)	^{197}Au	197	196.966551	~100
^{176}Hf	176	175.941403	5.206(5)	^{196}Hg	196	195.965814	0.15(1)
^{177}Hf	177	176.943220	18.606(4)	^{198}Hg	198	197.966752	9.97(8)
^{178}Hf	178	177.943698	27.297(4)	^{199}Hg	199	198.968262	16.87(10)
^{179}Hf	179	178.945815	13.629(6)	^{200}Hg	200	199.968309	23.10(16)
^{180}Hf	180	179.946549	35.100(7)	^{201}Hg	201	200.970285	13.18(8)
^{180}Ta	180	179.947466	0.012(2)	^{202}Hg	202	201.970625	29.86(20)
^{181}Ta	181	180.947996	99.988(2)	^{204}Hg	204	203.973475	6.87(4)
^{180}W	180	179.946706	0.120(1)	^{203}Tl	203	202.972329	29.524(14)
^{182}W	182	181.948205	26.498(29)	^{205}Tl	205	204.974412	70.476(14)
^{183}W	183	182.950224	14.314(4)	^{204}Pb	204	203.973028	1.4(1)
^{184}W	184	183.950932	30.642(8)	^{206}Pb	206	205.974449	24.1(1)
^{186}W	186	185.954362	28.426(37)	^{207}Pb	207	206.975880	22.1(1)
^{185}Re	185	184.952955	37.40(2)	^{208}Pb	208	207.976636	52.4(1)
^{187}Re	187	186.955751	62.60(2)	^{209}Bi	209	208.980384	~100
^{184}Os	184	183.952491	0.020(3)	^{232}Th	232	232.038050	~100
^{186}Os	186	185.953838	1.58(10)	^{234}U	234	234.040945	0.0055(5)
^{187}Os	187	186.955748	1.6(1)	^{235}U	235	235.043922	0.720(1)
^{188}Os	188	187.955836	13.3(2)	^{238}U	238	238.050784	99.2745(15)
^{189}Os	189	188.958145	16.1(3)				

APPENDIX 4
Intensities of Molecular Ions for Compounds Containing Bromine, Chlorine and Combinations of Bromine and Chlorine

		Relative intensity		
Halogen	M	M + 2	M + 4	M + 6
Br	1.0	0.98	—	—
Br_2	1.0	1.95	0.95	—
Br_3	1.0	2.93	2.86	0.93
Cl	1.0	0.33	—	—
Cl_2	1.0	0.65	0.10	—
Cl_3	1.0	0.98	0.32	0.03
BrCl	1.0	1.30	0.32	—
Br_2Cl	1.0	2.28	1.59	0.32
$BrCl_2$	1.0	1.63	0.74	0.01

APPENDIX 5
Mass-to-Charge Ratios of McLafferty Rearrangement Ions for Common Carbonyl Compounds

m/e of ion formed

Functional group	X	*m/e*
Aldehyde	–H	44
Methyl ketone	–CH_3	58
Amide	–NH_2	59
Carboxylic acid	–OH	60
Ethyl ketone	–CH_2CH_3	72
Methyl ester	–OCH_3	74
Ethyl ester	–OCH_2CH_3	88

APPENDIX 6
Primary Fragmentations Associated With Some Common Functional Groups

Functional group	Fragmentation
Amine	$$\left[R_1(CH_2)_n\underset{R_1}{N}-CH_2R_3 \right]^{+\bullet} \longrightarrow R_1(CH_2)_n\underset{R_2}{\overset{+}{N}}{=}CH_2 + R_3^\bullet \longrightarrow \underset{R_2}{\overset{+}{HN}}{=}CH_2 + Alkene$$
Aldehydes (R_2=H) and ketones	$$\left[\underset{R_1}{\overset{O}{\|}}R_2 \right]^{+\bullet} \longrightarrow R_1-C\equiv\overset{+}{O} + R_2^\bullet \longrightarrow R_1 + CO$$
Halides	$$[R{-}X]^{+\bullet} \longrightarrow R^+ + X\bullet$$
Acyl compounds	$$\left[\underset{R_1}{\overset{O}{\|}}X \right]^{+\bullet} \longrightarrow R-C\equiv\overset{+}{O} + X^\bullet$$
Alcohols and thiols	$$\left[\underset{R_2}{R_1CH_2-XH} \right]^{+\bullet} \longrightarrow R_1CH_2{=}\overset{+}{X}H + R_2^\bullet$$

APPENDIX 7
Fragments Commonly Lost From Molecular Ions

Mass	Group	Mass	Group
15	CH_3	32	CH_3OH
16	NH_2	44	C_3H_5
17	OH	42	CH_2CO
18	H_2O	42	C_3H_6
19	F	43	C_3H_7
20	HF	43	CH_3CO
26	C_2H_2	44	CO_2
29	CHO	44	C_3H_8
29	CH_2CH_3	45	CO_2H
30	CH_2O	45	OCH_2CH_3
31	OCH_3	46	CH_3CH_2OH

APPENDIX 8
Change in Mass Associated With Possible Metabolic Reaction

Metabolic reaction	Change in mass	Metabolic reaction	Change in mass
Methylation	14	Demethylation	−14
Hydroxylation	16	Acetylation	42
Epoxidation	16	Desulphurization	32
Decarboxylation	44	Hydration	18
Dehydration	16		

APPENDIX 9
Masses and Possible Structures of Common Fragment Ions

m/e	Associated structures	m/e	Associated structures
29	CHO^+, $CH_3CH_2^+$	30	$H_2C=NH_2^+$
31	$H_2C=OH^+$	41	$H_2C=CH-CH_2^+$
42	$H_2C=CH-CH_3^{+\bullet}$	43	$CH_3C\equiv O^+$, $CH_3^+CHCH_3$
44	$CH_3CHNH_2^+$, $CO_2^{+\bullet}$, $O=C=NH_2^+$, $C_3H_3^{+\bullet}$, $H_2C=CHOH^{+\bullet}$		
45	$H_2C=\overset{+}{O}CH_3$, $CH_3CH=OH^{+\bullet}$	54	$H_2C=CH-CH=CH_2^{+\bullet}$
57	$C_4H_9^+$, $CH_3CH_2C\equiv O^+$		
58	$CH_3CH_2NH^+=CH_2$, $H_2C=C(OH)CH_2^{+\bullet}$, $CH_2CH_2CH=NH_2^+$		
59	$CH_3CH_2CH=OH^+$, $H_2C=C(OH)NH_2^{+\bullet}$, $H_2C=O^+-CH_2CH_3$, $CH_3O-C\equiv O^+$		
60	$H_2C=C(OH)OH^{+\bullet}$	65	$C_5H_5^+$, $H\overset{+}{—}\langle\text{cyclopentadienyl}\rangle$
66	$\overset{\bullet}{\underset{+}{\langle\text{cyclopentadiene}\rangle}}$	67	$+\langle\text{cyclopentenyl}\rangle$
71	$CH_3CH_2CH_2C\equiv O^+$ (and isomers)	72	$H_2C=C(OH)CH_2CH_3^{+\bullet}$
73	$^+O\equiv C-OCH_2CH_3$	74	$H_2C=C(OH)OCH_3^{+\bullet}$
76	$C_6H_4^{+\bullet}$	77	$+\langle\text{phenyl}\rangle$
78	$C_6H_5^{+\bullet}$	79 & 81	$Br^{+\bullet}$
80 & 82	$HBr^{+\bullet}$	85	$C_4H_9C\equiv O^+$
88	$H_2C=C(OH)OCH_2CH_3^{+\bullet}$	91	$+\langle\text{cycloheptatrienyl}\rangle$ (and isomers)
105	$^+O\equiv C-\langle\text{phenyl}\rangle$, $CH_3-\overset{+}{\langle\text{cycloheptatrienyl}\rangle}$ (and isomers)		

APPENDIX 10
Amino Acid Residue Masses and Modifying Groups

Residue	Code	Elemental composition	Monoisotopic mass	Average mass
Alanine	A	C_3H_5NO	71.03711	71.0788
Cysteine	C	C_3H_5NOS	103.00919	103.1448
Aspartic acid	D	$C_4H_5NO_3$	115.02694	115.0886
Glutamic acid	E	$C_5H_7NO_3$	129.04259	129.1155
Phenylalanine	F	C_9H_9NO	147.06841	147.1766
Glycine	G	C_2H_3NO	57.02146	57.0520
Histidine	H	$C_6H_7N_3O$	137.05891	137.1412
Isoleucine	I	$C_6H_{11}NO$	113.08406	113.1595
Lysine	K	$C_6H_{12}N_2O$	128.09496	128.1742
Leucine	L	$C_6H_{11}NO$	113.08406	113.1595
Methionine	M	C_5H_9NOS	131.04049	131.1986
Asparagine	N	$C_4H_6N_2O_2$	114.04293	114.1039
Proline	P	C_5H_7NO	97.05276	97.1167
Glutamine	Q	$C_5H_8N_2O_2$	128.05858	128.1742
Arginine	R	$C_6H_{12}N_4O$	156.10111	156.1876
Serine	S	$C_3H_5NO_2$	87.03203	87.0782
Threonine	T	$C_4H_7NO_2$	101.04768	101.1051
Valine	V	C_5H_9NO	99.06841	99.1326
Tryptophan	W	$C_{11}H_{10}N_2O$	186.07931	186.2133
Tyrosine	Y	$C_9H_9NO_2$	163.06333	163.1760

Modifying group	Elemental composition	Monoisotopic mass	Average mass
Hydrogen	H	1.00782	1.0079
Methyl	CH_3	15.02347	15.0348
Formyl	CHO	29.00274	29.0183
Acetyl	C_2H_3O	43.01839	43.0452
t-Butyl	C_4H_9	57.07042	57.1154
Hydroxyl	HO	17.00274	17.0073

APPENDIX 11
Monosaccharide Residue Masses

Residue	Abbreviation	Monoisotopic mass	Average mass
Pentoses	Ara Rib Xyl	132.04226	132.1161
Deoxyhexoses	Fuc Rha	146.05791	146.1430
Hexosamines	GalN GlcN	161.06881	161.1577
Hexoses	Fru Gal Glc Man	162.05282	162.1424
Glucuronic acid	HexA	176.0321	176.1259
N-Acetylhexosamines	GalNAc GlcNAc	203.07937	203.1950
N-Acetylneuraminic acid	NeuAc	291.09542	291.2579
N-Glycolylneuraminic acid	NeuGc	307.09033	307.2573

APPENDIX 12
Isotopic Abundances for Various Elemental Compositions

		M + 1	M + 2			M + 1	M + 2
12	C	1.11	0.00	33	HO_2	0.10	0.40
13	CH	1.13	0.00		H_3NO	0.46	0.20
14	N	0.37	0.00	34	H_2O_2	0.11	0.40
	CH_2	1.14	0.00	36	C_3	3.33	0.04
15	HN	0.39	0.00	37	C_3H	3.35	0.04
	CH_3	1.16	0.00	38	C_2N	2.59	0.02
16	O	0.04	0.20		C_3H_2	3.36	0.04
	H_2N	0.40	0.00	39	C_2HN	2.61	0.02
	CH_4	1.17	0.00		C_3H_3	3.38	0.04
17	HO	0.06	0.20	40	CN_2	1.85	0.01
	H_3N	0.42	0.00		C_2O	2.26	0.21
18	H_2O	0.07	0.20		C_2H_2N	2.62	0.02
24	C_2	2.22	0.01		C_3H_4	3.39	0.04
25	C_2H	2.24	0.01	41	CHN_2	1.87	0.01
26	CN	1.48	0.00		C_2HO	2.28	0.21
	C_2H_2	2.25	0.01		C_2H_3N	2.64	0.02
27	CHN	1.50	0.00		C_3H_5	3.41	0.04
	C_2H_3	2.27	0.01	42	N_3	1.11	0.00
28	N_2	0.74	0.00		CNO	1.52	0.21
	CO	1.15	0.20		CH_2N_2	1.88	0.01
	CH_2N	1.51	0.00		C_2H_2O	2.29	0.21
	C_2H_4	2.28	0.01		C_2H_4N	2.65	0.02
29	HN_2	0.76	0.00		C_3H_6	3.42	0.04
	CHO	1.17	0.20	43	HN_3	1.13	0.00
	CH_3N	1.53	0.00		CHNO	1.54	0.21
	C_2H_5	2.30	0.01		CH_3N_2	1.90	0.01
30	NO	0.41	0.20		C_2H_3O	2.31	0.21
	H_2N_2	0.77	0.00		C_2H_5N	2.67	0.02
	CH_2O	1.18	0.20		C_3H_7	3.44	0.04
	CH_4N	1.54	0.01	44	N_2O	0.78	0.20
	C_2H_6	2.31	0.01		H_2N_3	1.14	0.00
31	HNO	0.43	0.20		CO_2	1.19	0.40
	H_3N_2	0.79	0.00		CH_2NO	1.55	0.21
	CH_3O	1.20	0.20		CH_4N_2	1.91	0.01
	CH_5N	1.56	0.01		C_2H_4O	2.32	0.21
32	O_2	0.08	0.40		C_2H_6N	2.68	0.02
	H_2NO	0.44	0.20		C_3H_8	3.45	0.04
	H_4N_2	0.80	0.00	45	HN_2O	0.80	0.20
	CH_4O	1.21	0.20		H_3N_3	1.16	0.00

(Contd.)

		M + 1	M + 2			M + 1	M + 2
	CHO_2	1.21	0.40		C_4H_6	4.53	0.08
	CH_3NO	1.57	0.21	55	CHN_3	2.24	0.02
	CH_5N_2	1.93	0.01		C_2HNO	2.65	0.22
	C_2H_5O	2.34	0.21		$C_2H_3N_2$	3.01	0.03
	C_2H_7N	2.70	0.02		C_3H_3O	3.42	0.24
46	NO_2	0.45	0.40		C_3H_5N	3.78	0.05
	H_2N_2O	0.81	0.20		C_4H_7	4.55	0.08
	H_4N_3	1.17	0.01	56	N_4	1.48	0.01
	CH_2O_2	1.22	0.40		CN_2O	1.89	0.21
	CH_4NO	1.58	0.21		CH_2N_3	2.25	0.02
	CH_6N_2	1.94	0.01		C_2O_2	2.30	0.41
	C_2H_6O	2.35	0.22		C_2H_2NO	2.66	0.22
47	HNO_2	0.47	0.40		$C_2H_4N_2$	3.02	0.03
	H_3N_2O	0.83	0.20		C_3H_4O	3.43	0.24
	H_5N_3	1.19	0.01		C_3H_6N	3.79	0.05
	CH_3O_2	1.24	0.40		C_4H_8	4.56	0.08
	CH_5NO	1.60	0.21	57	HN_4	1.50	0.01
48	O_3	0.12	0.60		CHN_2O	1.91	0.21
	H_2NO_2	0.48	0.40		CH_3N_3	2.27	0.02
	H_4N_2O	0.84	0.20		C_2HO_2	2.32	0.41
	CH_4O_2	1.25	0.40		C_2H_3NO	2.68	0.22
	C_4	4.44	0.07		$C_2H_5N_2$	3.04	0.03
49	HO_3	0.14	0.60		C_3H_5O	3.45	0.24
	H_3NO_2	0.50	0.40		C_3H_7N	3.81	0.05
	C_4H	4.46	0.07		C_4H_9	4.59	0.08
50	H_2O_3	0.15	0.60	58	N_3O	1.15	0.20
	C_3N	3.70	0.05		H_2N_4	1.51	0.01
	C_4H_2	4.47	0.07		CNO_2	1.56	0.41
51	C_3HN	3.72	0.05		CH_2N_2O	1.92	0.21
	C_4H_3	4.49	0.08		CH_4N_3	2.28	0.02
52	C_2N_2	2.96	0.03		$C_2H_2O_2$	2.33	0.42
	C_3O	3.37	0.24		C_2H_4NO	2.69	0.22
	C_3H_2N	3.73	0.05		$C_2H_6N_2$	3.05	0.03
	C_4H_4	4.50	0.08		C_3H_6O	3.46	0.24
53	C_2HN_2	2.98	0.03		C_3H_8N	3.82	0.05
	C_3HO	3.39	0.24		C_4H_{10}	4.59	0.08
	C_3H_3N	3.75	0.05	59	HN_3O	1.17	0.20
	C_4H_5	4.52	0.08		H_3N_4	1.53	0.01
54	CN_3	2.22	0.02		$CHNO_2$	1.58	0.41
	C_2NO	2.63	0.22		CH_3N_2O	1.94	0.21
	$C_2H_2N_2$	2.98	0.03		CH_5N_3	2.30	0.02
	C_3H_2O	3.40	0.24		$C_2H_3O_2$	2.35	0.42
	C_3H_4N	3.76	0.05		C_2H_5NO	2.71	0.22

(Contd.)

		M + 1	M + 2			M + 1	M + 2
	$C_2H_7N_2$	3.07	0.03		H_2NO_3	0.52	0.60
	C_3H_7O	3.48	0.24		$H_4N_2O_2$	0.88	0.40
	C_3H_9N	3.84	0.05		CH_4O_3	1.29	0.60
60	N_2O_2	0.82	0.40		C_3N_2	4.07	0.06
	H_2N_3O	1.18	0.20		C_4O	4.48	0.27
	H_4N_4	1.54	0.01		C_4H_2N	4.84	0.09
	CO_3	1.23	0.60		C_5H_4	5.61	0.12
	CH_2NO_2	1.59	0.41	65	HO_4	0.18	0.80
	CH_4N_2O	1.95	0.21		H_3NO_3	0.54	0.60
	CH_6N_3	2.31	0.02		C_3HN_2	4.09	0.06
	$C_2H_4O_2$	2.36	0.42		C_4HO	4.50	0.27
	C_2H_6NO	2.72	0.22		C_4H_3N	4.86	0.09
	$C_2H_8N_2$	3.08	0.03		C_5H_5	5.63	0.12
	C_3H_8O	3.49	0.24	66	H_2O_4	0.19	0.80
	C_5	5.55	0.12		C_2N_3	3.33	0.04
61	HN_2O_2	0.84	0.40		C_3NO	3.74	0.25
	H_3N_3O	1.20	0.21		$C_3H_2N_2$	4.10	0.06
	H_5N_4	1.56	0.01		C_4H_4N	4.87	0.09
	CHO_3	1.25	0.60		C_5H_6	5.64	0.12
	CH_3NO_2	1.61	0.41	67	C_2HN_3	3.35	0.04
	CH_5N_2O	1.97	0.21		C_3HNO	3.76	0.25
	CH_7N_3	2.33	0.02		$C_3H_3N_2$	4.12	0.06
	$C_2H_5O_2$	2.38	0.42		C_4H_3O	4.53	0.27
	C_2H_7NO	2.74	0.22		C_4H_5N	4.89	0.09
	C_5H	5.57	0.12		C_5H_7	5.66	0.12
62	NO_3	0.49	0.60	68	CN_4	2.59	0.02
	$H_2N_2O_2$	0.85	0.40		C_2N_2O	3.00	0.23
	H_4N_3O	1.21	0.42		$C_2H_2N_3$	3.36	0.04
	H_6N_4	1.57	0.01		C_3O_2	3.41	0.44
	CH_2O_3	1.26	0.60		C_3H_2NO	3.77	0.25
	CH_4NO_2	1.62	0.41		$C_3H_4N_2$	4.13	0.06
	CH_6N_2O	1.98	0.21		C_4H_4O	4.54	0.28
	$C_2H_6O_2$	2.39	0.42		C_4H_6N	4.90	0.09
	C_4N	4.81	0.09		C_5H_8	5.67	0.13
	C_5H_2	5.58	0.12	69	CHN_4	2.61	0.03
63	HNO_3	0.51	0.60		C_2HN_2O	3.02	0.23
	$H_3N_2O_2$	0.87	0.40		$C_2H_3N_3$	3.38	0.04
	H_5N_3O	1.23	0.21		C_3HO_2	3.43	0.44
	CH_3O_3	1.28	0.60		C_3H_3NO	3.79	0.25
	CH_5NO_2	1.64	0.41		$C_3H_5N_2$	4.15	0.06
	C_4HN	4.83	0.09		C_4H_5O	4.56	0.28
	C_5H_3	5.60	0.12		C_4H_7N	4.92	0.09
64	O_4	0.16	0.80		C_5H_5	5.63	0.12

(Contd.)

		M + 1	M + 2			M + 1	M + 2
	C_5H_9	5.69	0.13		$C_2H_3NO_2$	2.72	0.42
70	CN_3O	2.26	0.22		$C_2H_5N_2O$	3.08	0.23
	CH_2N_4	2.62	0.03		$C_2H_7N_3$	3.44	0.04
	C_2NO_2	2.67	0.42		$C_3H_5O_2$	3.49	0.44
	$C_2H_2N_2O$	3.03	0.23		C_3H_7NO	3.85	0.25
	$C_2H_4N_3$	3.39	0.04		$C_3H_9N_2$	4.21	0.07
	$C_3H_2O_2$	3.44	0.44		C_4H_9O	4.62	0.28
	C_3H_4NO	3.80	0.25		$C_4H_{11}N$	4.98	0.09
	$C_3H_6N_2$	4.16	0.07		C_6H	6.68	0.18
	C_4H_6O	4.57	0.28	74	N_3O_2	1.19	0.41
	C_4H_8N	4.93	0.09		H_2N_2O	1.55	0.21
	C_5H_{10}	5.70	0.13		CNO_3	1.60	0.61
71	CHN_3O	2.28	0.22		$CH_2N_2O_2$	1.96	0.41
	CH_3N_4	2.64	0.03		CH_4N_3O	2.32	0.22
	C_2HNO_2	2.69	0.42		CH_6N_4	2.68	0.03
	$C_2H_3N_2O$	3.05	0.23		$C_2H_2O_3$	2.37	0.62
	$C_2H_5N_3$	3.41	0.04		$C_2H_4NO_2$	2.73	0.42
	$C_3H_3O_2$	3.46	0.44		$C_2H_6N_2O$	3.09	0.23
	C_3H_5NO	3.82	0.25		$C_2H_8N_3$	3.45	0.05
	$C_3H_7N_2$	4.18	0.07		$C_3H_6O_2$	3.50	0.44
	C_4H_7O	4.59	0.28		C_3H_8NO	3.86	0.25
	C_4H_9N	4.95	0.10		$C_3H_{10}N_2$	4.22	0.07
	C_5H_{11}	5.72	0.13		$C_4H_{10}O$	4.63	0.28
72	N_4O	1.52	0.21		C_5N	5.92	0.14
	CN_2O_2	1.93	0.41		C_6H_2	6.69	0.18
	CH_2N_3O	2.29	0.22	75	HN_3O_2	1.21	0.41
	CH_4N_4	2.65	0.03		H_3N_4O	1.57	0.21
	C_2O_3	2.34	0.62		$CHNO_3$	1.62	0.61
	$C_2H_2NO_2$	2.70	0.42		$CH_3N_2O_2$	1.98	0.41
	$C_2H_4N_2O$	3.06	0.23		CH_5N_3O	2.34	0.22
	$C_2H_6N_3$	3.42	0.04		CH_7N_4	2.70	0.03
	$C_3H_4O_2$	3.47	0.44		$C_2H_3O_3$	2.39	0.62
	C_3H_6NO	3.83	0.25		$C_2H_5NO_2$	2.75	0.43
	$C_3H_8N_2$	4.19	0.07		$C_2H_7N_2O$	3.11	0.23
	C_4H_8O	4.60	0.28		$C_2H_9N_3$	3.47	0.05
	$C_4H_{10}N$	4.96	0.09		$C_3H_7O_2$	3.52	0.44
	C_5H_{12}	5.73	0.13		C_3H_9NO	3.88	0.25
	C_6	6.66	0.18		C_5HN	5.94	0.14
73	HN_4O	1.54	0.21		C_6H_3	6.71	0.18
	CHN_2O_2	1.95	0.41	76	N_2O_3	0.86	0.60
	CH_3N_3O	2.31	0.22		$H_2N_3O_2$	1.22	0.41
	CH_5N_4	2.67	0.03		H_4N_4O	1.58	0.21
	C_2HO_3	2.36	0.62		CO_4	1.27	0.80

(Contd.)

		M + 1	M + 2			M + 1	M + 2
	CH_2NO_3	1.63	0.61		CH_5NO_3	1.68	0.61
	$CH_4N_2O_2$	1.99	0.41		C_3HN_3	4.46	0.08
	CH_6N_3O	2.35	0.22		C_4HNO	4.87	0.29
	CH_8N_4	2.71	0.03		$C_4H_3N_2$	5.23	0.11
	$C_2H_4O_3$	2.40	0.62		C_5H_3O	5.64	0.32
	$C_2H_6NO_2$	2.76	0.43		C_5H_5N	6.00	0.14
	$C_2H_8N_2O$	3.12	0.24		C_6H_7	6.77	0.19
	$C_3H_8O_2$	3.53	0.44	80	H_2NO_4	0.56	0.80
	C_4N_2	5.18	0.10		$H_4N_2O_3$	0.92	0.60
	C_5O	5.59	0.32		CH_4O_4	1.33	0.80
	C_5H_2N	5.95	0.14		C_2N_4	3.70	0.05
	C_6H_4	6.72	0.19		C_3N_2O	4.11	0.26
77	HN_2O_3	0.88	0.60		$C_3H_2N_3$	4.47	0.08
	$H_3N_3O_2$	1.24	0.41		C_4O_2	4.52	0.47
	H_5N_4O	1.60	0.21		C_4H_2NO	4.88	0.29
	CHO_4	1.29	0.80		$C_4H_4N_2$	5.24	0.11
	CH_3NO_3	1.65	0.61		C_5H_4O	5.65	0.32
	$CH_5N_2O_2$	2.01	0.41		C_5H_6N	6.01	0.14
	CH_7N_3O	2.37	0.22		C_6H_8	6.78	0.19
	$C_2H_5O_3$	2.42	0.62	81	H_3NO_4	0.58	0.80
	$C_2H_7NO_2$	2.78	0.43		C_2NH_4	3.72	0.05
	C_4HN_2	5.20	0.11		C_3HN_2O	4.13	0.26
	C_5HO	5.61	0.32		$C_3H_3N_3$	4.49	0.08
	C_5H_3N	5.97	0.15		C_4HO_2	4.54	0.48
	C_6H_5	6.74	0.19		C_4H_3NO	4.90	0.29
78	NO_4	0.53	0.80		$C_4H_5N_2$	6.26	0.11
	$H_2N_2O_3$	0.89	0.60		C_5H_5O	5.67	0.32
	$H_4N_3O_2$	1.25	0.41		C_5H_7N	6.02	0.14
	H_6N_4O	1.61	0.21		C_6H_9	6.80	0.19
	CH_2O_4	1.30	0.80	82	C_2N_3O	3.37	0.24
	CH_4NO_3	1.66	0.61		$C_2H_2N_4$	3.73	0.05
	$CH_6N_2O_2$	2.02	0.41		C_3NO_2	3.78	0.45
	$C_2H_6O_3$	2.43	0.62		$C_3H_2N_2O$	4.14	0.26
	C_3N_3	4.44	0.08		$C_3H_4N_3$	4.50	0.08
	C_4NO	4.85	0.29		$C_4H_2O_2$	4.55	0.48
	$C_4H_2N_2$	5.21	0.11		C_4H_4NO	4.91	0.29
	C_5H_2O	5.62	0.32		$C_4H_6N_2$	5.27	0.11
	C_5H_4N	5.98	0.14		C_5H_6O	5.68	0.32
	C_6H_6	6.75	0.19		C_5H_8N	6.04	0.14
79	HNO_4	0.55	0.80		C_6H_{10}	6.81	0.19
	$H_3N_2O_3$	0.91	0.60	83	C_2HN_3O	3.39	0.24
	$H_5N_3O_2$	1.27	0.41		$C_2H_3N_4$	3.75	0.06
	CH_3O_4	1.32	0.80		C_3HNO_2	3.80	0.45

(Contd.)

		M + 1	M + 2			M + 1	M + 2
	$C_3H_3N_2O$	4.16	0.27		$C_2H_6N_4$	3.79	0.06
	$C_3H_5N_3$	4.52	0.08		$C_3H_2O_3$	3.48	0.64
	$C_4H_3O_2$	4.57	0.48		$C_3H_4NO_2$	3.84	0.45
	C_4H_5NO	4.93	0.29		$C_3H_6N_2O$	4.20	0.27
	$C_4H_7N_2$	5.29	0.11		$C_3H_8N_3$	4.56	0.08
	C_5H_7O	5.70	0.33		$C_4H_6O_2$	4.61	0.48
	C_5H_9N	6.06	0.15		C_4H_8NO	4.97	0.30
	C_6H_{11}	6.83	0.19		$C_4H_{10}N_2$	5.33	0.11
84	CN_4O	2.63	0.23		$C_5H_{10}O$	5.74	0.33
	$C_2H_2N_3O$	3.40	0.24		$C_5H_{12}N$	6.10	0.16
	$C_2N_2O_2$	3.04	0.43		C_6H_{14}	6.87	0.21
	$C_2H_4N_4$	3.76	0.06		C_6N	7.03	0.21
	C_3O_3	3.45	0.64		C_7H_2	7.80	0.26
	$C_3H_2NO_2$	3.81	0.45	87	CHN_3O_2	2.32	0.42
	$C_3H_4N_2O$	4.17	0.27		CH_3N_4O	2.68	0.23
	$C_3H_6N_3$	4.53	0.08		C_2HNO_3	2.73	0.62
	$C_4H_4O_2$	4.58	0.48		$C_2H_3N_2O_2$	3.09	0.43
	C_4H_6NO	4.94	0.29		$C_2H_5N_3O$	3.45	0.25
	$C_4H_8N_2$	5.30	0.11		$C_2H_7N_4$	3.81	0.06
	C_5H_8O	5.71	0.33		$C_3H_3O_3$	3.50	0.64
	$C_5H_{10}N$	6.07	0.15		$C_3H_5NO_2$	3.86	0.45
	C_6H_{12}	6.84	0.19		$C_3H_7N_2O$	4.22	0.27
	C_7	7.77	0.26		$C_3H_9N_3$	4.58	0.08
85	CHN_4O	2.65	0.23		$C_4H_7O_2$	4.63	0.48
	$C_2HN_2O_2$	3.06	0.43		C_4H_9NO	4.99	0.30
	$C_2H_3N_3O$	3.42	0.24		$C_4H_{11}N_2$	5.35	0.11
	$C_2H_5N_4$	3.78	0.06		$C_5H_{11}O$	5.76	0.33
	C_3HO_3	3.47	0.64		$C_5H_{13}N$	6.12	0.15
	$C_3H_3NO_2$	3.83	0.45		C_6HN	7.05	0.21
	$C_3H_5N_2O$	4.19	0.27		C_7H_3	7.82	0.26
	$C_3H_7N_3$	4.55	0.08	88	N_4O_2	1.56	0.41
	$C_4H_5O_2$	4.60	0.48		CN_2O_3	1.97	0.61
	C_4H_7NO	4.96	0.29		$CH_2N_3O_2$	2.33	0.42
	$C_4H_9N_2$	5.32	0.11		CH_4N_4O	2.69	0.23
	C_5H_9O	5.73	0.33		C_2O_4	2.38	0.82
	$C_5H_{11}N$	6.09	0.16		$C_2H_2NO_3$	2.74	0.63
	C_6H_{13}	6.86	0.20		$C_2H_4N_2O_2$	3.10	0.43
	C_7H	7.79	0.26		$C_2H_6N_3O$	3.46	0.25
86	CN_3O_2	2.30	0.41		$C_2H_8N_4$	3.82	0.06
	CH_2N_4O	2.66	0.21		$C_3H_4O_3$	3.51	0.64
	C_2NO_3	2.71	0.62		$C_3H_6NO_2$	3.87	0.45
	$C_2H_2N_2O_2$	3.07	0.43		$C_3H_8N_2O$	4.23	0.27
	$C_2H_4N_3O$	3.43	0.24		$C_3H_{10}N_5$	4.59	0.08

(Contd.)

	M + 1	M + 2			M + 1	M + 2
$C_4H_8O_2$	4.64	0.48		C_5NO	5.96	0.34
$C_4H_{10}NO$	5.00	0.30		$C_5H_2N_2$	6.32	0.17
$C_4H_{12}N_2$	5.36	0.11		C_6H_2O	6.73	0.38
$C_5H_{12}O$	5.77	0.33		C_6H_4N	7.09	0.21
C_5N_2	6.29	0.16		C_7H_6	7.86	0.26
C_6O	6.70	0.38	91	HN_3O_3	1.25	0.60
C_6H_2N	7.06	0.21		$H_3N_4O_2$	1.61	0.41
C_7H_4	7.83	0.26		$CHNO_4$	1.66	0.81
89 HN_4O_2	1.58	0.41		$CH_3N_2O_3$	2.02	0.61
CHN_2O_3	1.99	0.61		$CH_5N_3O_2$	2.38	0.42
$CH_3N_3O_2$	2.35	0.42		CH_7N_4O	2.74	0.23
CH_5N_4O	2.71	0.23		$C_2H_3O_4$	2.43	0.82
C_2HO_4	2.40	0.82		$C_2H_5NO_3$	2.79	0.63
$C_2H_3NO_3$	2.76	0.63		$C_2H_7N_2O_2$	3.15	0.44
$C_2H_5N_2O_2$	3.12	0.44		$C_2H_9N_3O$	3.51	0.25
$C_2H_7N_3O$	3.48	0.25		$C_3H_7O_3$	3.56	0.64
$C_2H_9N_4$	3.84	0.06		$C_3H_9NO_2$	3.92	0.46
$C_3H_5O_3$	3.53	0.64		C_4HN_3	5.57	0.13
$C_3H_7NO_2$	3.89	0.46		C_5HNO	5.98	0.34
$C_3H_9N_2O$	4.25	0.27		$C_5H_3N_2$	6.34	0.17
$C_3H_{11}N_3$	4.61	0.08		C_6H_3O	6.75	0.38
$C_4H_9O_2$	4.66	0.48		C_6H_5N	7.22	0.21
$C_4H_{11}NO$	5.02	0.30		C_7H_7	7.88	0.26
C_5HN_2	6.31	0.16	92	N_2O_4	0.90	0.80
C_6HO	6.72	0.38		$H_2N_3O_3$	1.26	0.60
C_6H_3N	7.08	0.21		$H_4N_4O_2$	1.62	0.41
C_7H_5	7.85	0.26		CH_2NO_4	1.67	0.81
90 N_3O_3	1.23	0.60		$CH_4N_2O_3$	2.03	0.61
$H_2N_4O_2$	1.59	0.40		$CH_6N_3O_2$	2.39	0.42
CNO_4	1.64	0.80		CH_8N_4O	2.75	0.23
$CH_2N_2O_3$	2.00	0.61		$C_2H_4O_4$	2.44	0.82
$CH_4N_3O_2$	2.36	0.42		$C_2H_6NO_3$	2.80	0.63
CH_6N_4O	2.72	0.23		$C_2H_8N_2O_2$	3.16	0.44
$C_2H_2O_4$	2.41	0.82		$C_3H_8O_3$	3.57	0.64
$C_2H_4NO_3$	2.77	0.63		C_3N_4	4.81	0.09
$C_2H_6N_2O_2$	3.13	0.44		C_4N_2O	5.22	0.31
$C_2H_8N_3O$	3.49	0.25		$C_4H_2N_3$	5.58	0.13
$C_2H_{10}N_4$	3.85	0.06		C_5O_2	5.63	0.52
$C_3H_6O_3$	3.54	0.64		C_5H_2NO	5.99	0.34
$C_3H_8NO_2$	3.90	0.46		$C_5H_4N_2$	6.35	0.17
$C_3H_{10}N_2O$	4.26	0.27		C_6H_4O	5.76	0.38
$C_4H_{10}O_2$	4.67	0.48		C_6H_6N	7.12	0.21
C_4N_3	5.55	0.13		C_7H_8	7.89	0.27

(Contd.)

		M + 1	M + 2			M + 1	M + 2
93	HN_2O_4	0.92	0.80		C_5H_5NO	6.04	0.35
	$H_3N_3O_3$	1.28	0.60		$C_5H_7N_2$	6.40	0.17
	$H_5N_4O_2$	1.64	0.41		C_6H_7O	6.81	0.39
	CH_3NO_4	1.69	0.81		C_6H_9N	7.17	0.22
	$CH_5N_2O_3$	2.05	0.61		C_7H_{11}	7.94	0.27
	$CH_7N_3O_2$	2.41	0.42	96	$H_4N_2O_4$	0.96	0.80
	$C_2H_5O_4$	2.46	0.82		C_2N_4O	3.74	0.26
	$C_2H_7NO_3$	2.82	0.63		$C_3N_2O_2$	4.15	0.47
	C_3HN_4	4.83	0.09		$C_3H_2N_3O$	4.51	0.28
	C_4HN_2O	5.24	0.31		$C_3H_4N_4$	4.87	0.10
	$C_4H_3N_3$	5.60	0.13		C_4O_3	4.56	0.67
	C_5HO_2	5.65	0.52		$C_4H_2NO_2$	4.92	0.49
	C_5H_3NO	6.01	0.35		$C_4H_4N_2O$	5.28	0.31
	$C_5H_5N_2$	6.37	0.17		$C_4H_6N_3$	5.64	0.13
	C_6H_5O	6.78	0.38		$C_5H_4O_2$	5.69	0.53
	C_6H_7N	7.14	0.22		C_5H_6NO	6.05	0.35
	C_7H_9	7.91	0.27		$C_5H_8N_2$	6.41	0.17
94	$H_2N_2O_4$	0.93	0.80		C_6H_8O	6.82	0.39
	$H_4N_3O_3$	1.29	0.61		$C_6H_{10}N$	7.18	0.22
	$H_6N_4O_2$	1.65	0.41		C_7H_{12}	7.95	0.27
	CH_4NO_4	1.70	0.81		C_8	8.88	0.34
	$CH_6N_2O_3$	2.06	0.62	97	C_2HN_4O	3.76	0.26
	$C_2H_6O_4$	2.47	0.82		$C_3HN_2O_2$	4.17	0.47
	C_3N_3O	4.48	0.28		$C_3H_3N_3O$	4.53	0.28
	$C_3H_2N_4$	4.84	0.09		$C_3H_5N_4$	4.89	0.10
	C_4NO_2	4.89	0.49		C_4HO_3	4.58	0.68
	$C_4H_2N_2O$	5.25	0.31		$C_4H_3NO_2$	4.94	0.49
	$C_4H_4N_3$	5.61	0.13		$C_4H_5N_2O$	5.30	0.31
	$C_5H_2O_2$	5.66	0.52		$C_4H_7N_3$	5.66	0.13
	C_5H_4NO	6.02	0.35		$C_5H_5O_2$	5.71	0.53
	$C_5H_6N_2$	6.38	0.17		C_5H_7NO	6.07	0.35
	C_6H_6O	6.79	0.38		$C_5H_9N_2$	6.43	0.17
	C_6H_8N	7.15	0.22		C_6H_9O	6.84	0.39
	C_7H_{10}	7.92	0.27		$C_6H_{11}N$	7.20	0.22
95	$H_3N_2O_4$	0.95	0.80		C_7H_{13}	7.97	0.27
	$H_5N_3O_3$	1.31	0.60		C_8H	8.90	0.34
	CH_5NO_4	1.72	0.81	98	$C_2N_3O_2$	3.41	0.44
	C_3HN_3O	4.50	0.28		$C_2H_2N_4O$	3.77	0.26
	$C_3H_3N_4$	4.86	0.10		C_3NO_3	3.82	0.65
	C_4HNO_2	4.91	0.49		$C_3H_2N_2O_2$	4.18	0.47
	$C_4H_3N_2O$	5.27	0.31		$C_3H_4N_3O$	4.54	0.28
	$C_4H_5N_3$	5.63	0.13		$C_3H_6N_4$	4.90	0.10
	$C_5H_3O_2$	5.68	0.52		$C_4H_2O_3$	4.59	0.68

(Contd.)

	M + 1	M + 2			M + 1	M + 2
$C_4H_4NO_2$	4.95	0.49		$C_5H_{10}NO$	6.11	0.35
$C_4H_6N_2O$	5.31	0.31		$C_5H_{12}N_2$	6.47	0.18
$C_4H_8N_3$	5.67	0.13		$C_6H_{12}O$	6.88	0.39
$C_5H_6O_2$	5.72	0.53		$C_6H_{14}N$	7.24	0.22
C_5H_8NO	6.08	0.35		C_6N_2	7.40	0.23
$C_5H_{10}N_2$	6.44	0.17		C_7H_{16}	8.01	0.28
$C_6H_{10}O$	6.85	0.39		C_7O	7.81	0.46
$C_6H_{12}N$	7.21	0.21		C_7H_2N	8.17	0.29
C_7H_{14}	7.98	0.26		C_8H_4	8.94	0.35
C_7N	8.14	0.27	101	CHN_4O_2	2.701	0.428
C_8H_2	8.91	0.33		$C_2HN_2O_3$	3.057	0.635
99 $C_2HN_3O_2$	3.43	0.44		$C_2H_3N_3O_2$	3.432	0.446
$C_2H_3N_4O$	3.79	0.25		$C_2H_5N_4O$	3.806	0.258
C_3HNO_3	3.84	0.65		C_3HO_4	3.414	0.842
$C_3H_3N_2O_2$	4.20	0.47		$C_3H_3NO_3$	3.789	0.654
$C_3H_5N_3O$	4.56	0.28		$C_3H_5N_2O_2$	4.163	0.468
$C_3H_7N_4$	4.92	0.10		$C_3H_7N_3O$	4.537	0.283
$C_4H_3O_3$	4.61	0.68		$C_3H_9N_4$	4.912	0.100
$C_4H_5NO_2$	4.97	0.49		$C_4H_5O_3$	4.520	0.680
$C_4H_7H_2O$	5.33	0.31		$C_4H_7NO_2$	4.894	0.496
$C_4H_9N_3$	5.69	0.13		$C_4H_9N_2O$	5.268	0.314
$C_5H_7O_2$	5.74	0.53		$C_4H_{11}N_3$	5.643	0.137
C_5H_9NO	6.11	0.35		$C_5H_9O_2$	5.625	0.529
$C_5H_{11}N_2$	6.46	0.17		$C_5H_{11}NO$	6.000	0.350
$C_6H_{11}O$	6.86	0.39		$C_5H_{13}N_2$	6.374	0.172
$C_6H_{13}N$	7.23	0.22		C_6HN_2	7.263	0.227
C_7H_{15}	8.00	0.27		$C_6H_{13}O$	6.731	0.391
C_7HN	8.16	0.29		$C_6H_{15}N$	7.105	0.216
C_8H_3	8.93	0.35		C_7HO	7.619	0.449
100 CN_4O_2	2.67	0.43		C_7H_3N	7.994	0.277
$C_2N_2O_3$	3.08	0.63		C_8H_5	8.725	0.333
$C_2H_2N_3O_2$	3.44	0.45	102	$CH_2N_4O_2$	2.717	0.429
$C_2H_4N_4O$	3.80	0.26		$C_2H_2N_2O_3$	3.073	0.635
C_3O_4	3.45	0.84		$C_2H_4N_3O_2$	3.448	0.446
$C_3H_2NO_3$	3.85	0.65		$C_2H_6N_4O$	3.822	0.258
$C_3H_4N_2O_2$	4.21	0.47		$C_3H_2O_4$	3.430	0.843
$C_3H_6N_3O$	4.57	0.28		$C_3H_4NO_3$	3.805	0.655
$C_3H_8N_4$	4.94	0.10		$C_3H_6N_2O_2$	4.179	0.470
$C_4H_4O_3$	4.62	0.68		$C_3H_8N_3O$	4.553	0.284
$C_4H_6NO_2$	4.98	0.49		$C_3H_{10}N_4$	4.928	0.101
$C_4H_8N_2O$	5.34	0.31		$C_4H_6O_3$	4.536	0.681
$C_4H_{10}N_3$	5.70	0.13		$C_4H_8NO_2$	4.910	0.497
$C_5H_8O_2$	5.76	0.53		$C_4H_{10}N_2O$	5.284	0.315

(Contd.)

		M + 1	M + 2			M + 1	M + 2
	$C_4H_{12}N_3$	5.659	0.135		$C_4H_8O_3$	4.568	0.682
	$C_5H_{10}O_2$	5.641	0.531		$C_4H_{10}NO_2$	4.942	0.499
	$C_5H_{12}NO$	6.016	0.351		$C_4H_{12}N_2O$	5.316	0.317
	$C_5H_{14}N_2$	6.390	0.173		$C_5H_2N_3$	6.579	0.185
	$C_6H_2N_2$	7.279	0.228		$C_5H_{12}O_2$	5.673	0.532
	$C_6H_{14}O$	6.747	0.393		C_6H_2NO	6.936	0.405
	C_7H_2O	7.635	0.451		$C_6H_4N_2$	7.311	0.230
	C_7H_4N	8.010	0.279		C_7H_4O	7.667	0.453
	C_8H_6	8.741	0.333		C_7H_6N	8.042	0.281
103	CHN_3O_3	2.358	0.621		C_8H_8	8.773	0.338
	$CH_3N_4O_2$	2.733	0.429	105	CHN_2O_4	2.016	0.814
	C_2HNO_4	2.715	0.826		$CH_3N_3O_3$	2.390	0.622
	$C_2H_3N_2O_3$	3.089	0.636		$CH_5N_4O_2$	2.765	0.430
	$C_2H_5N_3O_2$	3.464	0.447		$C_2H_3NO_4$	2.747	0.827
	$C_2H_7N_4O$	3.838	0.260		$C_2H_5N_2O_3$	3.121	0.637
	$C_3H_3O_4$	3.446	0.844		$C_2H_7N_3O_2$	3.496	0.448
	$C_3H_3NO_3$	3.821	0.656		$C_2H_9N_4O$	3.870	0.260
	$C_3H_7N_2O_2$	4.195	0.470		$C_3H_5O_4$	3.478	0.844
	$C_3H_9N_3O$	4.569	0.285		$C_3H_7NO_3$	3.853	0.657
	$C_3H_{11}N_4$	4.944	0.102		$C_3H_9N_2O_2$	4.227	0.471
	$C_4H_7O_3$	4.552	0.681		$C_3H_{11}N_3O$	4.601	0.286
	$C_4H_9NO_2$	4.926	0.498		C_4HN_4	5.864	0.145
	$C_4H_{11}N_2O$	5.300	0.316		$C_4H_9O_3$	4.584	0.683
	$C_4H_{13}N_3$	5.675	0.135		$C_4H_{11}NO_2$	4.958	0.499
	C_5HN_3	6.563	0.184		C_5HN_2O	6.221	0.363
	$C_5H_{11}O_2$	5.657	0.531		$C_5H_3N_3$	6.595	0.186
	$C_5H_{13}NO$	6.032	0.352		C_6HO_2	6.578	0.582
	C_6HNO	6.920	0.404		C_6H_3NO	6.952	0.406
	$C_6H_3N_2$	7.295	0.229		$C_6H_5N_2$	7.327	0.231
	C_7H_3O	7.651	0.452		C_7H_5O	7.683	0.454
	C_7H_5N	8.026	0.280		C_7H_7N	8.058	0.283
	C_8H_7	8.757	0.336		C_8H_9	8.789	0.339
104	$CH_2N_3O_3$	2.374	0.621	106	$CH_2N_2O_4$	2.032	0.815
	$CH_4N_4O_2$	2.749	0.430		$CH_4N_3O_3$	2.406	0.622
	$C_2H_2NO_4$	2.731	0.826		$CH_6N_4O_2$	2.781	0.431
	$C_2H_4N_2O_3$	3.105	0.636		$C_2H_4NO_4$	2.763	0.828
	$C_2H_6N_3O_2$	3.480	0.447		$C_2H_6N_2O_3$	3.137	0.638
	$C_2H_8N_4O$	3.854	0.260		$C_2H_8N_3O_2$	3.512	0.449
	$C_3H_4O_4$	3.462	0.844		$C_2H_{10}N_4O$	3.886	0.261
	$C_3H_6NO_3$	3.837	0.656		$C_3H_6O_4$	3.494	0.845
	$C_3H_8N_2O_2$	4.211	0.470		$C_3H_8NO_3$	3.869	0.658
	$C_3H_{10}N_3O$	4.585	0.285		$C_3H_{10}N_2O_2$	4.243	0.472
	$C_3H_{12}N_4$	4.960	0.102		$C_4H_2N_4$	5.880	0.146

(Contd.)

		M + 1	M + 2			M + 1	M + 2
	$C_4H_{10}O_3$	4.600	0.683		$C_7H_{10}N$	8.106	0.286
	$C_5H_2N_2O$	6.237	0.364		C_8H_{12}	8.837	0.343
	$C_5H_4N_3$	6.611	0.187	109	$CH_5N_2O_4$	2.080	0.816
	$C_6H_2O_2$	6.594	0.583		$CH_7N_3O_3$	2.454	0.624
	C_6H_4NO	6.968	0.407		$C_2H_7NO_4$	2.811	0.829
	$C_6H_6N_2$	7.343	0.233		C_3HN_4O	4.823	0.296
	C_7H_6O	7.699	0.456		$C_4HN_2O_2$	5.179	0.510
	C_7H_8N	8.074	0.284		$C_3H_3N_3O$	5.554	0.329
	C_8H_{10}	8.805	0.341		$C_4H_5N_4$	5.928	0.149
107	$CH_3N_2O_4$	2.048	0.816		C_5HO_3	5.536	0.725
	$CH_5N_3O_3$	2.422	0.623		$C_5H_3NO_2$	5.911	0.545
	$CH_7N_4O_2$	2.797	0.431		$C_5H_5N_2O$	6.285	0.367
	$C_2H_5NO_4$	2.779	0.828		$C_5H_7N_3$	6.659	0.190
	$C_2H_7N_2O_3$	3.153	0.638		$C_6H_5O_2$	6.642	0.587
	$C_2H_9N_3O_2$	3.528	0.449		C_6H_7NO	7.016	0.411
	$C_3H_7O_4$	3.510	0.846		$C_6H_9N_2$	7.391	0.237
	$C_3H_9NO_3$	3.885	0.659		C_7H_9O	7.747	0.460
	C_4HN_3O	5.522	0.327		$C_7H_{11}N$	8.122	0.288
	$C_4H_3N_4$	5.896	0.148		C_8H_{13}	8.853	0.345
	C_5HNO_2	5.879	0.544		C_9H	9.741	0.421
	$C_5H_3N_2O$	6.253	0.365	110	$CH_6N_2O_4$	2.096	0.816
	$C_5H_5N_3$	6.627	0.188		$C_3H_2N_4O$	4.839	0.297
	$C_6H_3O_2$	6.610	0.584		$C_4H_2N_2O_2$	5.196	0.511
	C_6H_5NO	6.984	0.409		$C_4H_4N_3O$	5.570	0.330
	$C_6H_7N_2$	7.359	0.234		$C_4H_6N_4$	5.944	0.150
	C_7H_7O	7.715	0.457		$C_5H_2O_3$	5.552	0.726
	C_7H_9N	8.090	0.286		$C_5H_4NO_2$	5.927	0.546
	C_8H_{11}	8.821	0.342		$C_5H_6N_2O$	6.301	0.368
108	$CH_4N_2O_4$	2.064	0.815		$C_5H_8N_3$	6.675	0.191
	$CH_6N_3O_3$	2.438	0.623		$C_6H_6O_2$	6.658	0.587
	$CH_8N_4O_2$	2.813	0.432		C_6H_8NO	7.032	0.412
	$C_2H_6NO_4$	2.795	0.828		$C_6H_{10}N_2$	7.407	0.237
	$C_2H_8N_2O_3$	3.169	0.638		$C_7H_{10}O$	7.763	0.460
	$C_3H_8O_4$	3.526	0.846		$C_7H_{12}N$	8.138	0.289
	$C_4H_2N_3O$	5.538	0.328		C_8H_{14}	8.869	0.346
	$C_4H_4N_4$	5.912	0.148		C_9H_2	9.757	0.423
	$C_5H_2NO_2$	5.895	0.544	111	$C_3HN_3O_2$	4.480	0.481
	$C_5H_4N_2O$	6.269	0.366		$C_3H_3N_4O$	4.855	0.297
	C_5H_6Na	6.643	0.189		C_4HNO_3	4.837	0.694
	$C_6H_4O_2$	6.626	0.585		$C_4H_3N_2O_2$	5.212	0.511
	C_6H_6NO	7.000	0.409		$C_4H_5N_3O$	5.586	0.330
	$C_6H_8N_2$	7.375	0.235		$C_4H_7N_4$	5.960	0.151
	C_7H_8O	7.731	0.458		$C_5H_3O_3$	5.568	0.727

(Contd.)

		M + 1	M + 2			M + 1	M + 2
	$C_5H_5NO_2$	5.943	0.547		$C_6H_{11}NO$	7.080	0.415
	$C_5H_7N_2O$	6.317	0.369		$C_6H_{13}N_2$	7.455	0.241
	$C_5H_9N_3$	6.691	0.192		C_7HN_2	8.343	0.305
	$C_6H_7O_2$	6.674	0.588		$C_7H_{13}O$	7.811	0.464
	C_6H_9NO	7.048	0.413		$C_7H_{15}N$	8.186	0.293
	$C_6H_{11}N_2$	7.423	0.239		C_8HO	8.700	0.532
	$C_7H_{11}O$	7.779	0.462		C_8H_3N	9.074	0.364
	$C_7H_{13}N$	8.154	0.291		C_8H_{17}	8.917	0.350
	C_8HN	9.042	0.361		C_9H_5	9.805	0.428
	C_8H_{15}	8.885	0.348	114	$C_2H_2N_4O_2$	3.797	0.458
	C_9H_3	9.773	0.425		$C_3H_2N_2O_3$	4.154	0.668
112	$C_3H_2N_3O_2$	4.496	0.482		$C_3H_4N_3O_2$	4.528	0.483
	$C_3H_4N_4O$	4.871	0.298		$C_3H_6N_4O$	4.903	0.300
	$C_4H_2NO_3$	4.853	0.695		$C_4H_2O_4$	4.511	0.880
	$C_4H_4N_2O_2$	5.228	0.512		$C_4H_4NO_3$	4.885	0.696
	$C_4H_6N_3O$	5.602	0.331		$C_4H_6N_2O_2$	5.260	0.514
	$C_4H_8N_4$	5.976	0.152		$C_4H_8N_3O$	5.634	0.336
	$C_5H_4O_3$	5.584	0.728		$C_4H_{10}N_4$	6.008	0.154
	$C_5H_6NO_2$	5.959	0.548		$C_5H_6O_3$	5.616	0.729
	$C_5H_8N_2O$	6.333	0.370		$C_5H_8NO_2$	5.991	0.550
	$C_5H_{10}N_3$	6.707	0.193		$C_5H_{10}N_2O$	6.365	0.372
	$C_6H_8O_2$	6.690	0.589		$C_5H_{12}N_3$	6.739	0.195
	$C_6H_{10}NO$	7.064	0.414		$C_6H_{10}O_2$	6.722	0.591
	$C_6H_{12}N_2$	7.439	0.240		$C_6H_{12}NO$	7.096	0.416
	$C_7H_{12}O$	7.795	0.463		$C_6H_{14}N_2$	7.471	0.242
	$C_7H_{14}N$	8.170	0.292		$C_7H_2N_2$	8.359	0.307
	C_8H_2N	9.058	0.362		$C_7H_{14}O$	7.827	0.465
	C_8H_{16}	8.901	0.349		$C_7H_{16}N$	8.202	0.294
	C_9H_4	9.789	0.426		C_8H_2O	8.716	0.533
113	$C_2HN_4O_2$	3.781	0.458		C_8H_4N	9.090	0.365
	$C_3HN_2O_3$	4.138	0.668		C_8H_{18}	8.933	0.352
	$C_3H_3N_3O_2$	4.512	0.483		C_9H_6	9.821	0.429
	$C_3H_5N_4O$	4.887	0.299	115	$C_2HN_3O_3$	3.439	0.647
	C_4HO_4	4.495	0.880		$C_2H_3N_4O_2$	3.813	0.459
	$C_4H_3NO_3$	4.869	0.696		C_3HNO_4	3.796	0.856
	$C_4H_5N_2O_2$	5.244	0.514		$C_3H_3N_2O_3$	4.170	0.669
	$C_4H_7N_3O$	5.618	0.333		$C_3H_5N_3O_2$	4.544	0.485
	$C_4H_9N_4$	5.992	0.153		$C_3H_7N_4O$	4.919	0.301
	$C_5H_5O_3$	5.600	0.729		$C_4H_3O_4$	4.527	0.881
	$C_5H_7NO_2$	5.975	0.550		$C_4H_5NO_3$	4.901	0.697
	$C_5H_9N_2O$	6.349	0.371		$C_4H_7N_2O_2$	5.276	0.515
	$C_5H_{11}N_3$	6.723	0.194		$C_4H_9N_3O$	5.650	0.335
	$C_6H_9O_2$	6.706	0.590		$C_4H_{11}N_4$	6.024	0.155

(Contd.)

	M + 1	M + 2			M + 1	M + 2
	$C_5H_7O_3$	5.632	0.731	$C_3H_3NO_4$	3.828	0.857
	$C_5H_9NO_2$	6.007	0.551	$C_3H_5N_2O_3$	4.202	0.671
	$C_5H_{11}N_2O$	6.381	0.373	$C_3H_7N_3O_2$	4.576	0.486
	$C_5H_{13}N_3$	6.755	0.197	$C_3H_9N_4O$	4.951	0.303
	C_6HN_3	7.644	0.255	$C_4H_5O_4$	4.560	0.882
	$C_6H_{11}O_2$	6.738	0.593	$C_4H_7NO_3$	4.933	0.699
	$C_6H_{13}NO$	7.112	0.418	$C_4H_9N_2O_2$	5.308	0.517
	$C_6H_{15}N_2$	7.487	0.244	$C_4H_{11}N_3O$	5.682	0.336
	C_7HNO	8.001	0.479	$C_4H_{13}N_4$	6.056	0.157
	$C_7H_3N_2$	8.375	0.308	C_5HN_4	6.945	0.209
	$C_7H_{15}O$	7.843	0.467	$C_5H_9O_3$	5.664	0.733
	$C_7H_{17}N$	8.218	0.296	$C_5H_{11}NO_2$	6.039	0.553
	C_8H_3O	8.732	0.535	$C_5H_{13}N_2O$	6.413	0.375
	C_8H_5N	9.106	0.367	$C_5H_{15}N_3$	6.787	0.199
	C_9H_7	9.837	0.431	C_6HN_2O	7.302	0.431
116	$C_2H_2N_3O_3$	3.455	0.647	$C_6H_3N_3$	7.676	0.257
	$C_2H_4N_4O_2$	3.829	0.460	$C_6H_{13}O_2$	6.770	0.595
	$C_3H_2NO_4$	3.812	0.856	$C_6H_{15}NO$	7.144	0.420
	$C_3H_4N_2O_3$	4.186	0.670	C_7HO_2	7.658	0.653
	$C_3H_6N_3O_2$	4.560	0.485	C_7H_3NO	8.033	0.482
	$C_3H_8N_4O$	4.935	0.301	$C_7H_5N_2$	8.407	0.311
	$C_4H_4O_4$	4.543	0.881	C_8H_5O	8.764	0.538
	$C_4H_6NO_3$	4.917	0.698	C_8H_7N	9.138	0.370
	$C_4H_8N_2O_2$	5.292	0.516	C_9H_9	9.869	0.435
	$C_4H_{10}N_3O$	5.666	0.335	118 $C_2H_2N_3O_4$	3.113	0.837
	$C_4H_{12}N_4$	6.040	0.156	$C_2H_4N_3O_3$	3.487	0.648
	$C_5H_8O_3$	5.648	0.731	$C_2H_6N_4O_2$	3.861	0.461
	$C_5H_{10}NO_2$	6.023	0.552	$C_3H_4NO_4$	3.844	0.858
	$C_5H_{12}N_2O$	6.397	0.374	$C_3H_6N_2O_3$	4.218	0.671
	$C_5H_{14}N_3$	6.771	0.197	$C_3H_8N_3O_2$	4.592	0.487
	$C_6H_2N_3$	7.660	6.256	$C_3H_{10}N_4O$	4.967	0.303
	$C_6H_{12}O_2$	6.754	0.594	$C_4H_6O_4$	4.575	0.883
	$C_6H_{14}NO$	7.128	0.418	$C_4H_8NO_3$	4.949	0.700
	$C_6H_{16}N_2$	7.503	0.244	$C_4H_{10}N_2O_2$	5.324	0.518
	C_7H_2NO	8.017	0.480	$C_4H_{12}N_3O$	5.698	0.337
	$C_7H_4N_2$	8.391	0.309	$C_4H_{14}N_4$	6.072	0.158
	$C_7H_{16}O$	7.859	0.468	$C_5H_2N_4$	6.961	0.210
	C_8H_4O	8.748	0.536	$C_5H_{10}O_3$	5.680	0.734
	C_8H_6N	9.122	0.368	$C_5H_{12}NO_2$	6.055	0.554
	C_9H_8	9.853	0.432	$C_5H_{14}N_2O$	6.429	0.376
117	$C_2HN_2O_4$	3.097	0.837	$C_6H_2N_2O$	7.318	0.431
	$C_2H_3N_3O_3$	3.471	0.648	$C_6H_4N_3$	7.692	0.259
	$C_2H_5N_4O_2$	3.845	0.460	$C_6H_{14}O_2$	6.786	0.596

(Contd.)

		M + 1	M + 2			M + 1	M + 2
	$C_7H_2O_2$	7.674	0.655		$C_6H_2NO_2$	6.975	0.608
	C_7H_4NO	8.049	0.483		$C_6H_4N_2O$	7.350	0.434
	$C_7H_6N_2$	8.423	0.312		$C_6H_6N_3$	7.724	0.261
	C_8H_6O	8.780	0.539		$C_7H_4O_2$	7.706	0.657
	C_8H_8N	9.154	0.372		C_7H_6NO	8.081	0.485
	C_9H_{10}	9.885	0.436		$C_7H_8N_2$	8.455	0.315
119	$C_2H_3N_2O_4$	3.129	0.838		C_8H_8O	8.812	0.542
	$C_2H_5N_3O_3$	3.503	0.649		$C_8H_{10}N$	9.186	0.374
	$C_2H_7N_4O_2$	3.877	0.462		C_9H_{12}	9.917	0.439
	$C_3H_5NO_4$	3.860	0.858	121	$C_2H_5N_2O_4$	3.161	0.839
	$C_3H_7N_2O_3$	4.234	0.672		$C_2H_7N_3O_3$	3.535	0.650
	$C_3H_9N_3O_2$	4.608	0.487		$C_2H_9N_4O_2$	3.909	0.463
	$C_3H_{11}N_4O$	4.983	0.304		$C_3H_7NO_4$	3.892	0.859
	$C_4H_7O_4$	4.591	0.884		$C_3H_9N_2O_3$	4.266	0.673
	$C_4H_9NO_3$	4.965	0.701		$C_3H_{11}N_3O_2$	4.640	0.489
	$C_4H_{11}N_2O_2$	5.340	0.519		C_4HN_4O	5.903	0.349
	$C_4H_{13}N_3O$	5.714	0.338		$C_4H_9O_4$	4.623	0.885
	C_5HN_3O	6.602	0.387		$C_4H_{11}NO_3$	4.997	0.702
	$C_5H_3N_4$	6.977	0.211		$C_5HN_2O_2$	6.260	0.566
	$C_5H_{11}O_3$	5.696	0.735		$C_5H_3N_3O$	6.635	0.389
	$C_5H_{13}NO_2$	6.071	0.555		$C_5H_5N_4$	7.009	0.214
	C_6HNO_2	6.959	0.607		C_6HO_3	6.617	0.785
	$C_6H_3N_2O$	7.334	0.433		$C_6H_3NO_2$	6.991	0.610
	$C_6H_5N_3$	7.708	0.260		$C_6H_5N_2O$	7.366	0.435
	$C_7H_3O_2$	7.690	0.656		$C_6H_7N_3$	7.740	0.262
	C_7H_5NO	8.065	0.484		$C_7H_5O_2$	7.722	0.658
	$C_7H_7N_2$	8.439	0.314		C_7H_7NO	8.097	0.487
	C_8H_7O	8.796	0.541		$C_7H_9N_2$	8.471	0.317
	C_8H_9N	9.170	0.373		C_8H_9O	8.828	0.543
	C_9H_{11}	9.901	0.438		$C_8H_{11}N$	9.202	0.376
120	$C_2H_4N_2O_4$	3.145	0.838		C_9H_{13}	9.933	0.441
	$C_2H_6N_3O_3$	3.519	0.650		$C_{10}H$	10.822	0.527
	$C_2H_8N_4O_2$	3.893	0.462	122	$C_2H_6N_2O_4$	3.177	0.839
	$C_3H_6NO_4$	3.876	0.859		$C_2H_8N_3O_3$	3.551	0.651
	$C_3H_8N_2O_3$	4.250	0.673		$C_2H_{10}N_4O_2$	3.925	0.463
	$C_3H_{10}N_3O_2$	4.624	0.488		$C_3H_8NO_4$	3.908	0.860
	$C_3H_{12}N_4O$	4.999	0.305		$C_3H_{10}N_2O_3$	4.282	0.674
	$C_4H_8O_4$	4.607	0.885		$C_4H_2N_4O$	5.919	0.349
	$C_4H_{10}NO_3$	4.981	0.701		$C_4H_{10}O_4$	4.639	0.886
	$C_4H_{12}N_2O_2$	5.356	0.520		$C_5H_2N_2O_2$	6.276	0.567
	$C_5H_2N_3O$	6.618	0.388		$C_5H_4N_3O$	6.651	0.390
	$C_5H_4N_4$	6.993	0.212		$C_5H_6N_4$	7.025	0.215
	$C_5H_{12}O_3$	5.712	0.735		$C_6H_2O_3$	6.633	0.786

(Contd.)

	M + 1	M + 2			M + 1	M + 2
$C_6H_4NO_2$	7.00	0.611		$C_7H_{10}NO$	8.145	0.491
$C_6H_6N_2O$	7.382	0.436		$C_7H_{12}N_2$	8.519	0.321
$C_6H_8N_3$	7.756	0.264		$C_8H_{12}O$	8.876	0.548
$C_7H_6O_2$	7.738	0.660		$C_8H_{14}N$	9.250	0.380
C_7H_8NO	8.113	0.488		C_9H_2N	10.139	0.461
$C_7H_{10}N_2$	8.487	0.318		C_9H_{16}	9.981	0.446
$C_8H_{10}O$	8.844	0.545		$C_{10}H_4$	10.870	0.532
$C_8H_{12}N$	9.218	0.377	125	$C_3HN_4O_2$	4.862	0.499
C_9H_{14}	9.949	0.442		$C_4HN_2O_3$	5.219	0.713
$C_{10}H_2$	10.838	0.529		$C_4H_3N_3O_2$	5.593	0.532
123 $C_2H_7N_2O_4$	3.193	0.840		$C_4H_5N_4O$	5.967	0.352
$C_2H_9N_3O_3$	3.567	0.651		C_5HO_4	5.575	0.928
$C_3H_9NO_4$	3.924	0.861		$C_5H_3NO_3$	5.950	0.749
$C_4HN_3O_2$	5.561	0.530		$C_5H_5N_2O_2$	6.324	0.570
$C_4H_3N_4O$	5.935	0.350		$C_5H_7N_3O$	6.699	0.393
C_5HNO_3	5.918	0.747		$C_5H_9N_4$	7.073	0.218
$C_5H_3N_2O_2$	6.292	0.568		$C_6H_5O_3$	6.681	0.790
$C_5H_5N_3O$	6.667	0.391		$C_6H_7NO_2$	7.055	0.614
$C_5H_7N_4$	7.041	0.216		$C_6H_9N_2O$	7.430	0.440
$C_6H_3O_3$	6.649	0.787		$C_6H_{11}N_3$	7.804	0.267
$C_6H_5NO_2$	7.023	0.612		$C_7H_9O_2$	7.786	0.663
$C_6H_7N_2O$	7.398	0.438		$C_7H_{11}NO$	8.161	0.492
$C_6H_9N_3$	7.772	0.265		$C_7H_{13}N_2$	8.535	0.322
$C_7H_7O_2$	7.754	0.661		C_8HN_2	9.424	0.396
C_7H_9NO	8.129	0.489		$C_8H_{13}O$	8.892	0.549
$C_7H_{11}N_2$	8.503	0.319		$C_8H_{15}N$	9.266	0.382
$C_8H_{11}O$	8.860	0.546		C_9HO	9.781	0.626
$C_8H_{13}N$	9.234	0.379		C_9H_3N	10.155	0.462
C_9HN	10.123	0.459		C_9H_{17}	9.998	0.447
C_9H_{15}	9.965	0.444		$C_{10}H_5$	10.886	0.534
$C_{10}H_3$	10.854	0.531	126	$C_3H_2N_4O_2$	4.878	0.500
124 $C_2H_8N_2O_4$	3.209	0.840		$C_4H_2N_2O_3$	5.235	0.714
$C_4H_2N_3O_2$	5.577	0.531		$C_4H_4N_3O_2$	5.609	0.533
$C_4H_4N_4O$	5.951	0.351		$C_4H_6N_4O$	5.983	0.353
$C_5H_2NO_3$	5.934	0.748		$C_5H_2O_4$	5.591	0.929
$C_5H_4N_2O_2$	6.308	0.569		$C_5H_4NO_3$	5.966	0.749
$C_5H_6N_3O$	6.683	0.392		$C_5H_6N_2O_2$	6.340	0.571
$C_5H_8N_4$	7.057	0.217		$C_5H_8N_3O$	6.715	0.350
$C_6H_4O_3$	6.665	0.789		$C_5H_{10}N_4$	7.089	0.219
$C_6H_6NO_2$	7.039	0.613		$C_6H_6O_3$	6.697	0.791
$C_6H_8N_2O$	7.414	0.439		$C_6H_8NO_2$	7.071	0.615
$C_6H_{10}N_3$	7.788	0.266		$C_6H_{10}N_2O$	7.446	0.441
$C_7H_8O_2$	7.770	0.662		$C_6H_{12}N_3$	7.820	0.268

(Contd.)

		M + 1	M + 2			M + 1	M + 2
	$C_7H_{10}O_2$	7.802	0.664		$C_5H_4O_4$	5.623	0.931
	$C_7H_{12}NO$	8.177	0.493		$C_5H_6NO_3$	5.998	0.751
	$C_7H_{14}N_2$	8.551	0.323		$C_5H_8N_2O_2$	6.372	0.573
	$C_8H_2N_2$	9.440	0.397		$C_5H_{10}N_3O$	6.747	0.397
	$C_8H_{14}O$	8.908	0.550		$C_5H_{12}N_4$	7.121	0.221
	$C_8H_{16}N$	9.282	0.383		$C_6H_8O_3$	6.729	0.793
	C_9H_2O	9.797	0.628		$C_6H_{10}NO_2$	7.103	0.618
	C_9H_4N	10.171	0.464		$C_6H_{12}N_2O$	7.478	0.444
	C_9H_{18}	10.013	0.448		$C_6H_{14}N_3$	7.852	0.271
	$C_{10}H_6$	10.902	0.535		$C_7H_2N_3$	8.741	0.339
127	$C_3HN_3O_3$	4.519	0.684		$C_7H_{12}O_2$	7.834	0.667
	$C_3H_3N_4O_2$	4.894	0.500		$C_7H_{14}NO$	8.209	0.496
	C_4HNO_4	4.876	0.897		$C_7H_{16}N_2$	8.583	0.326
	$C_4H_3N_2O_3$	5.251	0.715		C_8H_2NO	9.097	0.567
	$C_4H_5N_3O_2$	5.625	0.534		$C_8H_4N_2$	9.472	0.400
	$C_4H_7N_4O$	5.999	0.354		$C_8H_{16}O$	8.940	0.553
	$C_5H_3O_4$	5.607	0.930		$C_8H_{18}N$	9.314	0.386
	$C_5H_5NO_3$	5.982	0.750		C_9H_4O	9.829	0.631
	$C_5H_7N_2O_2$	6.356	0.572		C_9H_6N	10.203	0.467
	$C_5H_9N_3O$	6.731	0.396		C_9H_{20}	10.045	0.452
	$C_5H_{11}N_4$	7.105	0.220		$C_{10}H_8$	10.935	0.539
	$C_6H_7O_3$	6.713	0.792	129	$C_3HN_2O_4$	4.177	0.870
	$C_6H_9NO_2$	7.087	0.616		$C_3H_3N_3O_3$	4.551	0.685
	$C_6H_{11}N_2O$	7.462	0.442		$C_3H_5N_4O_2$	4.926	0.502
	$C_6H_{13}N_3$	7.836	0.270		$C_4H_3NO_4$	4.908	0.898
	C_7HN_3	8.725	0.338		$C_4H_5N_2O_3$	5.283	0.716
	$C_7H_{11}O_2$	7.818	0.666		$C_4H_7N_3O_2$	5.657	0.535
	$C_7H_{13}NO$	8.193	0.495		$C_4H_9N_4O$	6.031	0.356
	$C_7H_{15}N_2$	8.567	0.325		$C_5H_5O_4$	5.639	0.932
	C_8HNO	9.081	0.565		$C_5H_7NO_3$	6.014	0.752
	$C_8H_3N_2$	9.456	0.399		$C_5H_9N_2O_2$	6.388	0.574
	$C_8H_{15}O$	8.924	0.552		$C_5H_{11}N_3O$	6.763	0.398
	$C_8H_{17}N$	9.298	0.385		$C_5H_{13}N_4$	7.137	0.223
	C_9H_3O	9.813	0.629		C_6HN_4	8.025	0.284
	C_9H_5N	10.187	0.466		$C_6H_9O_3$	6.745	0.794
	C_9H_{19}	10.030	0.450		$C_6H_{11}NO_2$	7.119	0.619
	$C_{10}H_7$	10.918	0.548		$C_6H_{13}N_2O$	7.494	0.445
128	$C_3H_2N_3O_3$	4.535	0.685		$C_6H_{15}N_3$	7.868	0.272
	$C_3H_4N_4O_2$	5.910	0.501		C_7HN_2O	8.382	0.510
	$C_4H_2NO_4$	4.892	0.898		$C_7H_3N_3$	8.757	0.340
	$C_4H_4N_2O_3$	5.267	0.715		$C_7H_{13}O_2$	7.850	0.668
	$C_4H_6N_3O_2$	5.641	0.535		$C_7H_{15}NO$	8.225	0.497
	$C_4H_8N_4O$	6.015	0.355		$C_7H_{17}N_2$	8.599	0.327

(Contd.)

		M + 1	M + 2			M + 1	M + 2
	C_8HO_2	8.739	0.736		$C_4H_{11}N_4O$	6.063	0.358
	C_8H_3NO	9.113	0.568		$C_5H_7O_4$	5.671	0.933
	$C_8H_5N_2$	9.488	0.402		$C_5H_9NO_3$	6.046	0.754
	$C_8H_{17}O$	8.956	0.555		$C_5H_{11}N_2O_2$	6.420	0.576
	$C_8H_{19}N$	9.330	0.388		$C_5H_{13}N_3O$	6.795	0.399
	C_9H_5O	9.845	0.633		$C_5H_{15}N_4$	7.169	0.224
	C_9H_7N	10.219	4.688		C_6HN_3O	7.683	0.458
	$C_{10}H_9$	10.950	0.541		$C_6H_3N_4$	8.057	0.286
130	$C_3H_2N_2O_4$	4.193	0.871		$C_6H_{11}O_3$	6.777	0.796
	$C_3H_4N_3O_3$	4.567	0.686		$C_6H_{13}NO_2$	7.151	0.621
	$C_3H_6N_4O_2$	4.942	0.503		$C_6H_{15}N_2O$	7.526	0.447
	$C_4H_4NO_4$	4.924	0.899		$C_6H_{17}N_3$	7.900	0.274
	$C_4H_6N_2O_3$	5.299	0.717		C_7HNO_2	8.040	0.682
	$C_4H_8N_3O_2$	5.673	0.536		$C_7H_3N_2O$	8.414	0.512
	$C_4H_{10}N_4O$	6.047	0.357		$C_7H_5N_3$	8.789	0.343
	$C_5H_6O_4$	5.655	0.933		$C_{17}H_{15}O_2$	7.882	0.670
	$C_5H_8NO_3$	6.030	0.753		$C_7H_{17}NO$	8.257	0.499
	$C_5H_{10}N_2O_2$	6.404	0.575		$C_8H_3O_2$	8.771	0.738
	$C_5H_{12}N_3O$	6.779	0.399		C_8H_5NO	9.145	0.571
	$C_5H_{14}N_4$	7.153	0.224		$C_8H_7N_2$	9.520	0.405
	$C_6H_2N_4$	8.041	0.285		C_9H_7O	9.877	0.635
	$C_6H_{10}O_3$	6.761	0.795		C_9H_9N	10.251	0.472
	$C_6H_{12}NO_2$	7.135	0.620		$C_{10}H_{11}$	10.982	0.544
	$C_6H_{14}N_2O$	7.510	0.446	132	$C_3H_4N_2O_4$	4.225	0.872
	$C_6H_{16}N_3$	7.884	0.274		$C_3H_6N_3O_3$	4.599	0.688
	$C_7H_2N_2O$	8.398	0.511		$C_3H_8N_4O_2$	4.974	0.504
	$C_7H_4N_3$	8.773	0.342		$C_4H_6NO_4$	4.956	0.900
	$C_7H_{14}O_2$	7.866	0.670		$C_4H_8N_2O_3$	5.331	0.718
	$C_7H_{16}NO$	8.241	0.498		$C_4H_{10}N_3O_2$	5.705	0.538
	$C_7H_{18}N_2$	8.615	0.329		$C_4H_{12}N_4O$	6.079	0.359
	$C_8H_2O_2$	8.755	0.738		$C_5H_8O_4$	5.687	6.934
	C_8H_4NO	9.129	0.570		$C_5H_{10}NO_3$	6.062	0.755
	$C_8H_6N_2$	9.504	0.403		$C_5H_{12}N_2O_2$	6.436	0.577
	$C_8H_{18}O$	8.972	0.556		$C_5H_{14}N_3O$	6.811	0.401
	C_9H_6O	9.861	0.634		$C_5H_{16}N_4$	7.185	0.226
	C_9H_8N	10.235	0.471		$C_6H_2N_3O$	7.699	0.460
	$C_{10}H_{10}$	10.966	0.543		$C_6H_4N_4$	8.073	0.288
131	$C_3H_3N_2O_4$	4.209	0.972		$C_6H_{12}O_3$	6.793	0.797
	$C_3H_5N_3O_3$	4.583	0.687		$C_6H_{14}NO_2$	7.167	0.622
	$C_3H_7N_4O_2$	4.958	0.503		$C_6H_{16}N_2O$	7.542	0.448
	$C_4H_5NO_4$	4.940	0.900		$C_7H_2NO_2$	8.056	0.683
	$C_4H_7N_2O_2$	5.315	0.717		$C_7H_4N_2O$	8.430	0.514
	$C_4H_9N_3O_2$	5.689	0.537		$C_7H_6N_3$	8.805	0.345

(Contd.)

		M + 1	M + 2			M + 1	M + 2
	$C_7H_{16}O_2$	7.898	0.672		$C_5H_{10}O_4$	5.719	0.936
	$C_8H_4O_2$	8.787	0.740		$C_5H_{12}NO_3$	6.094	0.757
	C_8H_6NO	9.161	0.573		$C_5H_{14}N_2O_2$	6.468	0.580
	$C_8H_8N_2$	9.536	0.407		$C_6H_2N_2O_2$	7.357	0.635
	C_9H_8O	9.893	0.637		$C_6H_4N_3O$	7.731	0.462
	$C_9H_{10}N$	10.267	0.474		$C_6H_6N_4$	8.105	0.291
	$C_{10}H_{12}$	10.998	0.546		$C_6H_{14}O_3$	6.825	0.799
133	$C_3H_5N_2O_4$	4.241	0.8729		$C_7H_2O_3$	7.714	0.858
	$C_3H_7N_3O_3$	4.616	0.688		$C_7H_4NO_2$	8.088	0.686
	$C_3H_9N_4O_2$	4.990	0.505		$C_7H_6N_2O$	8.462	0.516
	$C_4H_7NO_4$	4.972	0.901		$C_7H_8N_3$	8.837	0.347
	$C_4H_9N_2O_3$	5.347	0.720		$C_8H_6O_2$	8.819	0.743
	$C_4H_{11}N_3O_2$	5.721	0.539		C_8H_8NO	9.193	0.576
	$C_4H_{13}N_4O$	6.095	0.360		$C_8H_{10}N_2$	9.568	0.410
	C_5HN_4O	6.984	0.412		$C_9H_{10}O$	9.925	0.640
	$C_5H_9O_4$	5.703	0.935		$C_9H_{12}N$	10.299	0.477
	$C_5H_{11}NO_3$	6.078	0.756		$C_{10}H_{14}$	11.030	0.550
	$C_5H_{13}N_2O_2$	6.452	5.784		$C_{11}H_2$	11.919	0.646
	$C_5H_{15}N_3O$	6.827	0.402	135	$C_3H_7N_2O_4$	4.273	0.874
	$C_6HN_2O_2$	7.341	0.634		$C_3H_9N_3O_3$	4.648	0.690
	$C_6H_3N_3O$	7.715	0.461		$C_3H_{11}N_4O_2$	5.022	0.507
	$C_6H_5N_4$	8.089	0.289		$C_4H_9NO_4$	5.004	0.903
	$C_6H_{13}O_3$	6.809	0.798		$C_4H_{11}N_2O_2$	5.379	0.721
	$C_6H_{15}NO_2$	7.183	0.623		$C_4H_{13}N_3O_2$	5.753	0.541
	C_7HO_3	7.698	0.857		$C_5HN_3O_2$	6.642	0.590
	$C_7H_3NO_2$	8.072	0.685		$C_5H_3N_4O$	7.016	0.415
	$C_7H_5N_2O$	8.446	0.515		$C_5H_{11}O_4$	5.735	0.937
	$C_7H_7N_3$	8.821	0.346		$C_5H_{13}NO_3$	6.110	0.758
	$C_8H_5O_2$	8.803	0.742		C_6HNO_3	6.998	0.810
	C_8H_7NO	9.177	0.574		$C_6H_3N_2O_2$	7.373	0.636
	$C_8H_9N_2$	9.552	0.408		$C_6H_5N_3O$	7.747	0.463
	C_9H_9O	9.909	0.638		$C_6H_7N_4$	8.121	0.291
	$C_9H_{11}N$	10.283	0.475		$C_7H_3O_3$	7.730	0.859
	$C_{10}H_{13}$	11.014	0.548		$C_7H_5NO_2$	8.104	0.687
	$C_{11}H$	11.903	0.644		$C_7H_7N_2O$	8.478	0.517
134	$C_3H_6N_2O_4$	4.257	0.874		$C_7H_9N_3$	8.853	0.348
	$C_3H_8N_3O_3$	4.632	0.689		$C_8H_7O_2$	8.835	0.744
	$C_3H_{10}N_4O_2$	5.006	0.506		C_8H_9NO	9.209	0.577
	$C_4H_8NO_4$	4.988	0.902		$C_8H_{11}N_2$	9.584	0.411
	$C_4H_{10}N_2O_3$	5.363	0.720		$C_9H_{11}O$	9.941	0.642
	$C_4H_{12}N_3O_2$	5.737	0.540		$C_9H_{13}N$	10.315	0.478
	$C_4H_{14}N_4O$	6.111	0.361		$C_{10}HN$	11.203	0.569
	$C_5H_2N_4O$	7.000	0.413		$C_{10}H_{15}$	11.046	0.552

(Contd.)

		M + 1	M + 2			M + 1	M + 2
	$C_{11}H_3$	11.935	0.648		$C_8H_{13}N_2$	9.616	0.414
136	$C_3H_8N_2O_4$	4.289	0.875		C_9HN_2	10.504	0.498
	$C_3H_{10}N_3O_3$	4.664	0.691		$C_9H_{13}O$	9.973	0.645
	$C_3H_{12}N_4O_2$	5.038	0.507		$C_9H_{15}N$	10.347	0.482
	$C_4H_{10}NO_4$	5.020	0.904		$C_{10}HO$	10.861	0.732
	$C_4H_{12}N_2O_3$	5.395	0.722		$C_{10}H_3N$	11.235	0.572
	$C_5H_2N_3O_2$	6.658	0.591		$C_{10}H_{17}$	11.078	0.555
	$C_5H_4N_4O$	7.032	0.416		$C_{11}H_5$	11.967	0.652
	$C_5H_{12}O_4$	5.751	0.938	138	$C_3H_{10}N_2O_4$	4.321	0.876
	$C_6H_2NO_3$	7.014	0.812		$C_4H_2N_4O_2$	5.958	0.552
	$C_6H_4N_2O_2$	7.389	0.637		$C_5H_2N_2O_3$	6.315	0.770
	$C_6H_6N_3O$	7.763	0.465		$C_5H_4N_3O_2$	6.690	0.593
	$C_6H_8N_4$	8.137	0.293		$C_5H_6N_4O$	7.064	0.418
	$C_7H_4O_3$	7.746	0.861		$C_6H_2O_4$	6.672	0.989
	$C_7H_6NO_2$	8.120	0.689		$C_6H_4NO_3$	7.046	0.814
	$C_7H_8N_2O$	8.494	0.519		$C_6H_6N_2O_2$	7.421	0.639
	$C_7H_{10}N_3$	8.869	0.350		$C_6H_8N_3O$	7.795	0.467
	$C_8H_8O_2$	8.851	0.746		$C_6H_{10}N_4$	8.169	0.295
	$C_8H_{10}NO$	9.225	0.579		$C_7H_6O_3$	7.778	0.863
	$C_8H_{12}N_2$	9.600	0.413		$C_7H_8NO_2$	8.152	0.691
	$C_9H_{12}O$	9.957	0.644		$C_7H_{10}N_2O$	8.526	0.521
	$C_9H_{14}N$	10.331	0.480		$C_7H_{12}N_3$	8.901	0.353
	$C_{10}H_2N$	11.219	0.570		$C_8H_{10}O_2$	8.883	0.748
	$C_{10}H_{16}$	11.062	0.553		$C_8H_{12}NO$	9.257	0.581
	$C_{11}H_4$	11.951	0.650		$C_8H_{14}N_2$	9.632	0.415
137	$C_3H_9N_2O_4$	4.305	0.876		$C_9H_2N_2$	10.520	0.499
	$C_3H_{11}N_3O_3$	4.680	0.691		$C_9H_{14}O$	9.989	0.646
	$C_4HN_4O_2$	5.942	0.551		$C_9H_{16}N$	10.363	0.483
	$C_4H_{11}NO_4$	5.036	0.905		$C_{10}H_2O$	10.877	0.733
	$C_5HN_2O_3$	6.299	0.769		$C_{10}H_4N$	11.251	0.573
	$C_5H_3N_3O_2$	6.674	0.592		$C_{10}H_{18}$	11.094	0.557
	$C_5H_5N_4O$	7.048	0.417		$C_{11}H_6$	11.983	0.653
	C_6HO_4	6.656	0.988	139	$C_4H_3N_3O_3$	5.600	0.733
	$C_6H_3NO_3$	7.030	0.813		$C_4H_3N_4O_2$	5.974	0.553
	$C_6H_5N_2O_2$	7.405	0.639		C_5HNO_4	5.957	0.949
	$C_6H_7N_3O$	7.779	0.466		$C_5H_3N_2O_3$	6.331	0.771
	$C_6H_9N_4$	8.153	0.294		$C_5H_5N_3O_2$	6.706	0.594
	$C_7H_5O_3$	7.762	0.862		$C_5H_7N_4O$	7.080	0.419
	$C_7H_7NO_2$	8.136	0.690		$C_6H_3O_4$	6.688	0.991
	$C_7H_9N_2O$	8.510	0.520		$C_6H_5NO_3$	7.062	0.815
	$C_7H_{11}N_3$	8.885	0.352		$C_6H_7N_2O_2$	7.437	0.641
	$C_8H_9O_2$	8.867	0.747		$C_6H_9N_3O$	7.811	0.468
	$C_8H_{11}NO$	9.241	0.580		$C_6H_{11}N_4$	8.185	0.297

(Contd.)

	M + 1	M + 2			M + 1	M + 2
$C_7H_7O_3$	7.794	0.864	141	$C_4HN_2O_4$	5.258	0.915
$C_7H_9NO_2$	8.168	0.692		$C_4H_3N_3O_3$	5.632	0.735
$C_7H_{11}N_2O$	8.542	0.523		$C_4H_5N_4O_2$	6.006	0.555
$C_7H_{13}N_3$	8.917	0.354		$C_5H_3NO_4$	5.989	0.951
C_8HN_3	9.805	0.431		$C_5H_5N_2O_3$	6.363	0.773
$C_8H_{11}O_2$	8.899	0.750		$C_5H_7N_3O_2$	6.738	0.597
$C_8H_{13}NO$	9.273	0.583		$C_5H_9N_4O$	7.112	0.421
$C_8H_{15}N_2$	9.648	0.417		$C_6H_5O_4$	6.720	0.993
C_9HNO	10.162	0.663		$C_6H_7NO_3$	7.094	0.817
$C_9H_3N_2$	10.536	0.501		$C_6H_9N_2O_2$	7.469	0.643
$C_9H_{15}O$	10.005	0.648		$C_6H_{11}N_3O$	7.843	0.471
$C_9H_{17}N$	10.379	0.485		$C_6H_{13}N_4$	8.217	0.300
$C_{10}H_3O$	10.893	0.735		C_7HN_4	9.106	0.371
$C_{10}H_5N$	11.267	0.575		$C_7H_9O_3$	7.826	0.867
$C_{10}H_{19}$	11.110	0.558		$C_7H_{11}NO_2$	8.200	0.696
$C_{11}H_7$	11.999	0.655		$C_7H_{13}N_2O$	8.574	0.526
140 $C_4H_2N_3O_3$	5.616	0.734		$C_7H_{15}N_3$	8.949	0.357
$C_4H_4N_4O_2$	5.990	0.554		C_8HN_2O	9.463	0.600
$C_5H_2NO_4$	5.973	0.950		$C_8H_3N_3$	9.837	0.435
$C_5H_4N_2O_3$	6.347	0.772		$C_8H_{13}O_2$	8.931	0.753
$C_5H_6N_3O_2$	6.722	0.596		$C_8H_{15}NO$	9.305	0.586
$C_5H_8N_4O$	7.096	0.420		$C_8H_{17}N_2$	9.680	0.420
$C_6H_4O_4$	6.704	0.992		C_9HO_2	9.820	0.830
$C_6H_6NO_3$	7.078	0.816		C_9H_3NO	10.194	0.666
$C_6H_8N_2O_2$	7.453	0.642		$C_9H_5N_2$	10.568	0.504
$C_6H_{10}N_3O$	7.827	0.470		$C_9H_{17}O$	10.037	0.651
$C_6H_{12}N_4$	8.201	9.298		$C_9H_{19}N$	10.411	0.488
$C_7H_8O_3$	7.810	0.866		$C_{10}H_5O$	10.925	0.738
$C_7H_{10}NO_2$	8.184	0.694		$C_{10}H_7N$	11.299	0.579
$C_7H_{12}N_2O$	8.558	0.524		$C_{10}H_{21}$	11.142	0.562
$C_7H_{14}N_3$	8.933	0.356		$C_{11}H_9$	12.031	0.659
$C_8H_2N_3$	9.821	0.433	142	$C_4H_2N_2O_4$	5.274	0.916
$C_8H_{12}O_2$	8.915	0.751		$C_4H_4N_3O_3$	5.648	0.735
$C_8H_{14}NO$	9.289	0.584		$C_4H_6N_4O_2$	6.022	0.556
$C_8H_{16}N_2$	9.664	0.418		$C_5H_4NO_4$	6.005	0.952
C_9H_2NO	10.178	0.665		$C_5H_6N_2O_3$	6.379	0.774
$C_9H_4N_2$	10.552	0.502		$C_5H_8N_3O_2$	6.754	0.598
$C_9H_{16}O$	10.021	0.650		$C_5H_{10}N_4O$	7.128	0.422
$C_9H_{18}N$	10.395	0.487		$C_6H_6O_4$	6.736	0.994
$C_{10}H_4O$	10.909	0.737		$C_6H_8NO_3$	7.110	0.818
$C_{10}H_6N$	11.283	0.577		$C_6H_{10}N_2O_2$	7.485	0.645
$C_{10}H_{20}$	11.126	0.560		$C_6H_{12}N_3O$	7.859	0.472
$C_{11}H_8$	12.015	0.657		$C_6H_{14}N_4$	8.233	0.301

(Contd.)

		M + 1	M + 2			M + 1	M + 2
	$C_7H_2N_4$	9.122	0.372		$C_9H_3O_2$	9.852	0.834
	$C_7H_{10}O_3$	7.842	0.868		C_9H_5NO	10.226	0.670
	$C_7H_{12}NO_2$	8.216	0.697		$C_9H_7N_2$	10.600	0.508
	$C_7H_{14}N_2O$	8.590	0.527		$C_9H_{19}O$	10.069	0.655
	$C_7H_{16}N_3$	8.965	0.359		$C_9H_{21}N$	10.443	0.492
	$C_8H_2N_2O$	9.479	0.602		$C_{10}H_7O$	10.957	0.742
	$C_8H_4N_3$	9.853	0.437		$C_{10}H_9N$	11.331	0.582
	$C_8H_{14}O_2$	8.947	0.755		$C_{11}H_{11}$	12.063	0.663
	$C_8H_{16}NO$	9.321	0.588	144	$C_4H_9N_2O_4$	5.306	0.917
	$C_8H_{18}N_2$	9.696	0.422		$C_4H_6N_3O_3$	5.680	0.737
	$C_9H_2O_2$	9.836	0.832		$C_4H_8N_4O_2$	6.054	0.558
	C_9H_4NO	10.210	0.668		$C_5H_6NO_4$	6.037	0.954
	$C_9H_6N_2$	10.584	0.506		$C_5H_8N_2O_2$	6.411	0.776
	$C_9H_{18}O$	10.053	0.653		$C_5H_{10}N_3O_2$	6.786	0.599
	$C_9H_{20}N$	10.427	0.490		$C_5H_{12}N_4O$	7.160	0.424
	$C_{10}H_6O$	10.941	0.741		$C_6H_8O_4$	6.768	0.995
	$C_{10}H_8N$	11.315	0.581		$C_7H_{10}NO_3$	7.142	0.820
	$C_{10}H_{22}$	11.158	0.564		$C_6H_{12}N_2O_2$	7.517	0.647
	$C_{11}H_{10}$	12.047	0.661		$C_6H_{14}N_3O$	7.891	0.474
143	$C_4H_3N_2O_4$	5.290	0.917		$C_6H_{16}N_4$	8.265	0.303
	$C_4H_5N_3O_3$	5.664	0.736		$C_7H_2N_3O$	8.780	0.542
	$C_4H_7N_4O_2$	6.038	0.557		$C_7H_4N_4$	9.154	0.375
	$C_5H_5NO_4$	6.021	0.953		$C_7H_{12}O_3$	7.874	0.870
	$C_5H_7N_2O_3$	6.395	0.775		$C_7H_{14}NO_2$	8.248	0.699
	$C_5H_9N_3O_2$	6.770	0.599		$C_7H_{16}N_2O$	8.622	0.529
	$C_5H_{11}N_4O$	7.144	0.424		$C_7H_{18}N_3$	8.997	0.361
	$C_6H_7O_4$	6.752	0.995		$C_8H_2NO_2$	9.136	0.770
	$C_6H_9NO_3$	7.126	0.820		$C_8H_4N_2O$	9.511	0.604
	$C_6H_{11}N_2O_2$	7.501	0.646		$C_8H_6N_3$	9.885	0.439
	$C_6H_{13}N_3O$	7.875	0.473		$C_8H_{16}O_2$	8.979	0.757
	$C_6H_{15}N_4$	8.249	0.302		$C_8H_{18}NO$	9.353	0.590
	C_7HN_3O	8.764	0.541		$C_8H_{20}N_2$	9.728	0.425
	$C_7H_3N_4$	9.138	0.374		$C_9H_4O_2$	9.868	0.835
	$C_7H_{11}O_3$	7.858	0.869		C_9H_6NO	10.242	0.671
	$C_7H_{13}NO_2$	8.232	0.698		$C_9H_8N_2$	10.616	0.509
	$C_7H_{15}N_2O$	8.606	0.529		$C_9H_{20}O$	10.085	0.656
	$C_7H_{17}N_3$	8.981	0.360		$C_{10}H_8O$	10.973	0.744
	C_8HNO_2	9.120	0.769		$C_{10}H_{10}N$	11.347	0.584
	$C_8H_3N_2O$	9.495	0.603		$C_{11}H_{12}$	12.079	0.665
	$C_8H_5N_3$	9.869	0.438	145	$C_4H_5N_2O_4$	5.322	0.918
	$C_8H_{15}O_2$	8.963	0.755		$C_4H_7N_3O_3$	5.696	0.738
	$C_8H_{17}NO$	9.337	0.589		$C_4H_9N_4O_2$	6.070	0.558
	$C_8H_{19}N_2$	9.712	0.423		$C_5H_7NO_4$	6.053	0.955

(Contd.)

	M + 1	M + 2		M + 1	M + 2
$C_5H_9N_2O_3$	6.427	0.777	$C_7H_4N_3O$	8.812	0.546
$C_5H_{11}N_3O_2$	6.802	0.600	$C_7H_6N_4$	9.186	0.378
$C_5H_{13}N_4O$	7.176	0.425	$C_7H_{14}O_3$	7.906	0.873
C_6HN_4O	8.065	0.487	$C_7H_{16}NO_2$	8.280	0.702
$C_6H_9O_4$	6.784	0.997	$C_7H_{18}N_2O$	8.654	0.533
$C_6H_{11}NO_3$	7.158	0.821	$C_8H_2O_3$	8.794	0.941
$C_6H_{13}N_2O_2$	7.533	0.648	$C_8H_4NO_2$	9.168	0.773
$C_6H_{15}N_3O$	7.907	0.475	$C_8H_6N_2O$	9.543	0.602
$C_6H_{17}N_4$	8.282	0.305	$C_8H_8N_3$	9.917	0.443
$C_7HN_2O_2$	8.421	0.713	$C_8H_{18}O_2$	9.011	0.760
$C_7H_3N_3O$	8.796	0.544	$C_9H_6O_2$	9.900	0.838
$C_7H_5N_4$	9.170	0.376	C_9H_8NO	10.274	0.675
$C_7H_{13}O_3$	7.890	0.871	$C_9H_{10}N_2$	10.648	0.513
$C_7H_{15}NO_2$	8.264	0.701	$C_{10}H_{10}O$	11.005	0.748
$C_7H_{17}N_2O$	8.638	0.531	$C_{10}H_{12}N$	11.379	0.588
$C_7H_{19}N_3$	9.013	0.363	$C_{11}H_{14}$	12.111	0.669
C_8HO_3	8.778	0.940	$C_{12}H_2$	12.999	0.774
$C_8H_3NO_2$	9.152	0.772	147 $C_4H_7N_2O_4$	5.354	0.920
$C_8H_5N_2O$	9.527	0.606	$C_4H_9N_3O_3$	5.728	0.740
$C_8H_7N_3$	9.901	0.441	$C_4H_{11}N_4O_2$	6.102	0.561
$C_8H_{17}O_2$	8.995	0.759	$C_5H_9NO_4$	6.085	0.957
$C_8H_{19}NO$	9.369	0.592	$C_5H_{11}N_2O_3$	6.459	0.779
$C_9H_5O_2$	9.884	0.837	$C_5H_{13}N_3O_2$	6.834	0.603
C_9H_7NO	10.258	0.673	$C_5H_{15}N_4O$	7.208	0.428
$C_9H_9N_2$	10.632	0.511	$C_6HN_3O_2$	7.722	0.662
$C_{10}H_9O$	10.989	0.746	$C_6H_3N_4O$	8.097	0.490
$C_{10}H_{11}N$	11.363	0.587	$C_6H_{11}O_4$	6.816	0.999
$C_{11}H_{13}$	12.095	0.667	$C_6H_{13}NO_3$	7.190	0.824
$C_{12}H$	12.983	0.773	$C_6H_{15}N_2O_2$	7.565	0.651
146 $C_4H_6N_2O_4$	5.338	9.195	$C_6H_{17}N_3O$	7.939	0.478
$C_4H_8N_3O_3$	5.712	0.739	C_7HNO_3	8.079	0.886
$C_4H_{10}N_4O_2$	6.086	0.559	$C_7H_3N_2O_2$	8.453	0.716
$C_5H_8NO_4$	6.069	0.956	$C_7H_5N_3O$	8.828	0.547
$C_5H_{10}N_2O_3$	6.443	0.778	$C_7H_7N_4$	9.202	0.380
$C_5H_{12}N_3O_2$	6.818	0.602	$C_7H_{15}O_3$	7.922	0.874
$C_5H_{14}N_4O$	7.192	0.427	$C_{17}H_{17}NO_2$	8.296	0.703
$C_6H_2N_4O$	8.081	0.489	$C_8H_3O_3$	8.810	0.943
$C_6H_{10}O_4$	6.800	0.998	$C_8H_5NO_2$	9.185	0.775
$C_6H_{12}NO_3$	7.174	0.823	$C_8H_7N_2O$	9.559	0.609
$C_6H_{14}N_2O_2$	7.549	0.649	$C_8H_9N_3$	9.933	0.444
$C_6H_{16}N_3O$	7.923	0.477	$C_9H_7O_2$	9.916	0.840
$C_6H_{18}N_4$	8.298	0.306	C_9H_9NO	10.290	6.766
$C_7H_2N_2O_2$	8.437	0.715	$C_9H_{11}N_2$	10.664	0.515

(Contd.)

		M + 1	M + 2			M + 1	M + 2
	$C_{10}H_{11}O$	11.021	0.749		$C_6H_5N_4O$	8.129	0.493
	$C_{10}H_{13}N$	11.395	5.902		$C_6H_{13}O_4$	6.848	1.001
	$C_{11}HN$	12.284	0.690		$C_6H_{15}NO_3$	7.222	0.827
	$C_{11}H_{15}$	12.127	0.671		C_7HO_4	7.737	1.060
	$C_{12}H_3$	13.015	7.769		$C_7H_3NO_3$	8.111	0.889
148	$C_4H_8N_2O_4$	5.370	0.921		$C_7H_5N_2O_2$	8.485	0.719
	$C_4H_{10}N_3O_3$	5.744	0.741		$C_7H_7N_3O$	8.860	0.550
	$C_4H_{12}N_4O_2$	6.118	0.562		$C_7H_9N_4$	9.234	0.382
	$C_5H_{10}NO_4$	6.101	0.958		$C_8H_5O_3$	8.842	0.945
	$C_5H_{12}N_2O_3$	6.475	0.780		$C_8H_7NO_2$	9.217	0.778
	$C_5H_{14}N_3O_2$	6.850	0.604		$C_8H_9N_2O$	9.591	0.612
	$C_5H_{16}N_4O$	7.224	0.429		$C_8H_{11}N_3$	9.965	0.447
	$C_6H_2N_3O_2$	7.738	0.663		$C_9H_9O_2$	9.948	0.843
	$C_6H_4N_4O$	8.113	0.492		$C_9H_{11}NO$	10.322	0.679
	$C_6H_{12}O_4$	6.832	1.000		$C_9H_{13}N_2$	10.696	0.518
	$C_6H_{14}NO_3$	7.206	0.825		$C_{10}HN_2$	11.585	0.611
	$C_6H_{16}N_2O_2$	7.581	0.652		$C_{10}H_{13}O$	11.053	0.752
	$C_7H_2NO_3$	8.095	0.888		$C_{10}H_{15}N$	11.427	0.593
	$C_7H_4N_2O_2$	8.469	0.717		$C_{11}HO$	11.942	0.849
	$C_7H_6N_3O$	8.844	0.549		$C_{11}H_3N$	12.316	0.693
	$C_7H_8N_4$	9.218	0.381		$C_{11}H_{17}$	12.159	0.674
	$C_7H_{16}O_3$	7.938	0.876		$C_{12}H_5$	13.047	0.781
	$C_8H_4O_3$	8.826	0.944	150	$C_4H_{10}N_2O_4$	5.402	0.923
	$C_8H_6NO_2$	9.201	0.777		$C_4H_{12}N_3O_3$	5.776	0.743
	$C_8H_8N_2O$	9.575	0.611		$C_4H_{14}N_4O_2$	6.150	0.564
	$C_8H_{10}N_3$	9.949	0.446		$C_5H_2N_4O_2$	7.039	0.617
	$C_9H_8O_2$	9.932	0.842		$C_5H_{12}NO_4$	6.133	0.960
	$C_9H_{10}NO$	10.306	0.678		$C_5H_{14}N_2O_3$	6.507	0.783
	$C_9H_{12}N_2$	10.680	0.516		$C_6H_2N_2O_3$	7.396	0.838
	$C_{10}H_{12}O$	11.037	0.751		$C_6H_4N_3O_2$	7.770	0.666
	$C_{10}H_{14}N$	11.411	0.592		$C_6H_6N_4O$	8.145	0.494
	$C_{11}H_2N$	12.300	0.692		$C_6H_{14}O_4$	6.864	1.002
	$C_{11}H_{16}$	12.143	0.673		$C_7H_2O_4$	7.753	1.061
	$C_{12}H_4$	13.031	0.779		$C_7H_4NO_3$	8.127	0.890
149	$C_4H_9N_2O_4$	5.386	0.922		$C_7H_6N_2O_2$	8.501	0.720
	$C_4H_{11}N_3O_3$	5.760	0.742		$C_7H_8N_3O$	8.876	0.551
	$C_4H_{13}N_4O_2$	6.134	0.563		$C_7H_{10}N_4$	9.250	0.384
	$C_5HN_4O_2$	7.023	0.616		$C_8H_6O_3$	8.858	0.947
	$C_5H_{11}NO_4$	6.117	0.959		$C_8H_8NO_2$	9.233	0.780
	$C_5H_{13}N_2O_3$	6.491	0.781		$C_8H_{10}N_2O$	9.607	0.614
	$C_5H_{15}N_3O_2$	6.866	0.605		$C_8H_{12}N_3$	9.981	0.449
	$C_6HN_2O_3$	7.380	0.837		$C_9H_{10}O_2$	9.964	0.845
	$C_6H_3N_3O_2$	7.754	0.664		$C_9H_{12}NO$	10.338	0.682

(Contd.)

		M + 1	M + 2			M + 1	M + 2
	$C_9H_{14}N_2$	10.712	0.520		$C_6H_6N_3O_2$	7.802	0.668
	$C_{10}H_2N_2$	11.601	0.613		$C_6H_8N_4O$	8.177	0.497
	$C_{10}H_{14}O$	11.069	0.755		$C_7H_4O_4$	7.785	1.064
	$C_{10}H_{16}N$	11.443	0.596		$C_7H_6NO_3$	8.159	0.893
	$C_{11}H_2O$	11.958	0.851		$C_7H_8N_2O_2$	8.533	0.723
	$C_{11}H_4N$	12.332	0.695		$C_7H_{10}N_3O$	8.908	0.554
	$C_{11}H_{18}$	12.175	0.677		$C_7H_{12}N_4$	9.282	0.387
	$C_{12}H_6$	13.062	0.783		$C_8H_8O_3$	8.890	0.950
151	$C_4H_{11}N_2O_4$	5.418	0.924		$C_8H_{10}NO_2$	9.265	0.783
	$C_4H_{13}N_3O_3$	5.792	0.744		$C_8H_{12}N_2O$	9.639	0.617
	$C_5HN_3O_3$	6.681	0.793		$C_8H_{14}N_3$	10.013	0.452
	$C_5H_3N_4O_2$	7.055	0.618		$C_9H_2N_3$	10.902	0.540
	$C_5H_{13}NO_4$	6.149	0.961		$C_9H_{12}O_2$	9.996	0.848
	C_6HNO_3	7.037	1.013		$C_9H_{14}NO$	10.370	0.685
	$C_6H_3N_2O_3$	7.412	0.840		$C_9H_{16}N_2$	10.744	0.523
	$C_6H_5N_3O_2$	7.786	0.667		$C_{10}H_2NO$	11.259	0.775
	$C_6H_7N_4O$	8.161	0.496		$C_{10}H_4N_2$	11.633	0.617
	$C_7H_3O_4$	7.769	1.062		$C_{10}H_{16}O$	11.101	0.758
	$C_7H_5NO_3$	8.143	0.891		$C_{10}H_{18}N$	11.476	0.599
	$C_7H_7N_2O_2$	8.517	0.721		$C_{11}H_4O$	11.990	0.855
	$C_7H_9N_3O$	8.892	0.553		$C_{11}H_6N$	12.364	0.699
	$C_7H_{11}N_4$	9.266	0.386		$C_{11}H_{20}$	12.207	0.681
	$C_8H_7O_3$	8.874	0.949		$C_{12}H_8$	13.095	0.787
	$C_8H_9NO_2$	9.249	0.781	153	$C_5HN_2O_4$	6.338	0.972
	$C_8H_{11}N_2O$	9.623	0.615		$C_5H_3N_3O_3$	6.713	0.795
	$C_8H_{13}N_3$	9.997	0.451		$C_5H_5N_4O_2$	7.087	0.620
	C_9HN_3	10.866	0.538		$C_6H_3NO_4$	7.069	1.016
	$C_9H_{11}O_2$	9.980	0.846		$C_6H_5N_2O_3$	7.444	0.842
	$C_9H_{13}NO$	10.356	0.683		$C_6H_7N_3O_2$	7.818	0.669
	$C_9H_{15}N_2$	10.728	0.522		$C_6H_9N_4O$	8.193	0.498
	$C_{10}HNO$	11.243	0.773		$C_7H_5O_4$	7.801	1.065
	$C_{10}H_3N_2$	11.617	0.615		$C_7H_7NO_3$	8.175	0.894
	$C_{10}H_{15}O$	11.085	0.757		$C_7H_9N_2O_2$	8.549	0.724
	$C_{10}H_{17}N$	11.460	0.598		$C_7H_{11}N_3O$	8.924	0.556
	$C_{11}H_3O$	11.974	0.853		$C_7H_{13}N_4$	9.298	0.389
	$C_{11}H_5N$	12.348	0.697		C_8HN_4	10.187	0.469
	$C_{11}H_{19}$	12.191	0.679		$C_8H_9O_3$	8.906	0.951
	$C_{12}H_7$	13.079	0.785		$C_8H_{11}NO_2$	9.281	0.784
152	$C_4H_{12}N_2O_4$	5.434	0.925		$C_8H_{13}N_2O$	9.655	0.618
	$C_5H_2N_3O_3$	6.697	0.794		$C_8H_{15}N_3$	10.029	0.454
	$C_5H_4N_4O_2$	7.071	0.619		C_9HN_2O	10.543	0.702
	$C_6H_2NO_4$	7.053	1.015		$C_9H_3N_3$	10.918	0.541
	$C_6H_4N_2O_3$	7.428	0.841		$C_9H_{13}O_2$	10.012	0.850

(Contd.)

		M + 1	M + 2			M + 1	M + 2
	$C_9H_{15}NO$	10.386	0.687		$C_5H_5N_3O_3$	6.745	0.798
	$C_9H_{17}N_2$	10.760	0.525		$C_5H_7N_4O_2$	7.119	0.622
	$C_{10}HO_2$	10.900	0.937		$C_6H_5NO_4$	7.101	1.018
	$C_{10}H_3NO$	11.275	0.777		$C_6H_7N_2O_3$	7.476	0.844
	$C_{10}H_5N_2$	11.649	0.619		$C_6H_9N_3O_2$	7.850	0.672
	$C_{10}H_{17}O$	11.117	0.760		$C_6H_{11}N_4O$	8.225	0.501
	$C_{10}H_{19}N$	11.492	0.601		$C_7H_7O_4$	7.833	1.067
	$C_{11}H_5O$	12.006	0.857		$C_7H_9NO_3$	8.207	0.897
	$C_{11}H_7N$	12.380	0.701		$C_7H_{11}N_2O_2$	8.581	0.727
	$C_{11}H_{21}$	12.223	0.682		$C_7H_{13}N_3O$	8.956	0.558
	$C_{12}H_9$	13.111	0.790		$C_7H_{15}N_4$	9.330	0.392
154	$C_5H_2N_2O_4$	6.354	0.973		C_8HN_3O	9.844	0.636
	$C_5H_4N_3O_3$	6.729	0.796		$C_8H_3N_4$	10.219	0.473
	$C_5H_6N_4O_2$	7.103	0.621		$C_8H_{11}O_3$	8.938	0.954
	$C_6H_4NO_4$	7.085	1.017		$C_8H_{13}NO_2$	9.313	0.787
	$C_6H_6N_2O_3$	7.460	0.843		$C_8H_{15}N_2O$	9.687	0.622
	$C_6H_8N_3O_2$	7.834	0.671		$C_8H_{17}N_3$	10.061	0.457
	$C_6H_{10}N_4O$	8.209	0.499		C_9HNO_2	10.201	0.868
	$C_7H_6O_4$	7.817	1.067		$C_9H_3N_2O$	10.575	0.706
	$C_7H_8NO_3$	8.191	0.895		$C_9H_5N_3$	10.950	0.545
	$C_7H_{10}N_2O_2$	8.565	0.725		$C_9H_{15}O_2$	10.044	0.853
	$C_7H_{12}N_3O$	8.940	0.557		$C_9H_{17}NO$	10.418	0.690
	$C_7H_{14}N_4$	9.314	0.390		$C_9H_{19}N_2$	10.792	0.528
	$C_8H_2N_4$	10.203	0.471		$C_{10}H_3O_2$	10.932	0.940
	$C_8H_{10}O_3$	8.922	0.953		$C_{10}H_5NO$	11.307	0.781
	$C_8H_{12}NO_2$	9.297	0.786		$C_{10}H_7N_2$	11.681	0.622
	$C_8H_{14}N_2O$	9.671	0.620		$C_{10}H_{19}O$	11.149	0.764
	$C_8H_{16}N_3$	10.045	0.456		$C_{10}H_{21}N$	11.524	0.605
	$C_9H_2N_2O$	10.559	7.040		$C_{11}H_7O$	12.038	0.861
	$C_9H_4N_3$	10.934	0.543		$C_{11}H_9N$	12.412	0.705
	$C_9H_{14}O_2$	10.028	0.851		$C_{11}H_{23}$	12.255	0.687
	$C_9H_{16}NO$	10.402	0.688		$C_{12}H_{11}$	13.143	0.794
	$C_9H_{18}N_2$	10.776	0.527	156	$C_5H_4N_2O_4$	6.386	0.975
	$C_{10}H_2O_2$	10.916	0.938		$C_5H_6N_3O_3$	6.761	0.799
	$C_{10}H_4NO$	11.291	0.779		$C_5H_8N_4O_2$	7.135	0.623
	$C_{10}H_6N_2$	11.665	0.621		$C_6H_6NO_4$	7.117	1.019
	$C_{10}H_{18}O$	11.133	0.762		$C_6H_8N_2O_3$	7.492	0.846
	$C_{10}H_{20}N$	11.508	0.603		$C_6H_{10}N_3O_2$	7.866	0.673
	$C_{11}H_6O$	12.022	0.859		$C_6H_{12}N_4O$	8.241	0.502
	$C_{11}H_8N$	12.396	0.703		$C_7H_8O_4$	7.849	1.069
	$C_{11}H_{22}$	12.239	0.685		$C_7H_{10}NO_3$	8.223	0.898
	$C_{12}H_{10}$	13.127	0.792		$C_7H_{12}N_2O_2$	8.597	0.728
155	$C_5H_3N_2O_4$	6.370	0.974		$C_7H_{14}N_3O$	8.972	0.560

(Contd.)

		M + 1	M + 2			M + 1	M + 2
	$C_7H_{16}N_4$	9.346	0.393		$C_9H_3NO_2$	10.233	0.871
	$C_8H_2N_3O$	9.860	0.638		$C_9H_5N_2O$	10.607	0.709
	$C_8H_4N_4$	10.235	0.474		$C_9H_7N_3$	10.982	0.548
	$C_8H_{12}O_3$	8.954	0.956		$C_9H_{17}O_2$	10.076	0.856
	$C_8H_{14}NO_2$	9.329	0.789		$C_9H_{19}NO$	10.450	0.693
	$C_8H_{16}N_2O$	9.703	0.623		$C_9H_{21}N_2$	10.824	0.532
	$C_8H_{18}N_3$	10.077	0.459		$C_{10}H_5O_2$	10.964	0.944
	$C_9H_2NO_2$	10.217	0.870		$C_{10}H_7NO$	11.339	0.784
	$C_9H_4N_2O$	10.591	0.707		$C_{10}H_9N_2$	11.713	0.626
	$C_9H_6N_3$	10.966	0.547		$C_{10}H_{21}O$	11.181	0.767
	$C_9H_{16}O_2$	10.060	0.854		$C_{10}H_{23}N$	11.556	0.609
	$C_9H_{18}NO$	10.434	0.692		$C_{11}H_9O$	12.070	0.865
	$C_9H_{20}N_2$	10.808	0.530		$C_{11}H_{11}N$	12.444	0.709
	$C_{10}H_4O_2$	10.948	0.942		$C_{12}H_{13}$	13.175	7.979
	$C_{10}H_6NO$	11.323	0.782		$C_{13}H$	14.064	0.913
	$C_{10}H_8N_2$	11.697	0.624	158	$C_5H_6N_2O_4$	6.418	0.977
	$C_{10}H_{20}O$	11.165	0.765		$C_5H_8N_3O_3$	6.793	0.801
	$C_{10}H_{22}N$	11.540	0.606		$C_5H_{10}N_4O_2$	7.167	0.626
	$C_{11}H_8O$	12.054	0.862		$C_6H_8NO_4$	7.149	1.022
	$C_{11}H_{10}N$	12.428	0.707		$C_6H_{10}N_2O_3$	7.524	0.848
	$C_{11}H_{24}$	12.271	0.688		$C_6H_{12}N_3O_2$	7.898	0.676
	$C_{12}H_{12}$	13.159	0.795		$C_6H_{14}N_4O$	8.273	0.505
157	$C_5H_5N_2O_4$	6.402	0.976		$C_7H_2N_4O$	9.161	0.576
	$C_5H_7N_3O_3$	6.777	0.800		$C_7H_{10}O_4$	7.881	1.072
	$C_5H_9N_4O_2$	7.151	0.625		$C_7H_{12}NO_3$	8.255	0.901
	$C_6H_7NO_4$	7.133	1.021		$C_7H_{14}N_2O_2$	8.629	0.731
	$C_6H_9N_2O_3$	7.508	0.847		$C_7H_{16}N_3O$	9.004	0.563
	$C_6H_{11}N_3O_2$	7.882	0.674		$C_7H_{18}N_4$	9.378	0.396
	$C_6H_{13}N_4O$	8.257	0.503		$C_8H_2N_2O_2$	9.518	0.806
	C_7HN_4O	9.145	0.575		$C_8H_4N_3O$	9.892	0.641
	$C_7H_9O_4$	7.865	1.070		$C_8H_6N_4$	10.267	0.477
	$C_7H_{11}NO_3$	8.239	0.899		$C_8H_{14}O_3$	8.986	0.958
	$C_7H_{13}N_2O_2$	8.613	0.730		$C_8H_{16}NO_2$	9.361	0.792
	$C_7H_{15}N_3O$	8.988	0.561		$C_8H_{18}N_2O$	9.735	0.626
	$C_7H_{17}N_4$	9.362	0.394		$C_8H_{20}N_3$	10.109	0.462
	$C_8HN_2O_2$	9.502	0.804		$C_9H_2O_3$	9.875	1.036
	$C_8H_3N_3O$	9.876	0.639		$C_9H_4NO_2$	10.249	0.873
	$C_8H_5N_4$	10.251	0.476		$C_9H_6N_2O$	10.623	0.711
	$C_8H_{13}O_3$	8.970	0.957		$C_9H_8N_3$	10.998	0.550
	$C_8H_{15}NO_2$	9.345	0.790		$C_9H_{18}O_2$	10.092	0.858
	$C_8H_{17}N_2O$	9.719	0.625		$C_9H_{20}NO$	10.466	0.695
	$C_8H_{19}N_3$	10.093	0.461		$C_9H_{22}N_2$	10.840	0.534
	C_9HO_3	9.859	1.035		$C_{10}H_6O_2$	10.980	0.945

(Contd.)

	M + 1	M + 2			M + 1	M + 2
$C_{10}H_8NO$	11.355	0.786	160	$C_5H_8N_2O_4$	6.450	0.979
$C_{10}H_{10}N_2$	11.729	0.628		$C_5H_{10}N_3O_3$	6.825	0.803
$C_{10}H_{22}O$	11.197	0.769		$C_5H_{12}N_4O_2$	7.199	0.628
$C_{11}H_{10}O$	12.086	0.867		$C_6H_{10}NO_4$	7.181	1.024
$C_{11}H_{12}N$	12.460	0.711		$C_6H_{12}N_2O_3$	7.556	0.850
$C_{12}H_{14}$	13.191	0.800		$C_6H_{14}N_3O_2$	7.930	0.678
$C_{13}H_2$	14.080	0.915		$C_6H_{16}N_4O$	8.305	0.507
159 $C_5H_7N_2O_4$	6.434	0.978		$C_7H_2N_3O_2$	8.819	0.747
$C_5H_9N_3O_3$	6.809	0.802		$C_7H_4N_4O$	9.193	0.579
$C_5H_{11}N_4O_2$	7.183	0.627		$C_7H_{12}O_4$	7.913	1.074
$C_6H_9NO_4$	7.165	1.023		$C_7H_{14}NO_3$	8.287	0.903
$C_6H_{11}N_2O_3$	7.540	0.849		$C_7H_{16}N_2O_2$	8.661	0.734
$C_6H_{13}N_3O_2$	7.914	0.677		$C_7H_{18}N_3O$	9.036	0.566
$C_6H_{15}N_4O$	8.285	0.506		$C_7H_{20}N_4$	9.410	0.399
$C_7HN_3O_2$	8.803	0.745		$C_8H_2NO_3$	9.176	0.974
$C_7H_3N_4O$	9.177	0.578		$C_8H_4N_2O_2$	9.550	0.809
$C_7H_{11}O_4$	7.897	1.073		$C_8H_6N_3O$	9.924	0.644
$C_7H_{13}NO_3$	8.271	0.902		$C_8H_8N_4$	10.299	0.481
$C_7H_{15}N_2O_2$	8.645	0.732		$C_8H_{16}O_3$	9.018	0.961
$C_7H_{17}N_3O$	9.020	0.564		$C_8H_{18}NO_2$	9.393	0.795
$C_7H_{19}N_4$	9.394	0.397		$C_8H_{20}N_2O$	9.767	0.629
C_8HNO_3	9.160	0.974		$C_9H_4O_3$	9.907	1.040
$C_8H_3N_2O_2$	9.534	0.807		$C_9H_6NO_2$	10.281	0.876
$C_8H_5N_3O$	9.908	0.642		$C_9H_8N_2O$	10.655	0.714
$C_8H_7N_4$	10.283	0.479		$C_9H_{10}N_3$	11.030	0.554
$C_8H_{15}O_3$	9.002	0.960		$C_9H_{20}O_2$	10.124	0.861
$C_8H_{17}NO_2$	9.377	0.793		$C_{10}H_8O_2$	11.012	0.949
$C_8H_{19}N_2O$	9.751	0.628		$C_{10}H_{10}NO$	11.387	0.790
$C_8H_{21}N_3$	10.125	0.464		$C_{10}H_{12}N_2$	11.761	0.632
$C_9H_3O_3$	9.891	1.038		$C_{11}H_{12}O$	12.118	0.870
$C_9H_5NO_2$	10.265	0.875		$C_{11}H_{14}N$	12.492	0.715
$C_9H_7N_2O$	10.639	0.713		$C_{12}H_2N$	13.381	0.824
$C_9H_9N_3$	11.014	0.552		$C_{12}H_{16}$	13.223	0.804
$C_9H_{19}O_2$	10.108	0.859		$C_{13}H_4$	14.112	0.920
$C_9H_{21}NO$	10.482	0.697	161	$C_5H_9N_2O_4$	6.466	0.980
$C_{10}H_7O_2$	10.996	0.947		$C_5H_{11}N_3O_3$	6.841	0.804
$C_{10}H_9NO$	11.371	0.788		$C_5H_{13}N_4O_2$	7.215	0.629
$C_{10}H_{11}N_2$	11.745	0.630		$C_6HN_4O_2$	8.104	0.691
$C_{11}H_{11}O$	12.102	0.869		$C_6H_{11}NO_4$	7.197	1.025
$C_{11}H_{13}N$	12.476	0.713		$C_6H_{13}N_2O_3$	7.572	0.852
$C_{12}HN$	13.365	0.822		$C_6H_{15}N_3O_2$	7.946	0.679
$C_{12}H_{15}$	13.207	0.802		$C_6H_{17}N_4O$	8.321	0.509
$C_{13}H_3$	14.096	0.917		$C_7HN_2O_3$	8.460	0.917

(Contd.)

		M + 1	M + 2			M + 1	M + 2
	$C_7H_3N_3O_2$	8.835	0.748		$C_8H_6N_2O_2$	9.582	0.812
	$C_7H_5N_4O$	9.209	0.581		$C_8H_8N_3O$	9.956	0.647
	$C_7H_{13}O_4$	7.929	1.075		$C_8H_{10}N_4$	10.331	0.484
	$C_7H_{15}NO_3$	8.303	0.905		$C_8H_{18}O_3$	9.050	0.964
	$C_7H_{17}N_2O_2$	8.677	0.735		$C_9H_6O_3$	9.939	1.043
	$C_7H_{19}N_3O$	9.052	0.567		$C_9H_8NO_2$	10.313	0.880
	C_8HO_4	8.817	1.144		$C_9H_{10}N_2O$	10.687	0.718
	$C_8H_3NO_3$	9.192	0.976		$C_9H_{12}N_3$	11.062	0.557
	$C_8H_5N_2O_2$	9.566	0.810		$C_{10}H_{10}O_2$	11.044	0.952
	$C_8H_7N_3O$	9.940	0.646		$C_{10}H_{12}NO$	11.419	0.793
	$C_8H_9N_4$	10.315	0.482		$C_{10}H_{14}N_2$	11.793	0.636
	$C_8H_{17}O_3$	9.034	0.963		$C_{11}H_2N_2$	12.682	0.738
	$C_8H_{19}NO_2$	9.409	0.796		$C_{11}H_{14}O$	12.150	0.874
	$C_9H_5O_3$	9.923	1.041		$C_{11}H_{16}N$	12.524	0.719
	$C_9H_7NO_2$	10.297	0.878		$C_{12}H_2O$	13.038	0.980
	$C_9H_9N_2O$	10.671	0.716		$C_{12}H_4N$	13.413	0.829
	$C_9H_{11}N_3$	11.046	0.555		$C_{12}H_{18}$	13.255	0.808
	$C_{10}H_9O_2$	11.028	0.951		$C_{13}H_6$	14.144	0.924
	$C_{10}H_{11}NO$	11.403	0.791	163	$C_5H_{11}N_2O_4$	6.498	0.982
	$C_{10}H_{13}N_2$	11.777	0.634		$C_5H_{13}N_3O_3$	6.873	0.806
	$C_{11}HN_2$	12.666	0.736		$C_5H_{15}N_4O_2$	7.247	0.631
	$C_{11}H_{13}O$	12.134	0.872		$C_6HN_3O_3$	7.761	0.865
	$C_{11}H_{15}N$	12.508	0.717		$C_6H_3N_4O_2$	8.136	0.694
	$C_{12}HO$	13.022	0.978		$C_6H_{13}NO_4$	7.230	1.027
	$C_{12}H_3N$	13.397	0.827		$C_6H_{15}N_2O_3$	7.604	0.854
	$C_{12}H_{17}$	13.239	0.806		$C_6H_{17}N_3O_2$	7.978	0.682
	$C_{13}H_5$	14.128	0.922		C_7HNO_4	8.118	1.090
162	$C_5H_{10}N_2O_4$	6.482	0.981		$C_7H_3N_2O_3$	8.492	0.920
	$C_5H_{12}N_3O_3$	6.857	0.805		$C_7H_5N_3O_2$	8.867	0.751
	$C_5H_{14}N_4O_2$	7.231	0.630		$C_7H_7N_4O$	9.241	0.584
	$C_6H_2N_4O_2$	8.120	0.692		$C_7H_{15}O_4$	7.961	1.078
	$C_6H_{12}NO_4$	7.213	1.026		$C_7H_{17}NO_3$	8.335	0.907
	$C_6H_{14}N_2O_3$	7.588	0.853		$C_8H_3O_4$	8.849	1.147
	$C_6H_{16}N_3O_2$	7.962	0.681		$C_8H_5NO_3$	9.224	0.979
	$C_6H_{18}N_4O$	8.337	0.510		$C_8H_7N_2O_2$	9.598	0.813
	$C_7H_2N_2O_3$	8.476	0.918		$C_8H_9N_3O$	9.972	0.649
	$C_7H_4N_3O_2$	8.851	0.750		$C_8H_{11}N_4$	10.347	0.486
	$C_7H_6N_4O$	9.225	0.582		$C_9H_7O_3$	9.955	1.044
	$C_7H_{14}O_4$	7.945	1.077		$C_9H_9NO_2$	10.329	0.881
	$C_7H_{16}NO_3$	8.319	0.906		$C_9H_{11}N_2O$	10.703	0.719
	$C_7H_{18}N_2O_2$	8.693	0.737		$C_9H_{13}N_3$	11.078	0.559
	C_8H_2O4	8.833	1.145		$C_{10}HN_3$	11.966	0.655
	$C_8H_4NO_3$	9.208	0.978		$C_{10}H_{11}O_2$	11.060	0.954

(Contd.)

	M + 1	M + 2			M + 1	M + 2
$C_{10}H_{13}NO$	11.435	0.795	165	$C_5H_{13}N_2O_4$	6.530	0.985
$C_{10}H_{15}N_2$	11.809	0.637		$C_5H_{15}N_3O_3$	6.905	0.808
$C_{11}HNO$	12.323	0.895		$C_6HN_2O_4$	7.419	1.041
$C_{11}H_3N_2$	12.698	0.741		$C_6H_3N_3O_3$	7.793	0.868
$C_{11}H_{15}O$	12.166	0.876		$C_6H_5N_4O_2$	8.168	0.697
$C_{11}H_{17}N$	12.540	0.721		$C_6H_{15}NO_4$	7.262	1.030
$C_{12}H_3O$	13.054	0.983		$C_7H_3NO_4$	8.150	1.092
$C_{12}H_5N$	13.429	0.831		$C_7H_5N_2O_3$	8.524	0.922
$C_{12}H_{19}$	13.271	0.811		$C_7H_7N_3O_2$	8.899	0.754
$C_{13}H_7$	14.160	0.927		$C_7H_9N_4O$	9.273	0.587
164 $C_5H_{12}N_2O_4$	6.514	0.983		$C_8H_5O_4$	8.881	1.150
$C_5H_{14}N_3O_3$	6.889	0.807		$C_8H_7NO_3$	9.256	0.982
$C_5H_{16}N_4O_2$	7.263	0.633		$C_8H_9N_2O_2$	9.630	0.817
$C_6H_2N_3O_3$	7.777	0.867		$C_8H_{11}N_3O$	10.004	0.652
$C_6H_4N_4O_2$	8.152	0.695		$C_8H_{13}N_4$	10.379	0.489
$C_6H_{14}NO_4$	7.246	1.029		C_9HN_4	11.267	0.579
$C_6H_{16}N_2O_3$	7.620	0.855		$C_9H_9O_3$	9.987	1.048
$C_7H_2NO_4$	8.134	1.091		$C_9H_{11}NO_2$	10.361	0.884
$C_7H_4N_2O_3$	8.508	0.921		$C_9H_{13}N_2O$	10.735	0.723
$C_7H_6N_3O_2$	8.883	0.752		$C_9H_{15}N_3$	11.110	0.562
$C_7H_8N_4O$	9.257	0.585		$C_{10}HN_2O$	11.624	0.816
$C_7H_{16}O_4$	7.977	1.079		$C_{10}H_3N_3$	11.998	0.659
$C_8H_4O_4$	8.865	1.148		$C_{10}H_{13}O_2$	11.092	0.958
$C_8H_6NO_3$	9.240	0.981		$C_{10}H_{15}NO$	11.467	0.799
$C_8H_8N_2O_2$	9.614	0.815		$C_{10}H_{17}N_2$	11.841	0.641
$C_8H_{10}N_3O$	9.988	0.650		$C_{11}HO_2$	11.981	1.054
$C_8H_{12}N_4$	10.363	0.487		$C_{11}H_3NO$	12.355	0.899
$C_9H_8O_3$	9.971	1.046		$C_{11}H_5N_2$	12.730	0.745
$C_9H_{10}NO_2$	10.345	0.883		$C_{11}H_{17}O$	12.198	0.880
$C_9H_{12}N_2O$	10.719	0.721		$C_{11}H_{19}N$	12.572	0.725
$C_9H_{14}Na$	11.094	0.561		$C_{12}H_5O$	13.086	0.987
$C_{10}H_2N_3$	11.982	0.657		$C_{12}H_7N$	13.461	0.835
$C_{10}H_{12}O_2$	11.076	0.956		$C_{12}H_{21}$	13.303	0.815
$C_{10}H_{14}NO$	11.451	0.797		$C_{13}H_9$	14.192	0.931
$C_{10}H_{16}N_2$	11.825	0.639	166	$C_5H_{14}N_2O_4$	6.546	0.986
$C_{11}H_2NO$	12.339	0.897		$C_6H_2N_2O_4$	7.435	1.042
$C_{11}H_4N_2$	12.714	0.743		$C_6H_4N_3O_3$	7.809	0.869
$C_{11}H_{16}O$	12.182	0.878		$C_6H_6N_4O_2$	8.184	0.698
$C_{11}H_{18}N$	12.556	0.723		$C_7H_4NO_4$	8.166	1.094
$C_{12}H_4O$	13.070	0.985		$C_7H_6N_2O_3$	8.540	0.924
$C_{12}H_6N$	13.445	0.833		$C_7H_8N_3O_2$	8.915	0.755
$C_{12}H_{20}$	13.287	0.813		$C_7H_{10}N_4O$	9.289	0.588
$C_{13}H_8$	14.176	0.929		$C_8H_6O_4$	8.897	1.151

(Contd.)

	Formula	M + 1	M + 2
	$C_8H_8NO_3$	9.272	0.984
	$C_8H_{10}N_2O_2$	9.646	0.818
	$C_8H_{12}N_3O$	10.020	0.654
	$C_8H_{14}N_4$	10.395	0.491
	$C_9H_2N_4$	11.283	0.581
	$C_9H_{10}O_3$	10.003	1.049
	$C_9H_{12}NO_2$	10.377	0.886
	$C_9H_{14}N_2O$	10.751	0.724
	$C_9H_{16}N_3$	11.126	0.564
	$C_{10}H_2N_2O$	11.640	0.818
	$C_{10}H_4N_3$	12.014	0.661
	$C_{10}H_{14}O_2$	11.108	0.959
	$C_{10}H_{16}NO$	11.483	0.800
	$C_{10}H_{18}N_2$	11.857	0.643
	$C_{11}H_2O_2$	11.997	1.056
	$C_{11}H_4NO$	12.371	0.901
	$C_{11}H_6N_2$	12.746	0.747
	$C_{11}H_{18}O$	12.214	0.882
	$C_{11}H_{20}N$	12.588	0.727
	$C_{12}H_6O$	13.102	0.989
	$C_{12}H_8N$	13.477	0.837
	$C_{12}H_{22}$	13.319	0.817
	$C_{13}H_{10}$	14.208	0.933
167	$C_6H_3N_2O_4$	7.451	1.043
	$C_6H_5N_3O_3$	7.825	0.870
	$C_6H_7N_4O_2$	8.200	0.699
	$C_7H_5NO_4$	8.182	1.095
	$C_7H_7N_2O_3$	8.556	0.925
	$C_7H_9N_3O_2$	8.931	0.757
	$C_7H_{11}N_4O$	9.305	0.590
	$C_8H_7O_4$	8.913	1.152
	$C_8H_9NO_3$	9.288	0.985
	$C_8H_{11}N_2O_2$	9.662	0.820
	$C_8H_{13}N_3O$	10.036	0.655
	$C_8H_{15}N_4$	10.411	0.492
	C_9HN_3O	10.925	0.743
	$C_9H_3N_4$	11.299	0.583
	$C_9H_{11}O_3$	10.019	1.051
	$C_9H_{13}NO_2$	10.393	0.888
	$C_9H_{15}N_2O$	10.767	0.726
	$C_9H_{17}N_3$	11.142	0.566
	$C_{10}HNO_2$	11.282	0.978
	$C_{10}H_3N_2O$	11.656	8.199

	Formula	M + 1	M + 2
	$C_{10}H_5Na$	12.030	0.663
	$C_{10}H_{15}O_2$	11.124	0.961
	$C_{10}H_{17}NO$	11.499	0.802
	$C_{10}H_{19}N_2$	11.873	0.645
	$C_{11}H_3O_2$	12.013	1.058
	$C_{11}H_5NO$	12.387	0.903
	$C_{11}H_7N_2$	12.762	0.749
	$C_{11}H_{19}O$	12.230	0.884
	$C_{11}H_{21}N$	12.604	0.729
	$C_{12}H_7O$	13.118	0.991
	$C_{12}H_9N$	13.493	0.840
	$C_{12}H_{23}$	13.335	0.819
	$C_{13}H_{11}$	14.224	0.936
168	$C_6H_4N_2O_4$	7.467	1.044
	$C_6H_6N_3O_3$	7.841	0.872
	$C_8H_8N_4O_2$	8.216	0.700
	$C_7H_6NO_4$	8.198	1.096
	$C_7H_8N_2O_3$	8.572	0.927
	$C_7H_{10}N_3O_2$	8.947	0.758
	$C_7H_{12}N_4O$	9.321	0.591
	$C_8H_8O_4$	8.929	1.154
	$C_8H_{10}NO_3$	9.304	9.868
	$C_8H_{12}N_2O_2$	9.678	0.821
	$C_8H_{14}N_3O$	10.052	0.657
	$C_8H_{16}N_4$	10.427	0.494
	$C_9H_2N_3O$	10.941	0.744
	$C_9H_4N_4$	11.315	0.585
	$C_9H_{12}O_3$	10.035	1.052
	$C_9H_{14}NO_2$	10.409	0.889
	$C_9H_{16}N_2O$	10.783	0.728
	$C_9H_{18}N_3$	11.158	0.568
	$C_{10}H_2NO_2$	11.298	0.980
	$C_{10}H_4N_2O$	11.672	0.822
	$C_{10}H_6N_3$	12.046	0.665
	$C_{10}H_{16}O_2$	11.140	0.963
	$C_{10}H_{18}NO$	11.515	0.804
	$C_{10}H_{20}N_2$	11.889	0.647
	$C_{11}H_4O_2$	12.029	1.060
	$C_{11}H_6NO$	12.403	0.905
	$C_{11}H_8N_2$	12.778	0.751
	$C_{11}H_{20}O$	12.246	0.886
	$C_{11}H_{22}N$	12.620	0.731
	$C_{12}H_8O$	13.134	0.993

(Contd.)

		M + 1	M + 2			M + 1	M + 2
	$C_{12}H_{10}N$	13.509	0.842		$C_7H_8NO_4$	8.230	1.099
	$C_{12}H_{24}$	13.351	0.821		$C_7H_{10}N_2O_3$	8.604	0.929
	$C_{13}H_{12}$	14.240	0.938		$C_7H_{12}N_3O_2$	8.979	0.761
169	$C_6H_5N_2O_4$	7.483	1.045		$C_7H_{14}N_4O$	9.353	0.594
	$C_6H_7N_3O_3$	7.857	0.873		$C_8H_2N_4O$	10.242	0.675
	$C_6H_9N_4O_2$	8.232	0.702		$C_8H_{10}O_4$	8.961	1.157
	$C_7H_7NO_4$	8.214	1.098		$C_8H_{12}NO_3$	9.336	0.990
	$C_7H_9N_2O_3$	8.588	9.279		$C_8H_{14}N_2O_2$	9.710	0.824
	$C_7H_{11}N_3O_2$	8.963	0.760		$C_8H_{16}N_3O$	10.084	0.660
	$C_7H_{13}N_4O$	9.337	0.593		$C_8H_{18}N_4$	10.459	0.497
	C_8HN_4O	10.226	0.674		$C_9H_2N_2O_2$	10.598	0.909
	$C_8H_9O_4$	8.945	1.155		$C_9H_4N_3O$	10.973	0.748
	$C_8H_{11}NO_3$	9.320	0.988		$C_9H_6N_4$	11.347	0.588
	$C_8H_{13}N_2O_2$	9.694	0.823		$C_9H_{14}O_3$	10.067	1.056
	$C_8H_{15}N_3O$	10.068	0.658		$C_9H_{16}NO_2$	10.441	0.893
	$C_8H_{17}N_4$	10.443	0.496		$C_9H_{18}N_2O$	10.815	0.731
	$C_9HN_2O_2$	10.582	0.907		$C_9H_{20}N_3$	11.190	0.571
	$C_9H_3N_3O$	10.957	0.746		$C_{10}H_2O_3$	10.955	1.143
	$C_9H_5N_4$	11.331	0.587		$C_{10}H_4NO_2$	11.330	0.984
	$C_9H_{13}O_3$	10.051	1.054		$C_{10}H_6N_2O$	11.704	0.826
	$C_9H_{15}NO_2$	10.425	0.891		$C_{10}H_8N_3$	12.078	0.669
	$C_9H_{17}N_2O$	10.799	0.730		$C_{10}H_{18}O_2$	11.172	0.967
	$C_9H_{19}N_3$	11.174	0.570		$C_{10}H_{20}NO$	11.547	0.808
	$C_{10}HO_3$	10.939	1.141		$C_{10}H_{22}N_2$	11.921	0.651
	$C_{10}H_3NO_2$	11.314	0.982		$C_{11}H_6O_2$	12.061	1.064
	$C_{10}H_5N_2O$	11.688	0.824		$C_{11}H_8NO$	12.435	0.909
	$C_{10}H_7N_3$	12.062	0.667		$C_{11}H_{10}N_2$	12.810	0.755
	$C_{10}H_{17}O_2$	11.156	0.965		$C_{11}H_{22}O$	12.278	0.890
	$C_{10}H_{19}NO$	11.531	0.806		$C_{11}H_{24}N$	12.652	0.735
	$C_{10}H_{21}N_2$	11.905	0.649		$C_{12}H_{10}O$	13.166	0.997
	$C_{11}H_5O_2$	12.045	1.062		$C_{12}H_{12}N$	13.541	0.846
	$C_{11}H_7NO$	12.419	0.907		$C_{12}H_{26}$	13.383	0.826
	$C_{11}H_9N_2$	12.794	0.753		$C_{13}H_{14}$	14.272	0.942
	$C_{11}H_{21}O$	12.262	0.888		$C_{14}H_2$	15.160	1.068
	$C_{11}H_{23}N$	12.636	0.733	171	$C_6H_7N_2O_4$	7.515	1.048
	$C_{12}H_9O$	13.150	0.995		$C_6H_9N_3O_3$	7.889	0.875
	$C_{12}H_{11}N$	13.525	0.844		$C_6H_{11}N_4O_2$	8.264	0.704
	$C_{12}H_{25}$	13.367	0.823		$C_7H_9NO_4$	8.246	1.100
	$C_{13}H_{13}$	14.256	0.940		$C_7H_{11}N_2O_3$	8.620	0.931
	$C_{14}H$	15.144	1.065		$C_7H_{13}N_3O_2$	8.995	0.762
170	$C_6H_6N_2O_4$	7.499	1.046		$C_7H_{15}N_4O$	9.369	0.596
	$C_6H_8N_3O_3$	7.873	0.874		$C_8HN_3O_2$	9.883	0.841
	$C_6H_{10}N_4O_2$	8.248	0.703		$C_8H_3N_4O$	10.258	0.677

(Contd.)

		M + 1	M + 2			M + 1	M + 2
	$C_8H_{11}O_4$	8.977	1.158		$C_8H_{20}N_4$	10.491	0.501
	$C_8H_{13}NO_3$	9.352	0.991		$C_9H_2NO_3$	10.256	1.074
	$C_8H_{15}N_2O_2$	9.726	0.826		$C_9H_4N_2O_2$	10.630	0.912
	$C_8H_{17}N_3O$	10.100	0.662		$C_9H_6N_3O$	11.005	0.751
	$C_8H_{19}N_4$	10.475	0.499		$C_9H_8N_4$	11.379	0.592
	C_9HNO_3	10.240	1.072		$C_9H_{16}O_3$	10.099	1.058
	$C_9H_3N_2O_2$	10.614	0.910		$C_9H_{18}NO_2$	10.473	0.896
	$C_9H_5N_3O$	10.989	0.750		$C_9H_{20}N_2O$	10.847	0.734
	$C_9H_7N_4$	11.363	0.590		$C_9H_{22}N_3$	11.222	0.575
	$C_9H_{15}O_3$	10.083	1.057		$C_{10}H_4O_3$	10.987	1.147
	$C_9H_{17}NO_2$	10.457	0.894		$C_{10}H_6NO_2$	11.362	0.987
	$C_9H_{19}N_2O$	10.831	0.733		$C_{10}H_8N_2O$	11.736	0.829
	$C_9H_{21}N_3$	11.206	0.573		$C_{10}H_{10}N_3$	12.110	0.673
	$C_{10}H_3O_3$	10.971	1.145		$C_{10}H_{20}O_2$	11.204	0.970
	$C_{10}H_5NO_2$	11.346	9.855		$C_{10}H_{22}NO$	11.579	0.812
	$C_{10}H_7N_2O$	11.720	0.827		$C_{10}H_{24}N_2$	11.953	0.655
	$C_{10}H_9N_3$	12.094	0.671		$C_{11}H_8O_2$	12.093	1.068
	$C_{10}H_{19}O_2$	11.188	0.968		$C_{11}H_{10}NO$	12.467	0.913
	$C_{10}H_{21}NO$	11.563	0.810		$C_{11}H_{12}N_2$	12.842	0.759
	$C_{10}H_{23}N_2$	11.937	0.653		$C_{11}H_{24}O$	12.310	0.894
	$C_{11}H_7O_2$	12.077	1.066		$C_{12}H_{12}O$	13.198	1.001
	$C_{11}H_9NO$	12.451	0.911		$C_{12}H_{14}N$	13.573	0.850
	$C_{11}H_{11}N_2$	12.826	0.757		$C_{13}H_2N$	14.461	0.969
	$C_{11}H_{23}O$	12.294	0.892		$C_{13}H_{16}$	14.304	0.947
	$C_{11}H_{25}N$	12.668	0.737		$C_{14}H_4$	15.192	1.072
	$C_{12}H_{11}O$	13.182	1.000	173	$C_6H_9N_2O_4$	7.547	1.050
	$C_{12}H_{13}N$	13.557	0.848		$C_6H_{11}N_3O_3$	7.921	0.878
	$C_{13}HN$	14.445	0.967		$C_6H_{13}N_4O_2$	8.296	0.707
	$C_{13}H_{15}$	14.288	0.945		$C_7HN_4O_2$	9.184	0.779
	$C_{14}H_3$	15.176	1.070		$C_7H_{11}NO_4$	8.278	1.103
172	$C_6H_8N_2O_4$	7.531	1.049		$C_7H_{13}N_2O_3$	8.652	0.933
	$C_9H_{10}N_3O_3$	7.905	0.877		$C_7H_{15}N_3O_2$	9.027	0.765
	$C_6H_{12}N_4O_2$	8.280	0.706		$C_7H_{17}N_4O$	9.401	0.599
	$C_7H_{10}NO_4$	8.262	1.102		$C_8HN_2O_3$	9.541	1.008
	$C_7H_{12}N_2O_3$	8.636	0.932		$C_8H_3N_3O_2$	9.915	8.436
	$C_7H_{14}N_3O_2$	9.011	0.764		$C_8H_5N_4O$	10.290	0.680
	$C_7H_{16}N_4O$	9.385	5.971		$C_8H_{13}O_4$	9.009	1.161
	$C_8H_2N_3O_2$	9.899	0.842		$C_8H_{15}NO_3$	9.384	0.994
	$C_8H_4N_4O$	10.274	0.679		$C_8H_{17}N_2O_2$	9.758	0.829
	$C_8H_{12}O_4$	8.993	1.160		$C_8H_{19}N_3O$	10.132	0.665
	$C_8H_{14}NO_3$	9.368	0.993		$C_8H_{21}N_4$	10.507	0.502
	$C_8H_{16}N_2O_2$	9.742	0.827		C_9HO_4	9.898	1.239
	$C_8H_{18}N_3O$	10.116	6.633		$C_9H_3NO_3$	10.272	1.076

(Contd.)

	M + 1	M + 2			M + 1	M + 2
$C_9H_5N_2O_2$	10.647	0.914		$C_9H_{10}N_4$	11.411	0.596
$C_9H_7N_3O$	11.021	0.753		$C_9H_{18}O_3$	10.131	1.062
$C_9H_9N_4$	11.395	0.594		$C_9H_{20}NO_2$	10.505	0.899
$C_9H_{17}O_3$	10.115	1.060		$C_9H_{22}N_2O$	10.879	0.738
$C_9H_{19}NO_2$	10.489	0.898		$C_{10}H_6O_3$	11.019	1.150
$C_9H_{21}N_2O$	10.863	0.737		$C_{10}H_8NO_2$	11.394	0.991
$C_9H_{23}N_3$	11.238	0.577		$C_{10}H_{10}N_2O$	11.768	0.833
$C_{10}H_5O_3$	11.003	1.148		$C_{10}H_{12}N_3$	12.142	0.677
$C_{10}H_7NO_2$	11.378	0.989		$C_{10}H_{22}O_2$	11.236	0.974
$C_{10}H_9N_2O$	11.752	0.831		$C_{11}H_{10}O_2$	12.125	1.072
$C_{10}H_{11}N_3$	12.126	0.675		$C_{11}H_{12}NO$	12.449	0.917
$C_{10}H_{21}O_2$	11.220	0.972		$C_{11}H_{14}N_2$	12.874	0.763
$C_{10}H_{23}NO$	11.595	0.814		$C_{12}H_2N_2$	13.762	0.876
$C_{11}H_9O_2$	12.109	1.070		$C_{12}H_{14}O$	13.230	1.006
$C_{11}H_{11}NO$	12.483	0.915		$C_{12}H_{16}N$	13.605	0.855
$C_{11}H_{13}N_2$	12.858	0.761		$C_{13}H_2O$	14.119	1.121
$C_{12}HN_2$	13.746	0.873		$C_{13}H_4N$	14.493	0.974
$C_{12}H_{13}O$	13.214	1.000		$C_{13}H_{18}$	14.336	0.952
$C_{12}H_{15}N$	13.589	0.852		$C_{14}H_6$	15.224	1.077
$C_{13}HO$	14.103	1.119	175	$C_6H_{11}N_2O_4$	7.579	1.052
$C_{13}H_3N$	14.477	0.971		$C_6H_{13}N_3O_3$	7.953	0.881
$C_{13}H_{17}$	14.320	0.949		$C_6H_{15}N_4O_2$	8.328	0.710
$C_{14}H_5$	15.208	1.075		$C_7HN_3O_3$	8.842	0.949
174 $C_6H_{10}N_2O_4$	7.563	1.051		$C_7H_3N_4O_2$	9.216	0.782
$C_6H_{12}N_3O_3$	7.937	0.879		$C_7H_{13}NO_4$	8.310	1.106
$C_6H_{14}N_4O_2$	8.312	0.708		$C_7H_{15}N_2O_3$	8.684	0.936
$C_7H_2N_4O_2$	9.200	0.780		$C_7H_{17}N_3O_2$	9.059	0.768
$C_7H_{12}NO_4$	8.294	1.104		$C_7H_{19}N_4O$	9.432	0.602
$C_7H_{14}N_2O_3$	8.668	0.935		C_8HNO_4	9.199	1.178
$C_7H_{16}N_3O_2$	9.043	0.767		$C_8H_3N_2O_3$	9.573	1.012
$C_7H_{18}N_4O$	9.417	0.600		$C_8H_5N_3O_2$	9.947	0.847
$C_8H_2N_2O_3$	9.557	1.010		$C_8H_7N_4O$	10.322	0.684
$C_8H_4N_3O_2$	9.931	0.845		$C_8H_{17}NO_3$	9.416	0.997
$C_8H_6N_4O$	10.306	0.682		$C_8H_{19}N_2O_2$	9.790	0.832
$C_8H_{14}O_4$	9.025	1.162		$C_8H_{21}N_3O$	10.164	0.668
$C_8H_{16}NO_3$	9.400	9.957		$C_8H_{15}O_4$	9.041	1.164
$C_8H_{18}N_2O_2$	9.774	0.830		$C_9H_3O_4$	9.930	1.242
$C_8H_{20}N_3O$	10.148	0.667		$C_9H_5NO_3$	10.304	1.079
$C_8H_{22}N_4$	10.523	0.504		$C_9H_7N_2O_2$	10.679	0.917
$C_9H_2O_4$	9.914	1.241		$C_9H_9N_3O$	11.053	7.566
$C_9H_4NO_3$	10.288	1.077		$C_9H_{11}N_4$	11.427	0.597
$C_9H_6N_2O_2$	10.663	0.915		$C_9H_{19}O_3$	10.147	1.064
$C_9H_8N_3O$	11.037	0.755		$C_9H_{21}NO_2$	10.521	0.901

(Contd.)

	M + 1	M + 2			M + 1	M + 2
$C_{10}H_7O_3$	11.035	1.151		$C_{11}H_{12}O_2$	12.157	1.076
$C_{10}H_9NO_2$	11.410	0.993		$C_{11}H_{14}NO$	12.531	0.921
$C_{10}H_{11}N_2O$	11.784	0.835		$C_{11}H_{16}N_2$	12.906	0.767
$C_{10}H_{13}N_3$	12.158	0.679		$C_{12}H_2NO$	13.420	1.030
$C_{11}HN_3$	13.047	0.785		$C_{12}H_4N_2$	13.794	0.880
$C_{11}H_{11}O_2$	12.141	1.073		$C_{12}H_{16}O$	13.262	1.010
$C_{11}H_{13}NO$	12.515	0.919		$C_{12}H_{18}N$	13.637	0.859
$C_{11}H_{15}N_2$	12.890	7.650		$C_{18}H_4O$	14.151	1.126
$C_{12}HNO$	13.404	1.028		$C_{13}H_6N$	14.525	0.978
$C_{12}H_3N_2$	13.778	0.878		$C_{13}H_{20}$	14.368	0.956
$C_{12}H_{15}O$	13.246	1.008		$C_{14}H_8$	15.256	1.082
$C_{12}H_{17}N$	13.621	0.857	177	$C_6H_{13}N_2O_4$	7.611	1.055
$C_{13}H_3O$	14.135	1.124		$C_6H_{15}N_3O_3$	7.985	0.883
$C_{13}H_5N$	14.509	0.976		$C_6H_{17}N_4O_2$	8.360	0.712
$C_{13}H_{19}$	14.352	0.954		$C_7HN_2O_4$	8.500	1.121
$C_{14}H_7$	15.240	1.080		$C_7H_3N_3O_3$	8.874	0.952
176 $C_6H_{12}N_2O_4$	7.595	1.054		$C_7H_5N_4O_2$	9.248	0.785
$C_6H_{14}N_3O_3$	7.969	0.882		$C_7H_{15}NO_4$	8.342	1.108
$C_6H_{16}N_4O_2$	8.344	0.711		$C_7H_{17}N_2O_3$	8.716	0.939
$C_7H_2N_3O_3$	8.858	0.951		$C_7H_{19}N_3O_2$	9.091	0.771
$C_7H_4N_4O_2$	9.232	0.783		$C_8H_3NO_4$	9.231	1.181
$C_7H_{14}NO_4$	8.326	1.107		$C_8H_5N_2O_3$	9.605	1.014
$C_7H_{16}N_2O_3$	8.700	0.938		$C_8H_7N_3O_2$	9.979	0.850
$C_7H_{18}N_3O_2$	9.075	0.770		$C_8H_9N_4O$	10.354	0.687
$C_7H_{20}N_4O$	9.449	0.603		$C_8H_{17}O_4$	9.073	1.166
$C_8H_2NO_4$	9.215	1.179		$C_8H_{19}NO_3$	9.448	1.000
$C_8H_4N_2O_3$	9.589	1.013		$C_9H_5O_4$	9.962	1.246
$C_8H_6N_3O_2$	9.963	0.848		$C_9H_7NO_3$	10.336	1.082
$C_8H_8N_4O$	10.338	0.685		$C_9H_9N_2O_2$	10.711	0.921
$C_8H_{16}O_4$	9.057	1.165		$C_9H_{11}N_3O$	11.085	0.760
$C_8H_{18}NO_3$	9.432	0.999		$C_9H_{13}N_4$	11.459	0.601
$C_8H_{20}N_2O_2$	9.806	0.834		$C_{10}HN_4$	12.348	0.701
$C_9H_4O_4$	9.946	1.244		$C_{10}H_9O_3$	11.067	1.155
$C_9H_6NO_3$	10.320	1.081		$C_{10}H_{11}NO_2$	11.442	0.996
$C_9H_8N_2O_2$	10.695	0.919		$C_{10}H_{13}N_2O$	11.816	0.839
$C_9H_{10}N_3O$	11.069	0.758		$C_{10}H_{15}N_3$	12.190	0.682
$C_9H_{12}N_4$	11.443	0.599		$C_{11}HN_2O$	12.705	0.942
$C_9H_{20}O_3$	10.163	1.065		$C_{11}H_3N_3$	13.079	0.789
$C_{10}H_8O_3$	11.051	1.154		$C_{11}H_{13}O_2$	12.173	1.078
$C_{10}H_{10}NO_2$	11.426	0.995		$C_{11}H_{15}NO$	12.547	0.923
$C_{10}H_{12}N_2O$	11.800	0.837		$C_{11}H_{17}N_2$	12.922	0.769
$C_{10}H_{14}N_3$	12.174	0.681		$C_{12}HO_2$	13.061	1.184
$C_{11}H_2N_3$	13.063	0.787		$C_{12}H_3NO$	13.436	1.032

(Contd.)

		M + 1	M + 2			M + 1	M + 2
	$C_{12}H_5N_2$	13.810	0.882		$C_{14}H_{10}$	15.288	1.087
	$C_{12}H_{17}O$	13.278	1.012	179	$C_6H_{15}N_2O_4$	7.643	1.058
	$C_{12}H_{19}N$	13.653	0.861		$C_6H_{17}N_3O_3$	8.017	0.886
	$C_{13}H_5O$	14.167	1.128		$C_7H_3N_2O_4$	8.532	1.124
	$C_{13}H_7N$	14.541	0.981		$C_7H_5N_3O_3$	8.906	0.955
	$C_{13}H_{21}$	14.384	0.959		$C_7H_7N_4O_2$	9.280	0.788
	$C_{14}H_9$	15.272	1.085		$C_7H_{17}NO_4$	8.374	1.111
178	$C_6H_{14}N_2O_4$	7.627	1.056		$C_8H_5NO_4$	9.263	1.183
	$C_6H_{16}N_3O_3$	8.001	0.884		$C_8H_7N_2O_3$	9.637	1.018
	$C_6H_{18}N_4O_2$	8.376	0.714		$C_8H_9N_3O_2$	10.011	0.853
	$C_7H_2N_2O_4$	8.516	1.122		$C_8H_{11}N_4O$	10.386	0.690
	$C_7H_4N_3O_3$	8.890	0.054		$C_9H_7O_4$	9.994	1.249
	$C_7H_6N_4O_2$	9.264	0.786		$C_9H_9NO_3$	10.368	1.086
	$C_7H_{16}NO_4$	8.358	1.110		$C_9H_{11}N_2O_2$	10.743	0.924
	$C_7H_{18}N_2O_3$	8.732	0.940		$C_9H_{13}N_3O$	11.117	0.764
	$C_8H_4NO_4$	9.247	1.182		$C_9H_{15}N_4$	11.491	0.605
	$C_8H_6N_2O_3$	9.621	1.016		$C_{10}HN_3O$	12.005	0.861
	$C_8H_8N_3O_2$	9.995	0.852		$C_{10}H_3N_4$	12.380	0.705
	$C_8H_{10}N_4O$	10.370	0.689		$C_{10}H_{11}O_3$	11.099	1.159
	$C_8H_{18}O_4$	9.089	1.168		$C_{10}H_{13}NO_2$	11.474	1.000
	$C_9H_6O_4$	9.978	1.247		$C_{10}H_{15}N_2O$	11.848	0.843
	$C_9H_8NO_3$	10.352	1.084		$C_{10}H_{17}N_3$	12.222	0.686
	$C_9H_{10}N_2O_2$	10.727	0.922		$C_{11}HNO_2$	12.362	1.001
	$C_9H_{12}N_3O$	11.101	0.762		$C_{11}H_3N_2O$	12.737	0.946
	$C_9H_{14}N_4$	11.475	0.603		$C_{11}H_5N_3$	13.111	0.793
	$C_{10}H_2N_4$	12.364	0.703		$C_{11}H_{15}O_2$	12.205	1.082
	$C_{10}H_{10}O_3$	11.083	1.157		$C_{11}H_{17}NO$	12.579	0.927
	$C_{10}H_{12}NO_2$	11.458	0.998		$C_{11}H_{19}N_2$	12.954	0.773
	$C_{10}H_{14}N_2O$	11.832	0.841		$C_{12}H_3O_2$	13.093	1.188
	$C_{10}H_{16}N_3$	12.206	0.684		$C_{12}H_5NO$	13.468	1.036
	$C_{11}H_2N_2O$	12.721	0.944		$C_{12}H_7N_2$	13.842	0.887
	$C_{11}H_4N_3$	13.095	0.791		$C_{12}H_{19}O$	13.310	1.016
	$C_{11}H_{14}O_2$	12.189	1.080		$C_{12}H_{21}N$	13.685	8.655
	$C_{11}H_{16}NO$	12.563	0.925		$C_{13}H_7O$	14.199	1.133
	$C_{11}H_{18}N_2$	12.938	0.771		$C_{13}H_9N$	14.573	0.985
	$C_{12}H_2O_2$	13.077	1.186		$C_{13}H_{23}$	14.416	0.963
	$C_{12}H_4NO$	13.452	1.034		$C_{14}H_{11}$	15.304	1.089
	$C_{12}H_6N_2$	13.826	0.884	180	$C_6H_{16}N_2O_4$	7.659	1.059
	$C_{12}H_{18}O$	13.294	1.014		$C_7H_4N_2O_4$	8.548	1.125
	$C_{12}H_{20}N$	13.669	0.863		$C_7H_6N_3O_3$	8.922	0.956
	$C_{13}H_6O$	14.183	1.130		$C_7H_8N_4O_2$	9.296	0.789
	$C_{13}H_8N$	14.557	0.983		$C_8H_6NO_4$	9.279	1.185
	$C_{13}H_{22}$	14.400	0.961		$C_8H_8N_2O_3$	9.653	1.019

(Contd.)

	M + 1	M + 2			M + 1	M + 2
$C_8H_{10}N_3O_2$	10.027	0.855		$C_{10}H_5N_4$	12.412	0.709
$C_8H_{12}N_4O$	10.402	0.692		$C_{10}H_{13}O_3$	11.131	1.163
$C_9H_8O_4$	10.010	1.250		$C_{10}H_{15}NO_2$	11.506	1.004
$C_9H_{10}NO_3$	10.384	1.087		$C_{10}H_{17}N_2O$	11.880	0.846
$C_9H_{12}N_2O_2$	10.759	0.926		$C_{10}H_{19}N_3$	12.254	0.690
$C_9H_{14}N_3O$	11.133	0.765		$C_{11}HO_3$	12.020	1.260
$C_9H_{16}N_4$	11.507	0.607		$C_{11}H_3NO_2$	12.394	1.104
$C_{10}H_2N_3O$	12.021	0.863		$C_{11}H_5N_2O$	12.769	0.950
$C_{10}H_4N_4$	12.396	0.707		$C_{11}H_7N_3$	13.143	0.797
$C_{10}H_{12}O_3$	11.115	1.161		$C_{11}H_{17}O_2$	12.237	1.085
$C_{10}H_{14}NO_2$	11.490	1.002		$C_{11}H_{19}NO$	12.611	0.931
$C_{10}H_{16}N_2O$	11.864	0.844		$C_{11}H_{21}N_2$	12.986	0.777
$C_{10}H_{18}N_3$	12.238	0.688		$C_{12}H_5O_2$	13.125	1.192
$C_{11}H_2NO_2$	12.378	1.102		$C_{12}H_7NO$	13.500	1.041
$C_{11}H_4N_2O$	12.753	0.948		$C_{12}H_9N_2$	13.874	0.891
$C_{11}H_6N_3$	13.127	0.795		$C_{12}H_{21}O$	13.342	1.021
$C_{11}H_{16}O_2$	12.221	1.083		$C_{12}H_{23}N$	13.717	0.870
$C_{11}H_{18}NO$	12.595	0.929		$C_{13}H_9O$	14.231	1.137
$C_{11}H_{20}N_2$	12.970	0.775		$C_{13}H_{11}N$	14.605	0.990
$C_{12}H_4O_2$	13.109	1.190		$C_{13}H_{25}$	14.448	0.968
$C_{12}H_6NO$	13.484	1.039		$C_{14}H_{13}$	15.337	1.094
$C_{12}H_8N_2$	13.858	0.889		$C_{15}H$	16.225	1.229
$C_{12}H_{20}O$	13.326	1.018	182	$C_7H_6N_2O_4$	8.580	1.127
$C_{12}H_{22}N$	13.701	0.868		$C_7H_8N_3O_3$	8.954	0.959
$C_{13}H_8O$	14.215	1.135		$C_7H_{10}N_4O_2$	9.328	0.792
$C_{13}H_{10}N$	14.589	0.988		$C_8H_8NO_4$	9.311	1.188
$C_{13}H_{24}$	14.432	0.966		$C_8H_{10}N_2O_3$	9.685	1.022
$C_{14}H_{12}$	15.321	1.092		$C_8H_{12}N_3O_2$	10.059	0.858
181	$C_7H_5N_2O_4$	8.564	1.126		$C_8H_{14}N_4O$	10.434	0.695
	$C_7H_7N_3O_3$	8.938	0.958		$C_8H_2N_4O$	11.322	0.786
	$C_7H_9N_4O_2$	9.312	0.791		$C_9H_{10}O_4$	10.042	1.254
	$C_8H_7NO_4$	9.295	1.186		$C_9H_{12}NO_3$	10.416	1.091
	$C_8H_9N_2O_3$	9.669	1.021		$C_9H_{14}N_2O_2$	10.791	0.929
	$C_8H_{11}N_3O_2$	10.043	0.856		$C_9H_{16}N_3O$	11.165	0.769
	$C_8H_{13}N_4O$	10.418	0.694		$C_9H_{18}N_4$	11.539	0.610
	C_9HN_4O	11.306	0.784		$C_{10}H_2N_2O_2$	11.679	1.023
	$C_9H_9O_4$	10.026	1.252		$C_{10}H_4N_3O$	12.053	0.866
	$C_9H_{11}NO_3$	10.400	1.089		$C_{10}H_6N_4$	12.428	0.711
	$C_9H_{13}N_2O_2$	10.775	0.927		$C_{10}H_{14}O_3$	11.147	1.164
	$C_9H_{15}N_3O$	11.149	0.767		$C_{10}H_{16}NO_2$	11.522	1.006
	$C_9H_{17}N_4$	11.523	0.609		$C_{10}H_{18}N_2O$	11.896	0.848
	$C_{10}HN_2O_2$	11.663	1.021		$C_{10}H_{20}N_3$	12.270	0.692
	$C_{10}H_3N_3O$	12.037	0.864		$C_{11}H_2O_3$	12.036	1.262

(Contd.)

	M + 1	**M + 2**			**M + 1**	**M + 2**
$C_{11}H_4NO_2$	12.410	1.106		$C_{11}H_{21}NO$	12.643	0.935
$C_{11}H_6N_2O$	12.785	0.952		$C_{11}H_{23}N_2$	13.018	0.782
$C_{11}H_8N_3$	13.159	0.799		$C_{12}H_7O_2$	13.157	1.197
$C_{11}H_{18}O_2$	12.253	1.087		$C_{12}H_9NO$	13.532	1.045
$C_{11}H_{20}NO$	12.627	0.933		$C_{12}H_{11}N_2$	13.906	0.895
$C_{11}H_{22}N_2$	13.002	0.780		$C_{12}H_{23}O$	13.374	1.025
$C_{12}H_6O_2$	13.141	1.194		$C_{12}H_{25}N$	13.749	0.874
$C_{12}H_8NO$	13.516	1.043		$C_{13}H_{11}O$	14.263	1.142
$C_{12}H_{10}N_2$	13.890	0.893		$C_{13}H_{13}N$	14.637	0.995
$C_{12}H_{22}O$	13.358	1.023		$C_{13}H_{27}$	14.480	0.972
$C_{12}H_{24}N$	13.733	0.872		$C_{14}HN$	15.526	1.123
$C_{13}H_{10}O$	14.247	1.139		$C_{14}H_{15}$	15.369	1.099
$C_{13}H_{18}N$	14.621	0.992		$C_{15}H_3$	16.257	1.234
$C_{13}H_{26}$	14.464	0.970	184	$C_7H_8N_2O_4$	8.612	1.130
$C_{14}H_{14}$	15.353	1.097		$C_7H_{10}N_3O_3$	8.986	0.962
$C_{15}H_2$	16.241	1.213		$C_7H_{12}N_4O_2$	9.360	0.795
183 $C_7H_7N_2O_4$	8.596	1.129		$C_8H_{10}NO_4$	9.343	1.191
$C_7H_8N_3O_3$	8.970	0.961		$C_8H_{12}N_2O_3$	9.717	1.025
$C_7H_{11}N_4O_2$	9.344	0.794		$C_8H_{14}N_3O_2$	10.091	0.861
$C_8H_9NO_4$	9.327	1.189		$C_8H_{16}N_4O$	10.466	0.699
$C_8H_{11}N_2O_3$	9.701	1.024		$C_9H_2N_3O_2$	10.980	0.949
$C_8H_{13}N_3O_2$	10.075	0.860		$C_9H_4N_4O$	11.354	0.790
$C_8H_{15}N_4O$	10.450	0.697		$C_9H_{12}O_4$	10.074	1.257
$C_9HN_3O_2$	10.964	0.947		$C_9H_{14}NO_3$	10.448	1.094
$C_9H_3N_4O$	11.338	0.788		$C_9H_{16}N_2O_2$	10.823	0.933
$C_9H_{11}O_4$	10.058	1.255		$C_9H_{18}N_3O$	11.197	0.773
$C_9H_{13}NO_3$	10.432	1.092		$C_9H_{20}N_4$	11.571	0.614
$C_9H_{15}N_2O_2$	10.807	0.931		$C_{10}H_2NO_3$	11.337	1.184
$C_9H_{17}N_3O$	11.181	0.771		$C_{10}H_4N_2O_2$	11.711	1.027
$C_9H_{19}N_4$	11.555	0.612		$C_{10}H_6N_3O$	12.085	0.870
$C_{10}HNO_3$	11.321	1.183		$C_{10}H_8N_4$	12.460	0.715
$C_{10}H_3N_2O_2$	11.695	1.025		$C_{10}H_{16}O_3$	11.179	1.168
$C_{10}H_5N_3O$	12.069	0.868		$C_{10}H_{18}NO_2$	11.554	1.009
$C_{10}H_7N_4$	12.444	0.713		$C_{10}H_{20}N_2O$	11.928	0.852
$C_{10}H_{15}O_3$	11.163	1.166		$C_{10}H_{22}N_3$	12.302	0.696
$C_{10}H_{17}NO_2$	11.538	1.007		$C_{11}H_4O_3$	12.068	1.265
$C_{10}H_{19}N_2O$	11.912	0.850		$C_{11}H_6NO_2$	12.442	1.110
$C_{10}H_{21}N_3$	12.286	0.694		$C_{11}H_8N_2O$	12.817	0.956
$C_{11}H_3O_3$	12.052	1.263		$C_{11}H_{10}N_3$	13.191	0.804
$C_{11}H_5NO_2$	12.426	1.108		$C_{11}H_{20}O_2$	12.285	1.091
$C_{11}H_7N_2O$	12.801	0.954		$C_{11}H_{22}NO$	12.659	0.937
$C_{11}H_9N_3$	13.175	0.802		$C_{11}H_{24}N_2$	13.034	0.784
$C_{11}H_{19}O_2$	12.269	1.089		$C_{12}H_8O_2$	13.173	1.199

(Contd.)

	M + 1	M + 2			M + 1	M + 2
$C_{12}H_{10}NO$	13.548	1.047		$C_{12}H_{11}NO$	13.564	1.050
$C_{12}H_{12}N_2$	13.922	0.898		$C_{12}H_{13}N_2$	13.938	0.900
$C_{12}H_{24}O$	13.390	1.027		$C_{12}H_{25}O$	13.406	1.029
$C_{12}H_{26}N$	13.765	0.877		$C_{12}H_{27}N$	13.781	0.879
$C_{13}H_{12}O$	14.279	1.144		$C_{13}HN_2$	14.827	1.022
$C_{13}H_{14}N$	14.653	0.997		$C_{13}H_{13}O$	14.295	1.146
$C_{13}H_{28}$	14.496	0.975		$C_{13}H_{15}N$	14.669	0.999
$C_{14}H_2N$	15.542	1.125		$C_{14}HO$	15.183	1.271
$C_{14}H_{16}$	15.384	1.102		$C_{14}H_3N$	15.558	1.128
$C_{15}H_4$	16.273	1.237		$C_{14}H_{17}$	15.400	1.104
185 $C_7H_9N_2O_4$	8.628	1.132		$C_{15}H_5$	16.289	1.239
$C_7H_{11}N_3O_3$	9.002	0.964	186	$C_7H_{10}N_2O_4$	8.644	1.133
$C_7H_{13}N_4O_2$	9.376	0.797		$C_7H_{12}N_3O_3$	9.018	0.965
$C_8HN_4O_2$	10.265	0.878		$C_7H_{14}N_4O_2$	9.392	0.798
$C_8H_{11}NO_4$	9.359	1.192		$C_8H_2N_4O_2$	10.281	0.880
$C_8H_{13}N_2O_3$	9.733	1.027		$C_8H_{12}NO_4$	9.375	1.194
$C_8H_{15}N_3O_2$	10.107	0.863		$C_8H_{14}N_2O_3$	9.749	1.029
$C_8H_{17}N_4O$	10.482	0.700		$C_8H_{16}N_3O_2$	10.123	0.865
$C_9HN_2O_3$	10.622	1.112		$C_8H_{18}N_4O$	10.498	0.702
$C_9H_3N_3O_2$	10.996	0.951		$C_9H_2N_2O_3$	10.638	1.113
$C_9H_8N_4O$	11.370	0.791		$C_9H_4N_3O_2$	11.012	0.953
$C_9H_{13}O_4$	10.090	1.258		$C_9H_6N_4O$	11.386	0.793
$C_9H_{15}NO_3$	10.464	1.096		$C_9H_{14}O_4$	10.106	1.260
$C_9H_{17}N_2O_2$	10.839	0.934		$C_9H_{16}NO_3$	10.480	1.097
$C_9H_{19}N_3O$	11.213	0.774		$C_9H_{18}N_2O_2$	10.855	0.936
$C_9H_{21}N_4$	11.587	0.616		$C_9H_{20}N_3O$	11.229	0.776
$C_{10}HO_4$	10.978	1.346		$C_9H_{22}N_4$	11.603	0.618
$C_{10}H_3NO_3$	11.353	1.187		$C_{10}H_2O_4$	10.994	1.347
$C_{10}H_5N_2O_2$	11.727	1.029		$C_{10}H_4NO_3$	11.369	1.189
$C_{10}H_7N_3O$	12.101	0.872		$C_{10}H_6N_2O_2$	11.743	1.031
$C_{10}H_9N_4$	12.476	0.717		$C_9H_{20}N_3O$	11.229	0.776
$C_{10}H_{17}O_3$	11.195	1.170		$C_9H_{22}N_4$	11.603	0.618
$C_{10}H_{19}NO_2$	11.570	1.011		$C_{10}H_4NO_3$	11.369	1.189
$C_{10}H_{21}N_2O$	11.944	0.854		$C_{10}H_6N_2O_2$	11.743	1.031
$C_{10}H_{23}N_3$	12.318	0.698		$C_{10}H_8N_3O$	12.117	0.874
$C_{11}H_5O_3$	12.084	1.267		$C_{10}H_{10}N_4$	12.492	0.719
$C_{11}H_7NO_2$	12.458	1.112		$C_{10}H_{18}O_3$	11.211	1.172
$C_{11}H_9N_2O$	12.833	0.958		$C_{10}H_{20}NO_2$	11.586	1.013
$C_{11}H_{11}N_3$	13.207	0.806		$C_{10}H_{22}N_2O$	11.960	0.856
$C_{11}H_{21}O_2$	12.301	1.093		$C_{10}H_{24}N_3$	12.334	0.700
$C_{11}H_{23}NO$	1.268	0.939		$C_{11}H_6O_3$	12.100	1.269
$C_{11}H_{25}N_2$	13.050	0.786		$C_{11}H_8NO_2$	12.474	1.114
$C_{12}H_9O_2$	13.189	1.201		$C_{11}H_{10}N_2O$	12.849	0.960

(Contd.)

		M + 1	M + 2			M + 1	M + 2
	$C_{11}H_{12}N_3$	13.223	0.808		$C_{11}H_9NO_2$	12.490	1.116
	$C_{11}H_{22}O_2$	12.317	1.095		$C_{11}H_{11}N_2O$	12.865	0.962
	$C_{11}H_{24}NO$	1.269	0.941		$C_{11}H_{13}N_3$	13.239	0.801
	$C_{11}H_{26}N_2$	13.066	0.788		$C_{11}H_{25}O_2$	12.333	10.972
	$C_{12}H_{10}O_2$	13.205	1.203		$C_{11}H_{25}NO$	12.707	0.943
	$C_{12}H_{12}NO$	13.580	1.052		$C_{12}HN_3$	14.128	0.926
	$C_{12}H_{14}N_2$	13.954	0.902		$C_{12}H_{11}O_2$	13.221	1.205
	$C_{12}H_{26}O$	13.422	1.031		$C_{12}H_{13}NO$	13.596	1.054
	$C_{13}H_2N_2$	14.843	1.024		$C_{12}H_{15}N_2$	13.970	0.904
	$C_{13}H_{14}O$	14.311	1.149		$C_{13}HNO$	14.484	1.173
	$C_{13}H_{16}N$	14.685	1.002		$C_{13}H_3N_2$	14.859	1.027
	$C_{14}H_2O$	15.199	1.274		$C_{13}H_{15}O$	14.327	1.151
	$C_{14}H_4N$	15.574	1.130		$C_{13}H_{17}N$	14.701	1.004
	$C_{14}H_{18}$	15.416	1.107		$C_{14}H_3O$	15.216	1.276
	$C_{15}H_6$	16.305	1.242		$C_{14}H_5N$	15.590	1.133
187	$C_7H_{11}N_2O_4$	8.660	1.135		$C_{14}H_{19}$	15.432	1.109
	$C_7H_{13}N_3O_3$	9.034	0.966		$C_{15}H_7$	16.321	1.244
	$C_7H_{15}N_4O_2$	9.408	0.800	188	$C_7H_{12}N_2O_4$	8.676	1.136
	$C_8HN_3O_3$	9.922	1.045		$C_7H_{14}N_3O_3$	9.050	0.968
	$C_8H_3N_4O_2$	10.297	0.881		$C_{17}H_{16}N_4O_2$	9.424	0.801
	$C_8H_{13}NO_4$	9.391	1.195		$C_8H_2N_3O_3$	9.938	1.046
	$C_8H_{15}N_2O_3$	9.765	1.030		$C_8H_4N_4O_2$	10.313	0.883
	$C_8H_{17}N_3O_2$	10.139	0.866		$C_8H_{14}NO_4$	9.407	1.197
	$C_8H_{19}N_4O$	10.514	0.704		$C_8H_{16}N_2O_3$	9.781	1.032
	C_9HNO_4	10.279	1.277		$C_8H_{18}N_3O_2$	10.155	0.868
	$C_9H_3N_2O_3$	10.654	1.115		$C_8H_{20}N_4O$	10.530	0.705
	$C_9H_5N_3O_2$	11.028	0.954		$C_9H_2NO_4$	10.295	1.279
	$C_9H_7N_4O$	11.402	0.795		$C_9H_4N_2O_3$	10.670	1.117
	$C_9H_{15}O_4$	10.122	1.262		$C_9H_6N_3O_2$	11.044	0.956
	$C_9H_{17}NO_3$	10.496	1.099		$C_9H_8N_4O$	11.418	0.797
	$C_9H_{19}N_2O_2$	10.871	0.938		$C_9H_{16}O_4$	10.138	1.263
	$C_9H_{21}N_3O$	11.245	0.778		$C_9H_{18}NO_3$	10.512	1.101
	$C_9H_{23}N_4$	11.619	0.620		$C_9H_{20}N_2O_2$	10.887	0.940
	$C_{10}H_3O_4$	11.010	1.349		$C_9H_{22}N_3O$	11.261	0.780
	$C_{10}H_5NO_3$	11.385	1.190		$C_9H_{24}N_4$	11.636	0.621
	$C_{10}H_7N_2O_2$	11.759	1.033		$C_{10}H_4O_4$	11.026	1.351
	$C_{10}H_9N_3O$	12.133	0.876		$C_{10}H_6NO_3$	11.401	1.192
	$C_{10}H_{11}N_4$	12.508	0.721		$C_{10}H_8N_2O_2$	11.775	1.034
	$C_{10}H_{19}O_3$	11.227	1.173		$C_{10}H_{10}N_3O$	12.149	0.978
	$C_{10}H_{21}NO_2$	11.602	1.015		$C_{10}H_{12}N_4$	12.524	0.723
	$C_{10}H_{23}N_2O$	11.976	0.858		$C_{10}H_{20}O_3$	11.243	1.175
	$C_{10}H_{25}N_3$	12.350	0.702		$C_{10}H_{22}NO_2$	11.618	1.017
	$C_{11}H_7O_3$	12.116	1.271		$C_{10}H_{24}N_2O$	11.992	0.860

(Contd.)

		M + 1	M + 2			M + 1	M + 2
	$C_{11}H_8O_3$	12.132	1.273		$C_{11}H_9O_3$	12.148	1.275
	$C_{11}H_{10}NO_2$	12.506	1.118		$C_{11}H_{11}NO_2$	12.522	1.120
	$C_{11}H_{12}N_2O$	12.881	0.964		$C_{11}H_{13}N_2O$	12.897	0.966
	$C_{11}H_{14}N_3$	13.255	0.812		$C_{11}H_{15}N_3$	13.271	0.814
	$C_{11}H_{24}O_2$	12.349	1.099		$C_{12}HN_2O$	13.785	1.079
	$C_{12}H_2N_3$	14.144	0.928		$C_{12}H_3N_3$	14.160	0.930
	$C_{12}H_{12}O_2$	13.237	1.207		$C_{12}H_{13}O_2$	13.253	1.209
	$C_{12}H_{14}NO$	13.612	1.056		$C_{12}H_{15}NO$	13.628	1.058
	$C_{12}H_{16}N_2$	13.986	0.907		$C_{12}H_{17}N_2$	14.002	0.909
	$C_{13}H_2NO$	14.500	1.175		$C_{13}HO_2$	14.142	1.325
	$C_{13}H_4N_2$	14.875	1.029		$C_{13}H_3NO$	14.516	1.178
	$C_{13}H_{16}O$	14.343	1.153		$C_{13}H_5N_2$	14.891	1.031
	$C_{13}H_{18}N$	14.717	1.006		$C_{13}H_{17}O$	14.359	1.155
	$C_{14}H_4O$	15.232	1.279		$C_{13}H_{19}N$	14.733	1.009
	$C_{14}H_6N$	15.606	1.135		$C_{14}H_5O$	15.248	1.281
	$C_{14}H_{20}$	15.448	1.112		$C_{14}H_7N$	15.622	1.133
	$C_{15}H_8$	16.337	1.247		$C_{14}H_{21}$	15.464	1.114
189	$C_7H_{13}N_2O_4$	8.692	1.137		$C_{15}H_9$	16.353	1.250
	$C_7H_{15}N_3O_3$	9.066	0.969	190	$C_7H_{14}N_2O_4$	8.708	1.139
	$C_7H_{17}N_4O_2$	9.440	0.803		$C_7H_{16}N_3O_3$	9.082	0.971
	$C_8HN_2O_4$	9.580	1.213		$C_7H_{18}N_4O_2$	9.456	0.804
	$C_8H_3N_3O_3$	9.954	1.048		$C_8H_2N_2O_4$	9.596	1.214
	$C_8H_5N_4O_2$	10.329	0.885		$C_8H_4N_3O_3$	9.970	1.050
	$C_8H_{15}NO_4$	9.423	1.198		$C_8H_6N_4O_2$	10.345	0.886
	$C_8H_{17}N_2O_3$	9.797	1.033		$C_8H_{16}NO_4$	9.439	1.200
	$C_8H_{19}N_3O_2$	10.171	0.869		$C_8H_{18}N_2O_3$	9.813	1.035
	$C_8H_{21}N_4O$	10.546	0.707		$C_8H_{20}N_3O_2$	10.187	0.871
	$C_9H_3NO_4$	10.311	1.280		$C_8H_{22}N_4O$	10.562	0.709
	$C_9H_5N_2O_3$	10.686	1.118		$C_9H_4NO_4$	10.327	1.282
	$C_9H_7N_3O_2$	11.060	0.958		$C_9H_6N_2O_3$	10.702	1.120
	$C_9H_9N_4O$	11.434	0.799		$C_9H_8N_3O_2$	1.108	0.960
	$C_9H_{17}O_4$	10.154	1.265		$C_9H_{10}N_4O$	11.450	0.801
	$C_9H_{19}NO_3$	10.528	1.102		$C_9H_{18}O_4$	10.170	1.267
	$C_9H_{21}N_2O_2$	10.903	0.941		$C_9H_{20}NO_3$	10.544	1.104
	$C_9H_{23}N_3O$	11.277	0.782		$C_9H_{22}N_2O_2$	10.919	0.943
	$C_{10}H_5O_4$	11.042	1.353		$C_{10}H_6O_4$	11.058	1.355
	$C_{10}H_7NO_3$	11.417	1.194		$C_{10}H_8NO_3$	11.433	1.196
	$C_{10}H_9N_2O_2$	11.791	1.036		$C_{10}H_{10}N_2O_2$	11.807	1.028
	$C_{10}H_{11}N_3O$	12.165	0.880		$C_{10}H_{12}N_3O$	12.181	0.882
	$C_{10}H_{13}N_4$	12.540	0.725		$C_{10}H_{14}N_4$	12.556	0.727
	$C_{10}H_{21}O_3$	11.259	1.177		$C_{10}H_{22}O_3$	11.275	1.179
	$C_{10}H_{23}NO_2$	11.634	1.019		$C_{11}H_2N_4$	13.444	0.837
	$C_{11}HN_4$	13.428	0.835		$C_{11}H_{10}O_3$	12.164	1.277

(Contd.)

	M + 1	M + 2		M + 1	M + 2
$C_{11}H_{12}NO_2$	12.538	1.122	$C_{12}HNO_2$	13.443	1.234
$C_{11}H_{14}N_2O$	12.913	0.969	$C_{12}H_3N_2O$	13.817	1.084
$C_{11}H_{16}N_3$	13.287	0.816	$C_{12}H_5N_3$	14.192	0.935
$C_{12}H_2N_2O$	13.801	1.081	$C_{12}H_{15}O_2$	13.285	1.213
$C_{12}H_4N_3$	14.176	0.933	$C_{12}H_{17}NO$	13.660	1.063
$C_{12}H_{14}O_2$	13.269	1.211	$C_{12}H_{19}N_2$	14.034	0.913
$C_{12}H_{16}NO$	13.644	1.060	$C_{13}H_3O_2$	14.174	1.330
$C_{12}H_{18}N_2$	14.018	0.911	$C_{13}H_5NO$	14.548	1.182
$C_{13}H_2O_2$	14.158	1.327	$C_{13}H_7N_2$	14.923	1.036
$C_{13}H_4NO$	14.532	1.180	$C_{13}H_{19}O$	14.391	1.160
$C_{13}H_6N_2$	14.907	1.034	$C_{13}H_{21}N$	14.765	1.013
$C_{13}H_{18}O$	14.375	1.158	$C_{14}H_7O$	15.280	1.286
$C_{13}H_{20}N$	14.749	1.011	$C_{14}H_9N$	15.654	1.143
$C_{14}H_6O$	15.264	1.284	$C_{14}H_{23}$	15.496	1.119
$C_{14}H_8N$	15.638	1.140	$C_{15}H_{11}$	16.385	1.255
$C_{14}H_{22}$	15.480	1.117	**192** $C_7H_{16}N_2O_4$	8.470	1.141
$C_{15}H_{10}$	16.369	1.252	$C_7H_{18}N_3O_3$	9.114	0.974
191 $C_7H_{15}N_2O_4$	8.724	1.140	$C_7H_{20}N_4O_2$	9.488	0.807
$C_7H_{17}N_3O_3$	9.098	0.972	$C_8H_4N_2O_4$	9.628	1.217
$C_7H_{19}N_4O_2$	9.472	0.806	$C_8H_6N_3O_3$	10.002	1.053
$C_8H_3N_2O_4$	9.612	1.216	$C_8H_8N_4O_2$	10.377	0.890
$C_8H_5N_3O_3$	9.986	1.051	$C_8H_{18}NO_4$	9.471	1.203
$C_8H_7N_4O_2$	10.361	0.888	$C_8H_{20}N_2O_3$	9.845	1.038
$C_8H_{17}NO_4$	9.455	1.201	$C_9H_6NO_4$	10.359	1.285
$C_8H_{19}N_2O_3$	9.829	1.036	$C_9H_8N_2O_3$	10.734	1.124
$C_8H_{21}N_3O_2$	10.203	0.873	$C_9H_{10}N_3O_2$	11.108	0.963
$C_9H_5NO_4$	10.343	1.284	$C_9H_{12}N_4O$	11.482	0.804
$C_9H_7N_2O_3$	10.718	1.122	$C_9H_{20}O_4$	10.202	1.270
$C_9H_9N_3O_2$	11.092	0.961	$C_{10}H_8O_4$	11.090	1.359
$C_9H_{11}N_4O$	11.466	0.802	$C_{10}H_{10}NO_3$	11.465	1.200
$C_9H_{19}O_4$	10.186	1.268	$C_{10}H_{12}N_2O_2$	11.839	1.042
$C_9H_{21}NO_3$	10.560	1.106	$C_{10}H_{14}N_3O$	12.213	0.886
$C_{10}H_7O_4$	11.074	1.357	$C_{10}H_{16}N_4$	12.588	0.731
$C_{10}H_9NO_3$	11.449	1.198	$C_{11}H_2N_3O$	13.102	0.992
$C_{10}H_{11}N_2O_2$	11.823	1.040	$C_{11}H_4N_4$	13.476	0.841
$C_{10}H_{13}N_3O$	12.197	0.884	$C_{11}H_{12}O_3$	12.196	1.281
$C_{10}H_{15}N_4$	12.572	0.729	$C_{11}H_{14}NO_2$	12.570	1.126
$C_{11}HN_3O$	13.086	0.990	$C_{11}H_{16}N_2O$	12.945	0.973
$C_{11}H_3N_4$	13.460	0.839	$C_{11}H_{18}N_3$	13.319	0.821
$C_{11}H_{11}O_3$	12.180	1.280	$C_{12}H_2NO_2$	13.459	1.236
$C_{11}H_{13}NO_2$	12.554	1.124	$C_{12}H_4N_2O$	13.833	1.086
$C_{11}H_{15}N_2O$	12.929	0.971	$C_{12}H_6N_3$	14.208	0.937
$C_{11}H_{17}N_3$	13.303	0.818	$C_{12}H_{16}O_2$	13.301	1.216

(Contd.)

		M + 1	M + 2			M + 1	M + 2
	$C_{12}H_{18}NO$	13.676	1.065		$C_{13}H_9N_2$	14.955	1.041
	$C_{12}H_{20}N_2$	14.050	0.916		$C_{13}H_{21}O$	14.423	1.165
	$C_{13}H_4O_2$	14.190	1.332		$C_{13}H_{23}N$	14.797	1.018
	$C_{13}H_6NO$	14.564	1.184		$C_{14}H_9O$	15.312	1.291
	$C_{13}H_8N_2$	14.939	1.039		$C_{14}H_{11}N$	15.686	1.148
	$C_{13}H_{20}O$	14.407	1.162		$C_{14}H_{25}$	15.528	1.124
	$C_{13}H_{22}N$	14.781	1.016		$C_{15}H_{13}$	16.417	1.260
	$C_{14}H_8O$	15.296	1.289		$C_{16}H$	17.306	1.404
	$C_{14}H_{10}N$	15.670	1.145	194	$C_7H_{18}N_2O_4$	8.772	1.144
	$C_{14}H_{24}$	15.512	1.121		$C_8H_6N_2O_4$	9.660	1.220
	$C_{15}H_{12}$	16.401	1.257		$C_8H_8N_3O_3$	10.034	1.056
193	$C_7H_{17}N_2O_4$	8.756	1.143		$C_8H_{10}N_4O_2$	10.409	0.893
	$C_7H_{19}N_3O_3$	9.130	0.975		$C_9H_8NO_4$	10.391	1.289
	$C_8H_5N_2O_4$	9.644	1.219		$C_9H_{10}N_2O_3$	10.766	1.127
	$C_8H_7N_3O_3$	10.018	1.054		$C_9H_{12}N_3O_2$	11.140	0.967
	$C_8H_9N_4O_2$	10.393	0.891		$C_9H_{14}N_4O$	11.514	0.808
	$C_8H_{19}NO_4$	9.487	1.204		$C_{10}H_2N_4O$	12.403	0.908
	$C_9H_7NO_4$	10.375	1.287		$C_{10}H_{10}O_4$	11.122	1.362
	$C_9H_9N_2O_3$	10.750	1.125		$C_{10}H_{12}NO_3$	11.497	1.203
	$C_9H_{11}N_3O_2$	11.124	0.965		$C_{10}H_{14}N_2O_2$	11.871	1.046
	$C_9H_{13}N_4O$	11.498	0.806		$C_{10}H_{16}N_3O$	12.245	0.890
	$C_{10}HN_4O$	12.387	0.906		$C_{10}H_{18}N_4$	12.620	0.735
	$C_{10}H_9O_4$	11.106	1.360		$C_{11}H_2N_2O_2$	12.760	1.149
	$C_{10}H_{11}NO_3$	11.481	1.201		$C_{11}H_4N_3O$	13.134	0.997
	$C_{10}H_{13}N_2O_2$	11.855	1.044		$C_{11}H_6N_4$	13.508	0.845
	$C_{10}H_{15}N_3O$	12.229	0.888		$C_{11}H_{14}O_3$	12.228	1.285
	$C_{10}H_{17}N_4$	12.604	0.733		$C_{11}H_{16}NO_2$	12.602	1.130
	$C_{11}HN_2O_2$	12.744	1.147		$C_{11}H_{18}N_2O$	12.977	0.977
	$C_{11}H_3N_3O$	13.118	0.995		$C_{11}H_{20}N_3$	13.351	0.825
	$C_{11}H_5N_4$	13.492	0.843		$C_{12}H_2O_3$	13.116	1.392
	$C_{11}H_{13}O_3$	12.212	1.283		$C_{12}H_4NO_2$	13.491	1.240
	$C_{11}H_{15}NO_2$	12.586	1.128		$C_{12}H_6N_2O$	13.865	1.090
	$C_{11}H_{17}N_2O$	12.961	0.975		$C_{12}H_8N_3$	14.240	0.942
	$C_{11}H_{19}N_3$	13.335	0.823		$C_{12}H_{18}O_2$	13.333	1.220
	$C_{12}HO_3$	13.100	1.390		$C_{12}H_{20}NO$	13.708	1.069
	$C_{12}H_3NO_2$	13.475	1.238		$C_{12}H_{22}N_2$	14.082	0.920
	$C_{12}H_5N_2O$	13.849	1.088		$C_{13}H_6O_2$	14.222	1.336
	$C_{12}H_7N_3$	14.224	0.939		$C_{13}H_8NO$	14.596	1.189
	$C_{12}H_{17}O_2$	13.317	1.218		$C_{13}H_{10}N_2$	14.971	1.043
	$C_{12}H_{19}NO$	13.692	1.067		$C_{13}H_{22}O$	14.439	1.167
	$C_{12}H_{21}N_2$	14.066	0.918		$C_{13}H_{24}N$	14.813	1.021
	$C_{13}H_5O_2$	14.206	1.334		$C_{14}H_{10}O$	15.328	1.293
	$C_{13}H_7NO$	14.580	1.187		$C_{14}H_{12}N$	15.702	1.150

(Contd.)

		M + 1	**M + 2**			**M + 1**	**M + 2**
	$C_{14}H_{26}$	15.544	1.126	196	$C_8H_8N_2O_4$	9.692	1.223
	$C_{15}H_{14}$	16.433	1.263		$C_8H_{10}N_3O_3$	10.066	1.059
	$C_{16}H_2$	17.322	1.407		$C_8H_{12}N_4O_2$	10.441	0.896
195	$C_8H_7N_2O_4$	9.676	1.222		$C_9H_{10}NO_4$	10.423	1.292
	$C_8H_9N_3O_3$	10.050	1.058		$C_9H_{12}N_2O_3$	10.798	1.131
	$C_8H_{11}N_4O_2$	10.425	0.895		$C_9H_{14}N_3O_2$	11.172	0.970
	$C_9H_9NO_4$	10.407	1.290		$C_9H_{16}N_4O$	11.546	0.812
	$C_9H_{11}N_2O_3$	10.782	1.129		$C_{10}H_2N_3O_2$	12.061	1.068
	$C_9H_{13}N_3O_2$	11.156	0.969		$C_{10}H_4N_4O$	12.435	0.912
	$C_9H_{15}N_4O$	11.530	0.810		$C_{10}H_{12}O_4$	11.154	1.366
	$C_{10}HN_3O_2$	12.045	1.066		$C_{10}H_{14}NO_3$	11.529	1.207
	$C_{10}H_3N_4O$	12.419	0.910		$C_{10}H_{16}N_2O_2$	11.903	1.050
	$C_{10}H_{11}O_4$	11.138	1.364		$C_{10}H_{18}N_3O$	12.277	0.894
	$C_{10}H_{13}NO_3$	11.513	1.205		$C_{10}H_{20}N_4$	12.652	0.739
	$C_{10}H_{15}N_2O_2$	11.887	1.048		$C_{11}H_2NO_3$	12.417	1.307
	$C_{10}H_{17}N_3O$	12.261	0.892		$C_{11}H_4N_2O_2$	12.792	1.153
	$C_{10}H_{19}N_4$	12.636	0.737		$C_{11}H_6N_3O$	13.166	1.001
	$C_{11}HNO_3$	12.401	1.305		$C_{11}H_8N_4$	13.540	0.850
	$C_{11}H_3N_2O_2$	12.776	1.151		$C_{11}H_{16}O_3$	12.260	1.289
	$C_{11}H_5N_3O$	13.150	0.999		$C_{11}H_{18}NO_2$	12.634	1.134
	$C_{11}H_7N_4$	13.524	0.848		$C_{11}H_{20}N_2O$	13.009	0.981
	$C_{11}H_{15}O_3$	12.244	1.287		$C_{11}H_{22}N_3$	13.383	0.829
	$C_{11}H_{17}NO_2$	12.618	1.132		$C_{12}H_4O_3$	13.148	1.396
	$C_{11}H_{19}N_2O$	12.993	0.979		$C_{12}H_6NO_2$	13.523	1.245
	$C_{11}H_{21}N_3$	13.367	0.827		$C_{12}H_8N_2O$	13.897	1.095
	$C_{12}H_3O_3$	13.132	1.394		$C_{12}H_{10}N_3$	14.272	0.946
	$C_{12}H_5NO_2$	13.507	1.242		$C_{12}H_{20}O_2$	13.365	1.224
	$C_{12}H_7N_2O$	13.881	1.092		$C_{12}H_{22}NO$	13.740	1.074
	$C_{12}H_9N_3$	14.256	0.944		$C_{12}H_{24}N_2$	14.114	0.925
	$C_{12}H_{19}O_2$	13.349	1.222		$C_{13}H_8O_2$	14.254	1.341
	$C_{12}H_{21}NO$	13.724	1.071		$C_{13}H_{10}NO$	14.628	1.194
	$C_{12}H_{23}N_2$	14.098	0.922		$C_{13}H_{12}N_2$	15.003	1.048
	$C_{13}H_7O_2$	14.238	1.339		$C_{13}H_{24}O$	14.471	1.172
	$C_{13}H_9NO$	14.612	1.191		$C_{13}H_{26}N$	14.845	1.025
	$C_{13}H_{11}N_2$	14.987	1.046		$C_{14}H_{12}O$	15.360	1.298
	$C_{13}H_{23}O$	14.455	1.169		$C_{14}H_{14}N$	15.734	1.155
	$C_{13}H_{25}N$	14.829	1.023		$C_{14}H_{28}$	15.576	1.131
	$C_{14}H_{11}O$	15.344	1.296		$C_{15}H_2N$	16.622	1.293
	$C_{14}H_{13}N$	15.718	1.153		$C_{15}H_{16}$	16.465	1.268
	$C_{14}H_{27}$	15.560	1.129		$C_{16}H_4$	17.354	1.412
	$C_{15}HN$	16.606	1.291	197	$C_8H_9N_2O_4$	9.708	1.225
	$C_{15}H_{15}$	16.449	1.265		$C_8H_{11}N_3O_3$	10.082	1.061
	$C_{16}H_3$	17.338	1.410		$C_8H_{13}N_4O_2$	10.457	0.898

(Contd.)

	M + 1	M + 2			M + 1	M + 2
$C_9HN_4O_2$	11.345	0.989	198	$C_8H_{10}N_2O_4$	9.724	1.227
$C_9H_{11}NO_4$	10.439	1.294		$C_8H_{12}N_3O_3$	10.098	1.062
$C_9H_{13}N_2O_3$	10.814	1.132		$C_8H_{14}N_4O_2$	10.473	0.890
$C_9H_{15}N_3O_2$	11.188	0.972		$C_9H_2N_4O_2$	11.361	0.991
$C_9H_{17}N_4O$	11.562	0.813		$C_9H_{12}NO_4$	10.455	1.295
$C_{10}HN_2O_3$	11.702	1.226		$C_9H_{14}N_2O_3$	10.830	1.134
$C_{10}H_3N_3O_2$	12.077	1.070		$C_9H_{16}N_3O_2$	11.204	0.974
$C_{10}H_5N_4O$	12.451	0.914		$C_9H_{18}N_4O$	11.578	0.815
$C_{10}H_{13}O_4$	11.170	1.367		$C_{10}H_2N_2O_3$	11.718	1.228
$C_{10}H_{15}NO_3$	11.545	1.209		$C_{10}H_4N_3O_2$	12.093	1.072
$C_{10}H_{17}N_2O_2$	11.919	1.051		$C_{10}H_6N_4O$	12.467	0.916
$C_{10}H_{19}N_3O$	12.294	0.896		$C_{10}H_{14}O_4$	11.186	1.369
$C_{10}H_{21}N_4$	12.668	0.741		$C_{10}H_{16}NO_3$	11.561	1.211
$C_{11}HO_4$	12.059	1.465		$C_{10}H_{18}N_2O_2$	11.935	1.053
$C_{11}H_3NO_3$	12.433	1.309		$C_{10}H_{20}N_3O$	122.310	0.898
$C_{11}H_5N_2O_2$	12.808	1.155		$C_{10}H_{22}N_4$	12.684	0.743
$C_{11}H_7N_3O$	13.182	1.003		$C_{11}H_2O_4$	12.075	14.667
$C_{14}H_9N_4$	13.556	0.852		$C_{11}H_4NO_3$	12.449	1.311
$C_{11}H_{17}O_3$	12.276	1.291		$C_{11}H_6N_2O_2$	12.824	1.158
$C_{11}H_{19}NO_2$	12.650	1.136		$C_{11}H_8N_3O$	13.198	1.005
$C_{11}H_{21}N_2O$	13.025	0.983		$C_{11}H_{10}N_4$	13.572	0.854
$C_{11}H_{23}N_3$	13.399	0.831		$C_{11}H_{18}O_3$	12.292	1.293
$C_{12}H_5O_3$	13.164	1.398		$C_{11}H_{20}NO_2$	12.666	1.138
$C_{12}H_7NO_2$	13.539	1.247		$C_{11}H_{22}N_2O$	13.041	0.985
$C_{12}H_9N_2O$	13.913	1.097		$C_{11}H_{24}N_3$	13.415	0.833
$C_{12}H_{11}N_3$	14.288	0.948		$C_{12}H_6O_3$	13.180	1.400
$C_{12}H_{21}O_2$	13.381	1.226		$C_{12}H_8NO_2$	13.555	1.249
$C_{12}H_{23}NO$	13.756	1.076		$C_{12}H_{10}N_2O$	13.929	1.099
$C_{12}H_{25}N_2$	14.130	0.927		$C_{12}H_{12}N_3$	14.304	0.951
$C_{13}H_9O_2$	14.270	1.343		$C_{12}H_{22}O_2$	13.397	1.228
$C_{13}H_{11}NO$	14.644	1.196		$C_{12}H_{24}NO$	13.772	1.078
$C_{13}H_{13}N_2$	15.019	1.050		$C_{12}H_{26}N_2$	14.146	0.929
$C_{13}H_{25}O$	14.487	1.174		$C_{13}H_{10}O_2$	14.286	1.346
$C_{13}H_{27}N$	14.861	1.028		$C_{13}H_{12}NO$	14.660	1.199
$C_{14}HN_2$	15.907	1.182		$C_{13}H_{14}N_2$	15.035	1.053
$C_{14}H_{13}O$	15.376	1.301		$C_{13}H_{26}O$	14.503	1.176
$C_{14}H_{15}N$	15.750	1.158		$C_{13}H_{28}N$	14.877	1.030
$C_{14}H_{29}$	15.592	1.134		$C_{14}H_2N_2$	15.923	1.485
$C_{15}HO$	16.264	1.436		$C_{14}H_{14}O$	15.392	1.303
$C_{15}H_3N$	16.638	1.296		$C_{14}H_{16}N$	15.766	1.160
$C_{15}H_{17}$	16.481	1.271		$C_{14}H_{30}$	15.608	1.136
$C_{16}H_5$	17.370	1.415		$C_{15}H_2O$	16.280	1.438
				$C_{15}H_4N$	16.654	1.299

(Contd.)

		M + 1	**M + 2**			**M + 1**	**M + 2**
	$C_{15}H_{18}$	16.497	1.273		$C_{14}H_3N_2$	15.939	1.187
	$C_{16}H_6$	17.386	1.418		$C_{14}H_{15}O$	15.408	1.306
199	$C_8H_{11}N_2O_4$	9.740	1.228		$C_{14}H_{17}N$	15.782	1.163
	$C_8H_{13}N_3O_3$	10.114	1.064		$C_{15}H_3O$	16.296	1.441
	$C_8H_{15}N_4O_2$	10.489	0.901		$C_{15}H_5N$	16.670	1.301
	$C_9HN_3O_3$	11.003	1.152		$C_{15}H_{19}$	16.513	1.276
	$C_9H_3N_4O_2$	11.377	0.993		$C_{16}H_7$	17.402	1.421
	$C_9H_{13}NO_4$	10.471	1.297	200	$C_8H_{12}N_2O_4$	9.756	1.230
	$C_9H_{15}N_2O_3$	10.846	1.136		$C_8H_{14}N_3O_3$	10.130	1.066
	$C_9H_{17}N_3O_2$	11.220	0.976		$C_8H_{16}N_4O_2$	10.505	0.903
	$C_9H_{19}N_4O$	11.594	0.817		$C_9H_2N_3O_3$	11.019	1.154
	$C_{10}HNO_4$	11.360	1.388		$C_9H_4N_4O_2$	11.393	0.995
	$C_{10}H_3N_2O_3$	11.734	1.230		$C_9H_{14}NO_4$	10.487	1.299
	$C_{10}H_5N_3O_2$	12.109	1.074		$C_9H_{16}N_2O_3$	10.862	1.137
	$C_{10}H_7N_4O$	12.483	0.918		$C_9H_{18}N_3O_2$	11.236	0.977
	$C_{10}H_{15}O_4$	11.202	1.371		$C_9H_{20}N_4O$	11.610	0.819
	$C_{10}H_{17}NO_3$	11.577	1.212		$C_{10}H_2NO_4$	11.376	1.390
	$C_{10}H_{19}N_2O_2$	11.951	1.055		$C_{10}H_4N_2O_3$	11.750	1.232
	$C_{10}H_{21}N_3O$	12.326	0.890		$C_{10}H_6N_3O_2$	12.125	1.075
	$C_{10}H_{23}N_4$	12.700	0.745		$C_{10}H_8N_3O$	12.499	0.920
	$C_{11}H_2O_4$	12.091	1.469		$C_{10}H_{16}O_4$	11.218	1.373
	$C_{11}H_5NO_3$	12.465	1.313		$C_{10}H_{18}NO_3$	11.593	1.214
	$C_{11}H_7N_2O_2$	12.840	1.160		$C_{10}H_{20}N_2O_2$	11.967	1.057
	$C_{11}H_9N_3O$	13.214	1.007		$C_{10}H_{22}N_3O$	12.342	0.901
	$C_{11}H_{11}N_4$	13.588	0.856		$C_{10}H_{24}N_4$	12.716	0.747
	$C_{11}H_{19}O_3$	12.308	1.295		$C_{11}H_4O_4$	12.107	1.471
	$C_{11}H_{21}NO_2$	12.682	1.140		$C_{11}H_6NO_3$	12.481	1.315
	$C_{11}H_{23}N_2O$	13.057	0.987		$C_{11}H_8N_2O_2$	12.856	1.162
	$C_{11}H_{25}N_3$	13.431	0.836		$C_{11}H_{10}N_3O$	13.230	1.009
	$C_{12}H_7O_3$	13.197	1.402		$C_{11}H_{12}N_4$	13.604	0.858
	$C_{12}H_9NO_2$	13.571	1.251		$C_{11}H_{20}O_3$	12.324	1.297
	$C_{12}H_{11}N_2O$	13.945	1.101		$C_{11}H_{22}NO_2$	12.698	1.142
	$C_{12}H_{13}N_3$	14.320	0.953		$C_{11}H_{24}N_2O$	13.073	0.989
	$C_{12}H_{23}O_2$	13.413	1.231		$C_{11}H_{26}N_3$	13.447	0.838
	$C_{12}H_{25}NO$	13.788	1.080		$C_{12}H_8O_3$	13.213	1.404
	$C_{12}H_{27}N_2$	14.162	0.931		$C_{12}H_{10}NO_2$	13.587	1.253
	$C_{13}HN_3$	15.208	1.078		$C_{12}H_{12}N_2O$	13.961	1.104
	$C_{13}H_{11}O_2$	14.302	1.348		$C_{12}H_{14}N_3$	13.336	0.955
	$C_{13}H_{13}NO$	14.676	1.201		$C_{12}H_{24}O_2$	13.429	1.233
	$C_{13}H_{15}N_2$	15.051	1.055		$C_{12}H_{26}NO$	13.804	1.082
	$C_{13}H_{27}O$	14.519	1.179		$C_{12}H_{28}N_2$	14.178	0.934
	$C_{13}H_{29}N$	14.893	1.032		$C_{13}H_2N_3$	15.224	1.081
	$C_{14}HNO$	15.565	1.329		$C_{13}H_{12}O_2$	14.318	1.350

(Contd.)

		M + 1	M + 2			M + 1	M + 2
	$C_{13}H_{14}NO$	14.692	1.203		$C_{12}H_{15}N_3$	14.352	0.958
	$C_{13}H_{16}N_2$	15.067	1.058		$C_{12}H_{25}O_2$	13.445	1.235
	$C_{13}H_{28}O$	14.535	1.181		$C_{12}H_{27}NO$	13.820	1.085
	$C_{14}H_2NO$	15.581	1.332		$C_{13}HN_2O$	14.866	1.228
	$C_{14}H_4N_2$	15.955	1.190		$C_{13}H_3N_3$	15.240	1.083
	$C_{14}H_{16}O$	15.424	1.308		$C_{13}H_{13}O_2$	14.334	1.352
	$C_{14}H_{18}N$	15.798	1.165		$C_{13}H_{15}NO$	14.708	1.206
	$C_{15}H_4O$	16.312	1.443		$C_{13}H_{17}N_2$	15.083	1.060
	$C_{15}H_6N$	16.686	1.304		$C_{14}HO_2$	15.223	1.478
	$C_{15}H_{20}$	16.529	1.279		$C_{14}H_3NO$	15.597	1.334
	$C_{18}H_8$	17.418	1.423		$C_{14}H_5N_2$	15.971	1.192
201	$C_8H_{13}N_2O_4$	9.772	1.231		$C_{14}H_{17}O$	15.440	1.311
	$C_8H_{15}N_3O_3$	10.146	1.067		$C_{14}H_{19}N$	15.814	1.168
	$C_8H_{17}N_4O_2$	10.521	0.905		$C_{15}H_5O$	16.328	1.446
	$C_9HN_2O_4$	10.661	1.316		$C_{15}H_7N$	16.702	1.307
	$C_9H_3N_3O_3$	11.035	1.156		$C_{15}H_{21}$	16.545	1.281
	$C_9H_5N_4O_2$	11.409	0.996		$C_{16}H_9$	17.434	1.426
	$C_9H_{15}NO_4$	10.503	1.300	202	$C_8H_{14}N_2O_4$	9.788	1.233
	$C_9H_{17}N_2O_3$	10.878	1.139		$C_8H_{16}N_3O_3$	10.162	1.069
	$C_9H_{19}N_3O_2$	11.252	0.979		$C_8H_{18}N_4O_2$	10.537	0.906
	$C_9H_{21}N_4O$	11.626	0.821		$C_9H_2N_2O_4$	10.677	1.318
	$C_{10}H_3NO_4$	11.392	1.392		$C_9H_4N_3O_3$	11.051	1.157
	$C_{10}H_5N_2O_3$	11.766	1.234		$C_9H_6N_4O_2$	11.425	0.998
	$C_{10}H_7N_3O_2$	12.141	1.077		$C_9H_{16}NO_4$	10.519	1.302
	$C_{10}H_9N_4O$	12.515	0.922		$C_9H_{18}N_2O_3$	10.894	1.141
	$C_{10}H_{17}O_4$	11.234	1.375		$C_9H_{20}N_3O_2$	11.268	0.981
	$C_{10}H_{19}NO_3$	11.609	1.216		$C_9H_{22}N_4O$	11.642	0.823
	$C_{10}H_{21}N_2O_2$	11.983	1.059		$C_{10}H_4NO_4$	11.408	1.394
	$C_{10}H_{23}N_3O$	12.358	0.903		$C_{10}H_6N_2O_3$	11.782	1.236
	$C_{10}H_{25}N_4$	12.732	0.749		$C_{10}H_8N_3O_2$	12.157	1.079
	$C_{11}H_5O_4$	12.123	1.473		$C_{10}H_{10}N_4O$	12.531	0.924
	$C_{11}H_7NO_3$	12.497	1.317		$C_{10}H_{18}O_4$	11.250	1.376
	$C_{11}H_9N_2O_2$	12.872	1.164		$C_{10}H_{20}NO_3$	11.625	1.218
	$C_{11}H_{11}N_3O$	13.246	1.011		$C_{10}H_{22}N_2O_2$	11.999	1.061
	$C_{11}H_{13}N_4$	13.620	0.860		$C_{10}H_{24}N_3O$	12.374	0.905
	$C_{11}H_{21}O_3$	12.340	1.299		$C_{10}H_{26}N_4$	12.748	0.751
	$C_{11}H_{23}NO_2$	12.714	1.144		$C_{11}H_6O_4$	12.139	1.474
	$C_{11}H_{25}N_2O$	13.089	0.991		$C_{11}H_8NO_3$	12.513	1.319
	$C_{11}H_{27}N_3$	13.463	0.840		$C_{11}H_{10}N_2O_2$	12.888	1.166
	$C_{12}HN_4$	14.509	0.980		$C_{11}H_{12}N_3O$	13.262	1.014
	$C_{12}H_9O_3$	13.229	1.406		$C_{11}H_{14}N_4$	13.636	0.863
	$C_{12}H_{11}NO_2$	13.603	1.255		$C_{11}H_{22}O_3$	12.356	1.301
	$C_{12}H_{13}N_2O$	13.977	1.106		$C_{11}H_{21}NO_2$	12.730	1.146

(Contd.)

	M + 1	M + 2		M + 1	M + 2
$C_{11}H_{26}N_2O$	13.105	0.993	$C_{11}H_{15}N_4$	13.652	0.865
$C_{12}H_2N_4$	14.525	0.982	$C_{11}H_{23}O_3$	13.372	1.303
$C_{12}H_{10}O_3$	13.245	1.408	$C_{11}H_{25}NO_2$	12.746	1.148
$C_{12}H_{12}NO_2$	13.619	1.258	$C_{12}HN_3O$	14.167	1.132
$C_{12}H_{14}N_2O$	13.993	1.108	$C_{12}H_3N_4$	14.541	0.984
$C_{12}H_{16}N_3$	14.368	0.960	$C_{12}H_{11}O_3$	13.261	1.411
$C_{12}H_{26}O_2$	13.461	1.237	$C_{12}H_{13}NO_2$	13.635	1.260
$C_{13}H_2N_2O$	14.882	1.231	$C_{12}H_{15}N_2O$	14.009	1.110
$C_{13}H_4N_3$	15.256	1.086	$C_{12}H_{17}N_3$	14.384	0.962
$C_{13}H_{14}O_2$	14.350	1.355	$C_{13}HNO_2$	14.523	1.379
$C_{13}H_{16}NO$	14.724	1.208	$C_{13}H_3N_2O$	14.898	1.233
$C_{13}H_{18}N_2$	15.099	1.063	$C_{13}H_5N_3$	15.272	1.088
$C_{14}H_2O_2$	15.239	1.480	$C_{13}H_{15}O_2$	14.366	1.357
$C_{14}H_4NO$	15.613	1.337	$C_{13}H_{17}NO$	14.740	1.210
$C_{14}H_6N_2$	15.987	1.195	$C_{13}H_{19}N_2$	15.115	1.065
$C_{14}H_{18}O$	15.456	1.313	$C_{14}H_3O_2$	15.255	1.483
$C_{14}H_{20}N$	15.830	1.170	$C_{14}H_5NO$	15.629	1.339
$C_{15}H_6O$	16.344	1.449	$C_{14}H_7N_2$	16.003	1.197
$C_{15}H_8N$	16.718	1.309	$C_{14}H_{19}O$	15.472	1.316
$C_{15}H_{22}$	16.561	1.284	$C_{14}H_{21}N$	15.486	1.173
$C_{16}H_{10}$	17.450	1.429	$C_{15}H_7O$	16.360	1.451
203 $C_8H_{15}N_2O_4$	9.804	1.234	$C_{15}H_9N$	16.734	1.312
$C_8H_{17}N_3O_3$	10.178	1.071	$C_{15}H_{23}$	16.577	1.286
$C_8H_{19}N_4O_2$	10.553	0.908	$C_{16}H_{11}$	17.466	1.432
$C_9H_3N_2O_4$	10.693	1.320	204 $C_8H_{16}N_2O_4$	9.820	1.236
$C_9H_5N_3O_3$	11.067	1.159	$C_8H_{18}N_3O_3$	10.194	1.072
$C_9H_7N_4O_2$	11.441	1.000	$C_8H_{20}N_4O_2$	10.569	0.910
$C_9H_{17}NO_4$	10.535	1.304	$C_9H_4N_2O_4$	10.709	1.321
$C_9H_{19}N_2O_3$	10.910	1.143	$C_9H_6N_3O_3$	11.083	1.161
$C_9H_{21}N_3O_2$	11.284	0.983	$C_9H_8N_4O_2$	11.457	1.002
$C_9H_{23}N_4O$	11.658	0.825	$C_9H_{18}NO_4$	10.551	1.305
$C_{10}H_5NO_4$	11.424	1.395	$C_9H_{20}N_2O_3$	10.926	1.144
$C_{10}H_7N_2O_3$	11.798	1.238	$C_9H_{22}N_3O_2$	11.300	0.985
$C_{10}H_9N_3O_2$	12.173	1.081	$C_9H_{24}N_4O$	11.674	0.826
$C_{10}H_{11}N_4O$	12.547	0.926	$C_{10}H_6NO_4$	11.440	1.397
$C_{10}H_{19}O_4$	11.266	1.378	$C_{10}H_8N_2O_3$	11.814	1.239
$C_{10}H_{21}NO_3$	11.641	1.220	$C_{10}H_{10}N_3O_2$	12.189	1.083
$C_{10}H_{23}N_2O_2$	12.015	1.063	$C_{10}H_{12}N_4O$	12.563	0.928
$C_{10}H_{25}N_3O$	12.390	0.907	$C_{10}H_{20}O_4$	11.282	1.380
$C_{11}H_7O_4$	12.155	1.476	$C_{10}H_{22}NO_3$	11.657	1.222
$C_{11}H_9NO_3$	12.529	1.321	$C_{10}H_{24}N_2O_2$	12.031	1.065
$C_{11}H_{11}N_2O_2$	12.904	1.168	$C_{11}H_8O_4$	12.171	1.478
$C_{11}H_{13}N_3O$	13.278	1.016	$C_{11}H_{10}NO_3$	12.545	1.323

(Contd.)

	M + 1	M + 2			M + 1	M + 2
	$C_{11}H_{12}N_2O_2$	12.920	1.170		$C_{11}H_{13}N_2O_2$	12.936	1.172
	$C_{11}H_{14}N_3O$	13.294	1.018		$C_{11}H_{15}N_3O$	13.310	1.020
	$C_{11}H_{16}N_4$	13.668	0.867		$C_{11}H_{17}N_4$	13.684	0.869
	$C_{11}H_{24}O_3$	12.388	1.305		$C_{12}HN_2O_2$	13.824	1.285
	$C_{12}H_2N_3O$	14.183	1.134		$C_{12}H_3N_3O$	14.199	1.136
	$C_{12}H_4N_4$	14.557	0.987		$C_{12}H_5N_4$	14.573	0.989
	$C_{12}H_{12}O_3$	13.277	1.413		$C_{12}H_{13}O_3$	13.293	1.415
	$C_{12}H_{14}NO_2$	13.651	1.262		$C_{12}H_{15}NO_2$	13.667	1.264
	$C_{12}H_{16}N_2O$	14.025	1.113		$C_{12}H_{17}N_2O$	14.041	1.148
	$C_{12}H_{18}N_3$	14.400	0.965		$C_{12}H_{19}N_3$	14.416	0.967
	$C_{13}H_2NO_2$	14.539	1.381		$C_{13}HO_3$	14.181	1.531
	$C_{13}H_4N_2O$	14.914	1.235		$C_{13}H_3NO_2$	14.555	1.384
	$C_{13}H_6N_3$	15.288	1.091		$C_{13}H_5N_2O$	14.930	1.238
	$C_{13}H_{16}O_2$	14.382	1.359		$C_{13}H_7N_3$	15.304	1.093
	$C_{13}H_{18}NO$	14.756	1.213		$C_{13}H_{17}O_2$	14.398	13.615
	$C_{13}H_{20}N_2$	15.131	1.067		$C_{13}H_{19}NO$	14.772	1.215
	$C_{14}H_4O_2$	15.271	1.485		$C_{13}H_{21}N_2$	15.147	1.070
	$C_{14}H_6NO$	15.645	1.642		$C_{14}H_5O_2$	15.287	1.488
	$C_{14}H_8N_2$	16.019	1.200		$C_{14}H_7NO$	15.287	1.488
	$C_{14}H_{20}O$	15.488	1.318		$C_{14}H_9N_2$	16.035	1.203
	$C_{14}H_{22}N$	15.862	1.176		$C_{14}H_{21}O$	15.504	1.321
	$C_{15}H_8O$	16.376	1.454		$C_{14}H_{23}N$	15.878	1.478
	$C_{15}H_{10}N$	16.750	1.315		$C_{15}H_9O$	16.392	1.456
	$C_{15}H_{24}$	16.593	1.289		$C_{15}H_{11}N$	16.766	1.317
	$C_{16}H_{12}$	17.482	1.437		$C_{15}H_{25}$	16.609	1.292
205	$C_8H_{17}N_2O_4$	9.836	1.237		$C_{16}H_{13}$	17.498	1.437
	$C_8H_{19}N_3O_3$	10.210	1.074		$C_{17}H$	18.386	1.591
	$C_8H_{21}N_4O_2$	10.585	0.912	206	$C_8H_{18}N_2O_4$	9.852	1.239
	$C_9H_5N_2O_4$	10.725	1.323		$C_8H_{20}N_3O_3$	10.226	1.075
	$C_9H_7N_3O_3$	11.099	1.163		$C_8H_{22}N_4O_2$	10.601	0.913
	$C_9H_9N_4O_2$	11.473	1.004		$C_9H_6N_2O_4$	10.741	1.325
	$C_9H_{19}NO_4$	10.567	1.307		$C_9H_8N_3O_3$	11.115	1.164
	$C_9H_{21}N_2O_3$	10.942	1.146		$C_9H_{10}N_4O_2$	11.489	1.006
	$C_9H_{23}N_3O_2$	11.316	0.987		$C_9H_{20}NO_4$	10.583	1.309
	$C_{10}H_7NO_4$	11.546	1.399		$C_9H_{22}N_2O_3$	10.958	1.148
	$C_{10}H_9N_2O_3$	11.830	1.241		$C_{10}H_8NO_4$	11.472	1.401
	$C_{10}H_{11}N_3O_2$	12.205	1.085		$C_{10}H_{10}N_2O_3$	11.846	1.243
	$C_{10}H_{13}N_4O$	12.579	0.930		$C_{10}H_{12}N_3O_2$	12.221	1.087
	$C_{10}H_{21}O_4$	11.298	1.382		$C_{10}H_{14}N_4O$	12.595	0.932
	$C_{10}H_{23}NO_3$	11.673	1.224		$C_{10}H_{22}O_4$	11.314	1.384
	$C_{11}HN_4O$	13.467	1.040		$C_{11}H_2N_4O$	13.483	1.042
	$C_{11}H_9O_4$	12.187	1.480		$C_{11}H_{10}O_4$	12.203	1.482
	$C_{11}H_{11}NO_3$	12.561	1.325		$C_{11}H_{12}NO_3$	12.577	1.327

(Contd.)

		M + 1	M + 2			M + 1	M + 2
	$C_{11}H_{14}N_2O_2$	12.952	1.174		$C_{11}H_{19}N_4$	13.716	0.874
	$C_{11}H_{16}N_3O$	13.326	1.022		$C_{12}HNO_3$	13.482	1.439
	$C_{11}H_{18}N_4$	13.700	0.871		$C_{12}H_3N_2O_2$	13.856	1.289
	$C_{12}H_2N_2O_2$	13.840	1.287		$C_{12}H_5N_3O$	14.231	1.141
	$C_{12}H_4N_3O$	14.215	1.139		$C_{12}H_7N_4$	14.605	0.994
	$C_{12}H_6N_4$	14.589	0.991		$C_{12}H_{15}O_3$	13.325	1.419
	$C_{12}H_{14}O_3$	13.309	1.417		$C_{12}H_{17}NO_2$	13.699	1.268
	$C_{12}H_{16}NO_2$	13.683	1.266		$C_{12}H_{19}N_2O$	14.073	1.119
	$C_{12}H_{18}N_2O$	14.057	1.117		$C_{12}H_{21}N_3$	14.448	0.971
	$C_{12}H_{20}N_3$	14.432	0.969		$C_{13}H_3O_3$	14.213	1.536
	$C_{13}H_2O_3$	14.197	1.533		$C_{13}H_5NO_2$	14.587	1.388
	$C_{13}H_4NO_2$	14.571	1.386		$C_{13}H_7N_2O$	14.962	1.242
	$C_{13}H_6N_2O$	14.946	1.240		$C_{13}H_9N_3$	15.336	1.098
	$C_{13}H_8N_3$	15.320	1.095		$C_{13}H_{19}O_2$	14.430	1.366
	$C_{13}H_{18}O_2$	14.414	1.364		$C_{13}H_{21}NO$	14.804	1.220
	$C_{13}H_{20}NO$	14.788	1.217		$C_{13}H_{23}N_2$	15.179	1.075
	$C_{13}H_{22}N_2$	15.163	1.072		$C_{14}H_7O_2$	15.319	1.493
	$C_{14}H_6O_2$	15.303	1.490		$C_{14}H_9NO$	15.693	1.349
	$C_{14}H_8NO$	15.677	1.347		$C_{14}H_{11}N_2$	16.067	1.208
	$C_{14}H_{10}N_2$	16.051	1.205		$C_{14}H_{23}O$	15.536	1.326
	$C_{14}H_{22}O$	15.520	1.323		$C_{14}H_{25}N$	15.910	1.183
	$C_{14}H_{24}N$	15.894	1.181		$C_{15}H_{11}O$	16.424	1.462
	$C_{15}H_{10}O$	16.408	1.459		$C_{15}H_{13}N$	16.798	1.323
	$C_{15}H_{12}N$	16.782	1.320		$C_{15}H_{27}$	16.641	1.297
	$C_{15}H_{26}$	16.625	1.294		$C_{16}HN$	17.687	1.470
	$C_{16}H_{14}$	17.514	1.440		$C_{16}H_{15}$	17.530	1.443
	$C_{17}H_2$	18.402	1.594		$C_{17}H_3$	18.418	1.597
207	$C_8H_{19}N_2O_4$	9.868	1.241	208	$C_8H_{20}N_2O_4$	9.884	1.242
	$C_8H_{21}N_3O_3$	10.242	1.077		$C_9H_8N_2O_4$	10.773	1.328
	$C_9H_7N_2O_4$	10.757	1.326		$C_9H_{10}N_3O_3$	11.147	1.168
	$C_9H_9N_3O_3$	11.131	1.166		$C_9H_{12}N_4O_2$	11.521	1.009
	$C_9H_{11}N_4O_2$	11.505	1.007		$C_{10}H_{10}NO_4$	11.504	1.405
	$C_9H_{21}NO_4$	10.599	1.310		$C_{10}H_{12}N_2O_3$	11.878	1.247
	$C_{10}H_9NO_4$	11.488	1.403		$C_{10}H_{14}N_3O_2$	12.253	1.091
	$C_{10}H_{11}N_2O_3$	11.862	1.245		$C_{10}H_{16}N_4O$	12.627	0.936
	$C_{10}H_{13}N_3O_2$	12.237	1.089		$C_{11}H_2N_3O_2$	13.141	1.198
	$C_{11}H_{15}N_4O$	12.611	0.934		$C_{11}H_4N_4O$	13.516	1.047
	$C_{11}HN_3O_2$	13.125	1.196		$C_{11}H_{12}O_4$	12.235	1.486
	$C_{11}H_3N_4O$	13.500	1.045		$C_{11}H_{14}NO_3$	12.609	1.331
	$C_{11}H_{11}O_4$	12.219	1.484		$C_{11}H_{16}N_2O_2$	12.984	1.178
	$C_{11}H_{13}NO_3$	12.593	1.329		$C_{11}H_{18}N_3O$	13.358	1.026
	$C_{11}H_{15}N_2O_2$	12.968	1.176		$C_{11}H_{20}N_4$	13.732	0.876
	$C_{11}H_{17}N_3O$	13.342	1.024		$C_{12}H_2NO_3$	13.498	1.442

(Contd.)

	M + 1	M + 2			M + 1	M + 2
	$C_{12}H_4N_2O_2$	13.872	1.292		$C_{12}H_5N_2O_2$	13.888	1.294
	$C_{12}H_6N_3O$	14.247	1.143		$C_{12}H_7N_3O$	14.263	1.145
	$C_{12}H_8N_4$	14.621	0.996		$C_{12}H_9N_4$	14.637	0.998
	$C_{12}H_{16}O_3$	13.341	1.421		$C_{12}H_{17}O_3$	13.357	1.423
	$C_{12}H_{18}NO_2$	13.715	1.271		$C_{12}H_{19}NO_2$	13.731	1.273
	$C_{12}H_{20}N_2O$	14.089	1.122		$C_{12}H_{21}N_2O$	14.105	1.124
	$C_{12}H_{22}N_3$	14.464	0.974		$C_{12}H_{23}N_3$	14.480	0.976
	$C_{13}H_4O_3$	14.229	1.538		$C_{13}H_5O_3$	14.245	1.540
	$C_{13}H_6NO_2$	14.603	1.391		$C_{13}H_7NO_2$	14.619	1.393
	$C_{13}H_8N_2O$	14.978	1.245		$C_{13}H_9N_2O$	14.994	1.247
	$C_{13}H_{10}N_3$	15.352	1.100		$C_{13}H_{11}N_3$	15.368	1.103
	$C_{13}H_{20}O_2$	14.446	1.369		$C_{13}H_{21}O_2$	14.462	1.371
	$C_{13}H_{22}NO$	14.820	1.222		$C_{13}H_{23}NO$	14.836	1.224
	$C_{13}H_{24}N_2$	15.195	1.077		$C_{13}H_{25}N_2$	15.211	1.080
	$C_{14}H_8O_2$	15.335	1.495		$C_{14}H_9O_2$	15.351	1.497
	$C_{14}H_{10}NO$	15.709	1.652		$C_{14}H_{11}NO$	15.725	1.354
	$C_{14}H_{12}N_2$	16.083	1.210		$C_{14}H_{13}N_2$	16.099	1.213
	$C_{14}H_{24}O$	15.552	1.328		$C_{14}H_{25}O$	15.568	1.331
	$C_{14}H_{26}N$	15.926	1.186		$C_{14}H_{27}N$	15.942	1.188
	$C_{15}H_{12}O$	16.440	1.464		$C_{15}HN_2$	16.988	1.354
	$C_{15}H_{14}N$	16.814	1.325		$C_{15}H_{13}O$	16.456	1.467
	$C_{15}H_{28}$	16.657	1.300		$C_{15}H_{15}N$	16.830	1.328
	$C_{16}H_2N$	17.703	1.473		$C_{15}H_{29}$	16.673	1.302
	$C_{16}H_{16}$	17.456	1.446		$C_{16}HO$	17.345	1.611
	$C_{17}H_4$	18.434	1.599		$C_{16}H_3N$	17.719	1.476
209	$C_9H_9N_2O_4$	10.789	1.330		$C_{16}H_{17}$	17.562	1.449
	$C_9H_{11}N_3O_3$	11.163	1.170		$C_{17}H_5$	18.450	1.602
	$C_9H_{13}N_4O_2$	11.537	1.011	210	$C_9H_{10}N_2O_4$	10.805	1.332
	$C_{10}HN_4O_2$	12.426	1.412		$C_9H_{12}N_3O_3$	11.179	1.172
	$C_{10}H_{11}NO_4$	11.520	1.406		$C_9H_{14}N_4O_2$	11.553	1.013
	$C_{10}H_{13}N_2O_3$	11.894	1.249		$C_{10}H_2N_4O_2$	12.442	1.114
	$C_{10}H_{15}N_3O_2$	12.269	1.093		$C_{10}H_{12}NO_4$	11.536	1.408
	$C_{10}H_{17}N_4O$	12.643	0.938		$C_{10}H_{14}N_2O_3$	11.910	1.251
	$C_{11}HN_2O_3$	12.783	1.353		$C_{10}H_{16}N_3O_2$	12.285	1.095
	$C_{11}H_3N_3O_2$	13.157	1.200		$C_{10}H_{18}N_4O$	12.659	0.940
	$C_{11}H_5N_4O$	13.532	1.049		$C_{11}H_2N_2O_3$	12.799	1.355
	$C_{11}H_{13}O_4$	12.251	1.488		$C_{11}H_4N_3O_2$	13.173	1.202
	$C_{11}H_{15}NO_3$	12.625	1.333		$C_{11}H_6N_4O$	13.548	1.051
	$C_{11}H_{17}N_2O_2$	13.000	1.180		$C_{11}H_{14}O_4$	12.267	1.190
	$C_{11}H_{19}N_3O$	13.374	1.028		$C_{11}H_{16}NO_3$	12.641	1.336
	$C_{11}H_{21}N_4$	13.748	0.878		$C_{11}H_{18}N_2O_2$	13.016	1.182
	$C_{12}HO_4$	13.140	1.595		$C_{11}H_{20}N_3O$	13.390	1.031
	$C_{12}H_3NO_3$	13.514	1.444		$C_{11}H_{22}N_4$	13.764	0.880

(Contd.)

	M + 1	M + 2			M + 1	M + 2
$C_{12}H_2O_4$	13.156	1.597		$C_{11}H_{17}NO_3$	12.657	1.338
$C_{12}H_4NO_3$	13.530	1.446		$C_{11}H_{19}N_2O_2$	13.032	1.184
$C_{12}H_6N_2O_2$	13.904	1.296		$C_{11}H_{21}N_3O$	13.406	1.033
$C_{12}H_8N_3O$	14.279	1.148		$C_{11}H_{23}N_4$	13.780	0.882
$C_{12}H_{10}N_4$	14.653	1.001		$C_{12}H_3O_4$	13.172	1.599
$C_{12}H_{18}O_3$	13.373	1.426		$C_{12}H_5NO_3$	13.546	1.448
$C_{12}H_{20}NO_2$	13.747	1.275		$C_{12}H_7N_2O_2$	13.920	1.298
$C_{12}H_{22}N_2O$	14.121	1.126		$C_{12}H_9N_3O$	14.295	1.150
$C_{12}H_{24}N_3$	14.496	0.978		$C_{12}H_{11}N_4$	14.669	1.603
$C_{13}H_6O_3$	14.261	1.542		$C_{12}H_{19}O_3$	13.389	1.428
$C_{13}H_8NO_2$	14.635	1.395		$C_{12}H_{21}NO_2$	13.763	1.277
$C_{13}H_{10}N_2O$	15.010	1.250		$C_{12}H_{23}N_2O$	14.137	1.128
$C_{13}H_{12}N_3$	15.384	1.105		$C_{12}H_{25}N_3$	14.512	0.981
$C_{13}H_{22}O_2$	14.478	1.373		$C_{13}H_7O_3$	14.277	1.545
$C_{13}H_{24}NO$	14.852	1.227		$C_{13}H_9NO_2$	14.651	1.398
$C_{13}H_{26}N_2$	15.227	1.082		$C_{13}H_{11}N_2O$	15.026	1.252
$C_{14}H_{10}O_2$	15.367	1.500		$C_{13}H_{13}N_3$	15.400	1.108
$C_{14}H_{12}NO$	15.741	1.357		$C_{13}H_{23}O_2$	14.494	1.375
$C_{14}H_{14}N_2$	16.115	1.215		$C_{13}H_{25}NO$	14.868	1.229
$C_{14}H_{26}O$	15.584	1.333		$C_{13}H_{27}N_2$	15.243	1.084
$C_{14}H_{28}N$	15.958	1.191		$C_{14}HN_3$	16.289	1.243
$C_{15}H_2N_2$	17.004	1.357		$C_{14}H_{11}O_2$	15.383	1.502
$C_{15}H_{14}O$	16.472	1.470		$C_{14}H_{13}NO$	15.757	1.359
$C_{15}H_{16}N$	16.846	1.331		$C_{14}H_{15}N_2$	16.131	1.218
$C_{15}H_{30}$	16.689	1.305		$C_{14}H_{27}O$	15.600	1.335
$C_{16}H_2O$	17.361	1.614		$C_{14}H_{29}N$	15.974	1.193
$C_{16}H_4N$	17.735	1.479		$C_{15}HNO$	16.646	1.498
$C_{16}H_{18}$	17.578	1.451		$C_{15}H_3N_2$	17.020	1.359
$C_{17}H_6$	18.466	1.606		$C_{15}H_{15}O$	16.488	1.472
211 $C_9H_{11}N_2O_4$	10.821	1.333		$C_{15}H_{17}N$	16.863	1.333
$C_9H_{13}N_3O_3$	11.195	1.173		$C_{15}H_{31}$	16.705	1.308
$C_9H_{15}N_4O_2$	11.569	1.015		$C_{16}H_3O$	17.377	1.617
$C_{10}HN_3O_3$	12.084	1.271		$C_{16}H_5N$	17.751	1.481
$C_{10}H_3N_4O_2$	12.458	1.116		$C_{16}H_{19}$	17.594	1.454
$C_{10}H_{13}NO_4$	11.552	1.410		$C_{17}H_7$	18.482	1.609
$C_{10}H_{15}N_2O_3$	11.926	1.253	212	$C_9H_{12}N_2O_4$	10.837	1.335
$C_{10}H_{17}N_3O_2$	12.301	1.097		$C_9H_{14}N_3O_3$	11.211	1.175
$C_{10}H_{19}N_4O$	12.675	0.942		$C_9H_{16}N_4O_2$	11.585	1.017
$C_{11}HNO_4$	12.440	1.511		$C_{10}H_2N_3O_3$	12.100	1.273
$C_{11}H_3N_2O_3$	12.815	1.357		$C_{10}H_4N_4O_2$	12.474	1.118
$C_{11}H_5N_3O_2$	13.189	1.204		$C_{10}H_{14}NO_4$	11.568	1.412
$C_{11}H_7N_4O$	13.564	1.053		$C_{10}H_{16}N_2O_3$	11.942	1.255
$C_{11}H_{15}O_4$	12.283	1.492		$C_{10}H_{18}N_3O_2$	12.317	1.099

(Contd.)

		M + 1	M + 2		M + 1	M + 2
	$C_{10}H_{20}N_4O$	12.691	0.945	$C_9H_{17}N_4O_2$	11.601	1.018
	$C_{11}H_2NO_4$	12.456	1.513	$C_{10}HN_2O_4$	11.741	1.431
	$C_{11}H_4N_2O_3$	12.831	1.359	$C_{10}H_3N_3O_2$	12.116	1.275
	$C_{11}H_6N_3O_2$	13.205	1.206	$C_{10}H_5N_4O_2$	12.490	1.120
	$C_{11}H_8N_4O$	13.580	1.055	$C_{10}H_{15}NO_4$	11.584	1.414
	$C_{11}H_{16}O_4$	12.299	1.494	$C_{10}H_{17}N_2O_3$	11.958	1.257
	$C_{14}H_{18}NO_3$	12.673	1.340	$C_{10}H_{19}N_3O_2$	12.333	1.101
	$C_{11}H_{20}N_2O_2$	13.048	1.187	$C_{10}H_{21}N_4O$	12.707	0.947
	$C_{11}H_{22}N_3O$	13.422	1.035	$C_{11}H_3NO_4$	12.472	1.515
	$C_{11}H_{24}N_4$	13.796	0.885	$C_{11}H_5N_2O_3$	12.847	1.361
	$C_{12}H_4O_4$	13.188	1.601	$C_{11}H_7N_3O_2$	13.221	1.209
	$C_{12}H_6NO_3$	13.562	1.450	$C_{11}H_9N_4O$	13.596	1.058
	$C_{12}H_8N_2O_2$	13.936	1.301	$C_{11}H_{17}O_4$	12.315	1.496
	$C_{12}H_{10}N_3O$	14.311	1.152	$C_{11}H_{19}NO_3$	12.689	1.342
	$C_{12}H_{12}N_4$	14.685	1.005	$C_{11}H_{21}N_2O_2$	13.064	1.189
	$C_{12}H_{20}O_3$	13.405	1.430	$C_{11}H_{23}N_3O$	13.438	1.037
	$C_{12}H_{22}NO_2$	13.779	1.279	$C_{11}H_{25}N_4$	13.812	0.887
	$C_{12}H_{24}N_2O$	14.153	1.131	$C_{12}H_5O_4$	13.204	1.604
	$C_{12}H_{26}N_3$	14.528	0.983	$C_{12}H_7NO_3$	13.578	1.452
	$C_{13}H_8O_3$	14.293	1.547	$C_{12}H_9N_2O_2$	13.952	1.303
	$C_{13}H_{10}NO_2$	14.667	1.400	$C_{12}H_{11}N_3O$	14.327	1.155
	$C_{13}H_{12}N_2O$	15.042	1.254	$C_{12}H_{13}N_4$	14.701	1.008
	$C_{13}H_{14}N_3$	15.416	1.110	$C_{12}H_{21}O_3$	13.421	1.432
	$C_{13}H_{24}O_2$	14.510	1.378	$C_{12}H_{23}NO_2$	13.795	1.282
	$C_{13}H_{26}NO$	14.884	1.232	$C_{12}H_{25}N_2O$	14.169	1.133
	$C_{13}H_{28}N_2$	15.259	1.087	$C_{12}H_{27}N_3$	14.544	0.985
	$C_{14}H_2N_3$	16.305	1.245	$C_{13}HN_4$	15.590	1.136
	$C_{14}H_{12}O_2$	15.399	1.505	$C_{13}H_9N_3$	14.309	1.549
	$C_{14}H_{14}NO$	15.773	1.362	$C_{13}H_{11}NO_2$	14.683	1.402
	$C_{14}H_{16}N_2$	16.147	1.221	$C_{13}H_{13}N_2O$	15.058	1.257
	$C_{14}H_{28}O$	15.616	1.338	$C_{13}H_{15}N_3$	15.432	1.113
	$C_{14}H_{30}N$	15.990	1.196	$C_{13}H_{25}O_2$	14.526	1.380
	$C_{15}H_2NO$	16.662	1.500	$C_{13}H_{27}NO$	14.900	1.234
	$C_{15}H_4N_2$	17.036	1.362	$C_{13}H_{29}N_2$	15.275	1.089
	$C_{15}H_{16}O$	16.504	1.475	$C_{14}HN_2O$	15.946	1.389
	$C_{15}H_{18}N$	16.879	1.336	$C_{14}H_3N_3$	16.321	1.248
	$C_{15}H_{32}$	16.721	1.310	$C_{14}H_{13}O_2$	15.415	1.507
	$C_{16}H_4O$	17.393	1.620	$C_{14}H_{15}NO$	15.789	1.565
	$C_{16}H_6N$	17.767	1.484	$C_{14}H_{17}N_2$	16.163	1.223
	$C_{16}H_{20}$	17.610	1.457	$C_{14}H_{29}O$	15.632	1.340
	$C_{17}H_8$	18.498	1.612	$C_{14}H_{31}N$	16.006	1.198
213	$C_9H_{13}N_2O_4$	10.853	1.337	$C_{15}HO_2$	16.303	1.642
	$C_9H_{15}N_3O_3$	11.227	1.177	$C_{15}H_3NO$	16.678	1.503

(Contd.)

	M + 1	M + 2			M + 1	M + 2
$C_{15}H_5N_2$	17.052	1.565		$C_{14}H_2N_2O$	15.962	1.391
$C_{15}H_{17}O$	16.520	1.478		$C_{14}H_4N_3$	16.337	1.251
$C_{15}H_{19}N$	16.895	1.339		$C_{14}H_{14}O_2$	15.431	1.510
$C_{16}H_5O$	17.409	1.622		$C_{14}H_{16}NO$	15.805	1.367
$C_{16}H_7N$	17.783	1.487		$C_{14}H_{18}N_2$	16.179	1.226
$C_{16}H_{21}$	17.626	1.460		$C_{14}H_{30}O$	15.648	1.343
$C_{17}H_9$	18.514	1.615		$C_{15}H_2O_2$	16.319	1.645
214 $C_9H_{14}N_2O_4$	10.869	1.339		$C_{15}H_4NO$	16.694	1.506
$C_9H_{16}N_3O_3$	11.243	1.179		$C_{15}H_6N_2$	17.068	1.368
$C_9H_{18}N_4O_2$	11.617	1.020		$C_{15}H_{18}O$	16.536	1.480
$C_{10}H_2N_2O_4$	11.757	1.433		$C_{15}H_{20}N$	16.911	1.342
$C_{10}H_4N_3O_3$	12.132	1.277		$C_{16}H_6O$	17.425	1.625
$C_{10}H_6N_4O_2$	12.506	1.122		$C_{16}H_8N$	17.799	1.490
$C_{10}H_{16}NO_4$	11.600	1.416		$C_{16}H_{22}$	17.642	1.463
$C_{10}H_{18}N_2O_3$	11.974	1.258		$C_{17}H_{20}$	18.530	1.618
$C_{10}H_{20}N_3O_2$	12.349	1.103	215	$C_9H_{15}N_2O_4$	10.885	1.340
$C_{10}H_{22}N_4O$	12.723	0.949		$C_9H_{17}N_3O_3$	11.259	1.181
$C_{11}H_4NO_4$	12.488	1.517		$C_9H_{19}N_4O_2$	11.633	1.022
$C_{11}H_6N_2O_3$	12.863	1.363		$C_{10}H_3N_2O_4$	11.773	1.435
$C_{11}H_8N_3O_2$	13.237	1.211		$C_{10}H_5N_3O_3$	12.148	1.279
$C_{11}H_{10}N_4O$	13.612	1.060		$C_{10}H_7N_4O_2$	12.522	1.124
$C_{11}H_{18}O_4$	12.331	1.498		$C_{10}H_{17}NO_4$	11.616	1.417
$C_{11}H_{20}NO_3$	12.705	1.344		$C_{10}H_{19}N_2O_3$	11.990	1.260
$C_{11}H_{22}N_2O_2$	13.080	1.191		$C_{10}H_{21}N_3O_2$	12.365	1.105
$C_{11}H_{24}N_3O$	13.454	1.039		$C_{10}H_{23}N_4O$	12.739	0.951
$C_{11}H_{26}N_4$	13.828	0.889		$C_{11}H_5NO_4$	12.504	1.519
$C_{12}H_6O_4$	13.220	1.606		$C_{11}H_7N_2O_3$	12.879	1.365
$C_{12}H_8NO_3$	13.594	1.455		$C_{11}H_9N_3O_2$	13.253	1.213
$C_{12}H_{10}N_2O_2$	13.968	1.305		$C_{11}H_{11}N_4O$	13.628	1.062
$C_{12}H_{12}N_3O$	14.343	1.157		$C_{11}H_{19}O_4$	12.347	1.500
$C_{12}H_{14}N_4$	14.717	1.010		$C_{11}H_{21}NO_3$	12.721	1.346
$C_{12}H_{22}O_3$	13.437	1.434		$C_{11}H_{23}N_2O_2$	13.096	1.193
$C_{12}H_{24}NO_2$	13.811	1.284		$C_{11}H_{25}N_3O$	13.470	1.041
$C_{12}H_{26}N_2O$	14.185	1.135		$C_{11}H_{27}N_4$	13.844	0.891
$C_{12}H_{28}N_3$	14.560	0.988		$C_{12}H_7O_4$	13.236	1.608
$C_{13}H_2N_4$	15.606	1.139		$C_{12}H_9NO_3$	13.610	1.457
$C_{13}H_{10}O_3$	14.325	1.552		$C_{12}H_{11}N_2O_2$	13.984	1.307
$C_{13}H_{12}NO_2$	14.699	1.405		$C_{12}H_{13}N_3O$	14.359	1.159
$C_{13}H_{14}N_2O$	15.074	1.259		$C_{12}H_{15}N$	14.733	1.012
$C_{13}H_{16}N_3$	15.448	1.151		$C_{12}H_{23}O_3$	13.453	14.362
$C_{13}H_{26}O_2$	14.542	1.382		$C_{12}H_{25}NO_2$	13.827	1.286
$C_{13}H_{28}NO$	14.916	1.236		$C_{12}H_{27}N_2O$	14.201	1.137
$C_{13}H_{30}N_2$	15.291	1.092		$C_{12}H_{29}N_3$	14.576	0.990

(Contd.)

	M + 1	M + 2		M + 1	M + 2
$C_{13}HN_3O$	15.247	1.285	$C_{12}H_{10}NO_3$	13.626	1.459
$C_{13}H_3N_4$	15.622	1.141	$C_{12}H_{12}N_2O_2$	14.000	1.310
$C_{13}H_{11}O_3$	14.341	1.554	$C_{12}H_{14}N_3O$	14.375	1.161
$C_{13}H_{13}NO_2$	14.715	1.407	$C_{12}H_{16}N_4$	14.749	1.015
$C_{13}H_{15}N_2O$	15.090	1.262	$C_{12}H_{24}O_3$	13.469	1.438
$C_{13}H_{17}N_3$	15.464	1.118	$C_{12}H_{26}NO_2$	13.843	1.288
$C_{13}H_{27}O_2$	14.558	1.385	$C_{12}H_{28}N_2O$	14.217	1.140
$C_{13}H_{29}NO$	14.932	1.239	$C_{13}H_2N_2O$	15.263	1.287
$C_{14}HNO_2$	15.604	1.536	$C_{13}H_4N_4$	15.638	1.144
$C_{14}H_3N_2O$	15.978	1.394	$C_{13}H_{12}O_3$	14.357	1.556
$C_{14}H_5N_3$	16.353	1.253	$C_{13}H_{14}NO_2$	14.731	1.409
$C_{14}H_{15}O_2$	15.447	1.512	$C_{13}H_{16}N_2O$	15.106	1.264
$C_{14}H_{17}NO$	15.821	1.370	$C_{13}H_{18}N_3$	15.480	1.120
$C_{14}H_{19}N_2$	16.195	1.228	$C_{13}H_{28}O_2$	14.574	1.387
$C_{15}H_3O_2$	16.335	1.648	$C_{14}H_2NO_2$	15.620	1.538
$C_{15}H_5NO$	16.710	1.508	$C_{14}H_4N_2O$	15.994	1.396
$C_{15}H_7N_2$	17.084	1.370	$C_{14}H_6N_3$	16.369	1.256
$C_{15}H_{19}O$	16.552	1.483	$C_{14}H_{16}O_2$	15.463	1.515
$C_{15}H_{21}N$	16.927	1.344	$C_{14}H_{18}NO$	15.837	1.372
$C_{17}H_7O$	17.441	1.628	$C_{14}H_{20}N_2$	16.211	1.231
$C_{16}H_9N$	17.815	1.493	$C_{15}H_4O_2$	16.351	1.650
$C_{16}H_{23}$	17.658	1.466	$C_{15}H_6NO$	16.726	1.511
$C_{17}H_{11}$	18.546	1.621	$C_{15}H_8N_2$	17.100	1.373
216 $C_9H_{16}N_2O_4$	10.901	1.342	$C_{15}H_{20}O$	16.568	1.485
$C_9H_{18}N_3O_3$	11.275	1.182	$C_{15}H_{22}N$	16.943	1.347
$C_9H_{20}N_4O_2$	11.649	1.024	$C_{16}H_8O$	17.457	1.631
$C_{10}H_4N_2O_4$	11.789	1.437	$C_{16}H_{10}N$	17.831	1.496
$C_{10}H_6N_3O_3$	12.164	1.281	$C_{16}H_{24}$	17.674	1.468
$C_{10}H_8N_4O_2$	12.538	1.126	$C_{17}H_{12}$	18.562	1.624
$C_{10}H_{18}NO_4$	11.632	1.419	217 $C_9H_{17}N_2O_4$	10.917	1.344
$C_{10}H_{20}N_2O_3$	12.006	1.262	$C_9H_{19}N_3O_3$	11.291	1.184
$C_{10}H_{22}N_3O_2$	12.381	1.107	$C_9H_{21}N_4O_2$	11.665	1.026
$C_{10}H_{24}N_4O$	12.755	0.953	$C_{10}H_5N_2O_4$	11.805	1.439
$C_{11}H_6NO_4$	12.520	1.521	$C_{10}H_7N_3O_3$	12.180	1.283
$C_{11}H_8N_2O_3$	12.895	1.367	$C_{10}H_9N_4O_2$	12.554	1.128
$C_{11}H_{10}N_3O_2$	13.269	1.215	$C_{10}H_{19}NO_4$	11.648	1.421
$C_{11}H_{12}N_4O$	13.644	1.064	$C_{10}H_{21}N_2O_3$	12.022	1.264
$C_{11}H_{20}O_4$	12.363	1.502	$C_{10}H_{23}N_3O_2$	12.397	1.109
$C_{11}H_{22}NO_3$	12.737	1.348	$C_{10}H_{25}N_4O$	12.771	0.955
$C_{11}H_{24}N_2O_2$	13.112	1.195	$C_{11}H_7NO_4$	12.536	1.523
$C_{11}H_{26}N_3O$	13.486	1.043	$C_{11}H_9N_2O_3$	12.911	1.369
$C_{11}H_{28}N_4$	13.860	0.893	$C_{11}H_{11}N_3O_2$	13.285	1.217
$C_{12}H_8O_4$	13.252	1.610	$C_{11}H_{13}N_4O$	13.660	1.066

(Contd.)

	M + 1	**M + 2**		**M + 1**	**M + 2**
$C_{11}H_{21}O_4$	12.379	1.504	$C_{10}H_{22}N_2O_3$	12.038	1.266
$C_{11}H_{23}NO_3$	12.753	1.350	$C_{10}H_{24}N_3O_2$	12.413	1.111
$C_{11}H_{25}N_2O_2$	13.128	1.197	$C_{10}H_{26}N_4O$	12.787	0.957
$C_{11}H_{27}N_3O$	13.502	1.046	$C_{11}H_8NO_4$	12.552	1.525
$C_{12}HN_4O$	14.548	1.186	$C_{11}H_{10}N_2O_3$	12.927	1.371
$C_{12}H_9O_4$	13.268	1.612	$C_{11}H_{12}N_3O_2$	13.301	1.219
$C_{12}H_{11}NO_3$	13.642	1.461	$C_{11}H_{14}N_4O$	13.676	1.068
$C_{12}H_{13}N_2O_2$	14.016	1.312	$C_{11}H_{22}O_4$	12.395	1.506
$C_{12}H_{15}N_3O$	14.391	1.164	$C_{11}H_{24}NO_3$	12.769	1.352
$C_{12}H_{17}N_4$	14.765	1.017	$C_{11}H_{26}N_2O_2$	13.144	1.199
$C_{12}H_{25}O_3$	13.485	1.441	$C_{12}H_2N_4O$	14.564	1.188
$C_{12}H_{27}NO_2$	13.859	1.291	$C_{12}H_{10}O_4$	13.284	1.614
$C_{13}HN_2O_2$	14.905	1.435	$C_{12}H_{12}NO_3$	13.658	1.463
$C_{13}H_3N_3O$	15.279	1.290	$C_{12}H_{14}N_2O_2$	14.032	1.314
$C_{13}H_5N_4$	15.654	1.146	$C_{12}H_{16}N_3O$	14.407	1.166
$C_{13}H_{13}O_3$	14.373	1.558	$C_{12}H_{18}N_4$	14.781	1.019
$C_{13}H_{15}NO_2$	14.747	1.412	$C_{12}H_{26}O_3$	13.501	1.443
$C_{13}H_{17}N_2O$	15.122	1.266	$C_{13}H_2N_2O_2$	14.921	1.437
$C_{13}H_{19}N_3$	15.496	1.123	$C_{13}H_4N_3O$	15.295	1.292
$C_{14}HO_3$	15.262	1.684	$C_{13}H_6N_4$	15.670	1.149
$C_{14}H_3NO_2$	15.636	1.541	$C_{13}H_{14}O_3$	14.389	1.561
$C_{14}H_5N_2O$	16.010	1.399	$C_{13}H_{16}NO_2$	14.763	1.414
$C_{14}H_7N_3$	16.385	1.258	$C_{13}H_{18}N_2O$	15.138	1.269
$C_{14}H_{17}O_2$	15.479	1.517	$C_{13}H_{20}N_3$	15.512	1.125
$C_{14}H_{19}NO$	15.583	1.375	$C_{14}H_2O_3$	15.278	1.687
$C_{14}H_{21}N_2$	16.227	1.233	$C_{14}H_4NO_2$	15.652	1.543
$C_{15}H_5O_2$	16.367	1.653	$C_{14}H_6N_2O$	16.026	1.402
$C_{15}H_7NO$	16.742	1.514	$C_{14}H_8N_3$	16.401	1.261
$C_{15}H_9N_2$	17.116	1.376	$C_{14}H_{18}O_2$	15.495	1.520
$C_{15}H_{21}O$	16.584	1.488	$C_{14}H_{20}NO$	15.869	1.377
$C_{15}H_{23}N$	16.959	1.350	$C_{14}H_{22}N_2$	16.243	1.236
$C_{16}H_9O$	17.473	1.634	$C_{15}H_6O_2$	16.383	1.655
$C_{16}H_{11}N$	17.847	1.498	$C_{15}H_8NO$	16.758	1.516
$C_{16}H_{25}$	17.690	1.471	$C_{15}H_{10}N_2$	17.132	1.379
$C_{17}H_{13}$	18.578	1.627	$C_{15}H_{22}O$	16.600	1.491
$C_{18}H$	19.467	1.790	$C_{15}H_{24}N$	16.975	1.352
218 $C_9H_{18}N_2O_4$	10.933	1.346	$C_{16}H_{10}O$	17.489	1.636
$C_9H_{20}N_3O_3$	11.307	1.186	$C_{16}H_{12}N$	17.863	1.501
$C_9H_{22}N_4O_2$	11.681	1.028	$C_{16}H_{26}$	17.706	1.474
$C_{10}H_6N_2O_4$	11.821	1.441	$C_{17}H_{14}$	18.594	1.630
$C_{10}H_8N_3O_3$	12.196	1.285	$C_{18}H_2$	19.483	1.793
$C_{10}H_{10}N_4O_2$	12.570	1.130	219 $C_9H_{19}N_2O_4$	10.949	1.347
$C_{10}H_{20}NO_4$	11.664	1.423	$C_9H_{21}N_3O_3$	11.323	1.188

(Contd.)

	M + 1	M + 2			M + 1	M + 2
$C_9H_{23}N_4O_2$	11.697	1.030		$C_{17}HN$	18.768	1.661
$C_{10}H_7N_2O_4$	11.837	1.443		$C_{17}H_{15}$	18.610	1.632
$C_{10}H_9N_3O_3$	12.212	1.287		$C_{18}H_3$	19.499	1.796
$C_{10}H_{11}N_4O_2$	12.586	1.132	220	$C_9H_{20}N_2O_4$	10.965	1.349
$C_{10}H_{21}NO_4$	11.680	1.425		$C_9H_{22}N_3O_3$	11.339	1.190
$C_{10}H_{23}N_2O_3$	12.054	1.268		$C_9H_{24}N_4O_2$	11.713	1.032
$C_{10}H_{25}N_3O_2$	12.429	1.113		$C_{10}H_8N_2O_4$	11.853	1.445
$C_{11}H_9NO_4$	12.568	1.527		$C_{10}H_{10}N_3O_3$	12.228	1.288
$C_{11}H_{11}N_2O_3$	12.943	1.373		$C_{10}H_{12}N_4O_2$	12.602	1.134
$C_{11}H_{13}N_3O_2$	13.317	1.221		$C_{10}H_{22}NO_4$	11.696	1.427
$C_{11}H_{15}N_4O$	13.692	1.071		$C_{10}H_{24}N_2O_3$	12.070	1.270
$C_{11}H_{23}O_4$	12.411	1.508		$C_{11}H_{10}NO_4$	12.584	1.529
$C_{11}H_{25}NO_3$	12.785	1.354		$C_{11}H_{12}N_2O_3$	12.959	1.375
$C_{12}HN_3O_2$	14.206	1.338		$C_{11}H_{14}N_3O_2$	13.333	1.234
$C_{12}H_3N_4O$	14.580	1.190		$C_{11}H_{16}N_4O$	13.708	1.073
$C_{12}H_{11}O_4$	13.300	1.616		$C_{11}H_{24}O_4$	12.427	1.510
$C_{12}H_{13}NO_3$	13.674	1.466		$C_{12}H_2N_3O_2$	14.222	1.340
$C_{12}H_{15}N_2O_2$	14.048	1.316		$C_{12}H_4N_4O$	14.596	1.193
$C_{12}H_{17}N_3O$	14.423	1.168		$C_{12}H_{12}O_4$	13.316	1.618
$C_{12}H_{19}N_4$	14.797	1.022		$C_{12}H_{14}NO_3$	13.690	1.468
$C_{13}HNO_3$	14.563	1.585		$C_{12}H_{16}N_2O_2$	14.064	1.318
$C_3H_3N_2O_2$	14.937	1.439		$C_{12}H_{18}N_3O$	14.439	1.171
$C_{13}H_5N_3O$	15.311	1.295		$C_{12}H_{20}N_4$	14.813	1.024
$C_{13}H_7N_4$	15.686	1.151		$C_{13}H_2NO_3$	14.579	1.588
$C_{13}H_{15}O_3$	14.405	1.563		$C_{13}H_4N_2O_2$	14.953	1.442
$C_{13}H_{17}NO_2$	14.779	1.417		$C_{13}H_6N_3O$	15.327	1.297
$C_{13}H_{19}N_2O$	15.154	1.271		$C_{13}H_8N_4$	15.702	1.154
$C_{13}H_{21}N_3$	15.528	1.128		$C_{13}H_{16}O_3$	14.421	1.565
$C_{14}H_3O_3$	15.294	1.689		$C_{13}H_{18}NO_2$	14.795	1.419
$C_{14}H_5NO_2$	15.668	1.546		$C_{13}H_{20}N_2O$	15.170	1.274
$C_{14}H_7N_2O$	16.042	1.404		$C_{13}H_{22}N_3$	15.544	1.130
$C_{14}H_9N_3$	16.417	1.264		$C_{14}H_4O_3$	15.310	1.692
$C_{14}H_{19}O_2$	15.511	1.522		$C_{14}H_6NO_2$	15.684	1.548
$C_{14}H_{21}NO$	15.885	1.380		$C_{14}H_8N_2O$	16.058	1.407
$C_{14}H_{23}N_2$	16.259	1.239		$C_{14}H_{10}N_3$	16.433	1.267
$C_{15}H_7O_2$	16.399	1.658		$C_{14}H_{20}O_2$	15.527	1.525
$C_{15}H_9NO$	16.774	1.519		$C_{14}H_{22}NO$	15.901	1.382
$C_{15}H_{11}N_2$	17.148	1.381		$C_{14}H_{24}N_2$	16.275	1.241
$C_{15}H_{23}O$	16.616	1.493		$C_{15}H_8O_2$	16.415	1.661
$C_{15}H_{25}N$	16.991	1.355		$C_{15}H_{10}NO$	16.790	1.522
$C_{16}H_{11}O$	17.505	1.639		$C_{15}H_{12}N_2$	17.164	1.384
$C_{16}H_{13}N$	17.879	1.504		$C_{15}H_{24}O$	16.632	1.496
$C_{16}H_{27}$	17.722	1.477		$C_{15}H_{26}N$	17.007	1.358

(Contd.)

		M + 1	M + 2			M + 1	M + 2
	$C_{16}H_{12}O$	17.521	1.642		$C_{15}H_{13}N_2$	17.180	1.387
	$C_{16}H_{14}N$	17.895	1.507		$C_{15}H_{25}O$	16.648	1.499
	$C_{16}H_{28}$	17.738	1.480		$C_{15}H_{27}N$	17.023	1.361
	$C_{17}H_2N$	18.784	1.664		$C_{16}HN_2$	18.069	1.538
	$C_{17}H_{16}$	18.626	1.635		$C_{16}H_{13}O$	17.537	1.645
	$C_{18}H_4$	19.515	1.799		$C_{16}H_{15}N$	17.911	1.510
221	$C_9H_{21}N_2O_4$	10.981	1.351		$C_{16}H_{29}$	17.754	1.483
	$C_9H_{23}N_3O_3$	11.355	1.191		$C_{18}HO$	18.425	1.799
	$C_{10}H_9N_2O_4$	11.869	1.446		$C_{18}H_3N$	18.800	1.667
	$C_{10}H_{11}N_3O_3$	12.244	1.290		$C_{18}H_5$	19.531	1.802
	$C_{10}H_{13}N_4O_2$	12.618	1.136	222	$C_9H_{22}N_2O_4$	10.997	1.353
	$C_{10}H_{23}NO_4$	11.712	1.429		$C_{10}H_{10}N_2O_4$	11.885	1.448
	$C_{11}HN_4O_2$	13.507	1.246		$C_{10}H_{12}N_3O_3$	12.260	1.292
	$C_{11}H_{11}NO_4$	12.600	1.531		$C_{10}H_{14}N_4O_2$	12.634	1.138
	$C_{11}H_{13}N_2O_3$	12.975	1.378		$C_{11}H_2N_4O_2$	13.523	1.248
	$C_{11}H_{15}N_3O_2$	13.349	1.226		$C_{11}H_{12}NO_4$	12.616	1.533
	$C_{11}H_{17}N_4O$	13.724	1.075		$C_{11}H_{14}N_2O_3$	12.991	1.380
	$C_{12}HN_2O_3$	13.863	1.491		$C_{11}H_{16}N_3O_2$	13.365	1.228
	$C_{12}H_3N_3O_2$	14.238	1.342		$C_{11}H_{18}N_4O$	13.740	1.077
	$C_{12}H_5N_4O$	14.612	1.195		$C_{12}H_2N_2O_3$	13.879	1.493
	$C_{12}H_{13}O_4$	13.332	1.621		$C_{12}H_4N_3O_2$	14.254	1.345
	$C_{12}H_{15}NO_3$	13.706	1.470		$C_{12}H_6N_4O$	14.628	1.197
	$C_{12}H_{17}N_2O_2$	14.080	1.321		$C_{12}H_{14}O_4$	13.348	1.623
	$C_{12}H_{19}N_3O$	14.455	1.173		$C_{12}H_{16}NO_3$	13.722	1.472
	$C_{12}H_{21}N_4$	14.829	1.027		$C_{12}H_{18}N_2O_2$	14.096	1.323
	$C_{13}HO_4$	14.220	1.737		$C_{12}H_{20}N_3O$	14.471	1.175
	$C_{13}H_3NO_3$	14.595	1.590		$C_{12}H_{22}N_4$	14.845	1.029
	$C_{13}H_5N_2O_2$	14.969	1.444		$C_{13}H_2O_4$	14.236	1.739
	$C_{13}H_7N_3O$	15.343	1.300		$C_{13}H_4NO_3$	14.611	1.592
	$C_{13}H_9N_4$	15.718	1.156		$C_{13}H_6N_2O_2$	14.985	1.446
	$C_{13}H_{17}O_3$	14.437	1.568		$C_{13}H_8N_3O$	15.359	1.302
	$C_{13}H_{19}NO_2$	14.811	1.421		$C_{13}H_{10}N_4$	15.734	1.159
	$C_{13}H_{21}N_2O$	15.186	1.276		$C_{13}H_{18}O_3$	14.453	1.570
	$C_{13}H_{23}N_3$	15.560	1.133		$C_{13}H_{20}NO_2$	14.827	1.424
	$C_{14}H_5O_3$	15.326	1.694		$C_{13}H_{22}N_2O$	15.202	1.279
	$C_{14}H_7NO_2$	15.700	1.551		$C_{13}H_{24}N_3$	15.576	1.135
	$C_{14}H_9N_2O$	16.074	1.409		$C_{14}H_6O_3$	15.342	1.697
	$C_{14}H_{11}N_3$	16.449	1.269		$C_{14}H_8NO_2$	15.716	1.553
	$C_{14}H_{21}O_2$	15.543	1.527		$C_{14}H_{10}N_2O$	16.090	1.412
	$C_{14}H_{23}NO$	15.917	1.385		$C_{14}H_{12}N_3$	16.465	1.272
	$C_{14}H_{25}N_2$	16.291	1.244		$C_{14}H_{22}O_2$	15.559	1.530
	$C_{15}H_9O_2$	16.431	1.663		$C_{14}H_{24}NO$	15.933	1.387
	$C_{15}H_{11}NO$	16.806	1.524		$C_{14}H_{26}N_2$	16.307	1.246

(Contd.)

		M + 1	M + 2			M + 1	M + 2
	$C_{15}H_{10}O_2$	16.447	1.666		$C_{14}H_{13}N_3$	16.481	1.274
	$C_{15}H_{12}NO$	16.822	1.527		$C_{14}H_{23}O_2$	15.575	1.532
	$C_{15}H_{14}N_2$	17.196	1.390		$C_{14}H_{25}NO$	15.949	1.390
	$C_{15}H_{26}O$	16.664	1.501		$C_{14}H_{27}N_2$	16.323	1.249
	$C_{15}H_{28}N$	17.039	1.363		$C_{15}HN_3$	17.369	1.419
	$C_{16}H_2N_2$	18.085	1.540		$C_{15}H_{11}O_2$	16.463	1.669
	$C_{16}H_{14}O$	17.553	1.648		$C_{15}H_{13}NO$	16.838	1.530
	$C_{16}H_{16}N$	17.927	1.513		$C_{15}H_{15}N_2$	17.212	1.392
	$C_{16}H_{30}$	17.770	1.485		$C_{15}H_{27}O$	16.680	1.504
	$C_{17}H_2O$	18.441	1.802		$C_{15}H_{29}N$	17.055	1.366
	$C_{17}H_4N$	18.816	1.670		$C_{16}HNO$	17.726	1.677
	$C_{17}H_{18}$	18.658	1.641		$C_{16}H_3N_2$	18.101	1.543
	$C_{18}H_6$	19.547	1.805		$C_{16}H_{15}O$	17.569	1.650
223	$C_{10}H_{11}N_2O_4$	11.901	1.450		$C_{16}H_{17}N$	17.943	1.516
	$C_{10}H_{13}N_3O_3$	12.276	1.294		$C_{16}H_{31}$	17.786	1.488
	$C_{10}H_{15}N_4O_2$	12.650	1.140		$C_{17}H_3O$	18.457	1.805
	$C_{11}HN_3O_3$	13.164	1.401		$C_{17}H_5N$	18.832	1.673
	$C_{11}H_3N_4O_2$	13.539	1.250		$C_{17}H_{19}$	18.674	1.644
	$C_{11}H_{13}NO_4$	12.632	1.535		$C_{18}H_7$	19.563	1.808
	$C_{11}H_{15}N_2O_3$	13.007	1.382	224	$C_{10}H_{12}N_2O_4$	11.917	1.452
	$C_{11}H_{17}N_3O_2$	13.381	1.230		$C_{10}H_{14}N_3O_3$	12.292	1.296
	$C_{11}H_{19}N_4O$	13.756	1.079		$C_{10}H_{16}N_4O_2$	12.666	1.142
	$C_{12}HNO_4$	13.521	1.645		$C_{11}H_2N_3O_3$	13.180	1.404
	$C_{12}H_3N_2O_3$	13.895	1.495		$C_{11}H_4N_4O_2$	13.555	1.252
	$C_{12}H_5N_3O_2$	14.270	1.347		$C_{11}H_{14}NO_4$	12.648	1.537
	$C_{12}H_7N_4O$	14.644	1.200		$C_{11}H_{16}N_2O_3$	13.023	1.384
	$C_{12}H_{15}O_4$	13.364	1.625		$C_{11}H_{18}N_3O_2$	13.397	1.232
	$C_{12}H_{17}NO_3$	13.738	1.474		$C_{11}H_{20}N_4O$	13.772	1.082
	$C_{12}H_{19}N_2O_2$	14.112	1.325		$C_{12}H_2NO_4$	13.537	1.647
	$C_{12}H_{21}N_3O$	14.487	1.178		$C_{12}H_4N_2O_3$	13.911	1.498
	$C_{12}H_{23}N_4$	14.861	1.031		$C_{12}H_6N_3O_2$	14.286	1.349
	$C_{13}H_3O_4$	14.252	1.742		$C_{12}H_8N_4O$	14.660	1.202
	$C_{13}H_5NO_3$	14.627	1.595		$C_{12}H_{16}O_4$	13.380	1.627
	$C_{13}H_7N_2O_2$	15.001	1.449		$C_{12}H_{18}NO_3$	13.754	1.477
	$C_{13}H_9N_3O$	15.375	1.304		$C_{12}H_{20}N_2O_2$	14.128	1.328
	$C_{13}H_{11}N_4$	15.750	1.161		$C_{12}H_{22}N_3O$	14.503	1.180
	$C_{13}H_{19}O_3$	14.469	1.572		$C_{12}H_{24}N_4$	14.877	1.034
	$C_{13}H_{21}NO_2$	14.843	1.426		$C_{13}H_4O_4$	14.268	1.744
	$C_{13}H_{23}N_2O$	15.218	1.281		$C_{13}H_6NO_3$	14.643	1.597
	$C_{13}H_{25}N_3$	15.592	1.138		$C_{13}H_8N_2O_2$	15.017	1.451
	$C_{14}H_7O_4$	15.358	1.699		$C_{13}H_{10}N_3O$	15.391	1.307
	$C_{14}H_9NO_2$	15.732	1.556		$C_{13}H_{12}N_4$	15.766	1.164
	$C_{14}H_{11}N_2O$	16.106	1.414		$C_{13}H_{20}O_3$	14.485	1.575

(Contd.)

	M + 1	M + 2			M + 1	M + 2
	$C_{13}H_{22}NO_2$	14.860	1.428		$C_{12}H_{25}N_4$	14.893	1.036
	$C_{13}H_{24}N_2O$	15.234	1.283		$C_{13}H_5O_4$	14.284	1.746
	$C_{13}H_{26}N_3$	15.608	1.140		$C_{13}H_7NO_2$	14.659	1.599
	$C_{14}H_8O_3$	15.374	1.701		$C_{13}H_9N_2O_2$	15.033	1.454
	$C_{14}H_{10}NO_2$	15.748	1.559		$C_{13}H_{11}N_3O$	15.407	1.309
	$C_{14}H_{12}N_2O$	16.122	1.417		$C_{13}H_{13}N_4$	15.782	1.167
	$C_{14}H_{14}N_3$	16.497	1.277		$C_{13}H_{21}O_3$	14.501	1.577
	$C_{14}H_{24}O_2$	15.591	1.535		$C_{13}H_{23}NO_2$	14.876	1.431
	$C_{14}H_{26}NO$	15.965	1.392		$C_{13}H_{25}N_2O$	15.250	1.286
	$C_{14}H_{28}N_2$	16.339	1.252		$C_{13}H_{27}N_3$	15.624	1.143
	$C_{15}H_2N_3$	17.385	1.422		$C_{14}HN_4$	16.670	1.305
	$C_{15}H_{12}O_2$	16.479	1.671		$C_{14}H_9O_3$	15.390	1.704
	$C_{15}H_{14}NO$	16.854	1.532		$C_{11}H_{11}NO_2$	15.764	1.561
	$C_{15}H_{16}N_2$	17.228	1.395		$C_{14}H_{13}N_2O$	16.138	1.420
	$C_{15}H_{28}O$	16.696	1.507		$C_{14}H_{15}N_3$	16.513	1.279
	$C_{15}H_{30}N$	17.071	1.369		$C_{14}H_{25}O_2$	15.607	1.537
	$C_{16}H_2NO$	17.742	1.680		$C_{14}H_{27}NO$	15.981	1.395
	$C_{16}H_4N_2$	18.117	1.546		$C_{14}H_{29}N_2$	16.355	1.254
	$C_{16}H_{16}O$	17.585	1.653		$C_{15}HN_2O$	17.027	1.561
	$C_{16}H_{18}N$	17.959	1.519		$C_{15}H_3N_3$	17.401	1.424
	$C_{16}H_{32}$	17.802	1.491		$C_{15}H_{13}O_2$	16.495	1.674
	$C_{17}H_4O$	18.473	1.808		$C_{15}H_{15}NO$	16.870	1.535
	$C_{17}H_6N$	18.848	1.676		$C_{15}H_{17}N_2$	17.244	1.398
	$C_{17}H_{20}$	18.690	1.647		$C_{15}H_{29}O$	16.712	1.509
	$C_{18}H_8$	19.579	1.182		$C_{15}H_{31}N$	17.087	1.371
225	$C_{10}H_{13}N_2O_4$	11.933	1.454		$C_{16}HO_2$	17.384	1.819
	$C_{10}H_{15}N_3O_3$	12.308	1.298		$C_{16}H_3NO$	17.758	1.683
	$C_{10}H_{17}N_4O_2$	12.682	1.144		$C_{16}H_5N_2$	18.133	1.549
	$C_{11}HN_2O_4$	12.822	1.558		$C_{16}H_{17}O$	17.601	1.656
	$C_{11}H_3N_3O_3$	13.196	1.406		$C_{16}H_{19}N$	17.975	1.521
	$C_{11}H_5N_4O_2$	13.571	1.255		$C_{16}H_{33}$	17.818	1.494
	$C_{11}H_{15}NO_4$	12.664	1.539		$C_{17}H_5O$	18.489	1.811
	$C_{11}H_{17}N_2O_3$	13.039	1.386		$C_{17}H_7N$	18.864	1.679
	$C_{11}H_{19}N_3O_2$	13.413	1.234		$C_{17}H_{21}$	18.706	1.650
	$C_{11}H_{21}N_4O$	13.788	1.084		$C_{18}H_9$	19.595	1.815
	$C_{12}H_3NO_4$	13.553	1.650	226	$C_{10}H_{14}N_2O_4$	11.949	1.456
	$C_{12}H_5N_2O_3$	13.927	1.500		$C_{10}H_{16}N_3O_3$	12.324	1.300
	$C_{12}H_7N_3O_2$	14.302	1.351		$C_{10}H_{18}N_4O_2$	12.698	1.146
	$C_{12}H_9N_4O$	14.676	1.204		$C_{11}H_2N_2O_4$	12.838	1.560
	$C_{12}H_{17}O_4$	13.396	1.629		$C_{11}H_4N_3O_3$	13.212	1.408
	$C_{12}H_{19}NO_3$	13.770	1.479		$C_{11}H_6N_4O_2$	13.587	1.257
	$C_{12}H_{21}N_2O_2$	14.144	1.330		$C_{11}H_{16}NO_4$	12.680	1.541
	$C_{12}H_{23}N_3O$	14.519	1.182		$C_{11}H_{18}N_2O_3$	13.055	1.388

(Contd.)

	M + 1	M + 2			M + 1	M + 2
$C_{11}H_{20}N_3O_2$	13.429	1.236		$C_{17}H_{22}$	18.722	1.653
$C_{11}H_{22}N_4O$	13.804	1.086		$C_{18}H_{10}$	19.611	1.818
$C_{12}H_4NO_4$	13.569	1.652	227	$C_{10}H_{15}N_2O_4$	11.965	1.458
$C_{12}H_6N_2O_3$	13.943	1.502		$C_{10}H_{17}N_3O_3$	12.340	1.302
$C_{12}H_8N_3O_2$	14.318	1.354		$C_{10}H_{19}N_4O_2$	12.714	1.148
$C_{12}H_{10}N_4O$	14.692	1.207		$C_{11}H_3N_2O_4$	12.854	1.562
$C_{12}H_{18}O_4$	13.412	1.631		$C_{11}H_5N_3O_3$	13.228	1.410
$C_{12}H_{20}NO_3$	13.786	1.481		$C_{11}H_7N_4O_2$	13.603	1.259
$C_{12}H_{22}N_2O_2$	14.160	1.332		$C_{11}H_{17}NO_4$	12.696	1.543
$C_{12}H_{24}N_3O$	14.535	1.185		$C_{11}H_{19}N_2O_3$	13.071	1.390
$C_{12}H_{26}N_4$	14.909	1.038		$C_{11}H_{21}N_3O_2$	13.445	1.238
$C_{13}H_6O_4$	14.300	1.749		$C_{11}H_{23}N_4O$	13.820	1.088
$C_{13}H_8NO_3$	14.675	1.602		$C_{12}H_5NO_4$	13.585	1.654
$C_3H_{10}N_2O_2$	15.049	1.456		$C_{12}H_7N_2O_3$	13.959	1.504
$C_{13}H_{12}N_3O$	15.423	1.312		$C_{12}H_9N_3O_2$	14.334	1.356
$C_{13}H_{14}N_4$	15.798	1.169		$C_{12}H_{11}N_4O$	14.708	1.209
$C_{13}H_{22}O_3$	14.517	1.579		$C_{12}H_{19}O_4$	13.428	1.633
$C_{13}H_{24}NO_2$	14.892	1.433		$C_{12}H_{21}NO_3$	13.802	1.483
$C_{13}H_{26}N_2O$	15.266	1.288		$C_{12}H_{23}N_2O_2$	14.176	1.334
$C_{13}H_{28}N_3$	15.640	1.145		$C_{12}H_{25}N_3O$	14.551	1.187
$C_{14}H_2N_4$	16.686	1.308		$C_{12}H_{27}N_4$	14.925	1.041
$C_{14}H_{10}O_3$	15.406	1.706		$C_{13}H_7O_4$	14.316	1.751
$C_{14}H_{12}NO_2$	15.780	1.564		$C_{13}H_9NO_3$	14.691	1.604
$C_{14}H_{14}N_2O$	16.154	1.422		$C_{13}H_{11}N_2O_2$	15.065	1.458
$C_{14}H_{16}N_3$	16.529	1.282		$C_{13}H_{13}N_3O$	15.439	1.314
$C_{14}H_{26}O_2$	15.623	1.540		$C_{13}H_{15}N_4$	15.814	1.172
$C_{14}H_{28}NO$	15.997	1.398		$C_{13}H_{23}O_3$	14.533	1.582
$C_{14}H_{30}N_2$	16.371	1.257		$C_{13}H_{25}NO_2$	14.908	1.436
$C_{15}H_2N_2O$	17.043	1.564		$C_{13}H_{27}N_2O$	15.282	1.291
$C_{15}H_4N_3$	17.417	1.427		$C_{13}H_{29}N_3$	15.656	1.148
$C_{15}H_{14}O_2$	16.511	1.677		$C_{14}HN_3O$	16.328	1.450
$C_{15}H_{16}NO$	16.886	1.538		$C_{14}H_3N_4$	16.702	1.310
$C_{15}H_{18}N_2$	17.260	1.401		$C_{14}H_{11}O_3$	15.422	1.709
$C_{15}H_{30}O$	16.728	1.512		$C_{14}H_{13}NO_2$	15.796	1.566
$C_{15}H_{32}N$	17.103	1.374		$C_{14}H_{15}N_2O$	16.170	1.425
$C_{16}H_2O_2$	17.400	1.821		$C_{14}H_{17}N_3$	16.545	1.285
$C_{16}H_4NO$	17.774	1.686		$C_{14}H_{27}O_2$	15.639	1.542
$C_{16}H_6N_2$	18.149	1.552		$C_{14}H_{29}NO$	16.013	1.400
$C_{16}H_{18}O$	17.617	1.659		$C_{14}H_{31}N_2$	16.387	1.260
$C_{16}H_{20}N$	17.991	1.524		$C_{15}HNO_2$	16.685	1.705
$C_{16}H_{34}$	17.834	1.497		$C_{15}H_3N_2O$	17.059	1.567
$C_{17}H_6O$	18.505	1.813		$C_{15}H_5N_3$	17.433	1.430
$C_{17}H_8N$	18.880	1.682		$C_{15}H_{15}O_2$	16.527	1.679

(Contd.)

	M + 1	**M + 2**		**M + 1**	**M + 2**
$C_{15}H_{17}NO$	16.902	1.541	$C_{14}H_{12}O_3$	15.438	1.711
$C_{15}H_{19}N_2$	17.276	1.403	$C_{14}H_{14}NO_2$	15.812	1.569
$C_{15}H_{31}O$	16.744	1.515	$C_{14}H_{16}N_2O$	16.186	1.427
$C_{15}H_{33}N$	17.119	1.377	$C_{14}H_{18}N_3$	16.561	1.287
$C_{16}H_3O_2$	17.416	1.824	$C_{14}H_{28}O_2$	15.655	1.545
$C_{16}H_5NO$	17.790	1.689	$C_{14}H_{30}NO$	16.029	1.403
$C_{16}H_7N_2$	18.165	1.555	$C_{14}H_{32}N_2$	16.403	1.262
$C_{16}H_{19}O$	17.633	1.662	$C_{15}H_2NO_2$	16.701	1.707
$C_{16}H_{21}N$	18.007	1.527	$C_{15}H_4N_2O$	17.075	1.569
$C_{17}H_7O$	18.521	1.816	$C_{15}H_6N_3$	17.449	1.433
$C_{17}H_9N$	18.896	1.685	$C_{15}H_{16}O_2$	16.543	1.682
$C_{17}H_{23}$	18.738	1.656	$C_{15}H_{18}NO$	16.918	1.543
$C_{18}H_{11}$	19.627	1.821	$C_{15}H_{20}N_2$	17.292	1.406
228 $C_{10}H_{16}N_2O_4$	11.981	1.460	$C_{15}H_{32}O$	16.760	1.517
$C_{10}H_{18}N_3O_3$	12.356	1.304	$C_{16}H_4O_2$	17.432	1.827
$C_{10}H_{20}N_4O_2$	12.730	1.150	$C_{16}H_6NO$	17.806	1.692
$C_{11}H_4N_2O_4$	12.870	1.564	$C_{16}H_8N_2$	18.181	1.558
$C_{11}H_6N_3O_3$	13.244	1.412	$C_{16}H_{20}O$	17.649	1.665
$C_{11}H_8N_4O_2$	13.619	1.261	$C_{16}H_{22}N$	18.023	1.530
$C_{11}H_{18}NO_4$	12.712	1.545	$C_{17}H_8O$	18.537	1.819
$C_{11}H_{20}N_2O_3$	13.087	1.392	$C_{17}H_{10}N$	18.912	1.688
$C_{11}H_{22}N_3O_2$	13.461	1.241	$C_{17}H_{24}$	18.754	1.659
$C_{11}H_{24}N_4O$	13.836	1.090	$C_{18}H_{12}$	19.643	1.824
$C_{12}H_6NO_4$	13.601	1.656	229 $C_{10}H_{17}N_2O_4$	11.997	1.462
$C_{12}H_8N_2O_3$	13.975	1.507	$C_{10}H_{19}N_3O_3$	12.372	1.306
$C_{12}H_{10}N_3O_2$	14.350	1.358	$C_{10}H_{21}N_4O_2$	12.746	1.152
$C_{12}H_{12}N_4O$	14.724	1.212	$C_{11}H_5N_2O_4$	12.886	1.566
$C_{12}H_{20}O_4$	13.444	1.636	$C_{11}H_7N_3O_3$	13.260	1.414
$C_{12}H_{22}NO_3$	13.818	1.485	$C_{11}H_9N_4O_2$	13.635	1.263
$C_{12}H_{24}N_2O_2$	14.192	1.337	$C_{11}H_{19}NO_4$	12.728	1.547
$C_{12}H_{26}N_3O$	14.567	1.189	$C_{11}H_{21}N_2O_3$	13.103	1.394
$C_{12}H_{28}N_4$	14.941	1.043	$C_{11}H_{23}N_3O_2$	13.477	1.243
$C_{13}H_8O_4$	14.332	1.753	$C_{11}H_{25}N_4O$	13.852	1.093
$C_{13}H_{10}NO_3$	14.707	1.606	$C_{12}H_7NO_4$	13.617	1.658
$C_{13}H_{12}N_2O_2$	15.081	1.461	$C_{12}H_9N_2O_3$	13.991	1.509
$C_{13}H_{14}N_3O$	15.455	1.317	$C_{12}H_{11}N_3O_2$	14.366	1.361
$C_{13}H_{16}N_4$	15.830	1.174	$C_{12}H_{13}N_4O$	14.740	1.214
$C_{13}H_{24}O_3$	14.549	1.584	$C_{12}H_{21}O_4$	13.460	1.638
$C_{13}H_{26}NO_2$	14.924	1.438	$C_{12}H_{23}NO_3$	13.834	1.488
$C_{13}H_{28}N_2O$	15.298	1.293	$C_{12}H_{25}N_2O_2$	14.208	1.339
$C_{13}H_{30}N_3$	15.672	1.150	$C_{12}H_{27}N_3O$	14.583	1.192
$C_{14}H_2N_3O$	16.344	1.452	$C_{12}H_{29}N_4$	14.957	1.046
$C_{14}H_4N_4$	16.718	1.313	$C_{13}HN_4O$	15.629	1.343

(Contd.)

	M + 1	M + 2		M + 1	M + 2
$C_{13}H_9O_4$	14.348	1.755	$C_{11}H_{24}N_3O_2$	13.493	1.245
$C_{13}H_{11}NO_3$	14.723	1.609	$C_{11}H_{26}N_4O$	13.868	1.095
$C_{13}H_{13}N_2O_2$	15.097	1.463	$C_{12}H_8NO_4$	13.633	1.660
$C_{13}H_{15}N_3O$	15.471	1.319	$C_{12}H_{10}N_2O_3$	14.007	1.511
$C_{13}H_{17}N_4$	15.846	1.177	$C_{12}H_{12}N_3O_2$	14.382	1.363
$C_{13}H_{25}O_3$	14.565	1.586	$C_{12}H_{14}N_4O$	14.756	1.216
$C_{13}H_{27}NO_2$	14.940	1.440	$C_{12}H_{22}O_4$	13.476	1.640
$C_{13}H_{29}N_2O$	15.314	1.296	$C_{12}H_{24}NO_3$	13.850	1.490
$C_{13}H_{31}N_3$	15.688	1.153	$C_{12}H_{26}N_2O_2$	14.224	1.341
$C_{14}HN_2O_2$	15.985	1.595	$C_{12}H_{28}N_3O$	14.599	1.194
$C_{14}H_3N_3O$	16.360	1.455	$C_{12}H_{30}N_4$	14.973	1.048
$C_{14}H_5N_4$	16.734	1.316	$C_{13}H_2N_4O$	15.645	1.345
$C_{14}H_{13}O_3$	15.454	1.714	$C_{13}H_{10}O_4$	14.364	1.758
$C_{14}H_{15}NO_2$	15.828	1.571	$C_{13}H_{12}NO_3$	14.739	1.611
$C_{14}H_{17}N_2O$	16.202	1.430	$C_{13}H_{14}N_2O_2$	15.113	1.466
$C_{14}H_{19}N_3$	16.577	1.290	$C_{13}H_{16}N_3O$	15.487	1.322
$C_{14}H_{29}O_2$	15.671	1.547	$C_{13}H_{18}N_4$	15.862	1.179
$C_{14}H_{31}NO$	16.045	1.405	$C_{13}H_{26}O_3$	14.581	1.589
$C_{15}HO_3$	16.342	1.849	$C_{13}H_{28}NO_2$	14.956	1.443
$C_{15}H_3NO_2$	16.717	1.710	$C_{13}H_{30}N_2O$	15.330	1.298
$C_{15}H_5N_2O$	17.091	1.572	$C_{14}H_2N_2O_2$	16.001	1.598
$C_{15}H_7N_3$	17.465	1.435	$C_{14}H_4N_3O$	16.376	1.457
$C_{15}H_{17}O_2$	16.559	1.684	$C_{14}H_6N_4$	16.750	1.318
$C_{15}H_{19}NO$	16.934	1.546	$C_{14}H_{14}O_3$	15.470	1.716
$C_{15}H_{21}N_2$	17.308	1.409	$C_{14}H_{16}NO_2$	15.844	1.574
$C_{16}H_5O_2$	17.448	1.830	$C_{14}H_{18}N_2O$	16.218	1.132
$C_{16}H_7NO$	17.822	1.695	$C_{14}H_{20}N_3$	16.593	1.293
$C_{16}H_9N_2$	18.197	1.561	$C_{14}H_{30}O_2$	15.687	1.550
$C_{16}H_{21}O$	17.665	1.667	$C_{15}H_2O_3$	16.358	1.852
$C_{16}H_{23}N$	18.039	1.533	$C_{15}H_4NO_2$	16.733	1.713
$C_{17}H_9O$	18.553	1.822	$C_{15}H_6N_2O$	17.107	1.575
$C_{17}H_{11}N$	18.928	1.691	$C_{15}H_8N_3$	17.481	1.438
$C_{17}H_{25}$	18.770	1.662	$C_{15}H_{18}O_2$	16.575	1.687
$C_{18}H_{13}$	19.659	1.827	$C_{15}H_{20}NO$	16.950	1.549
$C_{19}H$	20.547	2.000	$C_{15}H_{22}N_2$	17.324	1.412
230 $C_{10}H_{18}N_2O_4$	12.013	1.464	$C_{16}H_6O_2$	17.464	1.832
$C_{10}H_{20}N_3O_3$	12.388	1.308	$C_{16}H_8NO$	17.838	1.697
$C_{10}H_{22}N_4O_2$	12.762	1.154	$C_{16}H_{10}N_2$	18.213	1.564
$C_{11}H_6N_2O_4$	12.902	1.569	$C_{16}H_{22}O$	17.681	1.670
$C_{11}H_8N_3O_3$	13.276	1.116	$C_{16}H_{24}N$	18.055	1.536
$C_{11}H_{10}N_4O_2$	13.651	1.266	$C_{17}H_{10}O$	18.569	1.825
$C_{11}H_{20}NO_4$	12.744	1.549	$C_{17}H_{12}N$	18.944	1.694
$C_{11}H_{22}N_2O_3$	13.119	1.396	$C_{17}H_{26}$	18.786	1.665

(Contd.)

	M + 1	**M + 2**			**M + 1**	**M + 2**
	$C_{18}H_{14}$	19.675	1.830		$C_{15}H_{23}N_2$	17.340	1.414
	$C_{19}H_2$	20.563	2.003		$C_{16}H_7O_2$	17.480	1.835
231	$C_{10}H_{19}N_2O_4$	12.029	1.466		$C_{16}H_9NO$	17.854	1.700
	$C_{10}H_{21}N_3O_3$	12.404	1.310		$C_{16}H_{11}N_2$	18.229	1.567
	$C_{10}H_{23}N_4O_2$	12.778	1.456		$C_{16}H_{23}O$	17.697	1.673
	$C_{11}H_7N_2O_4$	12.918	1.571		$C_{16}H_{25}N$	18.071	1.539
	$C_{11}H_9N_3O_3$	13.292	1.418		$C_{17}H_{11}O$	18.585	1.828
	$C_{11}H_{11}N_4O_2$	13.667	1.268		$C_{17}H_{13}N$	18.960	1.697
	$C_{11}H_{21}NO_4$	12.760	1.551		$C_{17}H_{27}$	18.802	1.668
	$C_{11}H_{23}N_2O_3$	13.135	1.398		$C_{18}HN$	19.848	1.864
	$C_{11}H_{25}N_3O_2$	13.509	1.247		$C_{18}H_{15}$	19.691	1.834
	$C_{11}H_{27}N_4O$	13.884	1.097		$C_{19}H_3$	20.579	2.001
	$C_{12}H_9NO_4$	13.649	1.663	232	$C_{10}H_{20}N_2O_4$	12.045	1.468
	$C_{12}H_{11}N_2O_3$	14.023	1.513		$C_{10}H_{22}N_3O_3$	12.420	1.312
	$C_{12}H_{13}N_3O_2$	14.398	1.365		$C_{10}H_{24}N_4O_2$	12.974	1.458
	$C_{12}H_{15}N_4O$	14.772	1.219		$C_{11}H_8N_2O_4$	12.934	1.573
	$C_{12}H_{23}O_4$	13.492	1.642		$C_{11}H_{10}N_3O_3$	13.308	1.421
	$C_{12}H_{25}NO_3$	13.866	1.492		$C_{11}H_{12}N_4O_2$	13.683	1.270
	$C_{12}H_{27}N_2O_2$	14.240	1.343		$C_{11}H_{22}NO_4$	12.776	1.553
	$C_{12}H_{29}N_3O$	14.615	1.196		$C_{11}H_{24}N_2O_3$	13.151	1.400
	$C_{13}HN_3O_2$	15.286	1.491		$C_{11}H_{26}N_3O_2$	13.525	1.249
	$C_{13}H_3N_4O$	15.661	1.348		$C_{11}H_{28}N_4O$	13.900	1.099
	$C_{13}H_{11}O_4$	14.380	1.760		$C_{12}H_{10}NO_4$	13.665	1.665
	$C_{13}H_{13}NO_3$	14.755	1.613		$C_{12}H_{12}N_2O$	14.039	1.515
	$C_{13}H_{15}N_2O_2$	15.129	1.468		$C_{12}H_{14}N_3O_2$	14.414	1.368
	$C_{13}H_{17}N_3O$	15.503	1.324		$C_{12}H_{16}N_4O$	14.788	1.221
	$C_{13}H_{19}N_4$	15.878	1.182		$C_{12}H_{24}O_4$	13.508	1.644
	$C_{13}H_{27}O_3$	14.597	1.591		$C_{12}H_{26}NO_3$	13.882	1.494
	$C_{13}H_{29}NO_2$	14.972	1.445		$C_{12}H_{28}N_2O_2$	14.256	1.346
	$C_{14}HNO_3$	15.643	1.743		$C_{13}H_2N_3O_2$	15.302	1.494
	$C_{14}H_3N_2O_2$	16.017	1.501		$C_{13}H_4N_4O$	15.677	1.351
	$C_{14}H_5N_3O$	16.392	1.460		$C_{12}H_{12}O_4$	14.396	1.762
	$C_{14}H_7N_4$	16.766	1.321		$C_{13}H_{14}NO_3$	14.771	1.616
	$C_{14}H_{15}O_3$	15.486	1.719		$C_{13}H_{16}N_2O_2$	15.145	1.470
	$C_{14}H_{17}NO_2$	15.860	1.576		$C_{13}H_{18}N_3O$	15.519	1.327
	$C_{14}H_{19}N_2O$	16.234	1.435		$C_{13}H_{20}N_4$	15.894	1.484
	$C_{14}H_{21}N_3$	16.609	1.295		$C_{13}H_{28}O_3$	14.613	1.593
	$C_{15}H_3O_3$	16.374	1.854		$C_{14}H_2NO_3$	15.659	1.745
	$C_{15}H_5NO_2$	16.749	1.715		$C_{14}H_4N_2O_2$	16.033	1.603
	$C_{15}H_7N_2O$	17.123	1.577		$C_{14}H_6N_2O$	16.408	1.463
	$C_{15}H_9N_3$	17.497	1.441		$C_{14}H_8N_4$	16.782	1.324
	$C_{15}H_{19}O_2$	16.591	1.690		$C_{14}H_{16}O_3$	15.502	1.721
	$C_{15}H_{21}NO$	16.966	1.515		$C_{14}H_{18}NO_2$	15.876	1.579

(Contd.)

		M + 1	M + 2		M + 1	M + 2
	$C_{14}H_{20}N_2O$	16.250	1.438	$C_{13}H_{21}N_4$	15.910	1.187
	$C_{14}H_{22}N_3$	16.625	1.298	$C_{14}HO_4$	15.301	1.891
	$C_{15}H_4O_3$	16.390	1.857	$C_{14}H_3NO_3$	15.675	1.748
	$C_{15}H_6NO_2$	16.765	1.718	$C_{14}H_5N_2O_2$	16.050	1.606
	$C_{15}H_8N_2O$	17.139	1.580	$C_{14}H_7N_3O$	16.424	1.465
	$C_{15}H_{10}N_3$	17.513	1.444	$C_{14}H_9N_4$	15.798	1.326
	$C_{15}H_{20}O_2$	16.607	1.692	$C_{14}H_{17}O_3$	15.518	1.724
	$C_{15}H_{22}NO$	16.982	1.554	$C_{14}H_{19}NO_2$	15.892	1.581
	$C_{15}H_{24}N_2$	17.356	1.417	$C_{14}H_{21}N_2O$	16.266	1.440
	$C_{16}H_8O_2$	17.496	1.838	$C_{14}H_{23}N_3$	16.641	1.301
	$C_{16}H_{10}NO$	17.870	1.703	$C_{15}H_5O_3$	16.406	1.860
	$C_{16}H_{12}N_2$	18.245	1.569	$C_{15}H_7NO_2$	16.781	1.721
	$C_{16}H_{24}O$	17.713	1.676	$C_{15}H_9N_2O$	17.155	1.583
	$C_{16}H_{26}N$	18.087	1.542	$C_{15}H_{11}N_3$	17.529	1.447
	$C_{17}H_{12}O$	18.601	1.831	$C_{15}H_{21}O_2$	16.623	1.695
	$C_{17}H_{14}N$	18.976	1.700	$C_{15}H_{23}NO$	16.998	1.557
	$C_{17}H_{28}$	18.818	1.671	$C_{15}H_{25}N_2$	17.372	1.420
	$C_{18}H_2N$	19.864	1.867	$C_{16}H_9O_2$	17.512	1.841
	$C_{18}H_{16}$	19.707	1.837	$C_{16}H_{11}NO$	17.886	1.706
	$C_{19}H_4$	20.595	2.001	$C_{16}H_{13}N_2$	18.261	1.572
233	$C_{10}H_{21}N_2O_4$	12.061	1.469	$C_{16}H_{25}O$	17.729	1.679
	$C_{10}H_{23}N_3O_3$	12.436	1.314	$C_{16}H_{27}N$	18.103	1.544
	$C_{10}H_{25}N_4O_2$	12.810	1.160	$C_{17}HN_2$	19.149	1.733
	$C_{11}H_9N_2O_4$	12.950	1.575	$C_{17}H_{13}O$	18.617	1.834
	$C_{11}H_{11}N_3O_3$	13.324	1.423	$C_{17}H_{15}N$	18.992	1.703
	$C_{11}H_{13}N_4O_2$	13.699	1.272	$C_{17}H_{29}$	18.834	1.674
	$C_{11}H_{23}NO_4$	12.792	1.555	$C_{18}HO$	19.506	1.998
	$C_{11}H_{25}N_2O_3$	13.167	1.403	$C_{18}H_3N$	19.880	1.870
	$C_{11}H_{27}N_3O_2$	13.541	1.251	$C_{18}H_{17}$	19.723	1.840
	$C_{12}HN_4O_2$	14.587	1.392	$C_{19}H_5$	20.611	2.013
	$C_{12}H_{11}NO_4$	13.681	1.667	234 $C_{10}H_{22}N_2O_4$	12.077	1.471
	$C_{12}H_{13}N_2O_3$	14.055	1.518	$C_{10}H_{24}N_3O_3$	12.452	1.316
	$C_{12}H_{15}N_3O_2$	14.430	1.370	$C_{10}H_{26}N_4O_2$	12.826	1.162
	$C_{12}H_{17}N_4O$	14.804	1.223	$C_{11}H_{10}N_2O_4$	12.966	1.577
	$C_{12}H_{25}O_4$	13.524	1.646	$C_{11}H_{12}N_3O_3$	13.340	1.425
	$C_{12}H_{27}NO_3$	13.898	1.496	$C_{11}H_{14}N_4O_2$	13.715	1.274
	$C_{13}HN_2O_3$	14.944	1.641	$C_{11}H_{24}NO_4$	12.808	1.557
	$C_{13}H_3N_3O_2$	15.318	1.496	$C_{11}H_{26}N_2O_3$	13.183	1.405
	$C_{13}H_5N_4O$	15.693	1.353	$C_{12}H_2N_4O_2$	14.603	1.394
	$C_{13}H_{13}O_4$	14.412	1.765	$C_{12}H_{12}NO_4$	13.697	1.669
	$C_{13}H_{15}NO_3$	14.787	1.618	$C_{12}H_{14}N_2O_3$	14.071	1.520
	$C_{13}H_{17}N_2O_2$	15.161	1.473	$C_{12}H_{16}N_3O_2$	14.446	1.372
	$C_{13}H_{19}N_3O$	15.535	1.329	$C_{12}H_{18}N_4O$	14.820	1.226

(Contd.)

	M + 1	M + 2		M + 1	M + 2
$C_{12}H_{26}O_4$	13.540	1.648	$C_{11}H_{25}NO_4$	12.824	1.559
$C_{13}H_2N_2O_3$	14.960	1.643	$C_{12}HN_3O_3$	14.245	1.544
$C_{13}H_4N_3O_2$	15.334	1.499	$C_{12}H_3N_4O_2$	14.619	1.397
$C_{13}H_6N_4O$	15.709	1.356	$C_{12}H_{13}NO_4$	13.713	1.671
$C_{13}H_{14}O_4$	14.428	1.767	$C_{12}H_{15}N_2O_3$	14.087	1.522
$C_{13}H_{16}NO_3$	14.803	1.620	$C_{12}H_{17}N_3O_2$	14.462	1.374
$C_{13}H_{18}N_2O_2$	15.177	1.475	$C_{12}H_{19}N_4O$	14.836	1.228
$C_{13}H_{20}N_3O$	15.551	1.332	$C_{13}HNO_4$	14.602	1.791
$C_{13}H_{22}N_4$	15.926	1.189	$C_{13}H_3N_2O_3$	14.976	1.646
$C_{14}H_2O_4$	15.317	1.893	$C_{13}H_5N_3O_2$	15.350	1.501
$C_{14}H_4NO_3$	15.691	1.750	$C_{13}H_7N_4O$	15.725	1.358
$C_{14}H_6N_2O_2$	16.066	1.608	$C_{13}H_{15}O_4$	14.444	1.769
$C_{14}H_8N_3O$	16.440	1.468	$C_{13}H_{17}NO_3$	14.819	1.623
$C_{14}H_{10}N_4$	16.814	1.329	$C_{13}H_{19}N_2O_2$	15.193	1.478
$C_{14}H_{18}O_3$	15.534	1.726	$C_{13}H_{21}N_3O$	15.567	1.334
$C_{14}H_{20}NO_2$	15.908	1.584	$C_{13}H_{23}N_4$	15.942	1.192
$C_{14}H_{22}N_2O$	16.282	1.443	$C_{14}H_3O_4$	15.333	1.896
$C_{14}H_{24}N_3$	16.657	1.303	$C_{14}H_5NO_3$	15.707	1.753
$C_{15}H_6O_3$	16.422	1.862	$C_{14}H_7N_2O_2$	16.082	1.611
$C_{15}H_8NO_2$	16.797	1.723	$C_{14}H_9N_3O$	16.456	1.471
$C_{15}H_{10}N_2O$	17.171	1.586	$C_{14}H_{11}N_4$	16.830	1.332
$C_{15}H_{12}N_3$	17.545	1.449	$C_{14}H_{19}O_3$	15.550	1.729
$C_{15}H_{22}O_2$	16.639	1.698	$C_{14}H_{21}NO_2$	15.924	1.586
$C_{15}H_{24}NO$	17.014	1.560	$C_{14}H_{23}N_2O$	16.298	1.446
$C_{15}H_{26}N_2$	17.388	1.423	$C_{14}H_{25}N_3$	16.673	1.306
$C_{16}H_{10}O_2$	17.528	1.844	$C_{15}H_7O_3$	16.438	1.865
$C_{16}H_{12}NO$	17.902	1.709	$C_{15}H_9NO_2$	16.813	1.726
$C_{16}H_{14}N_2$	18.277	1.575	$C_{15}H_{11}N_2O$	17.187	1.588
$C_{16}H_{26}O$	17.745	1.681	$C_{15}H_{13}N_3$	17.561	1.442
$C_{16}H_{28}N$	18.119	1.547	$C_{15}H_{23}O_2$	16.655	1.700
$C_{17}H_2N_2$	19.165	1.736	$C_{15}H_{25}NO$	17.030	1.562
$C_{17}H_{14}O$	18.633	1.837	$C_{15}H_{27}N_2$	17.404	1.425
$C_{17}H_{16}N$	19.008	1.707	$C_{16}HN_3$	18.450	1.606
$C_{17}H_{30}$	18.850	1.677	$C_{16}H_{11}O_2$	17.544	1.846
$C_{18}H_2O$	19.522	2.001	$C_{16}H_{13}NO$	17.918	1.712
$C_{18}H_4N$	19.896	1.874	$C_{16}H_{15}N_2$	18.293	1.578
$C_{18}H_{18}$	19.739	1.843	$C_{16}H_{27}O$	17.761	1.684
$C_{19}H_6$	20.627	2.017	$C_{16}H_{29}N$	18.135	1.550
235 $C_{10}H_{23}N_2O_4$	12.093	1.473	$C_{17}HNO$	18.807	1.869
$C_{10}H_{25}N_3O_3$	12.468	1.318	$C_{17}H_3N_2$	19.181	1.739
$C_{11}H_{11}N_2O_4$	12.982	1.579	$C_{17}H_{15}O$	18.649	1.840
$C_{11}H_{13}N_3O_3$	13.356	1.427	$C_{17}H_{17}N$	19.024	1.710
$C_{11}H_{15}N_4O_2$	13.731	1.277	$C_{17}H_{31}$	18.866	1.680

(Contd.)

	M + 1	M + 2			M + 1	M + 2
$C_{18}H_3O$	19.538	2.004		$C_{16}H_{28}O$	17.777	1.687
$C_{18}H_3N$	19.912	1.877		$C_{16}H_{30}N$	18.151	1.553
$C_{18}H_{19}$	19.755	1.846		$C_{17}H_2NO$	18.823	1.872
$C_{19}H_7$	20.643	2.120		$C_{17}H_4N_2$	19.197	1.742
236 $C_{10}H_{24}N_2O_4$	12.109	1.475		$C_{17}H_{16}O$	18.665	1.843
$C_{11}H_{12}N_2O_4$	12.998	1.581		$C_{17}H_{18}N$	19.040	1.713
$C_{11}H_{14}N_3O_3$	13.372	1.429		$C_{17}H_{32}$	18.882	1.683
$C_{11}H_{16}N_4O_2$	13.747	1.279		$C_{18}H_4O$	19.554	2.007
$C_{12}H_{12}N_3O_3$	14.261	1.546		$C_{18}H_6N$	19.928	1.880
$C_{12}H_4N_4O_2$	14.635	1.399		$C_{18}H_{20}$	19.771	1.849
$C_{12}H_{14}NO_4$	13.729	1.674		$C_{19}H_8$	20.659	2.023
$C_{12}H_{16}N_2O_3$	14.103	1.524	237	$C_{11}H_{13}N_2O_4$	13.014	1.583
$C_{12}H_{18}N_3O_2$	14.478	1.377		$C_{11}H_{15}N_3O_3$	13.388	1.431
$C_{12}H_{20}N_4O$	14.852	1.230		$C_{11}H_{17}N_4O_2$	13.763	1.281
$C_{13}H_2NO_4$	14.618	1.794		$C_{12}HN_2O_4$	13.902	1.697
$C_{13}H_4N_2O_3$	14.992	1.648		$C_{12}H_3N_3O_3$	19.277	1.548
$C_{13}H_6N_3O_2$	15.366	1.504		$C_{12}H_5N_4O_2$	14.651	1.401
$C_{13}H_8N_4O$	15.741	1.361		$C_{12}H_{15}NO_4$	13.745	1.676
$C_{13}H_{16}O_4$	14.460	1.772		$C_{12}H_{17}N_2O_3$	14.119	1.527
$C_{13}H_{18}NO_3$	14.835	1.625		$C_{12}H_{19}N_3O_2$	14.494	1.379
$C_{13}H_{20}N_2O_2$	15.209	1.480		$C_{12}H_{21}N_4O$	14.868	1.233
$C_{13}H_{22}N_3O$	15.583	1.337		$C_{13}H_3NO_4$	14.634	1.796
$C_{13}H_{24}N_4$	15.958	1.494		$C_{13}H_5N_2O_3$	15.008	1.650
$C_{14}H_4O_4$	15.349	1.898		$C_{13}H_7N_3O_2$	15.382	1.506
$C_{14}H_6NO_3$	15.723	1.755		$C_{13}H_9N_4O$	15.757	1.363
$C_{14}H_8N_2O_2$	16.098	1.613		$C_{13}H_{17}O_4$	14.476	1.774
$C_{14}H_{10}N_3O$	16.472	1.473		$C_{13}H_{19}NO_3$	14.851	1.628
$C_{14}H_{12}N_4$	16.846	1.334		$C_{13}H_{21}N_2O_2$	15.225	1.483
$C_{14}H_{20}O_3$	15.566	1.731		$C_{13}H_{23}N_3O$	15.599	1.339
$C_{14}H_{22}NO_2$	15.940	1.589		$C_{13}H_{25}N_4$	15.974	1.197
$C_{14}H_{24}N_2O$	16.314	1.448		$C_{14}H_5O_4$	15.365	1.901
$C_{14}H_{26}N_3$	16.689	1.309		$C_{14}H_7NO_3$	15.739	1.758
$C_{15}H_8O_3$	16.454	1.868		$C_{14}H_9N_2O_2$	16.114	1.616
$C_{15}H_{10}NO_2$	16.829	1.729		$C_{14}H_{11}N_3O$	16.488	1.476
$C_{15}H_{12}N_2O$	17.203	1.591		$C_{14}H_{13}N_4$	16.862	1.337
$C_{15}H_{14}N_3$	17.577	1.455		$C_{14}H_{21}O_3$	15.582	1.734
$C_{15}H_{24}O_2$	16.671	1.703		$C_{14}H_{23}NO_2$	15.956	1.591
$C_{15}H_{26}NO$	17.046	1.565		$C_{14}H_{25}N_2O$	16.330	1.451
$C_{15}H_{28}N_2$	17.420	1.428		$C_{14}H_{27}N_3$	16.705	1.311
$C_{16}H_2N_3$	18.466	1.609		$C_{15}HN_4$	17.751	1.485
$C_{16}H_{12}O_2$	17.560	1.849		$C_{15}H_9O_3$	16.470	1.870
$C_{16}H_{14}NO$	17.934	1.715		$C_{15}H_{11}NO_2$	16.845	1.731
$C_{16}H_{16}N_2$	18.309	1.581		$C_{15}H_{13}N_2O$	17.219	1.594

(Contd.)

	M + 1	M + 2		M + 1	M + 2
C$_{15}$H$_{15}$N$_3$	17.593	1.458	C$_{14}$H$_{12}$N$_3$O	16.504	1.478
C$_{15}$H$_{25}$O$_2$	16.687	1.706	C$_{14}$H$_{14}$N$_4$	16.878	1.340
C$_{15}$H$_{27}$NO	17.062	1.568	C$_{14}$H$_{22}$O$_3$	15.598	1.736
C$_{15}$H$_{29}$N$_2$	17.436	1.431	C$_{14}$H$_{24}$NO$_2$	15.972	1.594
C$_{16}$HN$_2$O	18.108	1.745	C$_{14}$H$_{26}$N$_2$O	16.346	1.453
C$_{16}$H$_3$N$_3$	18.482	1.612	C$_{14}$H$_{28}$N$_3$	16.721	1.314
C$_{16}$H$_{13}$O$_2$	17.576	1.852	C$_{15}$H$_2$N$_4$	17.767	1.488
C$_{16}$H$_{15}$NO	17.950	1.717	C$_{15}$H$_{10}$O$_3$	16.486	1.873
C$_{16}$H$_{17}$N$_2$	18.325	1.584	C$_{15}$H$_{12}$NO$_2$	16.861	1.734
C$_{16}$H$_{29}$O	17.793	1.690	C$_{15}$H$_{14}$N$_2$O	17.235	1.597
C$_{16}$H$_{31}$N	18.167	1.556	C$_{15}$H$_{16}$N$_3$	17.609	1.461
C$_{17}$HO$_2$	18.464	2.006	C$_{15}$H$_{26}$O$_2$	16.703	1.708
C$_{17}$H$_3$NO	18.839	1.875	C$_{15}$H$_{28}$NO	17.078	1.570
C$_{17}$H$_5$N$_2$	19.213	1.745	C$_{15}$H$_{30}$N$_2$	17.452	1.434
C$_{17}$H$_{17}$O	18.681	1.846	C$_{16}$H$_2$N$_2$O	18.124	1.748
C$_{17}$H$_{19}$N	19.056	1.716	C$_{16}$H$_4$N$_3$	18.498	1.615
C$_{17}$H$_{33}$	18.898	1.686	C$_{16}$H$_{14}$O$_2$	17.592	1.855
C$_{18}$H$_5$O	19.570	2.010	C$_{16}$H$_{16}$NO	17.966	1.720
C$_{18}$H$_7$N	19.944	1.883	C$_{16}$H$_{18}$N$_2$	18.341	1.587
C$_{18}$H$_{21}$	19.787	1.853	C$_{16}$H$_{30}$O	17.809	1.693
C$_{19}$H$_9$	20.675	2.026	C$_{16}$H$_{32}$N	18.183	1.559
238 C$_{11}$H$_{14}$N$_2$O$_4$	13.030	15.851	C$_{17}$H$_2$O$_2$	18.480	2.009
C$_{11}$H$_{16}$N$_3$O$_3$	13.404	1.433	C$_{17}$H$_4$NO	18.855	1.878
C$_{11}$H$_{18}$N$_4$O$_2$	13.779	1.283	C$_{17}$H$_6$N$_2$	19.229	1.748
C$_{12}$H$_2$N$_2$O$_4$	13.918	1.699	C$_{17}$H$_{18}$O	18.697	1.849
C$_{12}$H$_4$N$_3$O$_3$	14.293	1.551	C$_{17}$H$_{20}$N	19.072	1.719
C$_{12}$H$_6$N$_4$O$_2$	14.667	1.404	C$_{17}$H$_{34}$	18.914	1.690
C$_{12}$H$_{16}$NO$_4$	13.761	1.678	C$_{18}$H$_6$O	19.586	2.013
C$_{12}$H$_{18}$N$_2$O$_3$	14.135	1.530	C$_{18}$H$_8$N	19.960	1.886
C$_{12}$H$_{20}$N$_3$O$_2$	14.510	1.381	C$_{18}$H$_{22}$	19.803	1.856
C$_{12}$H$_{22}$N$_4$O	14.884	1.235	C$_{10}$H$_{10}$	20.691	2.030
C$_{13}$H$_4$NO$_4$	14.650	1.798	239 C$_{11}$H$_{15}$N$_2$O$_4$	13.046	1.587
C$_{13}$H$_6$N$_2$O$_3$	15.024	1.653	C$_{11}$H$_{17}$N$_3$O$_3$	13.420	1.436
C$_{13}$H$_8$N$_3$O$_2$	15.398	1.508	C$_{11}$H$_{19}$N$_4$O$_2$	13.795	1.285
C$_{13}$H$_{10}$N$_4$O	15.773	1.366	C$_{12}$H$_3$N$_2$O$_4$	13.934	1.701
C$_{13}$H$_{18}$O$_4$	14.492	1.776	C$_{12}$H$_5$N$_3$O$_3$	14.309	1.553
C$_{13}$H$_{20}$NO$_3$	14.867	1.630	C$_{12}$H$_7$N$_4$O$_2$	14.683	1.406
C$_{13}$H$_{22}$N$_2$O$_2$	15.241	1.485	C$_{12}$H$_{17}$NO$_4$	13.777	1.680
C$_{13}$H$_{24}$N$_3$O	15.615	1.342	C$_{12}$H$_{19}$N$_2$O$_3$	14.151	1.531
C$_{13}$H$_{26}$N$_4$	15.990	1.200	C$_{12}$H$_{21}$N$_3$O$_2$	14.526	1.384
C$_{14}$H$_6$O$_4$	15.381	1.903	C$_{12}$H$_{23}$N$_4$O	14.900	1.238
C$_{14}$H$_8$NO$_3$	15.755	1.760	C$_{13}$H$_5$NO$_4$	14.666	1.801
C$_{14}$H$_{10}$N$_2$O$_2$	16.130	1.619	C$_{13}$H$_7$N$_2$O$_3$	15.040	1.655

(Contd.)

	M + 1	M + 2			M + 1	M + 2
$C_{13}H_9N_3O_2$	15.414	1.511	240	$C_{11}H_{16}N_2O_4$	13.062	1.589
$C_{13}H_{11}N_4O$	15.789	1.368		$C_{11}H_{18}N_3O_3$	13.436	1.438
$C_{13}H_{19}O_4$	14.508	1.778		$C_{11}H_{20}N_4O_2$	13.811	1.288
$C_{13}H_{21}NO_3$	14.883	1.632		$C_{12}H_4N_2O_4$	13.950	1.703
$C_{13}H_{23}N_2O_2$	15.257	1.487		$C_{12}H_6N_3O_3$	14.325	1.555
$C_{13}H_{25}N_3O$	15.631	1.344		$C_{12}H_8N_4O_2$	14.699	1.408
$C_{13}H_{27}N_4$	16.006	1.202		$C_{12}H_{18}NO_4$	13.793	1.682
$C_{14}H_7O_4$	15.397	1.906		$C_{12}H_{20}N_2O_3$	14.167	1.534
$C_{14}H_9NO_3$	15.771	1.763		$C_{12}H_{22}N_3O_2$	14.542	1.386
$C_{14}H_{11}N_2O_2$	16.146	1.621		$C_{12}H_{24}N_4O$	14.916	1.240
$C_{14}H_{13}N_3O$	16.520	1.481		$C_{13}H_6NO_4$	14.682	1.803
$C_{14}H_{15}N_4$	16.894	1.342		$C_{13}H_8N_2O_3$	15.056	1.658
$C_{14}H_{23}O_3$	15.614	1.739		$C_{13}H_{10}N_3O_2$	15.430	1.513
$C_{14}H_{25}NO_2$	15.988	1.597		$C_{13}H_{12}N_4O$	15.805	1.371
$C_{14}H_{27}N_2O$	16.362	1.456		$C_{13}{}^{H}{}_{20}O_4$	14.524	1.781
$C_{14}H_{29}N_3$	16.737	1.317		$C_{13}H_{22}NO_3$	14.899	1.635
$C_{15}HN_3O$	17.408	1.626		$C_{13}H_{24}N_2O_2$	15.273	1.490
$C_{15}H_3N_4$	17.783	1.491		$C_{13}H_{26}N_3O$	15.647	1.347
$C_{15}H_{11}O_3$	16.502	1.876		$C_{13}H_{28}N_4$	16.022	1.205
$C_{15}H_{13}NO_2$	16.877	1.737		$C_{14}H_8O_4$	15.413	1.908
$C_{15}H_{15}N_2O$	17.251	1.599		$C_{14}H_{10}NO_3$	15.787	1.765
$C_{15}H_{17}N_3$	17.625	1.464		$C_{14}H_{12}N_2O_2$	16.162	1.624
$C_{15}H_{27}O_2$	16.719	1.711		$C_{14}H_{14}N_3O$	16.536	1.484
$C_{15}H_{29}NO$	17.094	1.573		$C_{14}H_{16}N_4$	16.910	1.345
$C_{15}H_{31}N_2$	17.468	1.437		$C_{14}H_{24}O_3$	15.630	1.741
$C_{16}HNO_2$	17.765	1.885		$C_{14}H_{26}NO_2$	16.004	1.599
$C_{16}H_3N_2O$	18.140	1.751		$C_{14}H_{28}N_2O$	16.378	1.459
$C_{16}H_5N_3$	18.514	1.618		$C_{14}H_{20}N_3$	16.753	1.319
$C_{16}H_{15}O_2$	17.608	1.858		$C_{15}H_2N_3O$	17.424	1.629
$C_{16}H_{17}NO$	17.982	1.723		$C_{15}H_4N_4$	17.799	1.494
$C_{16}H_{19}N_2$	18.357	1.590		$C_{15}H_{12}O_3$	16.518	1.678
$C_{16}H_{31}O$	17.825	1.696		$C_{15}H_{14}NO_2$	16.893	1.739
$C_{16}H_{33}N$	18.199	1.562		$C_{15}H_{16}N_2O$	17.267	1.602
$C_{17}H_3O_2$	18.496	2.012		$C_{15}H_{18}N_3$	17.641	1.466
$C_{17}H_5NO$	18.871	1.881		$C_{15}H_{28}O_2$	16.735	1.714
$C_{17}H_7N_2$	19.245	1.751		$C_{15}H_{30}NO$	17.110	1.576
$C_{17}H_{19}O$	18.713	1.852		$C_{15}H_{32}N_2$	17.484	1.439
$C_{17}H_{21}N$	19.088	1.722		$C_{16}H_2NO_2$	17.781	1.688
$C_{17}H_{35}$	18.930	1.693		$C_{16}H_4N_2O$	18.156	1.754
$C_{18}H_7O$	19.602	2.017		$C_{16}H_6N_3$	18.530	1.621
$C_{18}H_9N$	19.976	1.889		$C_{16}H_{16}O_2$	17.624	1.861
$C_{18}H_{23}$	19.819	1.859		$C_{16}H_{18}NO$	17.998	1.726
$C_{19}H_{11}$	20.707	2.033		$C_{16}H_{20}N_2$	18.373	1.593

(Contd.)

	M + 1	M + 2			M + 1	M + 2
$C_{16}H_{32}O$	17.841	1.698		$C_{15}H_5N_4$	17.815	1.496
$C_{16}H_{34}N$	18.215	1.565		$C_{15}H_{13}O_3$	16.534	1.881
$C_{17}H_4O_2$	18.512	2.015		$C_{15}H_{15}NO_2$	16.909	1.742
$C_{17}H_6NO$	18.887	1.884		$C_{15}H_{17}N_2O$	17.283	1.605
$C_{17}H_8N_2$	19.261	1.754		$C_{15}H_{19}N_3$	17.657	1.469
$C_{17}H_{20}O$	18.729	1.855		$C_{15}H_{29}O_2$	16.751	1.716
$C_{17}H_{22}N$	19.104	1.725		$C_{15}H_{31}NO$	17.126	1.579
$C_{17}H_{36}$	18.946	1.696		$C_{15}H_{33}N_2$	17.500	1.442
$C_{18}H_8O$	19.618	2.020		$C_{16}HO_3$	17.423	2.026
$C_{18}H_{10}N$	19.992	1.693		$C_{16}H_3NO_2$	17.797	1.891
$C_{18}H_{24}$	19.835	1.862		$C_{16}H_5N_2O$	18.172	1.757
$C_{19}H_{12}$	20.723	2.036		$C_{16}H_7N_3$	18.546	1.624
241 $C_{11}H_{17}N_2O_4$	13.078	1.591		$C_{16}H_{17}O_2$	17.640	1.863
$C_{11}H_{19}N_3O_3$	13.452	1.140		$C_{16}H_{19}NO$	18.014	1.729
$C_{11}H_{21}N_4O_2$	13.827	1.290		$C_{16}H_{21}N_2$	18.389	1.596
$C_{12}H_5N_2O_4$	13.966	1.706		$C_{16}H_{33}O$	17.857	1.701
$C_{12}H_7N_3O_3$	14.341	1.558		$C_{16}H_{35}N$	18.231	1.568
$C_{12}H_9N_4O_2$	14.715	1.411		$C_{17}H_5O_2$	18.528	2.018
$C_{12}H_{19}NO_4$	13.809	1.685		$C_{17}H_7NO$	18.903	1.887
$C_{12}H_{21}N_2O_3$	14.183	1.536		$C_{17}H_9N_2$	19.277	1.757
$C_{12}H_{23}N_3O_2$	14.558	1.388		$C_{17}H_{21}O$	18.745	1.858
$C_{12}H_{25}N_4O$	14.932	1.242		$C_{17}H_{23}N$	19.120	1.728
$C_{13}H_7NO_4$	14.698	1.805		$C_{18}H_9O$	19.634	2.023
$C_{13}H_9N_2O_3$	15.072	1.660		$C_8H_{11}N$	20.008	1.896
$C_{13}H_{11}N_3O_2$	15.446	1.516		$C_{18}H_{25}$	19.851	1.865
$C_{13}H_{13}N_4O$	15.821	1.373		$C_{19}H_{13}$	20.739	2.040
$C_{13}H_{21}O_4$	14.540	1.783		$C_{20}H$	21.628	2.222
$C_{13}H_{23}NO_3$	14.915	1.637	242 $C_{11}H_{18}N_2O_4$	13.094	1.593
$C_{13}H_{25}N_2O_2$	15.289	1.492		$C_{11}H_{20}N_3O_3$	13.468	1.442
$C_{13}H_{27}N_3O$	15.663	1.349		$C_{11}H_{22}N_4O_2$	13.843	1.292
$C_{13}H_{29}N_4$	16.038	1.207		$C_{12}H_6N_2O_4$	13.982	1.708
$C_{14}HN_4O$	16.709	1.512		$C_{12}H_8N_3O_3$	14.357	1.560
$C_{14}H_9O_4$	15.429	1.910		$C_{12}H_{10}N_2O_2$	14.731	1.413
$C_{14}H_{11}NO_3$	15.803	1.768		$C_{12}H_{20}NO_4$	13.825	1.687
$C_{14}H_{13}N_2O_2$	16.178	1.626		$C_{12}H_{22}N_2O_3$	14.199	1.538
$C_{14}H_{15}N_3O$	16.552	1.486		$C_{12}H_{24}N_3O_2$	14.574	1.391
$C_{14}H_{17}N_4$	16.926	1.348		$C_{12}H_{26}N_4O$	14.948	1.245
$C_{14}H_{25}O_3$	15.646	1.744		$C_{13}H_8NO_4$	14.714	1.808
$C_{14}H_{27}NO_2$	16.020	1.602		$C_{13}H_{10}N_2O_3$	15.088	1.662
$C_{14}H_{29}N_2O$	16.394	1.461		$C_{13}H_{12}N_3O_2$	15.462	1.518
$C_{14}H_{31}N_3$	16.769	1.322		$C_{13}N_{14}N_4O$	15.837	1.676
$C_{15}HN_2O_2$	17.066	1.768		$C_{13}H_{22}O_4$	14.556	1.785
$C_{15}H_3N_3O$	17.440	1.632		$C_{13}H_{24}NO_3$	14.931	1.639

(Contd.)

	M + 1	M + 2		M + 1	M + 2
$C_{13}H_{26}N_2O_2$	15.305	1.495	$C_{11}H_{23}N_4O_2$	13.859	1.294
$C_{13}H_{28}N_3O$	15.679	1.652	$C_{12}H_7N_2O_4$	13.998	1.710
$C_{13}H_{30}N_4$	16.054	1.210	$C_{12}H_9N_3O_3$	14.373	1.562
$C_{14}H_2N_4O$	16.725	1.515	$C_{12}H_{11}N_4O_2$	14.747	1.415
$C_{14}H_{10}O_4$	15.445	1.913	$C_{12}H_{21}NO_4$	13.841	1.689
$C_{14}H_{12}NO_3$	15.819	1.770	$C_{12}H_{23}N_2O_3$	14.215	1.540
$C_{14}H_{14}N_2O_2$	16.194	1.629	$C_{12}H_{25}N_3O_2$	14.590	1.393
$C_{14}H_{16}N_3O$	16.568	1.489	$C_{12}H_{27}N_4O$	14.964	1.247
$C_{14}H_{18}N_4$	16.942	1.351	$C_{13}H_9NO_4$	14.730	1.810
$C_{14}H_{26}O_3$	15.662	1.746	$C_{13}H_{11}N_2O_3$	15.104	1.665
$C_{14}H_{28}NO_2$	16.036	1.604	$C_{13}H_{13}N_3O_2$	15.478	1.521
$C_{14}H_{30}N_2O$	16.410	1.464	$C_{13}H_{15}N_4O$	15.853	1.378
$C_{14}H_{32}N_3$	16.785	1.325	$C_{13}H_{23}O_4$	14.572	1.788
$C_{15}H_2N_2O_2$	17.082	1.771	$C_{13}H_{25}NO_3$	14.947	1.642
$C_{15}H_4N_3O$	17.456	1.634	$C_{13}H_{27}N_2O_2$	15.321	1.497
$C_{15}H_6N_4$	17.831	1.499	$C_{13}H_{29}N_3O$	15.695	1.354
$C_{15}H_{14}O_3$	16.550	1.883	$C_{13}H_{31}N_4$	16.070	1.212
$C_{15}H_{16}NO_2$	16.925	1.745	$C_{14}HN_3O_2$	16.367	1.656
$C_{15}H_{18}N_2O$	17.299	1.608	$C_{14}H_3N_4O$	16.741	1.517
$C_{15}H_{20}N_3$	17.673	1.472	$C_{14}H_{11}O_4$	15.461	1.915
$C_{15}H_{30}O_2$	16.767	1.719	$C_{14}H_{13}NO_3$	15.835	1.773
$C_{15}H_{32}NO$	17.142	1.581	$C_{14}H_{15}N_2O_2$	16.210	1.632
$C_{15}H_{34}N_2$	17.516	1.445	$C_{14}H_{17}N_3O$	16.584	1.492
$C_{16}H_2O_3$	17.439	2.029	$C_{14}H_{19}N_4$	16.958	1.353
$C_{16}H_4NO_2$	17.813	1.893	$C_{14}H_{27}O_3$	15.678	1.749
$C_{16}H_6N_2O$	18.188	1.760	$C_{14}H_{29}NO_2$	16.052	1.607
$C_{16}H_8N_3$	18.562	1.627	$C_{14}H_{31}N_2O$	16.426	1.466
$C_{16}H_{18}O_2$	17.656	1.866	$C_{14}H_{33}N_3$	16.801	1.327
$C_{16}H_{20}NO$	18.030	1.732	$C_{15}HNO_3$	16.724	1.912
$C_{16}H_{22}N_2$	18.405	1.599	$C_{15}H_3N_2O_2$	17.098	1.774
$C_{16}H_{34}O$	17.873	1.704	$C_{15}H_5N_3O$	17.472	1.637
$C_{17}H_6O_2$	18.544	2.021	$C_{15}H_7N_4$	17.847	1.502
$C_{17}H_8NO$	18.919	1.890	$C_{15}H_{15}O_3$	16.566	1.886
$C_{17}H_{10}N_2$	19.293	1.760	$C_{15}H_{17}NO_2$	16.941	1.748
$C_{17}H_{22}O$	18.761	1.861	$C_{15}H_{19}N_2O$	17.315	1.611
$C_{17}H_{24}N$	19.136	1.731	$C_{15}H_{21}N_3$	17.689	1.475
$C_{18}H_{10}O$	19.650	2.026	$C_{15}H_{31}O_2$	16.783	1.722
$C_{18}H_{12}N$	20.024	1.699	$C_{15}H_{33}NO$	17.158	1.584
$C_{18}H_{26}$	19.867	1.868	$C_{16}H_3O_3$	17.455	2.031
$C_{19}H_{14}$	20.755	2.043	$C_{16}H_5NO_2$	17.829	1.896
$C_{20}H_2$	21.644	2.226	$C_{16}H_7N_2O$	18.204	1.763
243 $C_{11}H_{19}N_2O_4$	13.110	1.596	$C_{16}H_9N_3$	18.578	1.630
$C_{11}H_{21}N_3O_3$	13.484	1.444	$C_{16}H_{19}O_2$	17.672	1.869

(Contd.)

	M + 1	M + 2			M + 1	M + 2
	$C_{16}H_{21}NO$	18.046	1.735	$C_{15}H_4N_2O_2$	17.114	1.776
	$C_{16}H_{23}N_2$	18.421	1.602	$C_{15}H_6N_3O$	17.488	1.640
	$C_{17}H_7O_2$	18.560	2.024	$C_{15}H_8N_4$	17.863	1.505
	$C_{17}H_9NO$	18.935	1.893	$C_{15}H_{16}O_3$	16.582	1.889
	$C_{17}H_{11}N_2$	19.309	1.764	$C_{15}H_{18}NO_2$	16.957	1.750
	$C_{17}H_{23}O$	18.777	1.864	$C_{15}H_{20}N_2O$	17.331	1.613
	$C_{17}H_{25}N$	19.152	1.734	$C_{15}H_{22}N_3$	17.705	1.478
	$C_{18}H_{11}O$	19.666	2.029	$C_{15}H_{32}O_2$	16.799	1.724
	$C_{18}H_{13}N$	20.040	1.902	$C_{16}H_4O_3$	17.471	2.034
	$C_{18}H_{27}$	19.883	1.872	$C_{16}H_6NO_2$	17.845	1.899
	$C_{19}HN$	20.929	2.078	$C_{16}H_8N_2O$	18.220	1.765
	$C_{19}H_{15}$	20.771	2.046	$C_{16}H_{10}N_3$	18.594	1.633
	$C_{20}H_3$	21.660	2.229	$C_{16}H_{20}O_2$	17.688	1.872
244	$C_{11}H_{20}N_2O_4$	13.126	1.598	$C_{16}H_{22}NO$	18.062	1.738
	$C_{11}H_{22}N_3O_3$	13.500	1.446	$C_{16}H_{24}N_2$	18.437	1.605
	$C_{11}H_{24}N_4O_2$	13.875	1.296	$C_{17}H_8O_2$	18.576	2.027
	$C_{12}H_8N_2O_4$	14.014	1.712	$C_{17}H_{10}NO$	18.951	1.896
	$C_{12}H_{10}N_3O_3$	14.389	1.564	$C_{17}H_{12}N_2$	19.325	1.767
	$C_{12}H_{12}N_4O_2$	14.763	1.418	$C_{17}H_{24}O$	18.793	1.867
	$C_{12}H_{22}NO_4$	13.857	1.691	$C_{17}H_{26}N$	19.168	1.737
	$C_{12}H_{24}N_2O_3$	14.231	1.543	$C_{18}H_{12}O$	19.682	2.032
	$C_{12}H_{26}N_3O_2$	14.606	1.395	$C_{18}H_{14}N$	20.056	1.905
	$C_{12}H_{28}N_4O$	14.980	1.250	$C_{18}H_{28}$	19.899	1.875
	$C_{13}H_{10}NO_4$	14.746	1.812	$C_{19}H_2N$	20.945	2.082
	$C_{13}H_{12}N_2O_3$	15.120	1.667	$C_{19}H_{16}$	20.787	2.050
	$C_{13}H_{14}N_3O_2$	15.494	1.523	$C_{20}H_4$	21.676	2.233
	$C_{13}H_{16}N_4O$	15.869	1.381	245 $C_{11}H_{21}N_2O_4$	13.142	1.600
	$C_{13}H_{24}O_4$	14.588	1.790	$C_{11}H_{23}N_3O_3$	13.516	1.448
	$C_{13}H_{26}NO_3$	14.963	1.644	$C_{11}H_{25}N_4O_2$	13.891	1.299
	$C_{13}H_{28}N_2O_2$	15.337	1.500	$C_{12}H_9N_2O_4$	14.030	1.715
	$C_{13}H_{30}N_3O$	15.711	1.357	$C_{12}H_{11}N_3O_3$	14.405	1.567
	$C_{13}H_{32}N_4$	16.086	1.215	$C_{12}H_{13}N_4O_2$	14.779	1.420
	$C_{14}H_2N_3O_2$	16.383	1.659	$C_{12}H_{23}NO_4$	13.873	1.693
	$C_{14}H_4N_4O$	16.757	1.520	$C_{12}H_{25}N_2O_3$	14.247	1.545
	$C_{14}H_{12}O_4$	15.477	1.918	$C_{12}H_{27}N_3O_2$	14.622	1.398
	$C_{14}H_{14}NO_3$	15.851	1.775	$C_{12}H_{29}N_4O$	14.996	1.252
	$C_{14}H_{16}N_2O_2$	16.226	1.634	$C_{13}HN_4O_2$	15.668	1.550
	$C_{14}H_{18}N_3O$	16.600	1.494	$C_{13}H_{11}NO_4$	14.762	1.815
	$C_{14}H_{20}N_4$	16.974	1.356	$C_{13}H_{13}N_2O_3$	15.136	1.670
	$C_{14}H_{28}O_3$	15.694	1.751	$C_{13}H_{15}N_3O_2$	15.510	1.526
	$C_{14}H_{30}NO_2$	16.068	1.609	$C_{13}H_{17}N_4O$	15.885	1.383
	$C_{14}H_{32}N_2O$	16.442	1.469	$C_{13}H_{25}O_4$	14.604	1.792
	$C_{15}H_2NO_3$	16.740	1.914	$C_{13}H_{27}NO_3$	14.979	1.647

(Contd.)

	M + 1	M + 2		M + 1	M + 2
$C_{13}H_{29}N_2O_2$	15.353	1.502	$C_{11}H_{26}N_4O_2$	13.907	1.301
$C_{13}H_{31}N_3O$	15.727	1.359	$C_{12}H_{10}N_2O_4$	14.047	1.717
$C_{14}HN_2O_3$	16.025	1.802	$C_{12}H_{12}N_3O_3$	14.421	1.569
$C_{14}H_3N_3O_2$	16.399	1.662	$C_{12}H_{14}N_2O_2$	14.795	1.422
$C_{14}H_5N_4O$	16.773	1.523	$C_{12}H_{24}NO_4$	13.889	1.696
$C_{14}H_{13}O_4$	15.493	1.920	$C_{12}H_{26}N_2O_3$	14.263	1.547
$C_{14}H_{15}NO_3$	15.867	1.778	$C_{12}H_{28}N_3O_2$	14.638	1.400
$C_{14}H_{17}N_2O_2$	16.242	1.637	$C_{12}H_{30}N_4O$	15.012	1.254
$C_{14}H_{19}N_3O$	16.616	1.497	$C_{13}H_2N_4O_2$	15.684	1.552
$C_{14}H_{21}N_4$	16.990	1.359	$C_{13}H_{12}NO_4$	14.778	1.817
$C_{14}H_{29}O_3$	15.710	1.754	$C_{13}H_{14}N_2O_3$	15.152	1.672
$C_{14}H_{31}NO_2$	16.084	1.612	$C_{13}H_{16}N_3O_2$	15.526	1.528
$C_{15}HO_4$	16.381	2.056	$C_{13}H_{18}N_4O$	15.901	1.386
$C_{15}H_3NO_3$	16.756	1.917	$C_{13}H_{26}O_4$	14.620	1.795
$C_{15}H_5N_2O_2$	17.130	1.779	$C_{13}H_{28}NO_3$	14.995	1.649
$C_{15}H_7N_3O$	17.504	1.643	$C_{13}H_{30}N_2O_2$	15.369	1.505
$C_{15}H_9N_4$	17.879	1.508	$C_{14}H_2N_2O_3$	16.041	1.805
$C_{15}H_{17}O_3$	16.598	1.891	$C_{14}H_4N_3O_2$	16.415	1.664
$C_{15}H_{19}NO_2$	16.973	1.753	$C_{14}H_6N_4O$	16.789	1.525
$C_{15}H_{21}N_2O$	17.347	1.616	$C_{14}H_{14}O_4$	15.509	1.923
$C_{15}H_{23}N_3$	17.721	1.481	$C_{14}H_{16}NO_3$	15.883	1.780
$C_{16}H_5O_3$	17.487	2.037	$C_{14}H_{18}N_2O_2$	16.258	1.639
$C_{16}H_7NO_2$	17.861	1.902	$C_{14}H_{20}N_3O$	16.632	1.500
$C_{16}H_9N_2O$	18.236	1.768	$C_{14}H_{22}N_4$	17.006	1.361
$C_{16}H_{11}N_3$	18.610	1.636	$C_{14}H_{30}O_3$	15.726	1.756
$C_{16}H_{21}O_2$	17.704	1.875	$C_{15}H_2O_4$	16.397	2.059
$C_{16}H_{23}NO$	18.078	1.740	$C_{15}H_4NO_3$	16.772	1.920
$C_{16}H_{25}N_2$	18.453	1.608	$C_{15}H_6N_2O_2$	17.146	1.782
$C_{17}H_9O_2$	18.592	2.030	$C_{15}H_8N_3O$	17.520	1.646
$C_{17}H_{11}NO$	18.967	1.899	$C_{15}H_{10}N_4$	17.895	1.511
$C_{17}H_{13}N_2$	19.341	1.770	$C_{15}H_{18}O_3$	16.614	1.894
$C_{17}H_{25}O$	18.809	1.870	$C_{15}H_{20}NO_2$	16.989	1.756
$C_{17}H_{27}N$	19.184	1.740	$C_{15}H_{22}N_2O$	17.363	1.619
$C_{18}HN_2$	20.230	1.940	$C_{15}H_{24}N_3$	17.737	1.483
$C_{18}H_{13}O$	19.698	2.035	$C_{16}H_6O_3$	17.503	2.040
$C_{18}H_{15}N$	20.072	1.909	$C_{16}H_8NO_2$	17.877	1.905
$C_{18}H_{29}$	19.915	1.878	$C_{16}H_{10}N_2O$	18.252	1.771
$C_{19}H_{17}$	20.803	2.053	$C_{16}H_{12}N_3$	18.626	1.639
$C_{19}HO$	20.586	2.209	$C_{16}H_{22}O_2$	17.720	1.878
$C_{19}H_3N$	20.961	2.085	$C_{16}H_{24}NO$	18.094	1.743
$C_{20}H_5$	21.692	2.236	$C_{16}H_{26}N_2$	18.469	1.611
246 $C_{11}H_{22}N_2O_4$	13.158	1.602	$C_{17}H_{10}O_2$	18.608	2.033
$C_{11}H_{24}N_3O_3$	13.532	1.451	$C_{17}H_{12}NO$	18.983	1.902

(Contd.)

	M + 1	M + 2			M + 1	M + 2
	$C_{17}H_{14}N_2$	19.357	1.773		$C_{15}H_{21}NO_2$	17.005	1.758
	$C_{17}H_{26}O$	18.825	1.873		$C_{15}H_{23}N_2O$	17.379	1.622
	$C_{17}H_{28}N$	19.200	1.743		$C_{15}H_{25}N_3$	17.753	1.486
	$C_{18}H_2N_2$	20.246	1.943		$C_{16}H_7O_3$	17.519	2.043
	$C_{18}H_{14}O$	19.714	2.039		$C_{16}H_9NO_2$	17.893	1.908
	$C_{18}H_{16}N$	20.088	1.912		$C_{16}H_{11}N_2O$	18.268	1.774
	$C_{18}H_{30}$	19.931	1.881		$C_{16}H_{13}N_3$	18.642	1.642
	$C_{19}H_2O$	20.602	2.212		$C_{16}H_{23}O_2$	17.736	1.880
	$C_{19}H_4N$	20.977	2.089		$C_{16}H_{25}NO$	18.110	1.746
	$C_{19}H_{18}$	20.819	2.056		$C_{16}H_{27}N_2$	18.485	1.614
	$C_{20}H_6$	21.708	2.239		$C_{17}HN_3$	19.531	1.806
247	$C_{11}H_{23}N_2O_4$	13.174	1.604		$C_{17}H_{11}O_2$	18.624	2.036
	$C_{11}H_{25}N_3O_3$	13.548	1.453		$C_{17}H_{13}NO$	18.999	1.905
	$C_{11}H_{27}N_4O_2$	13.923	1.303		$C_{17}H_{15}N_2$	19.373	1.776
	$C_{12}H_{11}N_2O_4$	14.063	1.719		$C_{17}H_{27}O$	18.841	1.876
	$C_{12}H_{13}N_3O_3$	14.437	1.571		$C_{17}H_{29}N$	19.216	1.746
	$C_{12}H_{15}N_4O_2$	14.811	1.425		$C_{18}HNO$	19.887	2.072
	$C_{12}H_{25}NO_4$	13.905	1.698		$C_{18}H_3N_2$	20.262	1.946
	$C_{12}H_{27}N_2O_3$	14.279	1.549		$C_{18}H_{15}O$	19.730	2.042
	$C_{12}H_{29}N_3O_2$	14.654	1.402		$C_{18}H_{17}N$	20.104	1.915
	$C_{13}HN_3O_3$	15.325	1.698		$C_{18}H_{31}$	19.947	1.884
	$C_{13}H_3N_4O_2$	15.700	1.555		$C_{19}H_3O$	20.619	2.215
	$C_{13}H_{13}NO_4$	14.794	1.820		$C_{19}H_5N$	20.993	2.092
	$C_{13}H_{15}N_2O_3$	15.168	1.674		$C_{19}H_{19}$	20.835	2.060
	$C_{13}H_{17}N_3O_2$	15.542	1.531		$C_{20}H_7$	21.724	2.243
	$C_{13}H_{19}N_4O$	15.917	1.388	248	$C_{11}H_{24}N_2O_4$	13.190	1.606
	$C_{13}H_{27}O_4$	14.636	1.797		$C_{11}H_{26}N_3O_3$	13.564	1.455
	$C_{13}H_{29}NO_3$	15.011	1.651		$C_{11}H_{28}N_4O_2$	13.939	1.305
	$C_{14}HNO_4$	15.682	1.949		$C_{12}H_{12}N_2O_4$	14.079	1.721
	$C_{14}H_3N_2O_3$	16.057	1.807		$C_{12}H_{14}N_3O_3$	14.453	1.574
	$C_{14}H_5N_3O_2$	16.431	1.667		$C_{12}H_{16}N_4O_2$	14.827	1.427
	$C_{14}H_7N_4O$	16.805	1.528		$C_{12}H_{26}NO_4$	13.921	1.700
	$C_{14}H_{15}O_4$	15.525	1.925		$C_{12}H_{28}N_2O_3$	14.295	1.552
	$C_{14}H_{17}NO_3$	15.899	1.783		$C_{13}H_2N_3O_3$	15.341	1.700
	$C_{14}H_{19}N_2O_2$	16.274	1.642		$C_{13}H_4N_4O_2$	15.716	1.557
	$C_{14}H_{21}N_3O$	16.648	1.502		$C_{13}H_{14}NO_4$	14.810	1.822
	$C_{14}H_{23}N_4$	17.022	1.364		$C_{13}H_{16}N_2O_3$	15.184	1.677
	$C_{15}H_3O_4$	16.413	2.061		$C_{13}H_{18}N_3O_2$	15.558	1.533
	$C_{15}H_5NO_3$	16.788	1.922		$C_{13}H_{20}N_4O$	15.933	1.391
	$C_{15}H_7N_2O_2$	17.162	1.785		$C_{13}H_{28}O_4$	14.652	1.799
	$C_{15}H_9N_3O$	17.536	1.648		$C_{14}H_2NO_4$	15.698	1.952
	$C_{15}H_{11}N_4$	17.911	1.514		$C_{14}H_4N_2O_3$	16.073	1.810
	$C_{15}H_{19}O_3$	16.630	1.897		$C_{14}H_6N_3O_2$	16.447	1.670

(Contd.)

		M + 1	M + 2		M + 1	M + 2
	$C_{14}H_8N_4O$	16.821	1.531	$C_{13}HN_2O_4$	14.983	1.847
	$C_{14}H_{16}O_4$	15.541	1.928	$C_{13}H_3N_3O_3$	15.357	1.703
	$C_{14}H_{18}NO_3$	15.915	1.785	$C_{13}H_5N_4O_2$	15.732	1.560
	$C_{14}H_{20}N_2O_2$	16.29	1.645	$C_{13}H_{15}NO_4$	14.826	1.824
	$C_{14}H_{22}N_3O$	16.664	1.505	$C_{13}H_{17}N_2O_3$	15.200	1.679
	$C_{14}H_{24}N_4$	17.038	1.367	$C_{13}H_{19}N_3O_2$	15.574	1.536
	$C_{15}H_4O_4$	16.429	2.064	$C_{13}H_{21}N_4O$	15.949	1.393
	$C_{15}H_6NO_3$	16.804	1.925	$C_{14}H_3NO_4$	15.714	1.954
	$C_{15}H_8N_2O_2$	17.178	1.787	$C_{14}H_5N_2O_3$	16.089	1.812
	$C_{15}H_{10}N_3O$	17.552	1.651	$C_{14}H_7N_3O_2$	16.463	1.672
	$C_{15}H_{12}N_4$	17.927	1.516	$C_{14}H_9N_4O$	16.837	1.533
	$C_{15}H_{20}O_3$	16.646	1.899	$C_{14}H_{17}O_4$	15.557	1.930
	$C_{15}H_{22}NO_2$	17.021	1.761	$C_{14}H_{19}NO_3$	15.931	1.788
	$C_{15}H_{22}N_2O$	17.395	1.624	$C_{14}H_{21}N_2O_2$	16.306	1.647
	$C_{15}H_{26}N_3$	17.769	1.489	$C_{14}H_{23}N_3O$	16.680	1.508
	$C_{16}H_8O_3$	17.535	2.045	$C_{14}H_{25}N_4$	17.054	1.370
	$C_{16}H_{10}NO_2$	17.909	1.911	$C_{15}H_5O_4$	16.445	2.067
	$C_{16}H_{12}N_2O$	18.284	1.777	$C_{15}H_7NO_3$	16.820	1.928
	$C_{16}H_{14}N_3$	18.658	1.645	$C_{15}H_9N_2O_2$	17.194	1.790
	$C_{16}H_{24}O_2$	17.752	1.883	$C_{15}H_{11}N_3O$	17.568	1.654
	$C_{16}H_{26}NO$	1.813	1.749	$C_{15}H_{13}N_4$	17.943	1.519
	$C_{16}H_{28}N_2$	18.501	1.617	$C_{15}H_{21}O_3$	16.662	1.902
	$C_{17}H_2N_3$	19.547	1.609	$C_{15}H_{23}NO_2$	17.037	1.764
	$C_{17}H_{12}O_2$	18.640	2.039	$C_{15}H_{25}N_2O$	17.411	1.627
	$C_{17}H_{14}NO$	19.015	1.908	$C_{15}H_{27}N_3$	17.785	1.492
	$C_{17}H_{16}N_2$	19.389	1.779	$C_{16}HN_4$	18.831	1.677
	$C_{17}H_{28}O$	18.857	1.879	$C_{16}H_9O_3$	17.551	2.048
	$C_{17}H_{30}N$	19.232	1.749	$C_{16}H_{11}NO_2$	17.925	1.913
	$C_{18}H_2NO$	19.903	2.075	$C_{16}H_{13}N_2O$	18.300	1.780
	$C_{18}H_4N_2$	20.278	1.949	$C_{16}H_{15}N_3$	18.674	1.648
	$C_{18}H_{16}O$	19.746	2.045	$C_{16}H_{25}O_2$	17.768	1.887
	$C_{18}H_{18}N$	20.120	1.918	$C_{16}H_{27}NO$	18.142	1.752
	$C_{18}H_{32}$	19.963	1.888	$C_{16}H_{29}N_2$	18.517	1.619
	$C_{19}H_4O$	20.635	2.219	$C_{17}HN_2O$	19.188	1.941
	$C_{19}H_6N$	21.009	2.095	$C_{17}H_3N_3$	19.563	1.812
	$C_{19}H_{20}$	20.851	2.063	$C_{17}H_{13}O_2$	18.656	2.042
	$C_{20}H_8$	21.740	2.246	$C_{17}H_{15}NO$	19.031	1.911
249	$C_{11}H_{25}N_2O_4$	13.206	1.608	$C_{17}H_{17}N_2$	19.405	1.782
	$C_{11}H_{27}N_3O_3$	13.580	1.457	$C_{17}H_{29}O$	18.873	1.882
	$C_{12}H_{13}N_2O_4$	14.095	1.724	$C_{17}H_{31}N$	19.248	1.752
	$C_{12}H_{15}N_3O_3$	14.469	1.576	$C_{18}HO_2$	19.545	2.206
	$C_{12}H_{17}N_4O_2$	14.843	1.430	$C_{128}H_3NO$	19.919	2.079
	$C_{12}H_{27}NO_4$	13.937	1.702	$C_{18}H_5N_2$	20.294	1.953

(Contd.)

		M + 1	**M + 2**		**M + 1**	**M + 2**
	$C_{18}H_{17}O$	19.762	2.048	$C_{15}H_{14}N_4$	17.959	1.522
	$C_{18}H_{19}N$	2.014	1.922	$C_{15}H_{22}O_3$	16.678	1.905
	$C_{18}H_{33}$	19.979	1.891	$C_{15}H_{24}NO_2$	17.053	1.767
	$C_{19}H_5O$	20.651	2.222	$C_{15}H_{26}N_2O$	17.427	1.630
	$C_{19}H_7N$	21.025	2.099	$C_{15}H_{28}N_3$	17.801	1.495
	$C_{19}H_{21}$	20.867	2.066	$C_{16}H_2N_4$	18.847	1.680
	$C_{20}H_9$	21.756	2.250	$C_{16}H_{10}O_3$	17.567	2.051
250	$C_{11}H_{26}N_2O_4$	13.222	1.610	$C_{16}H_{12}NO_2$	17.941	1.916
	$C_{12}H_{14}N_2O_4$	14.111	1.726	$C_{16}H_{14}N_2O$	18.31	1.783
	$C_{12}H_{16}N_3O_3$	14.485	1.578	$C_{16}H_{16}N_3$	18.690	1.651
	$C_{12}H_{18}N_4O_2$	14.859	1.432	$C_{16}H_{26}O_2$	17.784	1.889
	$C_{13}H_2N_2O_4$	14.999	1.849	$C_{16}H_{28}NO$	18.158	1.755
	$C_{13}H_4N_3O_3$	15.373	1.705	$C_{16}H_{30}N_2$	18.533	1.622
	$C_{13}H_6N_4O_2$	15.748	1.562	$C_{17}H_2N_2O$	19.204	1.944
	$C_{13}H_{16}NO_4$	14.842	1.827	$C_{17}H_4N_3$	19.579	1.815
	$C_{13}H_{18}N_2O_3$	15.216	1.682	$C_{17}H_{14}O_2$	18.672	2.045
	$C_{13}H_{20}N_3O_2$	15.590	1.538	$C_{17}H_{16}NO$	19.047	1.914
	$C_{13}H_{22}N_4O$	15.965	1.396	$C_{17}H_{18}N_2$	19.421	1.785
	$C_{14}H_4NO_4$	15.730	1.957	$C_{17}H_{30}O$	18.889	1.885
	$C_{14}H_6N_2O_3$	16.105	1.815	$C_{17}H_{32}N$	19.264	1.755
	$C_{13}H_8N_3O_2$	16.479	1.675	$C_{18}H_2O_2$	19.561	2.209
	$C_{14}H_{10}N_4O$	16.853	1.536	$C_{18}H_4NO$	19.935	2.082
	$C_{14}H_{18}N_4$	15.573	1.933	$C_{18}H_6N_2$	20.310	1.956
	$C_{14}H_{20}NO_3$	15.947	1.791	$C_{18}H_{18}O$	19.778	2.051
	$C_{14}H_{22}N_2O_2$	16.322	1.650	$C_{18}H_{20}N$	20.152	1.925
	$C_{14}H_{24}N_3O$	16.696	1.510	$C_{18}H_{34}$	19.995	1.894
	$C_{14}H_{26}N_4$	17.070	1.372	$C_{19}H_6O$	20.667	2.225
	$C_{15}H_6O_4$	16.461	2.069	$C_{19}H_8N$	21.041	2.102
	$C_{15}H_8NO_3$	16.836	1.930	$C_{19}H_{22}$	20.883	2.070
	$C_{15}H_{10}N_2O_2$	17.210	1.793	$C_{20}H_{10}$	21.772	2.253
	$C_{15}H_{12}N_3O$	17.584	1.657			

Exercise 1

Find out the chemical structure with the help of below given mass spectrum.

Molecular formula = $C_5H_{12}O$, molecular weight = 88.15

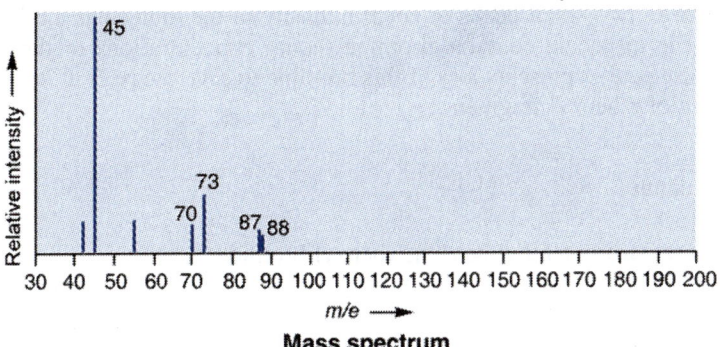

m/e ⟶

Mass spectrum

Answer

The spectrum shows a small molecular ion and a small M-1 peak, suggesting the presence of an alcohol (it cannot be an aldehyde since there are no degrees of unsaturation). The M-15 peak represents loss of a methyl group and the M-17 is consistent with loss of a hydroxy radical. For an alcohol, the base peak is often formed by expulsion of an alkyl chain to give the simple oxonium ion $R'CR''OH^+$; to generate the observed $m/e = 45$, R' must be CH_3 and R'' a H.

Chemical structure:

Chemical name: 2-Pentanol

MS fragments:

M^+	M-1	M-15	M-17	$m/e = 45$

Exercise 2

Give the structure to the compound, possessing molecular formula, $C_7H_{12}Br$ and molecular weight, 171.04. The mass spectrum is given below.

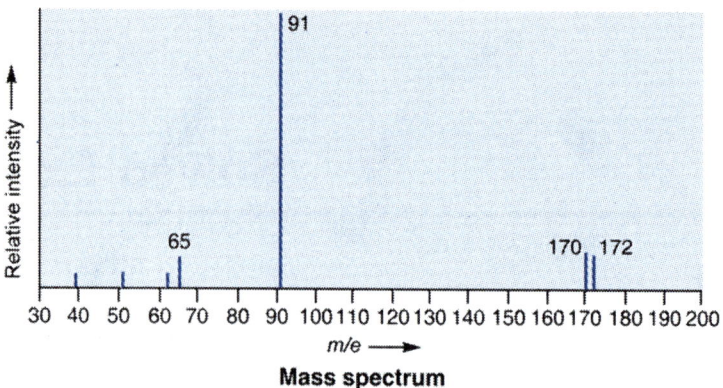

Mass spectrum

Answer

The spectrum shows two small peaks of equal intensity in the molecular ion region, strongly suggesting that the molecule contains bromine (equal concentrations of the ^{79}Br and ^{81}Br isotopes). The base peak represents loss of this bromine to give the peak at $m/e = 91$, which is highly suggestive of a benzyl fragment.

Chemical structure: [benzene ring]—CH_2Br

Chemical name: Bromomethyl benzene (benzyl bromide)

MS fragments:

[benzene ring]—CH_2Br [benzene ring]—CH_2 [Tropylium ion]

M^+ (170 and 172) m/e 91   Tropylium ion

Exercise 3

Give the structure to the compound having molecular formula, $C_9H_{10}O$ and molecular weight, 134.18. The mass spectrum is given below.

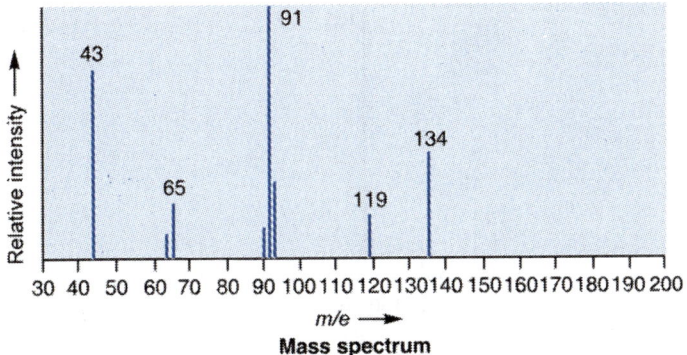

Mass spectrum

Answer

The spectrum shows a moderate molecular ion peak at *m/e* 134, and a peak at M-15, strongly suggesting the presence of a labile methyl group. The base peak occurs at *m/e* 91, which is highly suggestive of a benzyl fragment. The presence of an intense peak at *m/e* 43 is also suggestive of the presence of a methyl ketone, which can fragment to form the acylium ion.

Chemical structure:

Chemical name: Benzyl methyl ketone

MS fragments:

$M^{+\bullet}$	M-15	*m/e* 91	Tropylium ion	*m/e* 43

Exercise 4

Give the structure to the compound having molecular formula, $C_{11}H_{12}O_3$ and molecular weight, 192.21.The mass spectrum is given below.

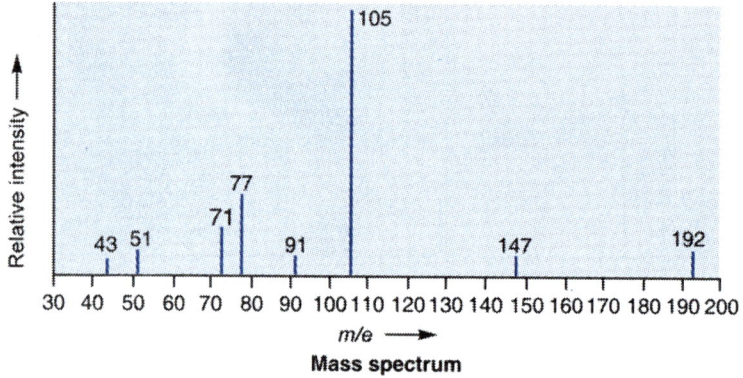

Mass spectrum

Answer

The spectrum shows a small molecular ion peak, and a peak at M-45, suggesting the presence of an ethoxy group ($-O-CH_2CH_3$). The very minor peaks at 91 and 43 suggest that the molecule does not contain a benzyl unit or a methyl ketone. The base peak occurs at *m/e* 105, and results from loss of a unit of *m/e* 73, which is also observed. The molecular ion at 105 is characteristic of a carbonyl bonded directly to an aromatic ring.

Chemical structure:

Chemical name: Ethyl 3-oxy-3-phenylpropanoate (ethyl benzoylacetate)

MS fragments:

| M⁺ | M-45 | *m/e* 105 | *m/e* 73 |

Exercise 5

Give the structure to the compound having molecular formula, $C_5H_8O_2$ and molecular weight, 100.12. The mass spectrum is given below.

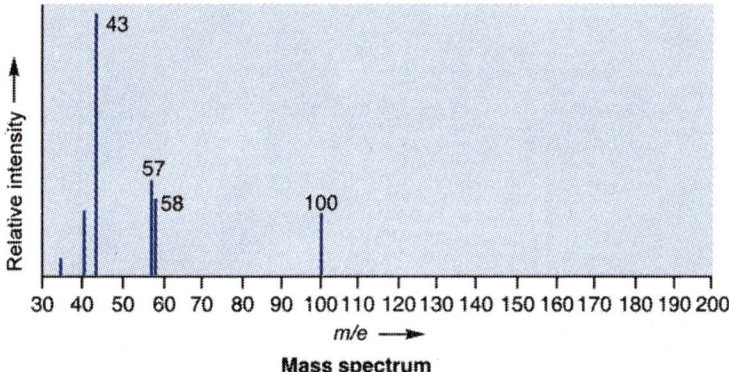

Mass spectrum

Answer

The spectrum shows a small molecular ion peak, and a pair of peaks at $m/e = 57$ and 58. The peak at $m/e = 57$ corresponds to loss of $m/e = 43$, which is the base peak and corresponds to the acylium ion (CH_3C-O^+). The $m/e = 57$ fragment corresponds to C_3H_5O, suggesting the original compound was an ester with this molecular formula.

Chemical structure:

Chemical name: 2-Propenyl ethanoate (allyl acetate)

MS fragments:

Exercise 6

One compound C_8H_8O gives the mass spectra presented in the following figure. Deduce the structure and the cleavage processes of main fragment ions.

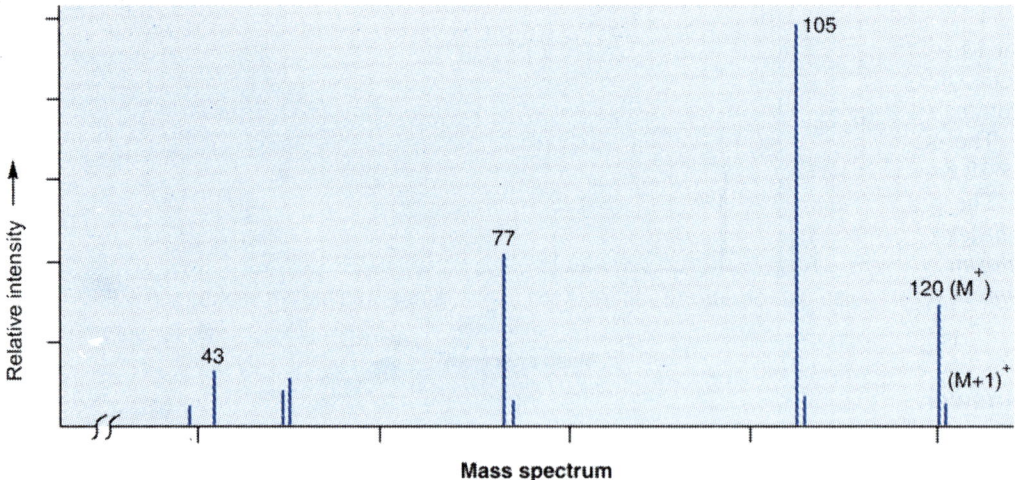

Mass spectrum

Answer

Acetophenone

MS fragments:

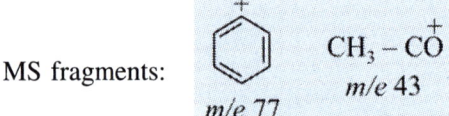

Exercise 7

Compound C_7H_8O, has important bands in its UV absorption shows typical of the absorption a benzene ring. Its IR at 3400 cm^{-1} (s), its NMR has bands at $\delta 4.3$, $\delta 5.1$, and $\delta 7.3$(s), the true ratios are 2:1:5. Assign a structure to the compound that is compatible with spectral data given above.

Answer

Benzyl alcohol

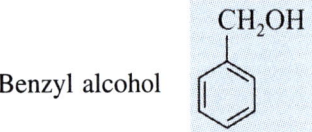

1. The formula is C_7H_8O, and there is, therefore, three double-bands equivalent, suggest the formation of the benzene ring.
2. The UV spectrum has λ_{max} 240–290 nm, the absorption is middle, and typical of the benzene ring group, the presence of an benzene ring group is immediately apparent in the UV.
3. The IR spectrum with its very strong band at 1600 cm^{-1}, is shown by C=C conjugated aryl group.
4. From low-field to high-field, the NMR spectrum has five aromatic protons, one hydroxyl protons, two methylene protons.

Exercise 8

The mass shows molecular ion peak at m/e 106 (100%), m/e 107 (8.91%) and m/e 10.5%).
Deduce the molecular formula.

Answer

The M + 1 value shows intensity of 8.91%, means there is a possibility of eight or nine carbon atoms (8.91 ÷ 1.1) in the compound. The M + 2 value does not indicate the presence of chlorine, bromine or sulphur atom.

The mass of M is 106 amu. If we assume that there are eight carbons, then we subtract 8×12 from 106, which gives 10 amu.

The possible molecular formula, is C_8H_{10}. It is also possible as per hydrogen rule. *The hydrogen rule states that for a molecule containing only hydrogen, carbon, oxygen, fluorine, chlorine, bromine, and iodine, the maximum number of monovalent atom (max H) for a given number of carbons (C) and nitrogen (N) is given by the equation.*

$$H = 2C + N + 2$$

To check the validity of the formula, $2(8) + 0 + 2 = 18$. The value '10' is well under 10. Hence, formula is valid.

Exercise 9

The McLafferty rearrangement of the molecular ion is responsible for the peak at m/e 74. Draw the structure.

Answer

Methyl butanoate m/e 74

Exercise 10

An unknown organic compound shows a molecular ion peak at m/e 170 with 100% relative intensity. M + 1 and M + 2 have peak intensities of 13.2 and 1.00, respectively. Deduce the molecular formula of unknown compound.

Answer

M^+ at m/e 170 indicates an even number of nitrogen or nitrogen is absent.

M^+ peak : 100% means it is a base peak
M + 1 peak : 13.2 means, $13.2/1.1 = 12$ carbon atoms
M + 2 peak : 1.00 of M^+ , means no S, Cl, Br

$17 - 12 \times 12 = 36$ $C_{12}H_{26}$ is possible
Or $170 - 12 \times 12 - 16 = 10$ $C_{12}H_{10}O$

Molecular formula of the compound

$C_{12}H_{10}O$ or $C_{12}H_{26}$

Exercise 11

The mass spectral data of an unknown liquid is given here. Deduce the molecular formula.

m/e	Intensity
78	23.6 ($M^{+\bullet}$)
79	0.79
80	7.55
81	0.25

Answer

M^+ at m/e 78 indicates the 0, 2, 4, ..., nitrogen atom.

$M^{+\bullet}$	23.6	100%	It is not the base peak
$M + 1$	0.79	3.35	3.35/1.1 = 3.05, 3 carbon atoms
$M + 2$	7.55	32%	Chlorine is present

$78 - (3 \times 12) - 35 = 7$

Molecular formula of the compound

C_3H_7Cl.

Exercise 12

Find out the expected molecular ion peak of 2-methylhexane. Give possible structures of the fragments at m/e 85, 57 and 43.

Answer

Molecular peak, m/e 100

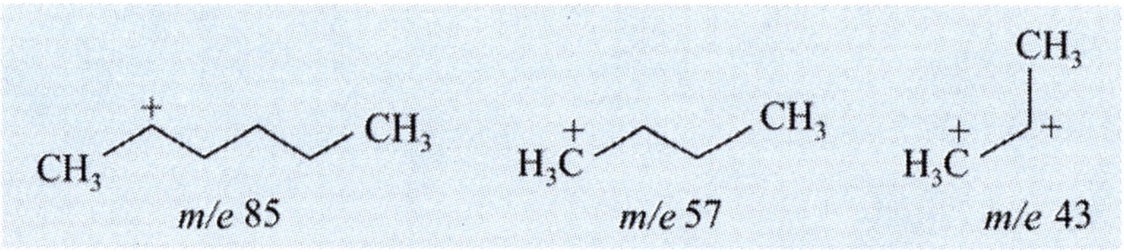

Exercise 13

Deduce the structure with the help of below given IR, ^{1}H-NMR and ^{13}C-NMR spectra.

Compound $C_7H_8O_2$
NMR solvent: $CDCl_3$
IR solvent: Neat

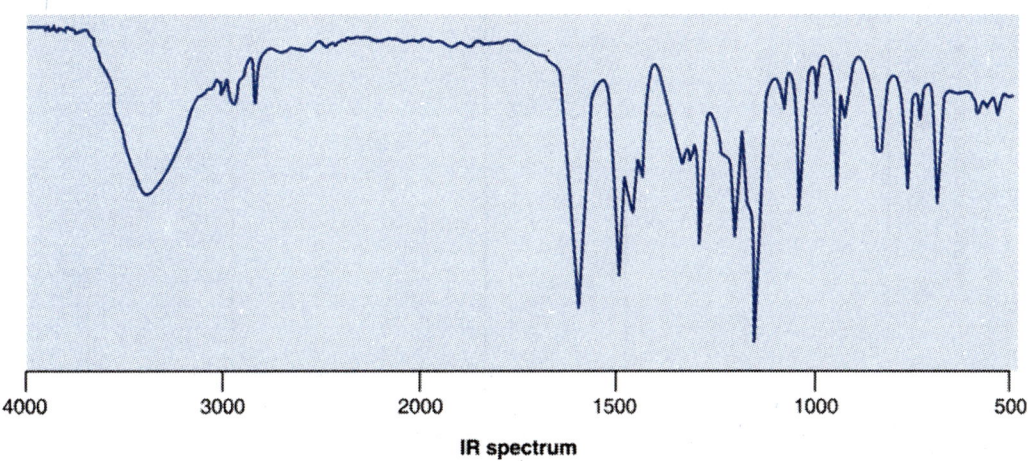

IR spectrum

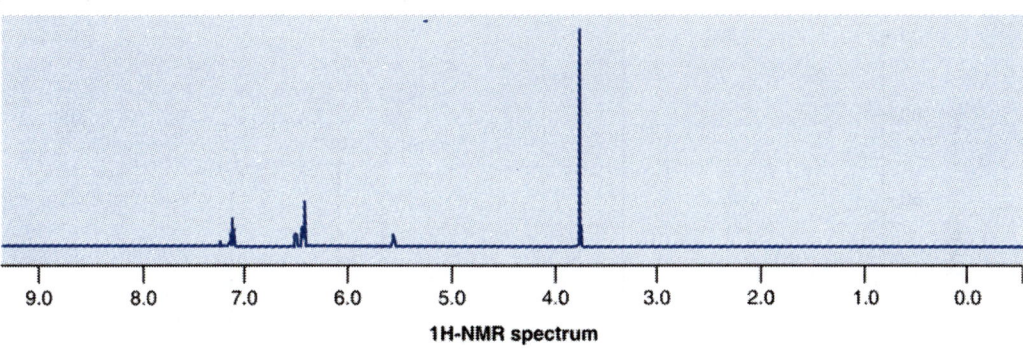

1H-NMR spectrum

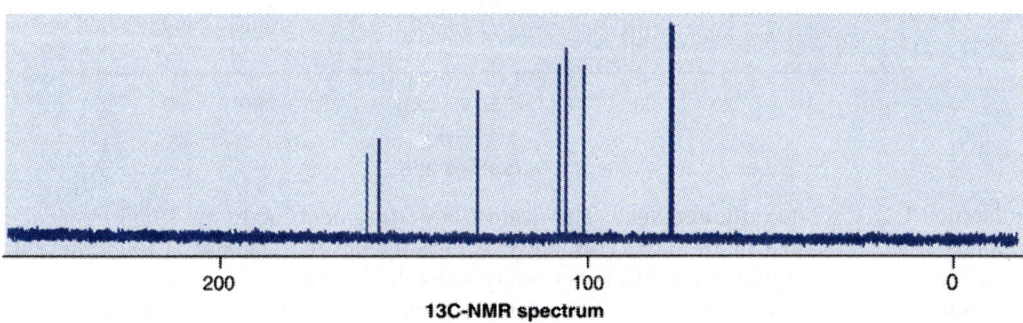

13C-NMR spectrum

Answer

3-Methoxyphenol

CH$_2$OH

OCH$_3$

Exercise 14

The mass spectra of two constitutional isomers are shown here. Both are gases at room temperature. The molecular ion is the small peak at *m/e* 102.

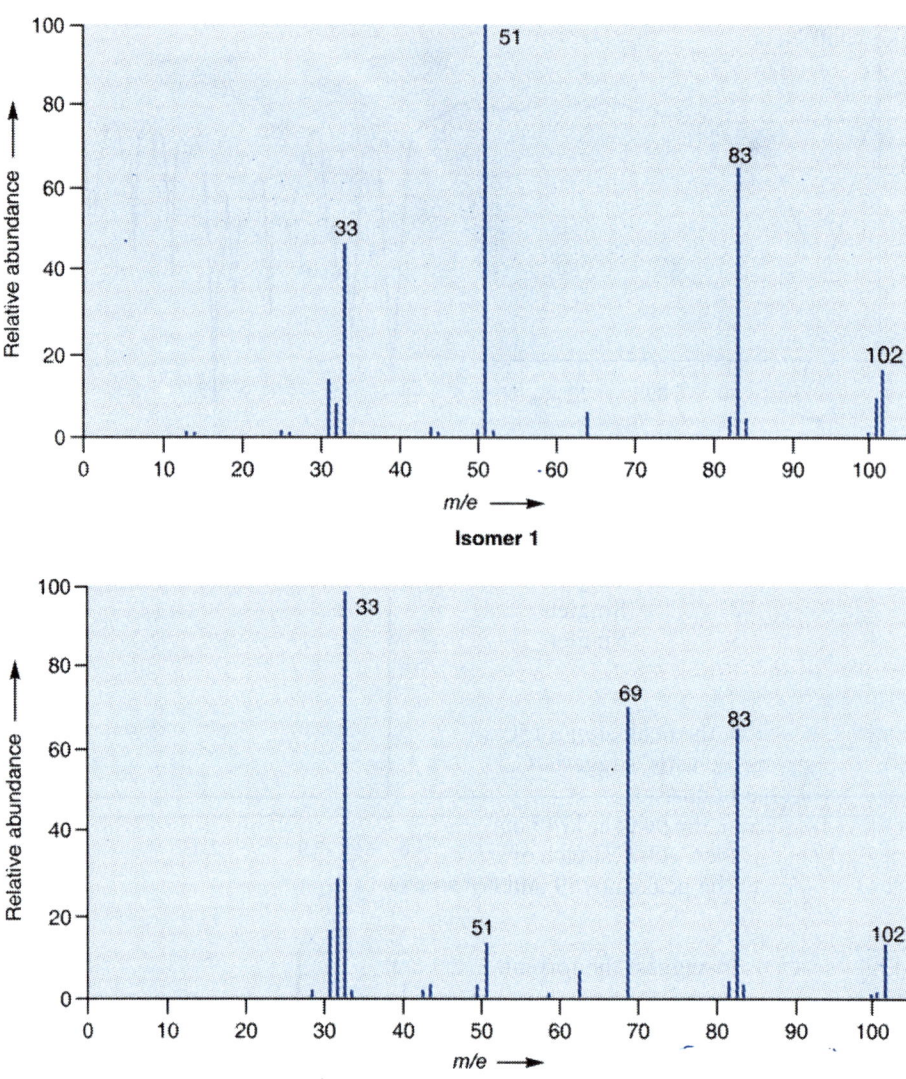

Isomer 1

Isomer 2

- **Isomer 1:** 1,1,2,2–tetrafluoroethane. The nearest large fragment ion to the small molecular ion is at *m/z* 83, a loss of 19 mass. This suggests loss of fluorine. The *m/e* 51 ion represents half the molecule. A hydrogen shift is necessary to explain the *m/e* 33 ion.
- **Isomer 2:** 1,1,1,2–tetrafluoroethane. Important differences from isomer 1 are that the *m/e* 51 ion is much smaller, *m/e* 33 is much larger, and a new strong ion at *m/e* 69 has appeared. Since fluorine is present, this may be assigned as a trifluoromethyl cation.

Exercise 15

Identify the organic compound that yields the following mass spectrum:

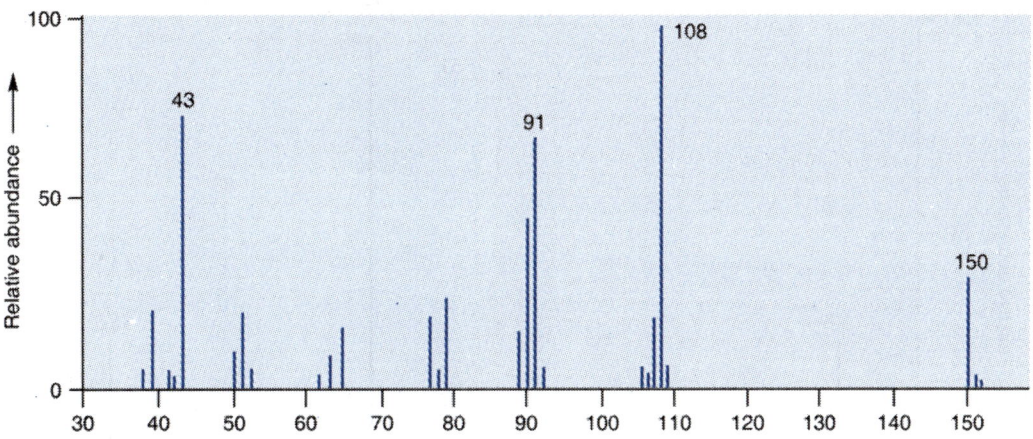

In IR, the main peaks appear at 3058, 2941, 1745, 1385, 1225, 1026, 749 and 697 cm^{-1}. In ^{1}H-NMR:

δ	Intensity	Multiplicity
7.22	5	Singlet
5.00	2	Singlet
1.96	3	Singlet

Answer

In the mass spectrum, the peak at m/e 150 can be the molecular peak (even number of nitrogen atoms). The isotopic cluster suggests $C_9H_{10}O_2$ as the most probable formula with aromatic rings and/or double bonds.

In the IR spectrum, the peak at 1745 cm^{-1} corresponds to the unconjugated C=O stretch, the peak at 1225 cm^{-1} is the C–O–C stretch of an acetal, the peak at 1026 cm^{-1} can be the asymmetric stretch of COC, and the peaks at 749 and 697 cm^{-1} point to a singly substituted benzene.

All this leads us to suggest the formula (benzyl acetate):

This is confirmed by the ^{1}H-NMR spectrum:

δ	Intensity	Assignment
7.22	5	Benzene protons
5.00	2	CH_2
1.96	3	CH_3

The main peaks in the mass spectrum are easily ascribed to the suggested formula:

150	Molecular peak
108	$COCH_3$ loss with H transfer
91	Tropylium ion
77, 78	Benzene ions
43	CH_3CO^+

Exercise 16

Identify the organic compound that yields the following mass spectrum:

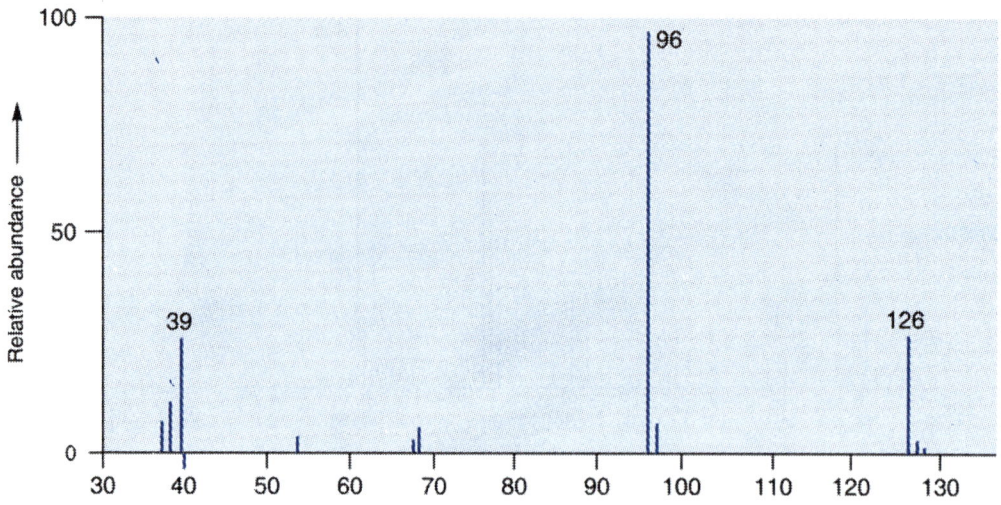

In IR, the main peaks appear at 3106, 2941, 1730, 1587, 1479, 1449, 1393, 1299, 1205, 1121 and 758 cm^{-1}. In ^{1}H-NMR:

δ	Intensity	Multiplicity
7.51	1	Quadruplet
7.02	1	Quadruplet
6.45	1	Quadruplet
3.80	3	Singlet

In off-resonance decoupled ^{13}C-NMR:

δ	Multiplicity
160	Singlet
146	Doublet
144	Singlet
118	Doublet
112	Doublet
51	Quadruplet

Answer

The mass spectrum shows a peak at m/z 126 which can be the molecular peak (even number of nitrogen atoms); the isotopic cluster corresponds to $C_6H_6O_3$ (the exact mass can be confirmed by high resolution) with four rings and two double bonds. The base peak at m/e 95 corresponds to M–OCH$_3$.

The IR spectrum shows a peak at 1730 cm^{-1} that can be ascribed to a conjugated C=O, peaks at 1205 and 1121 cm^{-1} that can be ascribed to an ester C–O stretch, the peak at 758 cm^{-1} indicates the presence of an aromatic ring, and the same is true for the peaks at 3106, 1587 and 1479 cm^{-1}.

The ^{1}H-NMR spectrum confirms the presence of six protons and the ^{13}C-NMR confirms the presence of six carbons. All these lead us to suggest the formula (carboxymethylfuran).

This structure is confirmed by the decoupled ^{13}C–NMR:

δ	Multiplicity	Number of H	Assignment
160	Singlet	0	C–5
146	Doublet	1	C–4
144	Singlet	0	C–1
118	Doublet	1	C–2
112	Doublet	1	C–3
51	Quadruplet	3	C–6

Exercise 17

Identify the organic compound that yields the following mass spectrum:

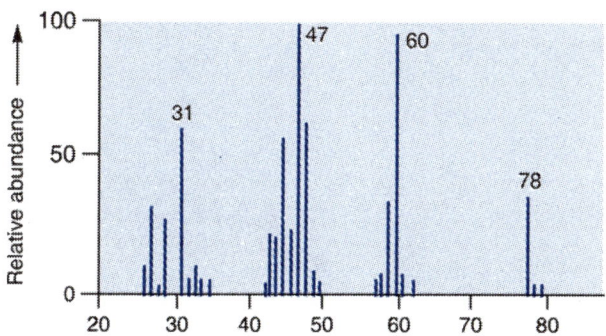

In IR, the main peaks appear at 3367, 3030, 2558, 1429, 1298, 1050 and 1020 cm^{-1}. In off-resonance decoupled ^{13}C-NMR:

δ	Multiplicity
28	Triplet
64	Triplet

Answer

The molecular peak at *m/e* 78 indicates a molecule with an even number of nitrogen atoms. The intensity of the M + 2 peak indicates the presence of S; the remainder (46) could be due to C_2H_6O.

The crude formula that is suggested is thus C_2H_6OS, with no ring or double bond.

The decoupled ^{13}C-NMR spectrum confirms the presence of two C; in the off-resonance spectrum, the triplets indicate that four out of the six hydrogens belong to neighbouring methylene groups. The structure that is proposed is thus 2-hydroxyethanethiol:

$$HOCH_2CH_2SH$$

The IR spectrum confirms such a structure:

3367 cm^{-1}: OH stretch
2558 cm^{-1}: SH stretch
1050 cm^{-1}: CH_2OH primary alcohol

The main peaks in the mass spectrum are easily assigned:

78: Molecular peak
60: H_2O loss
47: CH_2SH^+
31: CH_2OH^+

Index